NanoScience and Technology

NanoScience and Technology

Series Editors:
P. Avouris B. Bhushan D. Bimberg K. von Klitzing H. Sakaki R. Wiesendanger

The series NanoScience and Technology is focused on the fascinating nano-world, mesoscopic physics, analysis with atomic resolution, nano and quantum-effect devices, nanomechanics and atomic-scale processes. All the basic aspects and technology-oriented developments in this emerging discipline are covered by comprehensive and timely books. The series constitutes a survey of the relevant special topics, which are presented by leading experts in the field. These books will appeal to researchers, engineers, and advanced students.

Nano-Optoelectonics
Concepts, Physics and Devices
Editor: M. Grundmann

Noncontact Atomic Force Microscopy
Editors: S. Morita, R. Wiesendanger, E. Meyer

Nanoelectrodynamics
Electrons and Electromagnetic Fields in Nanometer-Scale Structures
Editor: H. Nejo

Single Organic Nanoparticles
Editors: H. Masuhara, H. Nakanishi, K. Sasaki

Epitaxy of Nanostructures
By V.A. Shchukin, N.N. Ledentsov and D. Bimberg

Applied Scanning Probe Methods I
Editors: B. Bhushan, H. Fuchs, S. Hosaka

Nanostructures
Theory and Modeling
By C. Delerue and M. Lannoo

Nanoscale Characterisation of Ferroelectric Materials
Scanning Probe Microscopy Approach
Editors: M. Alexe and A. Gruverman

Magnetic Microscopy of Nanostructures
Editors: H. Hopster and H.P. Oepen

Silicon Quantum Integrated Circuits
Silicon-Germanium Heterostructure Devices: Basics and Realisations
By E. Kasper, D.J. Paul

The Physics of Nanotubes
Fundamentals of Theory, Optics and Transport Devices
Editors: S.V. Rotkin and S. Subramoney

Single Molecule Chemistry and Physics
An Introduction
By C. Wang, C. Bai

Atomic Force Microscopy, Scanning Nearfield Optical Microscopy and Nanoscratching
Application to Rough and Natural Surfaces
By G. Kaupp

Applied Scanning Probe Methods II
Scanning Probe Microscopy Techniques
Editors: B. Bhushan, H. Fuchs

Applied Scanning Probe Methods III
Characterization
Editors: B. Bhushan, H. Fuchs

Applied Scanning Probe Methods IV
Industrial Application
Editors: B. Bhushan, H. Fuchs

Nanocatalysis
Editors: U. Heiz, U. Landman

U. Heiz U. Landman (Eds.)

Nanocatalysis

With 228 Figures, 5 in Color and 14 Tables

Springer

Professor Dr. Ulrich Heiz
Lehrstuhl für Physikalische Chemie
Technische Universität München
Lichtenbergstr. 4
85747 Garching
Germany
E-mail: ulrich.heiz@mytum.de

Professor Dr. Uzi Landman
Georgia Institute of Technology
School of Physics
Atlanta, GA 30332-0430, USA
E-mail: uzi.landman@physics.gatech.edu

Series Editors:

Professor Dr. Phaedon Avouris
IBM Research Division
Nanometer Scale Science & Technology
Thomas J. Watson Research Center
P.O. Box 218
Yorktown Heights, NY 10598, USA

Professor Dr. Bharat Bhushan
Ohio State University
Nanotribology Laboratory
for Information Storage
and MEMS/NEMS (NLIM)
Suite 255, Ackerman Road 650
Columbus, Ohio 43210, USA

Professor Dr. Dieter Bimberg
TU Berlin, Fakutät Mathematik/
Naturwissenschaften
Institut für Festkörperphyisk
Hardenbergstr. 36
10623 Berlin, Germany

Professor Dr., Dres. h.c. Klaus von Klitzing
Max-Planck-Institut
für Festkörperforschung
Heisenbergstr. 1
70569 Stuttgart, Germany

Professor Hiroyuki Sakaki
University of Tokyo
Institute of Industrial Science
4-6-1 Komaba, Meguro-ku
Tokyo 153-8505, Japan

Professor Dr. Roland Wiesendanger
Institut für Angewandte Physik
Universität Hamburg
Jungiusstr. 11
20355 Hamburg, Germany

ISSN 1434-4904
ISBN-10 3-540-32645-6 Springer Berlin Heidelberg New York
ISBN-13 978-3-540-32645-8 Springer Berlin Heidelberg New York

Library of Congress Control Number: 2006929219

Cover background image: "Single-wall nanotube METFET inverter: organic molecules wrap around a metallic tube channel and modify it to a semiconductor while the second tube is used for gating". Concept: S.V. Rotkin, Physics Department, Lehigh University, image: B. Grosser, Imaging Technology Group, Beckman Institute, UIUC.

Typesetting by the Authors and SPi using a Springer LaTeX macro package
Cover design: *design& production*, Heidelberg

Printed on acid-free paper SPIN: 11012795 57/3100/SPi - 5 4 3 2 1 0

Preface

Characterization and elucidation of the size-dependent evolution of the physical and chemical properties of finite materials aggregates, having discrete quantized energy level spectra, specific structural and morphological motifs and exhibiting unique dynamical characteristics, are among the outstanding challenges of modern materials science. Indeed, investigations of the unique properties of finite materials aggregates, and studies aiming at gaining deep insights into the development of materials characteristics from the molecular and cluster regimes to the bulk phase, are major themes of current interdisciplinary basic and applied research endeavors. In many instances, it is found that for larger materials aggregates the deviations of the properties from the bulk limit scale with the size of the aggregate. However, in many cases, at sufficiently small sizes, which most often lie in the nanoscale regime, the dependence of the material's property on size becomes non-scalable, and then small becomes different in an essential way, where the physical and chemical properties become emergent in nature, that is, they can no longer be deduced or extrapolated from those known for larger sizes.

Nanocatalysis is one of the most exciting subfields of nanoscience. Unlike the common practice in catalysis where the catalytic performance scales with the surface to volume ratio of the dispersed catalytic agent, nanocatalysts are distinguished by their unique and non-scaleable properties that originate from the highly reduced dimensions of the active catalytic aggregates. Consequently, the central aim of nanocatalysis is the promotion, enhancement, steering and control of chemical reactions by changing the size, dimensionality, chemical composition, morphology, or charge state of the catalyst or the reaction center, and/or by changing the kinetics through nanopatterning of the catalytic reaction centers. Since the aforementioned size-dependent non-scaleable, and often non-monotonic, evolution of materials' properties may occur when at least one of the material's dimensions is reduced to the nanoscale, nanocatalytic systems may appear as ultra-thin films, nanowires, or clusters. For these systems the chemical and physical properties are often controlled by quantum size effects and they present new opportunities for an atom-by-atom design, tuning and control of chemical activity, specificity, and selectivity.

This book is aimed at both individuals who wish to enter the field of nanocatalysis, as well as at scientists already active in this and related areas. It is the first book devoted to this subject, and by necessity it had to be limited to a selection of topics out of the broad range of issues pertaining to nanocatalysis. We intended to present the material in a pedagogical and methodological way, as well as to highlight specific illustrative examples of current research. The book covers various aspects of nanocatalysis, from basic research to applications. The various chapters attempt to provide both an introduction and reference material about fundamental studies of model systems, both experimental and theoretical, that reveal important concepts for understanding nanocatalysis, and about the new techniques and methodologies that are required for the preparation, characterization and probing of materials relevant to nanocatalysis.

The first two chapters of the book focus on nanocatalytic clusters consisting of up to a few tens of atoms. Here the size-dependent chemistry of free and supported clusters is discussed from both an experimental and theoretical points of view, and the various methodologies employed in studies of such systems are presented. Illustrative examples are described in great detail and concepts for understanding the remarkable size-dependent behavior are formulated. In particular, the role of the support material in changing the chemical and catalytic properties is highlighted. When larger nanoparticles, consisting of thousands of atoms, are considered, new phenomena emerge, and they are the focus of the third chapter. In Chap. 3 the specificity of nanoparticles is shown to originate from several factors, including: the presence of low coordinated sites, lattice distortions due to surface stress or accommodation with the substrate, the presence of different types of facets and the presence of the substrate. A challenge in nanocatalysis is the preparation of nanocatalytic materials. In Chap. 4 the use of lithographic techniques for the preparation of well-defined nanomaterials is reviewed. This chapter also describes the characterization of the prepared nanocatalysts and presents case studies of nanofabricated model catalysts. Theoretical considerations pertaining to formation of nanometer and sub-nanometer thin oxide films at surfaces of late transition metals and issues pertaining to the stability of surface oxides in an oxygen environment are discussed in Chap. 5. Gold serves as a useful paradigm in nanocatalysis because of the recently discovered unique chemical reactivity and catalytic properties of this metal that emerge when it is reduced to nanoscale dimensions, contrasting with the chemical inertness of bulk gold that has been known since antiquity. The "gold nanocatalysis paradigm" that is discussed already in the first chapter of the book is the topic of Chap. 6 where one can find a comprehensive review and discussion of catalytic applications of gold nanotechnology. Along with preparation methods, properties of gold particles, and reactions catalyzed by nanocatalytic gold and gold alloys, this chapter addresses issues related to the potential commercial applications for gold nanocatalysis.

Finally, we would like to thank all the authors that have contributed to this book for agreeing to take part in this endeavour and for their stimulating contributions which made this effort worthy. We would like to thank also our colleagues and co-workers who enhanced greatly our understanding of nanocatalysis and related areas. We are particularly pleased to acknowledge the US Air Force Office of Scientific Research, and in particular Dr. Michael Berman, as well as the German Science Foundation (DFG) for their continuous encouragement and for supporting the development of the nanocatalysis programs at the Georgia Institute of Technology and at the Technical University of Munich, respectively. The US Department of Energy, Dr. Dale Koelling in particular, is acknowledged for partial support of research activities at the Center for Computational Materials Science at Georgia Tech. Last but not least we thank our families for their support and patience.

München, Atlanta
July 2006

Ulrich Heiz
Uzi Landman

Contents

List of Contributors

Thorsten Bernhardt
Abteilung für Oberflächenchemie und Katalyse
Universität Ulm
89069 Ulm
Germany
thorsten.bernhardt@uni-ulm.de

Sónia A.C. Carabineiro
Centro de Química Estrutural
Departamento de Engenharia Química
Instituto Superior Técnico
Av. Rovisco Pais
1049-001 Lisboa
Portugal
sonia.carabineiro@ist.utl.pt

Ann W. Grant
Chemical Physics Group
Department of Applied Physics
Chalmers University of Technology
SE- 412 96 Gothenburg
Sweden
awgrant@fy.chalmers.se

Ulrich Heiz
Lehrstuhl für Physikalische Chemie
Technische Universität München
Lichtenbergstr. 4
85747 Garching
Germany
ulrich.heiz@mytum.de

Claude R. Henry
CRMCN-CNRS
Campus de Luminy, case 913
13288 Marseille cedex 09
France
henry@crmcn.univ-mrs.fr

Bengt Kasemo
Chemical Physics Group
Department of Applied Physics
Chalmers University of Technology
SE- 412 96 Gothenburg
Sweden
kasemo@fy.chalmers.se

Uzi Landman
School of Physics
Georgia Institute of Technology
Atlanta GA 30332-0430
USA
uzi.landman@physics.gatech.edu

Lars Österlund
Deputy Research Director
Department of Environment and Protection
FOI NBC-Defence
SE-901 82 UMEÅ
Sweden
lars.osterlund@foi.se

Gianfranco Pacchioni
Dipartimento di Scienza dei Materiali
Università degli Studi Milano-Bicocca
via R. Cozzi, 53 – 20125 Milano
Italy
gianfranco.pacchioni@unimib.it

Karsten Reuter
Abteilung Theorie
Fritz Haber Institut
Max Planck Gesellschaft
Faradayweg 4-6
14195 Berlin
Germany
reuter@fhi-berlin.mpg.de

David T. Thompson
Consultant to World Gold Council
'Newlands'
The Village
Whitchurch Hill
Reading RG8 7PN, UK
DTThompson@aol.com

1

Chemical and Catalytic Properties of Size-Selected Free and Supported Clusters

T.M. Bernhardt, U. Heiz, and U. Landman

1.1 Introduction

Clusters reveal unique physical and chemical properties as their dimensions are reduced to the nanoscale. Their properties are not scalable from those of bulk materials. Clusters are thus building blocks for nanocatalytic materials. Since the early days of cluster science, it has been speculated about strong size-dependent catalytic activity of metal clusters just a few or few tens of atoms in size. It was proposed that tuning reactivity and selectivity is feasible by simply changing cluster size [1] because intrinsic cluster properties relevant for catalytic behavior, like the electronic spectrum, the symmetry, or the spin multiplicities, can thus be tuned for obtaining the desired activation of the reactant molecules relevant in a catalytic process. It has been possible only in recent years, through the development of new experimental schemes, to show that model catalysts made of such small clusters indeed reveal variations in the catalytic activity when cluster size is changed just by a single atom. First results pointing toward such ability have been obtained using gas-phase clusters, where size-dependent adsorption of small molecules has been observed [2–6]. In another study, Shi and Ervin [7] even investigated a whole thermal catalytic cycle and reported on the size-dependent catalytic oxidation of CO on free Pt_n^- ($n = 3$–6) clusters in a molecular beam experiment. For supported clusters, Fayet et al. [8] and later Leisner et al. [9] showed that the latent image generation in the photographic process requires silver clusters of a critical size. Xu et al. [10] observed a size-dependent catalytic activity of supported metal clusters for the hydrogenation of toluene by using organometallic precursors. In another study, the dissociation of CO on size-selected supported nickel clusters has been shown to be strongly dependent on the exact cluster size [11] as previously demonstrated in the *gas phase* by Vajda et al. [12].

In this chapter, we focus on recent experimental and theoretical studies of chemical and catalytic properties of gas-phase metal clusters and metal clusters deposited onto support materials with sizes in the nonscalable regime, where each atom counts for the cluster reactive properties. By comparing the two cases, gas-phase and deposited clusters, it is not only possible to obtain information on the cluster–support interaction and its influence on the catalytic activity but also to elucidate the relevance of the size evolution of the intrinsic cluster properties. Some examples are presented to illustrate this approach, however, only very few cases exist where a direct comparison of free and deposited clusters is possible so far. Throughout this chapter, we focus on size-selected systems only. First, we give a brief description of the experimental techniques needed to carry out such studies in the gas phase and on surfaces. Second, computational techniques are reviewed, which are presently employed to gain insight into the molecular details of catalytic reactions on small free and deposited clusters. Third, concepts for understanding the catalytic properties in the nonscalable size regime are presented. In the last part, we review illustrative examples showing clusters in catalytic action.

1.2 Experimental Techniques

For the study of nanocatalysis in the nonscalable size regime, it is essential to control the preparation of the model systems with atomic precision because in this range the catalytic properties are changing atom by atom. In the gas phase, a size-selective detection scheme through mass spectrometry is commonly implemented and information about size-specific reactivity is readily obtained. In contrast, there are only few examples where perfect control of size has been achieved on surfaces by conventional methods like, e.g., self-assembly and growth or the use of coordinated cluster precursors. In none of these examples, however, it has been possible to control size over a large part of the nonscalable regime and thus to define concepts for understanding these interesting size effects. Recently, new methods have been introduced in which the clusters are produced in the gas phase and are then size-selected prior to deposition. In this approach, cluster sources developed during the last two decades are employed in combination with appropriate mass filters for selecting a single size from the cluster distribution present in the gas-phase cluster beam.

Thus, for studying free clusters in the gas phase and clusters on surfaces, the same cluster sources may be used, although for cluster deposition experiments higher cluster fluxes and narrow velocity distributions are required. This is necessary as the cluster-assembled materials must be prepared in short time to prevent adsorption from background gases and under soft-landing conditions. Normally, the soft-landing prerequisite is valid if clusters are deposited with less than 1 eV per atom kinetic energy onto solid surfaces. To illustrate the requirements on the cluster fluxes, the following example is illustrative: To study chemical and catalytic properties of supported clusters, typical surface areas are $\sim 0.1\,\mathrm{cm}^2$. In order to land the clusters well separated onto the surface, less than $\sim 1\%$ ML [1 monolayer (ML) $\sim 10^{15}$ clusters cm^{-2}] of clusters are deposited. For reasonable deposition times at background pressures of about 10^{-10} Torr (resulting in exposures smaller than 0.1 Langmuir), the flux of size-selected cluster beams consequently has to be in the range of 10^{10} clusters s^{-1}. If smaller areas can be probed (this depends on the analytical methods in use), the required cluster fluxes decrease linearly.

In Sect. 1.2.1 of the present chapter, we describe the most important cluster sources successfully used today. Section 1.2.2 introduces experimental techniques for mass-selecting single cluster sizes from the distribution generated by the cluster sources. In the gas phase as well as for clusters on surfaces, the densities are extremely low, thus only highly sensitive methods can be used for the characterization of the chemical and catalytic properties of the model systems. Some of the most commonly used techniques employed in gas-phase experiments are presented and discussed in Sect. 1.2.3, surface analysis techniques for cluster studies are presented in Sect. 1.2.4.

1.2.1 Cluster Sources

The following ablation thermalization principle is often used for producing atomic cluster beams. The material of interest is evaporated by means of intense laser pulses [13], electrical [14] and magnetron [15, 16] discharges, or highly energetic inert ions [17]. A buffer gas thermalizes the produced ultra-hot plasma or supersaturated atomic gas. The mixture then expands through a nozzle into the vacuum upon supersonic expansion. By this process, the formed clusters are cooled to cryogenic temperatures and generate well-defined molecular beams of neutral and charged clusters. The chemical reactivity of the clusters can then be studied in the gas phase after size-selection by various experimental techniques, including fast flow reactor kinetics in the post-vaporization expansion region of a laser evaporation source [18, 19], ion flow tube reactor kinetics of ionic clusters [20–23], ion cyclotron resonance [24–28], guided ion-beam [7, 29–31], and ion-trap experiments [32–34]. These cluster beams can also be deposited onto various support materials if the chemical and catalytic properties of size-selected clusters on surfaces are of interest.

At present there is no universal cluster source, which is capable of satisfying simultaneously all experimental requirements needed for gas-phase studies or deposition experiments. These are, typically, an intense beam with small energy spread for cluster sizes ranging from very small clusters of a few atoms to large clusters of a few hundreds of atoms at a desired, well-defined temperature, and, last but not least, for a large number of elements.

In a seeded supersonic expansion source [35–37], the material is heated in an oven and a mixture of gaseous material and a seed gas expands through a nozzle into the vacuum. This source produces a highly intense beam of small pure and mixed metal clusters, but it is limited to metals with a low boiling point (Li, Na, K, etc.). Laser vaporization sources [13, 38] are more widely used as they produce pure and mixed clusters of most elements and when operated at low frequency (1–10 Hz) they have been successfully used

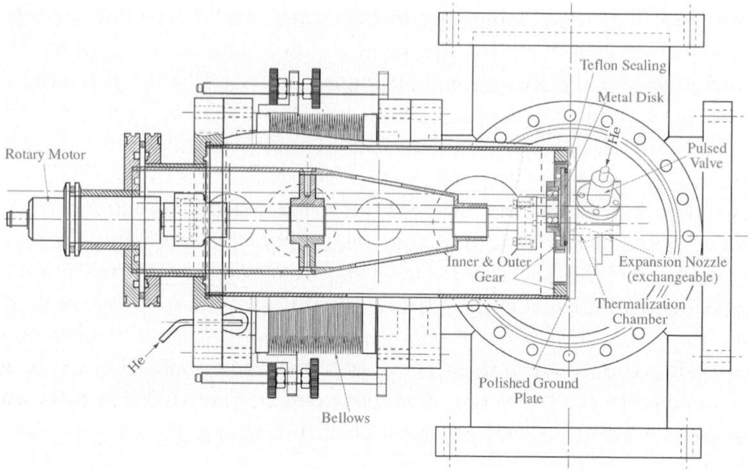

Fig. 1.1. Details of the high-frequency laser evaporation source. Shown are the rotary motor, which drives the planetary gear assembly for turning the target, and the thermalization chamber with exchangeable expansion nozzle. The laser-produced plasma expands into this thermalization chamber. A helium gas pulse is then introduced by a piezo-driven pulsed valve and synchronized with the laser pulse into the same volume. The metal–gas mixture then expands through the nozzle into the vacuum leading to cluster formation. In contrast to conventional sources, the laser beam is coaxial to the molecular beam axis. The bellow is used to align the source along the optical axis of the ion optics

in gas-phase cluster experiments for almost two decades. When increasing the repetition rate of the laser evaporation source, the cluster flux is substantially increased allowing this type of source to be used for cluster deposition experiments [39]. A typical design is shown in Fig. 1.1. A laser pulse is focused onto a rotating target disk driven by an integrated hypocycloidal planetary gear assembly. The produced plasma is thermalized by a helium gas pulse in an ellipsoidal thermalization chamber. Subsequent supersonic expansion of the helium–metal vapor through a nozzle leads to cold clusters with a narrow kinetic energy distribution. The generated clusters, after size-selection with a quadrupole mass spectrometer, are deposited with low kinetic ($E_{\mathrm{kin}} \approx 0.2\,\mathrm{eV}$ per atom) and internal energies at typical cluster ion currents of 0.5–10 nA.

Sputter sources have proven to be useful in producing small size-selected clusters in a continuous beam with high stability. But as the beam intensity falls off rapidly with size, these sources are not suitable to produce medium to large clusters in large enough quantities. Nevertheless, and in spite of a relatively large energy spread of the generated clusters, sputter sources are widely used for the production of small clusters. In a typical sputter source, a primary ion beam with energies between 0.1 and 15 keV and currents of 5–20 mA is focused onto a target producing neutrals and ions. The secondary ions are then drawn into ion optics and quadrupole mass spectrometers for mass-selection and beam transport [8,40–47]. With such a source (Fig. 1.2 [48,49]), for materials like Ag, Au, or Pd, beam currents for small clusters are typically in the range of a few to a few tens of nanoamperes in state-of-the-art experimental setups. Within the family of sputter sources, there has been a recent new development, which is a compact sputter source for the deposition of small size-selected clusters based on a cesium ion sputter source [50]. It has been designed to be compact and portable, enabling it to be attached to ultra-high vacuum (UHV) systems for use in deposition experiments. The sputter source, modified by introducing a beam extraction section into a high-energy ion beam source, produces a cluster beam of 1.5 kV energy. This allows mass-selection to be performed with a compact Wien velocity filter, greatly reducing the dimensions of the source. The source produces sufficient intensities of small cluster (1–15 atoms per cluster) for UHV deposition experiments, e.g., 0.002 ML min^{-1} for Al_7^- focused onto a 1 cm^2 area has been obtained. Cluster beams of many materials, including noble metals and carbon, have been generated. For deposition and ion-trap experiments, sputter sources are coupled to a "phase space compressor" [9,12,51–53] in order to reduce the kinetic energy distribution of the cluster beam (cf. Fig. 1.2a). In this device, inert gas collisions are utilized to reduce the ion's initial kinetic energy, allowing the production of clusters with relatively low kinetic energies. For example, clusters of Ag_{21} have been generated with kinetic energy distributions of 3.7 eV at full width half maximum (FWHM) and for smaller clusters, like Ag_4, energy distributions of only 1 eV FWHM have been obtained (Fig. 1.2b) [9].

A commercially available cluster source based on the original design of Haberland [15,16] combined with mass-selection is the NC200U nanocluster source by Oxford Applied Research. It is designed for use in an ultrahigh vacuum environment (Fig. 1.3). A magnetron discharge is used to generate the atomic vapor. Inside a liquid nitrogen cooled aggregation tube, a rare gas cools and sweeps the atoms and formed clusters from the aggregation region toward the aperture. By adjusting various parameters such as the power supplied to the magnetron, the aperture size, the rate of rare gas flow, type of collision gas, temperature and geometry of the aggregation region, cluster sizes can be adjusted. This source is combined with a quadrupole mass filter for mass-selection. The cluster source can produce metal and semiconductor clusters from sizes of a couple of atoms up to a couple of thousands of atoms.

Recently, a gas condensation source for production and deposition of size selected metal clusters has been developed showing remarkable features [54]. It operates on the principle of quenching of a hot metal vapor in a flowing stream of cool inert gas. The metal vapor condenses, producing clusters with a broad distribution of sizes. Clusters are usually extracted into a high vacuum region through a nozzle at the end of a condensation chamber. The gas condensation cluster source is capable of producing clusters containing from 2 to 10^6 atoms each [55]. Thus, this type of source is among the most flexible in terms of cluster sizes. Higher source temperatures can be achieved than are possible in

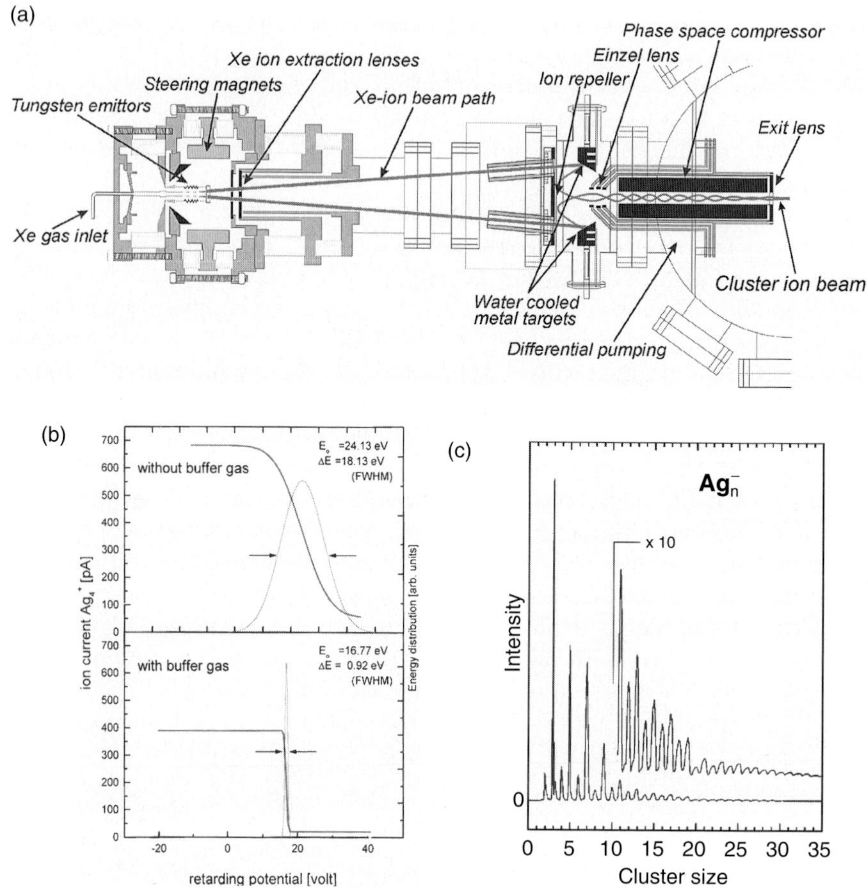

Fig. 1.2. (**a**) Schematic representation of an ion sputter cluster source. This particular source is used in both gas-phase and surface deposition experiments. The cross-section scheme shows two of the four primary xenon ion beams (bold lines) which are produced by the cold reflex discharge ion source (CORDIS) assembly on the left-hand side ([17]). These xenon ion beams are accelerated to kinetic energies of up to 15 keV before impinging on four metal targets. From this sputtering process, neutral as well as charged metal clusters are produced. The particular charge of interest is chosen by an appropriate bias voltage applied to the repeller plate opposite to the target holders. The cluster ion beams (thin lines) are thus deflected and subsequently focused into a quadrupole ion guide. This so-called "phase space compressor" is filled with helium buffer gas and serves to reduce the considerable initial energy spread of the cluster beam and to collimate the beam onto the center axis of the apparatus. (**b**) Kinetic energy distribution of a beam of $Ag_4{}^+$ clusters after passing a first phase space compressor (upper graph, see also **a**) and a second helium filled guiding quadrupole (lower graph, this compressor is not shown in **a**). (**c**) Silver cluster ion size distribution produced by this sputter source

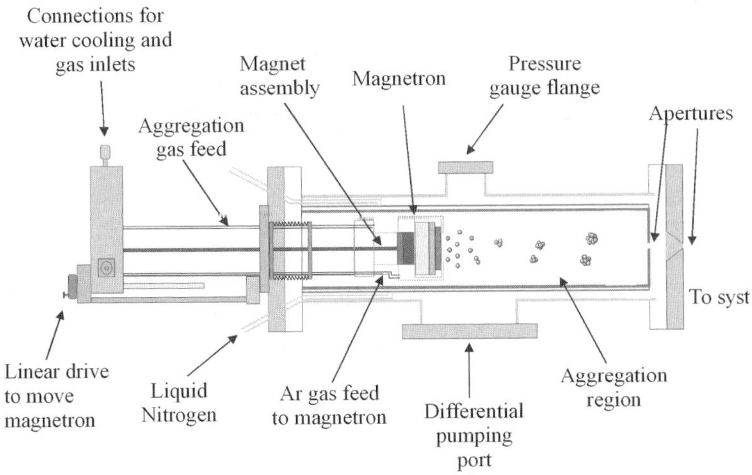

Fig. 1.3. Schematic representation of the NC200U nanocluster source designed at the University of Freiburg and distributed by Oxford Applied Research. Depending on magnetron power, aggregation length, aperture size, and rare gas flow rate, the clustering can be influenced. For Cu_n, the maximum cluster current can be changed from Cu_{20} up to Cu_{1500}

the seeded supersonic expansion source, so that silver and gold clusters are routinely produced by this method. With this source the size, energy, and deposition rate dependence of the deposition of silver clusters on a graphite surface have been determined, with cluster sizes ranging from 50 to 250 atoms and landing energies ranging from 25 to 800 eV [56, 57]. For Pb ($n = 2$–300) and Ag ($n = 20$–300), a density of size-selected clusters of about 10^{12} cm^{-2} in 10 min have been obtained.

Another cluster source is the pulsed arc discharge cluster ion source (PACIS), which generates ionized clusters directly but suffers from significant pulse-to-pulse variations in the beam intensity [14, 58–60] (Fig. 1.4). Recent improvements of the PACIS operate at considerably higher repetition rates providing a considerably enhanced cluster ion current for deposition experiments with mass-selected clusters [61]. Also the liquid metal ion source generates clusters of up to approximately 100 atoms. This latter cluster source is limited to metals with a fairly low melting point and produces cluster ions with a large kinetic energy spread, which is a problem for mass-selection and soft-landing [62, 63].

1.2.2 Mass-Selection and Soft Landing

All cluster sources presently employed generate a more or less broad distribution of clusters of different sizes. Therefore, in order to investigate and exploit the size-dependent cluster properties, the application of mass-selection

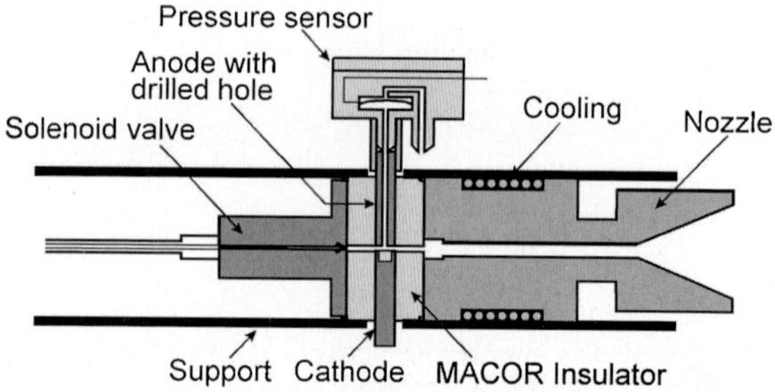

Fig. 1.4. Schematic drawing of a modified PACIS used in cluster deposition and surface scattering experiments [58, 59]. The cylindrical molybdenum electrodes are mounted in a MACOR insulator block with a center channel of rectangular cross section. The material to be evaporated during the electrical discharge is pressed into a hole in the cathode. A He seed gas pulse is introduced through a solenoid valve to provide cooling and cluster formation. For adjustment of the appropriate timing between valve opening and discharge ignition, the gas pressure inside the MACOR block can be monitored via a miniature pressure sensor attached to the bored anode. Supersonic expansion of the metal plasma through a cooled nozzle produces a wide range of cluster sizes

techniques is imperative to experimental studies concerned with clusters in the nonscalable size regime. The two most commonly used techniques for mass-separation of metal cluster beams are dynamic field quadrupole mass analyzers and time-of-flight mass spectrometers. Both methods require the clusters to be charged, which is accomplished either directly during the cluster generation process or by ionization with laser or electron beams. Radio frequency (rf) quadrupoles can be used to generate continuous beams of mass-separated clusters at high transmission; however, transmission is often a trade off with resolution and the mass range of these devices is inherently limited. Commercial quadrupole mass spectrometers are offered up to masses of some thousand atomic mass units (amu), which considerably restricts the size of the clusters that can be investigated, in particular for the catalytically interesting heavy metals like platinum (Pt_{20}, e.g., has already a mass of 3,900 amu). On the other hand, time-of-flight techniques offer the opportunity of fast, high duty cycle measurements in the microsecond range and of simultaneous monitoring of an essentially unlimited mass range. However, the time-of-flight mass analysis is inherently a pulsed technique, which determines a considerably lower transmission making it less suited for deposition experiments, where much larger cluster currents are required than in gas-phase experiments. Nevertheless, both techniques are employed in various experimental setups for cluster studies in the gas phase and at surfaces. The instrumental details of rf-quadrupole mass filters [30, 64] as well as of time-of-flight mass spectrometers [65–67] are well documented in the literature. Also experimental issues of their application to *gas-phase* metal cluster studies have been reviewed in several excellent contributions (see, e.g., [68–72])

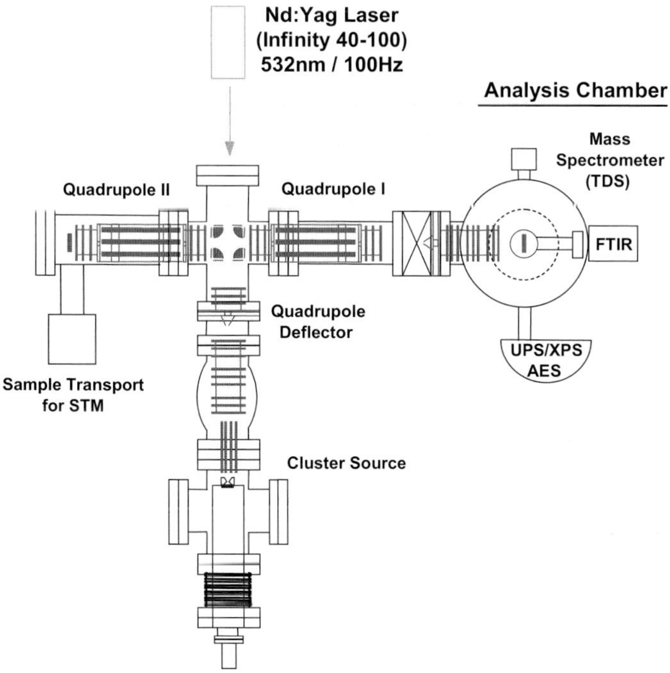

Fig. 1.5. Experimental setup of the high-frequency laser vaporization cluster ion source driven by a 100-Hz Nd:Yag laser for the production of ion clusters, ion optics with a quadrupole deflector, and quadrupole mass filter for size-selection and deposition; the analysis chamber with a mass spectrometer for thermal desorption spectroscopy (TDS), a Fourier transform infrared spectrometer, a spherical electron energy analyzer for Auger electron spectroscopy (AES) for in situ characterization of the clusters [73]

In the remainder of this section, we will therefore focus on the interaction of mass-selected cluster beams with solid surfaces.

A state-of-the-art experimental setup for cluster deposition and reaction experiments is depicted in Fig. 1.5 [73]. In this setup, the high-repetition rate laser vaporization source for cluster generation shown in Fig. 1.1 is combined with quadrupole mass filters to obtain continuous cluster beams with intensities in the nanoampere range for single cluster sizes. However, when depositing clusters, the deposition time should be as short as possible to prevent the degeneration of the clusters on the substrate. Therefore, to increase the cluster density on the substrates without increasing the deposition time, a new beam focusing element was implemented recently in this setup directly in front of the surface. For this purpose, an octopole ion guide of conical shape was constructed. The design is depicted in Fig. 1.6. With this construction, it is possible to focus 500 pA of an Ni_{20}^+ cluster ion beam from 9 onto 2 mm spot

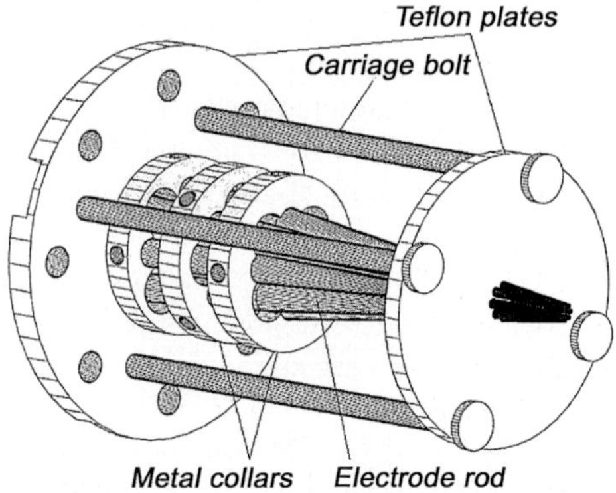

Fig. 1.6. Beam directioning during soft-landing: the focusing octopole ion guide. The schematics present a 3D view of the newly constructed conical octopole guide. Eight rods shaped as truncated cone (diameter rejuvenates from 3 to 0.5 mm) are arranged in a conical geometry. Two Teflon plates at the ends with four carriage bolts and two metal collars (in the cylindrical part) keep the conical geometry. The ion entrance orifice opens 9 mm in inner diameter and the exit focuses to around 2-mm spot size [74]

size in diameter with a transmittance up to about 70%. Thus, the cluster density increases by approximately a factor of 15 compared to the entrance orifice of the conical octopole [74].

A further less commonly employed but very classical method for mass-selection is the Wien-filter which combines an electrostatic and a magnetic field for mass-selective cluster deflection. This device is applied in particular with the purpose to obtain large cluster currents for cluster–surface interaction and deposition experiments at moderate resolution [75, 76]. Another approach for size-selection in combination with a sputter source [47] or a laser vaporization source [61, 77, 78] has been chosen recently. In both cases, a magnetic dipole field has achieved the size-selection. Subsequent deceleration of the size-selected clusters to less than 1 eV per atom enables soft-landing under UHV conditions [47]. Monodispersed chromium cluster beam densities range from 0.1 to 5 nA mm^{-2}, depending on the cluster size [47].

If pulsed time-of-flight techniques are employed in cluster–surface interaction experiments [79–81], a pulse of clusters of a single size is generated by a pulsed electrostatic deflection field in various geometric arrangements like, e.g., the interleaved comb mass-gate [82, 83] or a pulsed ion mirror [84, 85]. In this context, it is noteworthy to mention a new recent mass-selection technique, which allows to size-select particles from atoms to nanoparticles with practically unlimited size, providing a mass resolution of $m/\Delta m$ = 20–50 and a transmission of about 50% for the selected size, both independent of mass [57]. The method is in principle a time-of-flight method but differs fundamentally from the techniques normally used. The basic idea is to use time-limited high-voltage pulses to laterally displace a preaccelerated beam, without changing its direction or shape. As the movement of the ions perpendicular to their beam direction is independent of their forward velocity, mass resolution and calibration does not depend on the ion beam energy. A mass selector of this type has been implemented successfully into a cluster deposition experiment and has proven to be reliable and simple to operate [54, 86].

For mass-selection, the cluster ion beam is usually accelerated to a few tens of electron volts kinetic energy (quadrupole selection) or up to several thousands of electron volts (time-of-flight methods). The binding energy of metal clusters is in general around 1 eV per atom in the cluster. Thus, to provide soft-landing conditions, i.e., to avoid cluster fragmentation during the deposition on the surface, the clusters have to be decelerated to kinetic energies below about 1 eV per atom. This is achieved by applying an appropriate retarding potential to the substrate surface. However, the width of the energy distribution in the cluster beam determines, if all or only a fraction of the decelerated clusters do come in contact with the surface at the desired low collision energy. This issue of kinetic energy distribution in the cluster beam has already been briefly discussed in Sect. 1.2.1 in conjunction with the different cluster sources. In order to probe the processes that occur as a function of the kinetic energy of the clusters that approach the surface, the monitoring of the cluster ions that scatter back from the substrate is instructive. Figure 1.7a displays, for Sb_8^+ ions, the total yield of ions scattered back from the graphite surface during the deposition process [58, 59].

Already at a collision energy around 20 eV per cluster, about 10% of the impinging ions are recoiling from the surface. For silicon oxide as surface and larger clusters, this fraction of backscattered cluster ions can reach about 90% of the impinging cluster intensity [59]. Interestingly, in this particular case shown in Fig. 1.7, the yield of backscattered ions exhibits two separated maxima, indicating different mechanisms that prevail during cluster–surface interaction at the different collision energies. Figure 1.7b displays the major fragments that scatter back from the surface in this case. The change from Sb_4^+ as premier fragment ion at lower collision energies to Sb_3^+ at higher impact energies is found to be due to a transition from unimolecular decomposition of the impinging clusters between about 10 and 150 eV collision energy to a multifragmentation shattering above 150–300 eV. Below 10 eV, soft-landing with minimal decomposition can be achieved, whereas above 200 eV, increasing implantation of the colliding clusters into the surface leads to the observed decrease in scattered ion yield as evidenced from accompanying scanning tunneling microscopy (STM) investigations [58].

In this experiment, the analysis of the surface subsequent to the interaction with the clusters was performed in the same apparatus without breaking of the vacuum, thus providing valuable direct information about the fate of the clusters after the deposition process [88]. Figure 1.8a shows an STM image of a highly oriented pyrolytic graphite HOPG(0001) surface held at a temperature of 125 K after irradiation with Sb_{13}^+ clusters under soft-landing

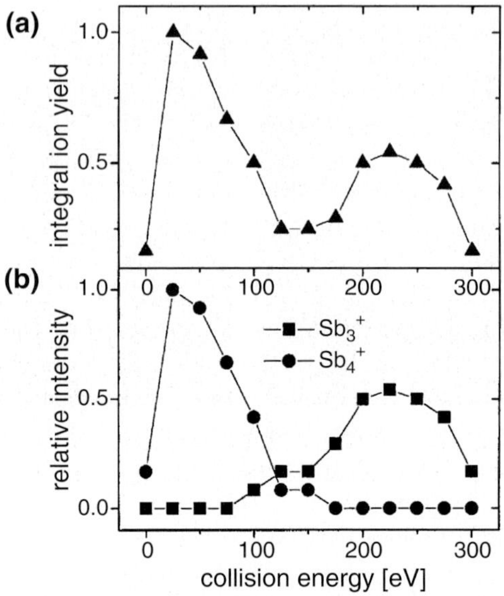

Fig. 1.7. Scattering of $Sb_8{}^+$ clusters from an HOPG surface. (**a**) Integral yield of the ions scattered off the surface and (**b**) relative intensity of the observed major fragment ion peaks as a function of the cluster-surface collision energy [58, 87]

conditions below 10 eV kinetic energy per cluster [89]. The arbitrary population of these clusters on the terraces indicates that surface diffusion is sufficiently slow at this substrate temperature and the clusters form stable nanodots due to the low substrate temperature. The number density amounts to 3.3×10^{13} cm^{-2}. By subsequent warming of the sample, surface diffusion is activated and the clusters start to diffuse from the terrace sites to more strongly binding step, edge, and kink sites. Figure 1.8b was obtained from the same sample at 170 K surface temperature. The clusters appear to have accumulated at the monoatomic step visible in the center of the STM image leaving behind the bare HOPG surface on the adjacent terraces. Thus, only the deposition at sufficiently low temperatures will avoid cluster diffusion and provide conditions to stably investigate isolated clusters on the substrate. The temperature required apparently depends on the substrate material. On the very weakly interacting HOPG surface, at room temperature, no clusters on terrace sites can be observed after soft-landing due to the discussed diffusion effects. However, if the clusters are deposited at much higher kinetic energy, stable nanodots can be imaged even at elevated temperatures as can be seen from Fig. 1.9a, which was obtained after deposition of $Sb_4{}^+$ with 230 eV kinetic energy. The locations of cluster impacts appear as hillocks. Figure 1.9b presents the enlarged image of a single protrusion observed after deposition of $Sb_4{}^+$ at 150 eV collision energy. Apparently, the atomic lattice structure of the HOPG(0001) surface is still intact on the location of the protrusion. Scanning

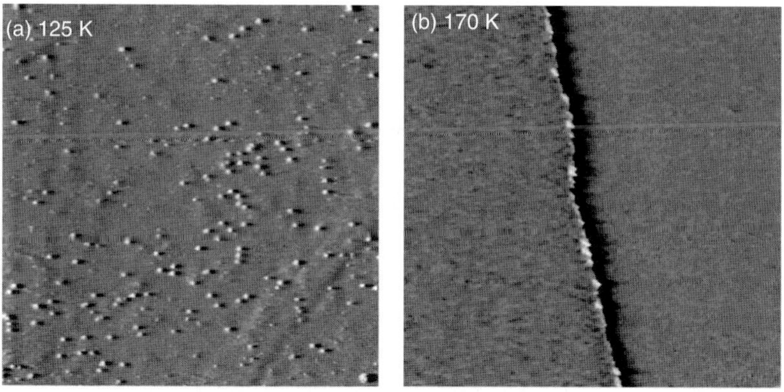

Fig. 1.8. Sb_{13} clusters deposited on HOPG under soft-landing conditions at 10 eV kinetic energy per cluster. (**a**) $2,480 \times 2,480$ Å^2 STM image obtained after deposition with the surface at 125 K. (**b**) $1,500 \times 1,500$ Å^2 STM image recorded from the same sample after warming of the substrate to 170 K [89]

tunneling spectroscopy (STS) reveals that also the electronic structure at this location is similar to the structure of the original carbon layers [90]. Therefore, it is concluded that the clusters are likely to be intercalated in between the graphite layers at these moderate collision energy. In contrast, the STM image shown in Fig. 1.9c was obtained after deposition of Sb_8^+ at 410 eV collision energy, i.e., under conditions where even no scattered ions are observed anymore (cf. Fig. 1.7). It is clearly visible that the surface lattice structure of the graphite plane has been destroyed. Strain-induced superstructures are observed next to the impact location. Also STS indicates a massive perturbation of the conductive properties of the graphite surface [90]. Therefore, at these energies the clusters are readily implanted into the graphite substrate leading to the known hillock formation resulting from energetic projectile collisions [58].

A variety of other experimental methods, including absorption spectroscopy and He-atom scattering, have been applied to investigate the interaction of mass-selected clusters with the surface during and after deposition. This work has been recently reviewed in detail [91]. A further versatile experimental method to probe the fate of mass-selected metal clusters on a solid surface is the femtosecond two-photon photoemission spectroscopy (2PPS) [92]. This method has been shown to be sensitive to the electronic surface structure upon cluster deposition [93]. For deposition experiments carried out with mass-selected silver clusters under soft-landing conditions at kinetic energies of 1–4 eV per cluster, the size dependence of the photoelectron spectra reveals a pronounced odd/even effect (see Fig. 1.10a), which is well known for gas-phase silver clusters [93]. This indicates that the deposited clusters retain their size and identity on the solid substrate. The lifetime of the photoexcitation rises with cluster size, which is attributed to an increasing electronic

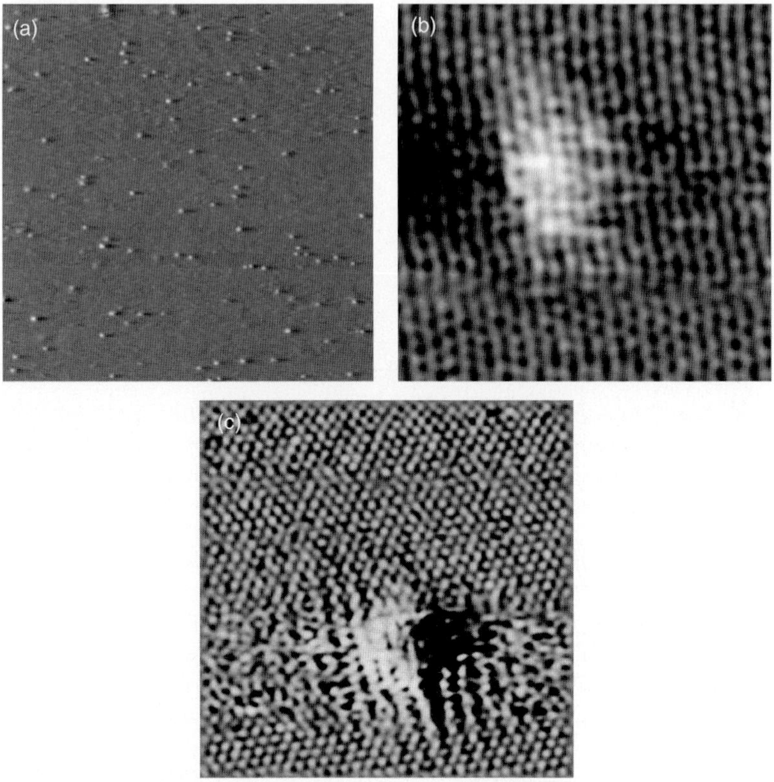

Fig. 1.9. Energetic impact of antimony clusters onto a graphite target. (**a**) 2,480 × 2,480 Å^2 STM image of an HOPG substrate after deposition of $Sb_4{}^+$ clusters with 230 eV kinetic energy at room temperature. Stable nanodot (hillock) formation is observed. (**b**) 47 × 47 Å^2 STM image of the impact location of an $Sb_4{}^+$ cluster with 150 eV collision energy. (**c**) 71 × 71 Å^2 STM image of the impact of an $Sb_8{}^+$ cluster with 410 eV collision energy [58, 87, 90]

density of states for larger clusters [94]. In addition, Fig. 1.10b reveals the influence of the deposition energy on the stability of the clusters during the deposition process. A series of measurements are plotted in which $Ag_9{}^+$ was deposited at different elevated kinetic energies in comparison to the result obtained under soft-landing conditions (3 eV per cluster). The average photoelectron energy rises abruptly from 0.9 eV at soft-landing conditions to about 1.3 eV at a deposition energy of 10 eV and then stays constant to higher deposition energies. This behavior clearly indicates a change in the topology of the cluster–surface system already at moderate collision energies of 10 eV or more [93].

The 2PPS method is also very powerful to probe the mobility of the clusters on the surface because the photoelectron yield is known to considerably increase as the clusters start to diffuse and agglomerate on the surface due to the development of a nanoparticle plasmon absorption. Figure 1.11 shows

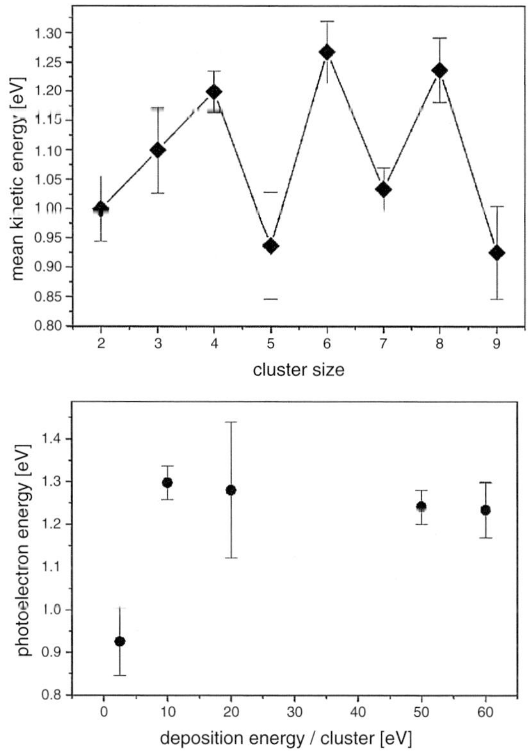

Fig. 1.10. The two graphs both show the mean photoelectron kinetic energy, defined as center of gravity of the recorded 2PPS spectra, plotted (**a**) as a function of the size of the clusters deposited under soft-landing conditions (3 eV per cluster) and (**b**) as a function of the deposition energy for Ag_9^+ clusters [93]. The corresponding photoelectron spectra were obtained with two 150 fs pulses of 3.2 eV photon energy each. The temperature in all experiments was 100 K. No diffusion of the clusters is observed at this temperature on HOPG

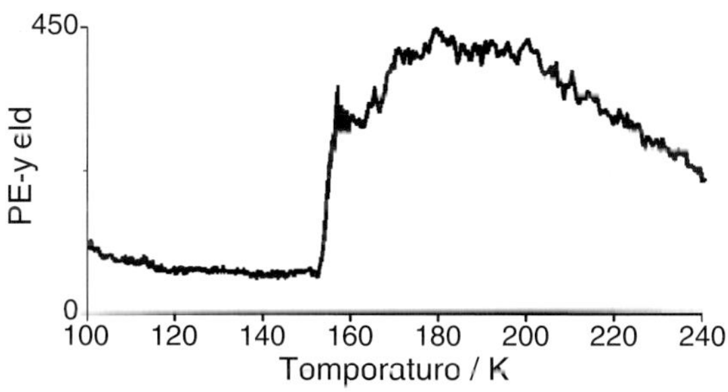

Fig. 1.11. Total photoelectron yield from Ag_6 clusters deposited on HOPG, integrated over all photoelectron energies, as a function of the substrate temperature [95]

the photoelectron yield from Ag_6 clusters deposited onto HOPG as a function of the substrate temperature. A steep increase of the photoelectron yield is apparent at 155 K. This increase indicates that the plasmon resonance of the growing nanoparticles begins to shift into the photon energy of 3.2 eV [95].

An innovative novel approach to the efficient realization of soft-landing conditions accompanied by the emerging of new and unexpected physical phenomena, like glass formation, has been put forward by Landman and coworkers [96–98]. Molecular dynamics simulations as shown in Fig. 1.12 demonstrate the dynamics and redistribution of energy during the impact of metal clusters [98] and nanocrystals [97] on solid surfaces with adsorbed liquid films of rare gases. Although, e.g., the impact of a 32-molecule NaCl cluster on an uncovered solid surface at 3 km s^{-1} (about 90 eV kinetic energy per cluster) leads to melting, disordering, fragmentation, and rebounding, the same size cluster colliding with a liquid neon film transfers its energy efficiently to the liquid for a controlled soft-landing (cf. Fig. 1.12). Impact on a higher density argon film leads to rapid attenuation of the cluster velocity accompanied by fast heating. Subsequent disordering, melting, and fast cooling by evaporation of argon quench the cluster to a glassy state as can be seen from Fig. 1.12e and f. Also the collision of a copper cluster with a Cu(111) surface was investigated [98]. While the impact of an ordered Cu_{147} icosahedral cluster on a bare copper surface at velocities larger than thermal leads to various outcomes including implantation, indentation, disordering, and spreading, deposition into a low-density liquid argon film results in efficient energy transfer to the liquid, which for incidence at velocities as high as 2–4 km s^{-1} can lead to controlled soft-landing of a crystalline cluster on the solid substrate. For an incident velocity of 2 km s^{-1}, the cluster does not melt maintaining its icosahedral structure, while for a velocity of 4 km s^{-1}, the cluster superheats and melts and subsequently recrystallizes after soft-landing on the Cu(111) surface. In collisions of the cluster with a higher density liquid (xenon), a larger fraction of the cluster translational energy is converted into internal energy of the cluster than in the case of the argon film. Such collisions lead to rapid attenuation of the cluster incident velocity, accompanied by ultrafast heating to high temperatures, superheating, and melting of the cluster. For certain impact velocities, e.g., 2 km s^{-1} into xenon, fast cooling via heat transport into the fluid can quench the metallic cluster into a glassy state [98].

This concept proposed on the basis of molecular dynamics simulations has been proven to be also experimentally extremely viable in the cluster deposition and STM investigations of Bromann et al. [99]. In this study, variable-temperature STM was used to monitor the effect of the cluster kinetic energy and rare-gas buffer layers on the deposition process of size-selected silver clusters on a platinum(111) surface. Clusters with impact energies smaller than 1 eV per atom in the cluster could be deposited nondestructively on the bare substrate, whereas at higher collision energies fragmentation and substrate damage were observed in agreement with the above discussion. Clusters at elevated impact energies could, however, be soft-landed via an argon buffer

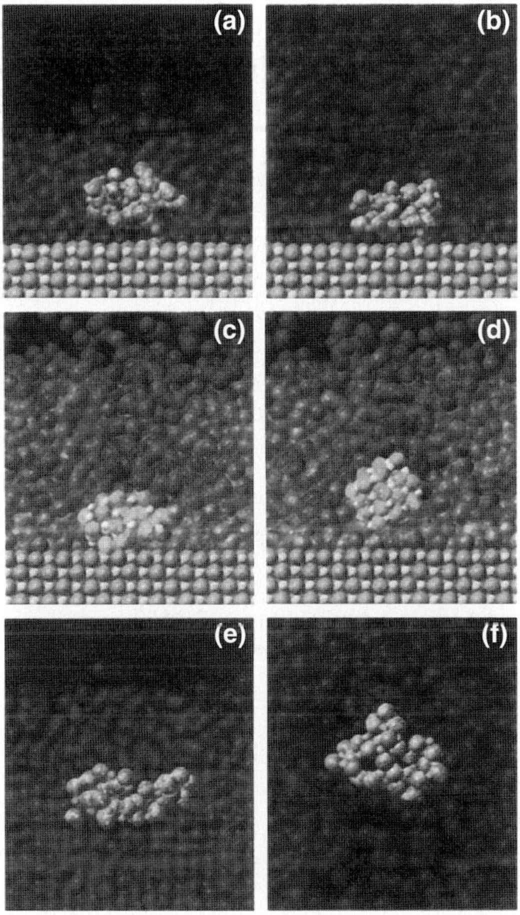

Fig. 1.12. Selected atomic configurations from simulations of collisions of an initially ordered NaCl nanocrystal (Na, small spheres; Cl, large spheres) with adsorbed Ar or Ne fluids. Disordered (glassy) $(NaCl)_{32}$ cluster at (**a**) $t = 2\,ps$ and (**b**) $t = 8\,ps$ in collision with an Ar film. Ordered cubic clusters from the limiting cases of initial orientation (**c**) face toward target surface and (**d**) corner toward surface, soft-landed on an Ne film ($t = 5\,ps$). The structural defect in (**c**) was caused by the initial impact with the liquid surface. Glassy nanocluster configurations, at (**e**) $t = 2\,ps$ and (**f**) $t = 22\,ps$, formed in a higher energy collision (center-of-mass velocity of $4\,km\,s^{-1}$) with a thick Ar film (only a portion of the system surrounding the cluster is shown). In all cases, the structures of the clusters at longer times remain virtually the same as the ones shown [97]

layer on the platinum substrate, which efficiently dissipated the kinetic energy. In this investigation, two similar experiments were compared: the deposition of Ag_7 with a kinetic energy of 20 eV on the clean surface and via an argon buffer layer. In the latter case, the clusters were deposited at 26 K substrate

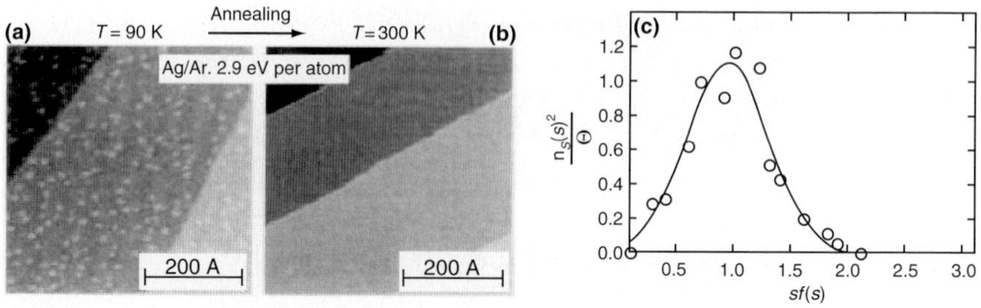

Fig. 1.13. Soft-landing of Ag_7 clusters (despite an incident kinetic energy of 2.9 eV per atom) through energy dissipation in an Ar buffer layer (10 ML) on Pt(111) at 26 K. The surface is imaged by STM after desorption of the Ar layer at 90 K (**a**) and after further annealing to 300 K (**b**). (**c**) The normalized island size distribution of the aggregates at 90 K. With n_S being the density of clusters of area s, $\langle s\rangle$ the average cluster area, and Θ the total Ag coverage, $n_S\langle s\rangle^2/\Theta$ is the normalized density of clusters of relative size $s/\langle s\rangle$. The standard deviation σ^* as a fraction of mean island size is 0.35, compared to $\sigma^* = 0.49$ for landing on the bare substrate. The line is a Gaussian fit that serves to guide the eye [99]

temperature into a preadsorbed argon layer of about 10 ML [99]. By subsequent annealing to 90 K, the argon was desorbed, and the clusters could be imaged on the platinum surface. The result is shown in Fig. 1.13a. In contrast to the bare substrate case, this landing procedure did not create any pinning centers and further annealing to 300 K resulted in cluster diffusion leaving behind the bare platinum surface (Fig. 1.13b). Furthermore, a sharp size distribution indicates only minor cluster fragmentation (Fig. 1.13c). This method is now also successfully applied by other groups to prepare materials based on monodisperse soft-landed metal clusters [47, 100].

1.2.3 Gas-Phase Analysis Techniques

Within the last three decades, highly sensitive experimental methods have been developed to characterize free metal clusters in the gas phase. First indications of the strongly size-dependent electronic cluster structure were obtained from ionization potential and electron affinity measurements [38]. More detailed information can be gained through photoelectron and absorption spectroscopy [71]. High-resolution mass-resolved data from these methods enable detailed comparison to high-level theoretical simulations, and this combination provided in particular in recent years unprecedented new insight into electronic as well as geometric metal cluster structures. Several illustrative examples will be presented in this section. In addition, novel experimental techniques have been advanced to find new and somewhat more direct approaches to the largely unknown metal cluster geometry. We will focus on representative recent examples obtained employing a free electron infrared laser, ion mobility measurements, and electron diffraction techniques. Finally, the chemical reactivity of *free* clusters is of major interest as it is expected to reveal the intrinsic size-dependent cluster properties without the influence of support and environment. The major developments and experimental techniques to study cluster gas-phase chemistry and catalytic activity will be reviewed in the last part of this section.

Geometric Structure of Gas-Phase Clusters

Atomic metal clusters represent an intermediate state between atoms and extended bulk. Therefore, they are also expected to exhibit new and different geometric arrangements that markedly depend on the exact number of atoms in the cluster and the cluster constituents, i.e., the type of metal. Revealing the geometric cluster structure inherently presents one of the largest challenges as metal clusters generally possess a large number of energetically close lying isomers. The number of isomers also rapidly increases with the number of atoms in the cluster [101]. Therefore, at finite temperature, a variety of isomers might be populated which makes structure determination difficult. On the other hand, the coexistence of different isomers also will be identified in Sect. 1.4 to be one of the reasons for the unique catalytic properties of metal clusters. The geometric cluster structure has thus an important impact on cluster reactivity, and it is highly desirable to measure metal cluster geometries as directly as possible under well-defined experimental conditions. Meeting these aspired prerequisites also ensures the most refined comparison to state-of-the-art ab initio theoretical simulations.

Four of the most powerful methods presently applied to elucidate metal cluster geometric structure will be presented in the following. These are mass-selected negative ion photoelectron spectroscopy, infrared vibrational spectroscopy made possible by very recent advances in free electron laser (FEL) technology, gas-phase ion chromatography (ion mobility measurements), and rf-ion trap electron diffraction of stored mass-selected cluster ions. All methods include mass-selection techniques as discussed in the previous section and efficient ion detection schemes which are customary in current gas-phase ion chemistry and physics [71].

Photoelectron Spectroscopy

Mass-selected negative ion photoelectron spectroscopy for the determination of metal cluster electronic structure was pioneered by Cheshnovsky and coworkers [102, 103], Lineberger and coworkers [104], as well as by Ganteför et al. [105]. Several other groups extended this work to, e.g., alloy clusters [106] and metal cluster oxides [107]. Figure 1.14 shows the schematic representation of a mass-selected negative ion photoelectron spectroscopy experiment [83] which was built based on earlier designs [102, 108, 109]. As in the majority of the presently employed setups, mass-selection is achieved in this case by a time-of-flight mass spectrometer comprising a pulsed electrostatic deflection

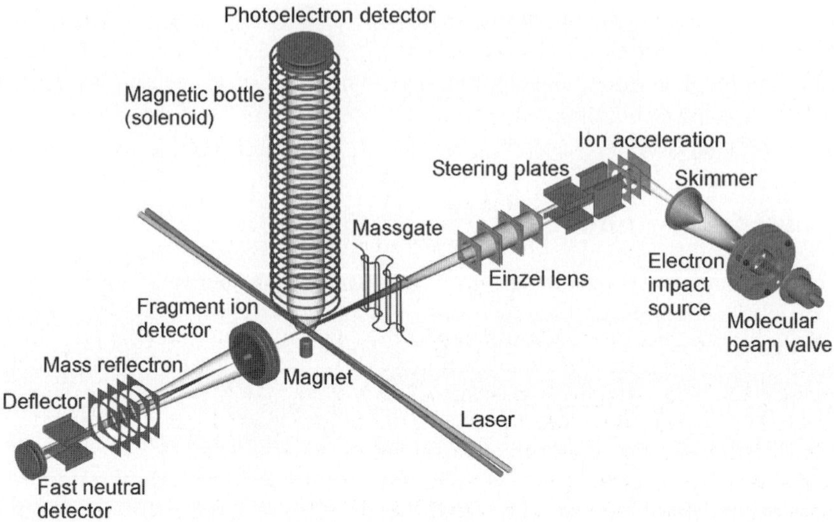

Fig. 1.14. Typical experimental setup for mass-selected anion photoelectron spectroscopy. This particular apparatus was constructed for femtosecond time-resolved anion photoelectron spectroscopy of mass-selected molecular clusters [83]. The cluster anions are produced by intersection of accelerated electrons with a pulsed molecular beam. The resulting plasma beam is skimmed and the negative ions are extracted perpendicular by a Wiley–McLaren type pulsed electrode arrangement [112]. The resulting ion pulses are steered, focused (Einzel lens), and subsequently separated spatially along the spectrometer axis according to their mass to charge ratio. The ion bunch of the clusters with the mass of interest is selected by deflecting all other ion bunches electrostatically by means of a pulsed high-voltage mass-gate [82]. The remaining ions reach the laser interaction region, and the kinetic energy of the detached electrons is analyzed with a magnetic bottle type time-of-flight photoelectron spectrometer. Additionally, in the depicted setup, the remaining neutral clusters and potential cluster fragments arising from the interaction with the laser can be analyzed by a linear reflectron mass spectrometer arrangement

mass-gate. The electrons released after interaction of the cluster anions with a laser pulse are collected and energy-analyzed by a magnetic bottle type photoelectron spectrometer. Instruments of this type combine almost 100% detection efficiency with a resolution of a few meV at 1 eV electron kinetic energy, if appropriate ion deceleration techniques are applied [102, 110, 111].

One very recent successful demonstration of the powerful capability of photoelectron spectroscopy to provide insight into electronic structure and, in combination with theoretical simulations, also into geometric cluster structure was given by the group of Wang [113]. Although the photoelectron spectra of negatively charged gold clusters have been measured already more than a decade ago [103, 104], only in a recent high-resolution study, the particular highly symmetric tetrahedral structure of the Au_{20} cluster could be revealed

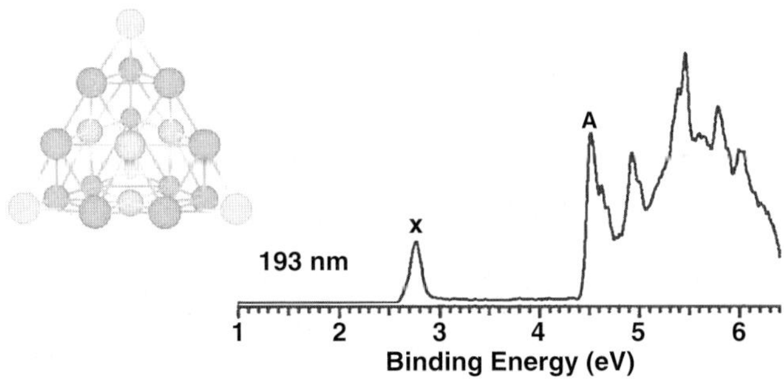

Fig. 1.15. Mass-selected negative ion photoelectron spectrum of Au_{20}^- (photoelectron intensity versus electron binding energy) obtained with 6.424 eV photon energy [113]. The extra electron in Au_{20}^- that enters the LUMO of neutral Au_{20} is removed upon photodetachment, yielding the neutral ground state signature marked X in the photoelectron spectrum. The feature A is assigned to the lowest triplet excited state of the neutral. The A–X separation, measured to be 1.77 eV, is an approximate experimental measure of the HOMO–LUMO gap of Au_{20}. This energy gap is about 0.2 eV larger than in the particularly stable C_{60} cluster (1.57 eV) [114–116]. The depicted tetrahedral structure of Au_{20} was found to be the most stable isomer in the relativistic DFT calculations [113]

through comparison to calculated lowest energy structures [113]. The photoelectron spectroscopic investigations showed that the 20-atom gold cluster has an extremely large energy gap between the lowest unoccupied molecular orbital (LUMO) and the highest occupied molecular orbital (HOMO) of the neutral cluster after photodetachment (Fig. 1.15). This HOMO–LUMO gap is even greater than that of C_{60}, and the electron affinity of Au_{20} is comparable with that of C_{60}. These observations already suggest the Au_{20} cluster to exhibit a prominent stability and chemical inertness. In conjunction with relativistic density functional theory (DFT) calculations, it was found that Au_{20} possesses a tetrahedral structure as depicted in Fig. 1.15. This Au_{20} structure has a calculated HOMO–LUMO gap of 1.8 eV, in excellent agreement with the experiment. Au_{20} therefore was claimed to represent a small piece of bulk gold. Each of the four faces represents a (111) surface of face centered cubic (fcc) gold. On the other hand, the chemical inertness and novel physical properties suggested by the observed large HOMO–LUMO gap indicate that Au_{20} is a unique cluster which, although exhibiting atomic packing similar to bulk gold, will show very different properties.

The structure of the Au_{20} cluster is, however, in marked contrast to the geometries of the smaller gold cluster ions. Only through the novel experimental technique of ion mobility measurements, which will be presented later, and new refined ab initio quantum calculations, the very particular structure of the small gold clusters was recognized recently [117–119]. The results of high-resolution photoelectron spectra of the gold clusters Au_n^- ($n = 4$–14) and relativistic DFT structure calculations confirm the structural planarity as

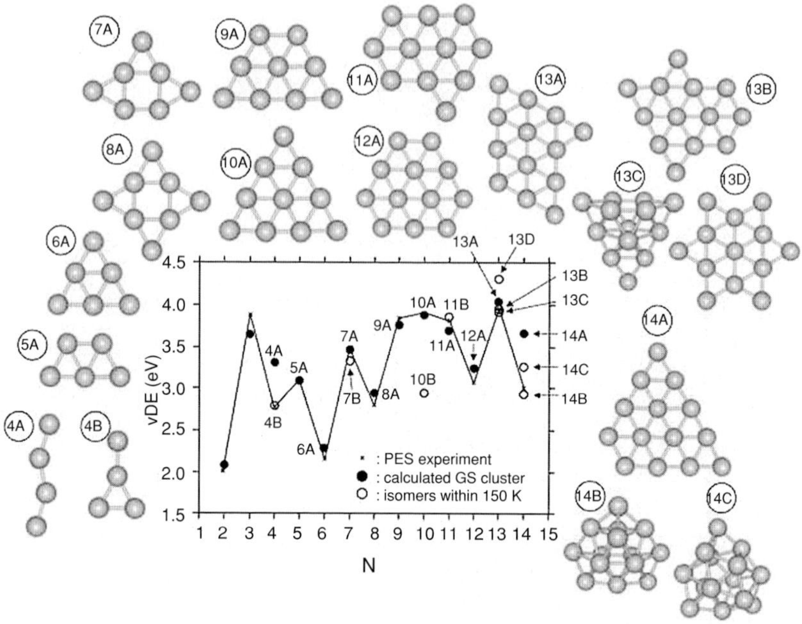

Fig. 1.16. The graph presents a comparison of theoretical and experimental vertical electron detachment energies (VDEs) for Au_n^- ($n = 4$–14). The optimized ground state structures (labeled "A" for each size) and close lying low-energy isomers are also displayed [120]. A change from planar geometry to 3D-structural motifs is apparent between Au_{12}^- and Au_{13}^-

geometric motif of gold clusters up to surprisingly large sizes [120] as indicated by the ion mobility data (see below) [118]. Figure 1.16 presents a comparison of calculated and measured vertical detachment energies from the photoelectron investigation [120]. Also displayed are the geometries of the ground state isomers. It is found that the main isomers observed experimentally indeed consist of planar clusters up to Au_{12}^-. For Au_{13}^-, a definite structural assignment cannot be done on the basis of the photoelectron data due to the existence of many low-lying two-dimensional (2D) and three-dimensional (3D) isomers. For Au_{14}^-, a 3D cagelike structure has been assigned (structure 14B in Fig. 1.16). These results therefore provide further support for the 2D to 3D structural transition at Au_{12}^-, in line with the conclusions from previous ion mobility experiments.

In a similar experiment, the group of von Issendorff measured photoelectron spectra of 55-atom silver and gold cluster anions [121]. Clusters consisting of 55 atoms have since long time been suspected to display highly symmetric structures because 55 constituents exactly built up a double shell icosahedron. Indeed, the photoelectron spectrum of Ag_{55}^- exhibits highly degenerate states, which is a direct consequence of its icosahedral symmetry (Fig. 1.17),

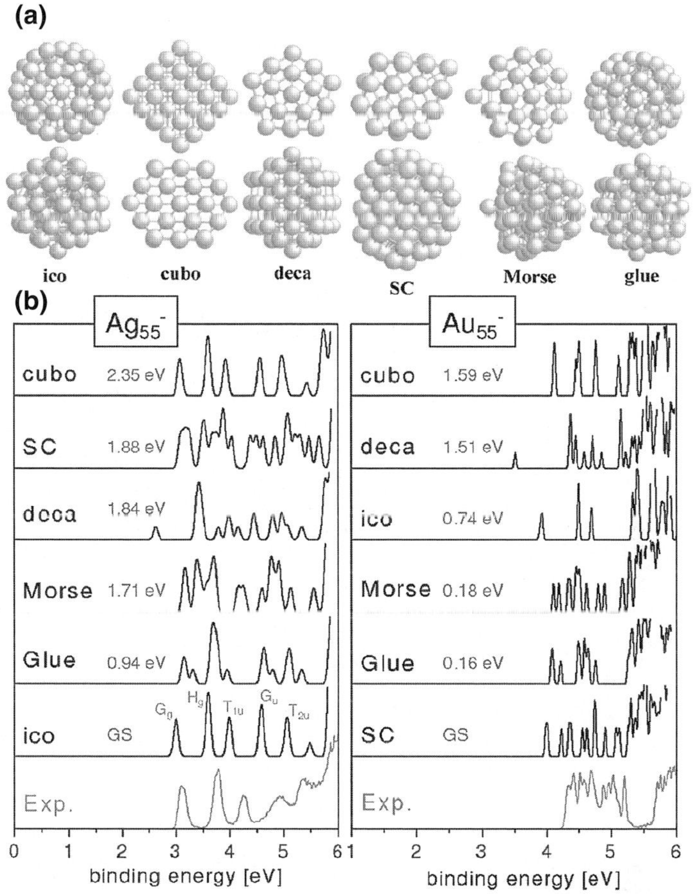

Fig. 1.17. Structures and density of states (DOS) of 55-atom silver and gold clusters obtained via DFT calculations in comparison to the experimentally determined photoelectron spectra (lowest panels). (**a**) Six candidate structures for Ag_{55}^- and Au_{55}^- representing different geometric motifs: closed atomic shell icosahedral (ICO), decahedral (DECA), and cuboctrahedral (CUBO) structures, and clusters optimized previously by classical Sutton-Chen (SC), Glue, and short-ranged Morse potentials; (**b**) DOS of the six structures (black curves) for Ag (left panel) and Au (right panel). The numbers denote the energy difference to the most stable ground state structure (GS). The experimental photoelectron spectrum of Ag_{55}^- points toward an icosahedral ground state structure with degenerate, clearly separated level structure, whereas the photoelectron spectrum of Au_{55}^- indicates a low-symmetry structure due to relativistic bonding effects in gold [121]

as confirmed by DFT calculations. A gold cluster of the same size, however, shows a completely different spectrum with almost no degeneracy, i.e., no separated state structure. This indicates that Au_{55}^- has a much lower symmetry.

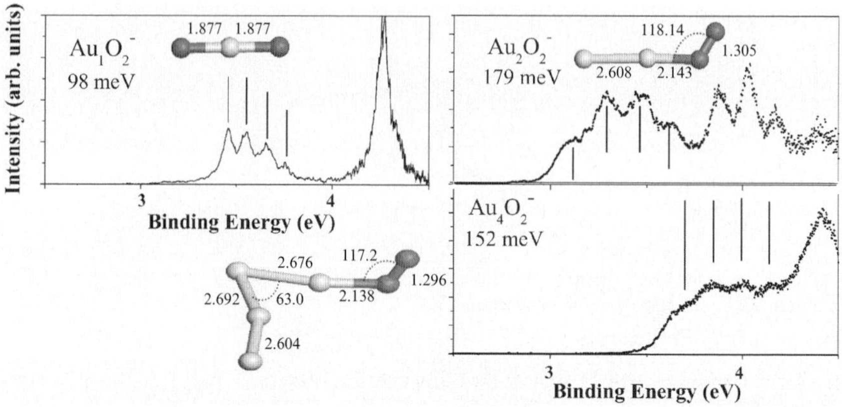

Fig. 1.18. Mass-selected negative ion photoelectron spectra for $Au_nO_2^-$ clusters ($n = 1, 2, 4$). The fine structures in the spectra correspond to the vibrational frequencies of the neutral state in the geometry of the anion. Indicated are also the geometric equilibrium structures of the cluster complexes obtained from DFT calculations [124, 125]

This marked difference of gold clusters from the homologous silver clusters is also found in experiments and calculations of smaller cluster sizes [119]. It is directly related to the strong relativistic bonding effects in gold [119, 122, 123]. A fictitious, nonrelativistic Au_{55}^- behaves very similar to Ag_{55}^-, possessing a clear icosahedral ground state and an almost identical density of states. Only in the scalar-relativistic calculation, the lower symmetry isomers are preferred, which is due to a change in the nature of the interatomic bonding originating from the relativistic effects [121].

As a final example for the capability of mass-selected anion photoelectron spectroscopy in conjunction with ab initio quantum chemical calculations to provide insight into cluster geometric structures, we present in Fig. 1.18 the photoelectron spectra of anionic gold cluster–oxygen complexes [124, 125]. These complexes are considered key reaction intermediates in the catalytic oxidation of carbon monoxide with molecular oxygen facilitated by small free gold clusters [33]. The details of the catalytic reaction mechanism will be presented in Sect. 1.5. Most importantly, the photoelectron spectra of $Au_2O_2^-$ and $Au_4O_2^-$ shown in Fig. 1.18 exhibit vibrational fine structures of 179 and 152 meV progressions, respectively. Considering that the vibrational structure in the photoelectron spectra reflects those of the final states, i.e., the neutral cluster with an anionic geometry, the observed vibrational frequencies can be assigned to oxygen molecules bound to the Au_2^- and Au_4^- clusters in a superoxo (O_2^-)- or peroxo (O_2^{2-})-like state. In this state, the molecular oxygen bond is clearly activated for further reaction as required for the catalytic cycle to proceed. In contrast, the observed vibrational frequency of the $Au_1O_2^-$

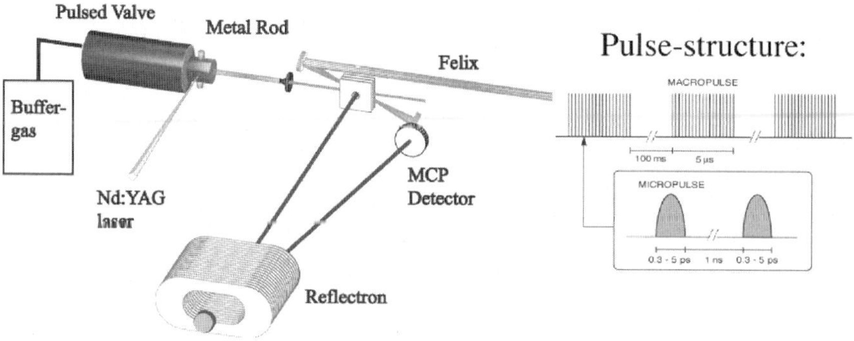

Fig. 1.19. Scheme of the experimental setup for infrared multiphoton ionization or dissociation of clusters or of metal clusters–rare gas complexes. The charged and neutral clusters are directly emitted from the laser vaporization/supersonic expansion source. The beam passes a skimmer and is subsequently crossed by the tightly focused beam of the FELIX. At some time after the FELIX pulse is over, the time-of-flight mass spectrometer acceleration plates are pulsed to high voltage, and a mass spectrum is recorded in a standard reflectron setup. Also schematically depicted is the particular pulse structure of the FELIX light [126, 127]

cluster (98 meV) is more likely to be assigned to atomic Au–O stretching, as also indicated by the calculated structure in Fig. 1.18 [125].

Vibrational Spectroscopy

Vibrationally resolved optical spectra of mass-selected free metal clusters have longtime been a dream of cluster scientists. However, the low number density of cluster beams and the nonavailability of versatile tunable light sources in the far infrared range of a few to a few tens of μm wavelength hampered this goal. The situation changed drastically with the advent of FELs operating in the far infrared region [126]. With respect to gas-phase clusters, the two key features of the FEL are tunability and high fluence on the microsecond time scale. The latter is particularly important to enable resonant excitation or even ionization via multiphoton absorption. In this regard, the pulse structure of the FEL facility FELIX (Free Electron Laser for Infrared eXperiments) in the Netherlands, which is depicted schematically on the right hand side in Fig. 1.19, is especially favorable. Trains of picosecond infrared micropulses are bunched to microsecond long macropulses. This pulse structure provides not only initial infrared excitation of molecules or clusters, but also, after dissipation of the initial excitation energy, fast subsequent multiphoton absorption within the macropulse which leads to further increase of the internal vibrational cluster energy and eventually to cluster fragmentation and/or to electron emission [126]. The ionized or charged cluster fragments are then detected mass-selectively in a high-resolution reflectron time-of-flight mass spectrometer. The experimental layout for cluster infrared spectroscopy with

FELIX is also shown schematically in Fig. 1.19. This method of infrared resonance enhanced multiphoton ionization (IR-REMPI) has been successfully applied to study fullerenes, metal carbide, metal oxide, and metal nitride clusters [126, 128–130] as well as metal–adsorbate complexes [131].

Very recently, this powerful method has now also been employed to obtain infrared spectra of pure metal clusters. For this purpose, the technique of rare gas atom tagging was applied. Charged complexes of vanadium clusters with one or two argon atoms $V_nAr_m{}^+$ ($n = 6$–$23, m = 1$–2) are irradiated by FELIX in the range between 140 and 580 cm^{-1} [127]. If the laser light is resonant with an IR active mode of the cluster, one or more photons can be absorbed by the cluster leading to desorption of the weakly bound argon atoms from the complexes. This results in a depletion of the mass-signal of the corresponding complex. IR depletion spectra are constructed in this way by recording the cluster ion intensities of the argon complexes as a function of the FELIX frequency. The result for the case of $V_8{}^+$ is shown in the lower panel of Fig. 1.20. The spectrum exhibits distinct sharp lines which can be compared to calculated infrared spectra obtained assuming different isomeric structures of the cluster. In the upper panels of Fig. 1.20, the calculated IR spectra of the four lowest energy isomers of $V_8{}^+$ are shown. The theoretical spectra are found to very sensitively depend on the geometric and electronic structure of the cluster [127]. The lowest energy structure A in Fig. 1.20 is a tetragonal bipyramid with two additional atoms on two faces. From Fig. 1.20, it is apparent that the spectra of isomers B and D are incompatible with the experimental findings. However, the spectra of the ground state isomer A and isomer C match the experimental spectrum quite well, also being very close in their geometric structure. Again, the combination of this new FEL-based infrared absorption technique with high-level quantum chemical simulations presents a powerful means to unravel the geometric structure of free mass-selected metal clusters.

Ion Mobility

In a conventional mass spectrometer, an ion M^+ is indistinguishable from an $M_2{}^{2+}$ ion because both possess the same mass to charge ratio. However, of course, the spatial extensions and the shapes of both ions differ significantly. To distinguish between structurally different species of identical mass to charge ratio in a gas-phase experiment, Bowers and coworkers pioneered a new technique similar to chromatography: the ion mobility measurement [132, 133]. In this technique, molecular or cluster ions are pulled through a rare gas filled chamber by a strong electrostatic field. On the pathway, the ions experience multiple collisions with the rare gas atoms. Thus, ions with a large geometric cross section are decelerated in comparison to ions with a smaller cross section. Under appropriate experimental conditions, this method is ideally suited to distinguish between isomers of cluster ions that exhibit even only small differences in their geometric effective cross sections [134]. By

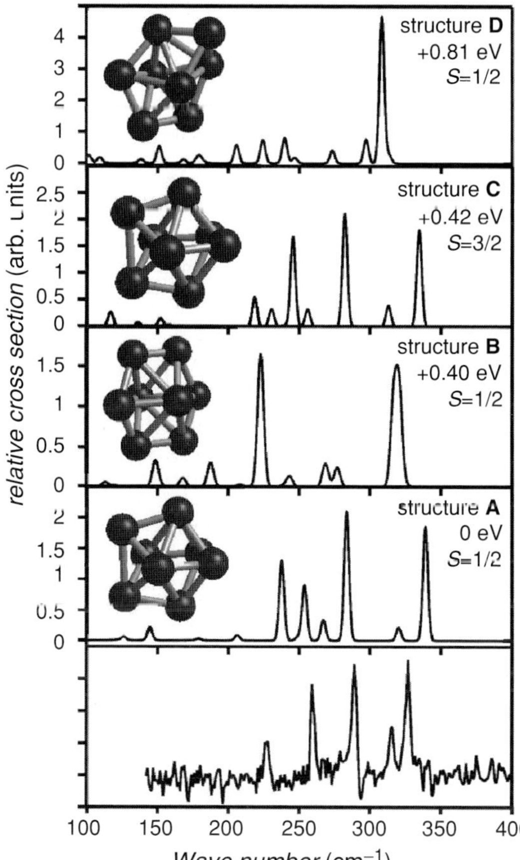

Fig. 1.20. Comparison of the experimental far infrared spectrum of $V_8{}^+$ with calculated IR absorption spectra of some geometric isomers. The experimental spectrum displayed in the lowest panel was obtained by resonant multiphoton dissociation of V_8Ar^+ with the FELIX light source and mass-selective detection of the V_8Ar^+-complex depletion during wavelength tuning of the laser [127]

comparing the measured ion mobilities with theoretically obtained geometric cross sections, a surprisingly accurate distinction between different structures is possible. This method is nowadays widely applied to, e.g., study the structure of large biomolecules [133, 135, 136]. A typical setup employed for the investigation of metal cluster ions is depicted schematically in Fig. 1.21.

As already mentioned above, this technique of geometric structure determination by ion mobility measurements has greatly contributed to unravel the surprising structural particularities of small charged gold clusters. The unexpected planar structure of negatively charged gold clusters with up to 12 atoms was first suggested based on ion mobility measurements [118]. Quantum

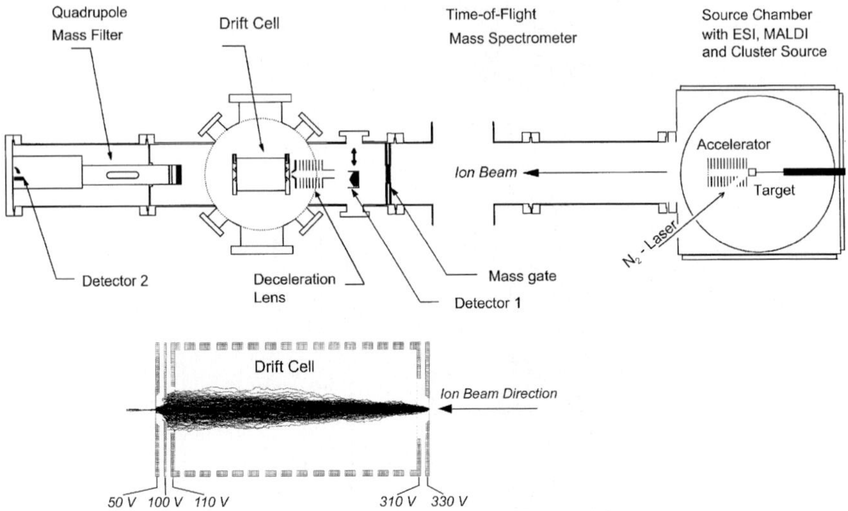

Fig. 1.21. Schematic layout of the ion mobility instrument employed in metal cluster ion studies. The setup consists of different cluster sources housed in a source chamber, a time-of-flight mass spectrometer, a helium filled drift cell, and a quadrupole mass filter for final ion detection (from right to left). Also displayed is an ion trajectory simulation of cluster ions of a mass of 500 amu drawn through the helium filled (7 mbar) drift cell at 300 K. The simulations show that under these conditions roughly 1% of the ions finally escape through the 0.5 mm diameter exit hole [137]

chemical calculations attribute these astounding geometries of the small gold clusters to relativistic effects pertinent to this heavy element [119]. Figure 1.22 presents a comparison of ion mobility measurements for positively and negatively charged gold clusters with up to 15 atoms per cluster [118]. Included in this figure is also a fit for cations (dashed line) that assumes near spherical shapes and that describes the larger gold cluster cations ($n > 12$) reasonably well. To facilitate the comparison, the cross sections of the anions are divided by this fit function (see caption of Fig. 1.22). It is apparent from this plot that for all cluster sizes, where a comparison can be made, the anions have significantly larger cross sections than the corresponding cations. Above Au_{10}^-, the anion cross sections rapidly approach the values of the cationic species. Two major reasons may be considered to explain this observations. First, they may originate from larger effective atomic radii of the anionic clusters (electron spillout). However, one would expect this effect to quickly decrease with increasing cluster size, which is not the case. Second, different structures may pertain for gold cluster cations and anions. Concurrent ab initio quantum chemical calculations confirm the structural differences between cationic and anionic gold clusters and reveal planar 2D structures with exceptionally large cross sections to be the origin of the measured mobility differences (cf. also

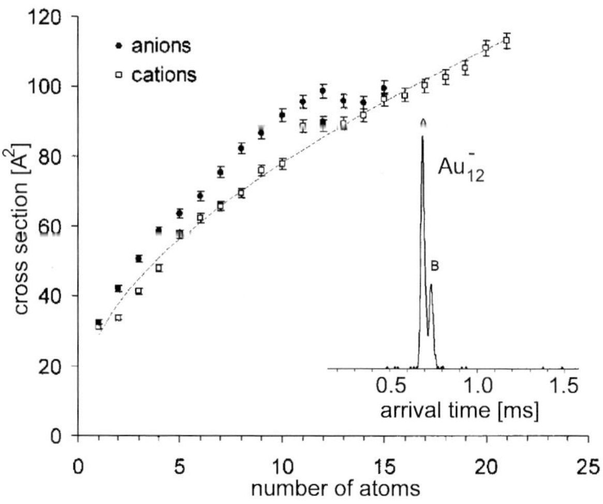

Fig. 1.22. The graph presents cross sections obtained from the ion mobility measurements for positively and negatively charged gold clusters. It is obvious that the small gold anions ($n < 13$) have much larger cross sections than the corresponding cations. The dashed line is a fit to the cationic cross sections with the function $\Omega(n) = 4/3\pi(n^{1/3}r_{Au} + r_{He})^2$. n is the number of atoms and the parameters r_{Au} and r_{He} are determined to be 1.47 and 1.15 Å, respectively. This fit function is used to normalize the calculated and experimental cross sections [118]. Also shown in the inset is the typical arrival distribution of Au_{12}^-. Au_{12}^- is the only cluster that was found to exhibit two peaks in the arrival time distribution in this experiment. This observation points toward two isomers (A and B) with largely different cross sections [118]

Fig. 1.16) [118]. The dodecamer Au_{12}^- is the only cluster size that shows two different peaks in the arrival time distribution as can be seen from the inset in Fig. 1.22. This indicates the presence of two isomers with largely different cross sections in this case. The cross section of isomer B (Fig. 1.22) is in line with a planar structure, while the cross section of isomer A is almost identical to that of Au_{12}^+, which interestingly represents a segment of the gold bulk structure [117], and therefore corresponds to a 3D geometry [118].

Electron Diffraction

The most direct approach to the geometric structures of molecules and also of clusters in the gas phase are diffraction methods, in particular the diffraction of an electron beam. Since an adequate cluster flux for such electron diffraction experiments has been so far only possible, if the full source output beam was sampled, the uncertainties in cluster size (no mass-selection) and internal energy prevent an unambiguous interpretation of electron diffraction patterns [138–145]. In a new development, a technique has been recently reported that relies upon an rf-Paul trap [146] to take advantage of the current

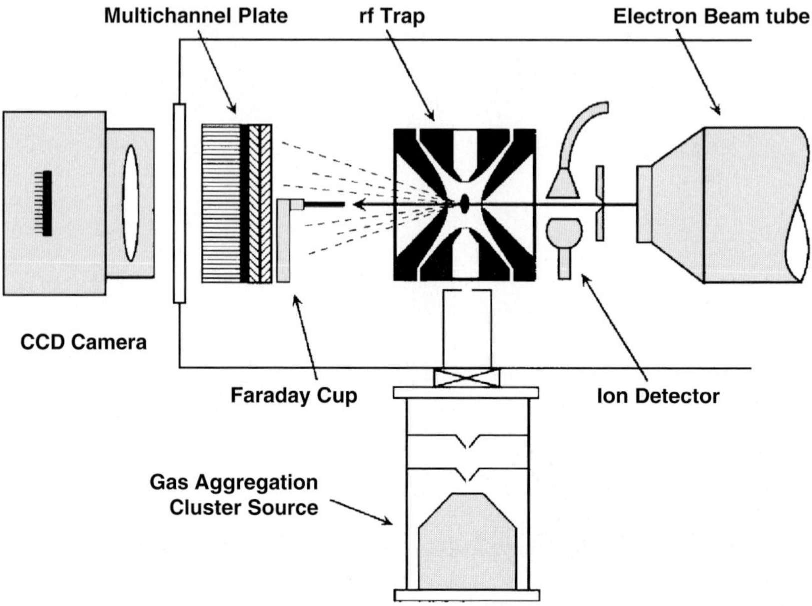

Fig. 1.23. The electron diffraction apparatus developed by Parks and coworkers includes an rf-ion trap, Faraday cup, and microchannel plate detector (MCP) and is structured to maintain a cylindrical symmetry around the electron beam axis [147]. The cluster aggregation source emits an ion beam that is injected into the trap through an aperture in the ring electrode. The electron beam passes through a trapped ion cloud producing diffracted electrons indicated by the dashed lines. The primary beam enters the Faraday cup and the diffracted electrons strike the MCP producing a ring pattern on the phosphor screen. This screen is imaged by a CCD camera mounted external to the UHV chamber. The distance from the trapped ion cloud to the MCP is approximately 10.5 cm in this experiment

source technologies yet avoiding the shortcomings of beam measurements. The rf-ion trap enables one to accumulate size-selected clusters, collisionally relax the vibrational energy distribution, and store the clusters for an adequate time to perform electron diffraction measurements. The components of the experimental apparatus are depicted schematically in Fig. 1.23. After interaction with the ion cloud inside the ion trap, the primary electron beam is captured in a Faraday cup. The diffracted electrons are detected by an image quality multichannel plate detector forming a pattern on the phosphor screen, which is then imaged and recorded on the charge-coupled device (CCD) pixel array. The diffraction pattern has the form of Debye–Scherrer rings similar to powder diffraction as a result of the orientational and spatial disorder of the trapped cluster ions. In the upper left corner of Fig. 1.24, a CCD image of such diffraction data obtained after subtracting the electron background contribution is shown [147].

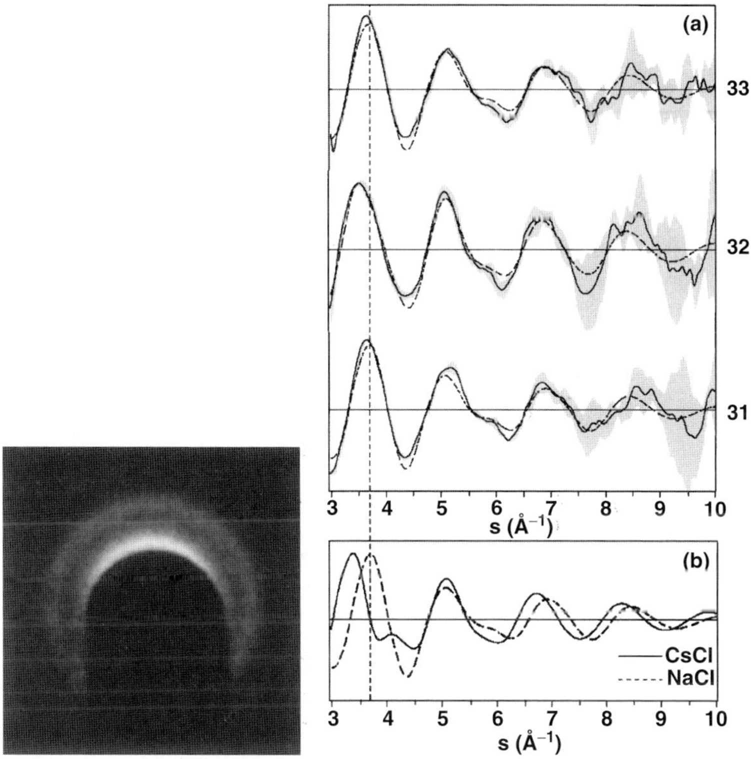

Fig. 1.24. The CCD image in the upper left corner displays the original diffraction data after subtraction of the electron background contribution. (**a**) The graphs compare the experimental (solid curve) and theoretical fit (dashed curve) of the molecular diffraction intensity at 300 K for cluster sizes $(CsI)_nCs^+$ with $n = 31$, 32, 33 averaged over multiple runs. The standard deviation $\pm 1\sigma$ is shown for each size (light gray band) characterizing the run to run reproducibility. (**b**) This graph shows the molecular diffraction intensity for the model NaCl and CsCl structures including broadening by vibrational motion at 300 K and by the finite electron beam size. The vertical dashed line which is aligned with the NaCl peak in (**b**) helps to highlight the phase shift for $n = 32$ in (**a**) [147]

As an example, an electron diffraction investigation of $(CsI)_nCs^+$ cluster structures ($n = 30$–39) are presented in Fig. 1.24. From the mass-resolved diffraction pattern contributions of both, rock salt (NaCl) and bulk cesium chloride lattice (CsCl) derived isomeric structures are observed at size $n = 32$. This particular size can form a closed shell rhombic dodecahedron corresponding to the CsI bulk structure. Interestingly, all other investigated sizes ($n \neq 32$) are dominated by the rock salt structure [147]. In Fig. 1.24a the molecular diffraction data and the best fits for $(CsI)_nCs^+$ cluster sizes $n = 31$, 32, and 33 obtained at 300 K are displayed. The plot in Fig. 1.24b shows the diffraction calculations for the DFT derived isomer structures used in the rock salt (NaCl) and the CsCl models.

The broadening introduced by temperature and finite electron beam size seriously reduces the differentiation between these structures. However, the phase shift of the diffraction peak near 3.5 Å^{-1} stands out as the most important signature for the assignment of predominant CsCl (for size 32) or NaCl (sizes 31 and 33) structural motifs [147]. Currently also geometry determinations for bare metal clusters of surprisingly small sizes are in progress in different groups which will fundamentally contribute to our understanding of the geometric factors that influence the size-dependent metal cluster chemical and catalytic properties [148].

Electronic Structure of Gas-Phase Clusters

This section will present two selected examples of electronic spectroscopy on mass-selected metal clusters in the gas phase. In the first example, time-resolved photoelectron spectroscopy is employed to monitor the real time evolution of an electronic excitation leading to the thermal desorption of an adsorbate molecule from a small gold cluster. In the second example, optical absorption–depletion spectroscopy in conjunction with first principles calculations provide insight into the excited state structure of mass-selected metal clusters.

Photoelectron Spectroscopy

Figure 1.25 displays pump–probe photoelectron data of the cluster complex Au_2CO^- [149] obtained in a mass-selective negative ion photoelectron spectroscopy setup. A pump photon of 1.5 eV energy excites the ground state of Au_2CO^-, and the fate of this electronic excitation is probed by time-delayed electron detachment with 3 eV probe photons. The kinetic energy distribution of the ejected electrons (plotted in Fig. 1.25 as electron binding energy = photon energy – electron kinetic energy) as a function of the pump–probe time-delay provides insight into the energy flow and nuclear dynamics of this model cluster–adsorbate system. From the spectra in Fig. 1.25, a rapid decay of the electronic population at low binding energies ($<$1 eV) is observed accompanied by a sharp peak evolving at 2 eV with increasing delay time. This emerging peak matches perfectly the photodetachment peak of unreacted Au_2^- (bottom trace of Fig. 1.25). Apparently, the photoexcited Au_2CO^- cluster dissociates into ground state Au_2^- and CO with an experimental dissociation time constant of about 470 fs. Besides this desorption dynamics, a fast (36 fs time constant) preceding electronic relaxation of the excited "hot" electrons below 1 eV binding energy is evident from the spectra in Fig. 1.25. This electron dynamics is explained in terms of inelastic electron scattering and thermalization due to electron–vibrational coupling. The thermalization between the electronic and the vibrational systems, which is observed to proceed about

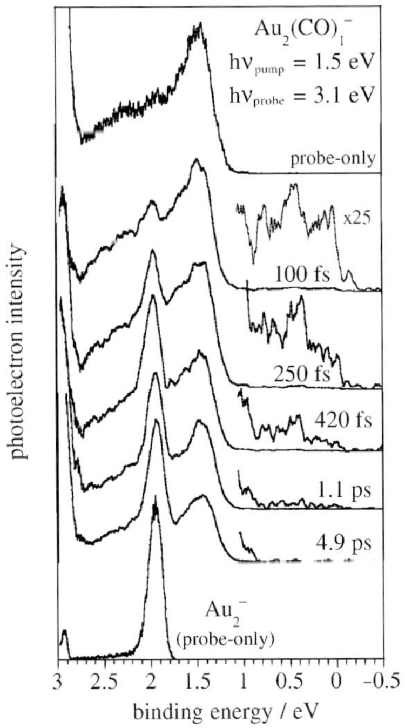

Fig. 1.25. Time-resolved pump–probe photoelectron spectra of Au_2CO^-. The topmost spectrum of Au_2CO^- is obtained with solely the probe laser beam (400 nm wavelength). The corresponding 400 nm probe-only photoelectron spectrum of Au_2^- is shown in the bottom trace [149]

100 times faster than the desorption of CO, is thus completely independent from the desorption process itself which therefore has to be regarded as purely thermal even if the initial excitation is clearly electronic [149].

Absorption–Depletion Spectroscopy

The absorption–depletion technique has already been presented in the context of vibrational spectroscopy on mass-selected metal clusters using a free electron far infrared laser. Visible or ultraviolet range lasers are more commonly available. Consequently, electronic absorption–depletion spectroscopy has been applied earlier to metal cluster system [35, 150–155]. The use of this indirect technique is necessary as direct electronic absorption spectroscopy is generally not applicable on mass-selected clusters due to the low densities even in not size-selected molecular cluster beams [156]. The absorption–depletion method was recently also used to investigate the optical spectrum in the visible wavelength range of Xe-tagged gold cluster anions (see Fig. 1.26) [157]. The absorption spectra of anionic metal clusters are particularly interesting as electronic charging is often identified as driving force in catalytic reaction

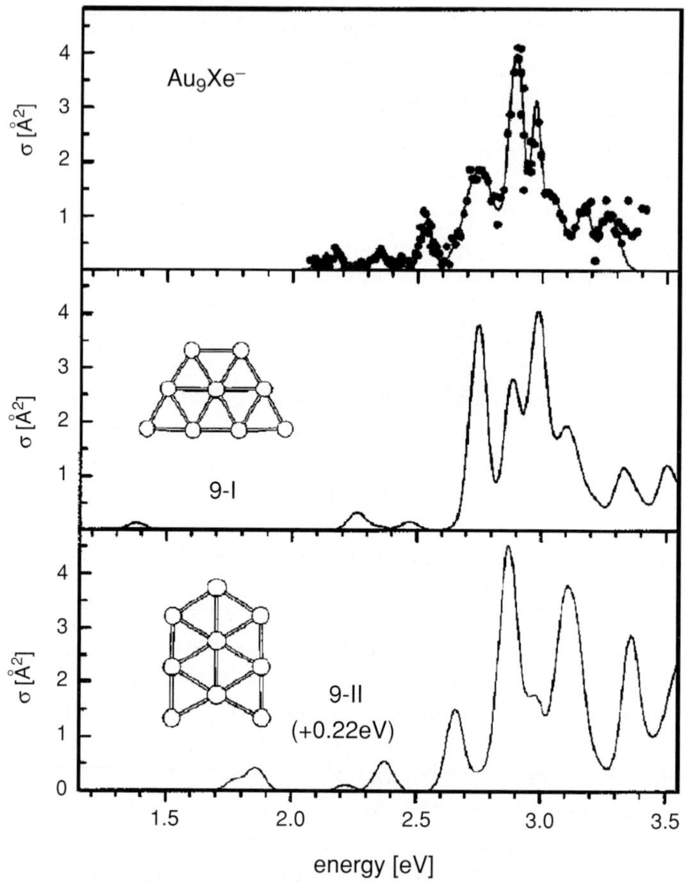

Fig. 1.26. The topmost graph shows the measured photon energy-dependent depletion cross section for Au_9Xe^-. The solid line in this plot results from fitting multiple Gaussian functions through the experimental datasets. The vertical electron detachment energy of the bare Au_9^- cluster (3.83 eV [120]) is larger than the wavelength range covered in the spectrum. The two lower panels show the TDDFT spectra for the two lowest energy 2D isomers 9-I and 9-II of Au_9^- for comparison [157]

mechanism. In addition, anions usually exhibit only rarely stable electronically excited states and very little is known about the nature of exited cluster anion states in general [158].

The absorption–depletion spectra obtained for the anionic gold clusters are highly structured with the narrowest features, which are assigned to individual electronic transitions, exhibiting bandwidths of less than 40 meV as can be seen from Fig. 1.26 for the case of Au_9^- [157]. Interestingly and in marked contrast to larger, near spherical gold colloids, which appear to be bright red in transmission, there was no indication of a characteristic transition near 530 nm assigned to the well-known surface plasmon resonance. On the contrary, but in agreement with the photoelectron spectra of gold clusters discussed earlier, the highly structured "molecular" absorption features as seen in Fig. 1.26 cannot simply be rationalized in terms of symmetry split components of a collective excitation comparable to simple s-electron metal clusters such as alkali and silver clusters [38]. Concurrent time-dependent density functional theory (TDDFT) calculations of optically allowed transitions for the most stable, planar isomers of the corresponding bare metal cluster anions are generally consistent with the experimental observations [157].

Chemical and Catalytic Properties of Gas-Phase Clusters

For almost two decades, isolated mass-selected metal clusters have been considered important model systems for the understanding of chemical reactions at macroscopic metal surface, in particular at catalytically active highly dispersed metal particles. First contributions concerned with the reactive properties of free metal clusters originate from the group of Cox and Kaldor at Exxon Research and Engineering Laboratories [159]. Several other groups, mainly in the United States, and also in Japan and Sweden, employed fast flow tube reactors to investigate the reactivity of transition metal clusters, building on this early work [20, 22, 70, 71, 160, 161]. The first catalytic reaction cycle involving a free metal cluster was discovered by Irion and coworkers employing an ion cyclotron mass spectrometer (FT-ICR). This group reported the cyclopolymerization of ethylene to yield benzene catalyzed by free Fe_4^+ clusters [25]. The different techniques presently utilized to investigate the chemical and catalytic activity of mass-selected metal clusters will be summarized in the following together with a few selected reaction examples that aim to demonstrate the applications of the various experimental approaches.

Flow Tube Reactor

This is the oldest method to investigate metal cluster reactions. In its most simple form, an extender is attached to a laser vaporization or discharge cluster source which permits the downstream addition of the reactive gases. Different variations of this principle have been realized [19, 20, 22, 160–162]. Depending on the exact design, in a flow tube reactor relatively high-reactant partial pressures might be reached. This means that (a) multiple collision conditions prevail, (b) thermal equilibration with the flow tube walls can be assumed, and (c) saturation coverages of adsorbates are expected to be observed at the largest reactant concentrations. In Fig. 1.27, a flow tube reactor attached to a pulsed arc cluster ion source (PACIS) is shown. Metal clusters are generated during the electrical discharge and during the subsequent flow through the cooled nozzle. The clusters are then drawn by the buffer gas flow into the reactor where they interact with reactive gases like, e.g., oxygen. The extension of the desired reaction is controlled by adjusting the reactant partial pressure in the reactor channel.

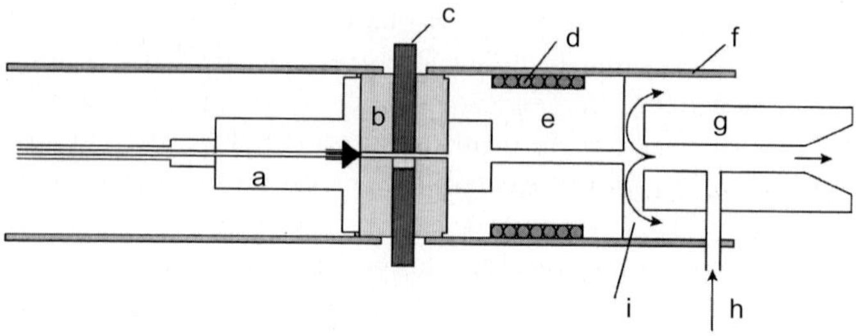

Fig. 1.27. Schematics of the PACIS with flow tube reactor attached to the source nozzle. (**a**) magnetic valve for buffer gas (He), (**b**) Macor insulator, (**c**) electrodes (the cluster material is contained in the cathode), (**d**) cooling coil, (**e**) nozzle, (**f**) source holder, (**g**) flow tube reactor, (**h**) reaction gas inlet, (**i**) pumping channels

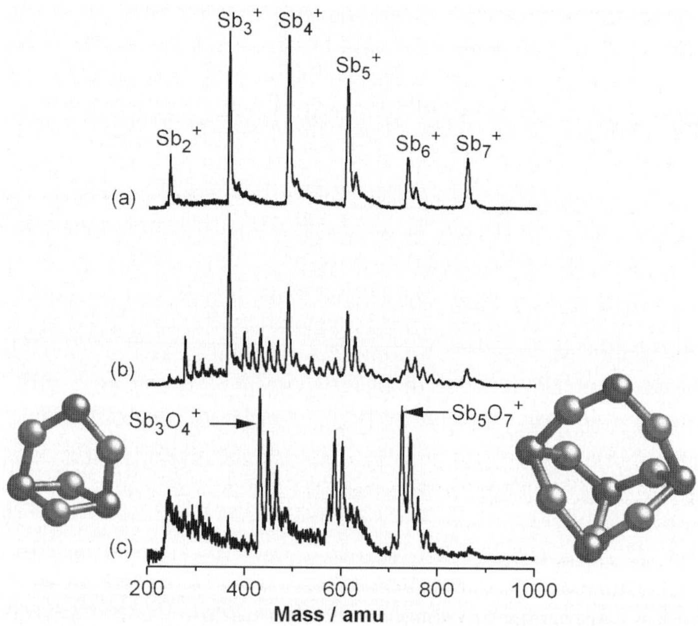

Fig. 1.28. Mass spectra of antimony and antimony oxide cluster distributions synthesized at different oxygen partial pressures in the flow tube reactor presented in Fig. 1.27. (**a**) oxygen traces only, (**b**) 4 mbar O_2, (**c**) 40 mbar O_2. The geometric structures shown for the most prominent antimony oxide cluster species in the mass spectra have been determined by quantum chemical calculations (dark *spheres*, antimony atoms; grey *spheres*, oxygen atoms) [23, 163]

Figure 1.28 presents an example for a cluster reaction study employing this experimental arrangement [23, 163]. Antimony cluster ions are generated and reacted with molecular oxygen at different oxygen partial pressures. With increasing oxygen pressure oxide peaks appear in the mass spectra that are sampled after the clusters emerged from the flow tube reactor. At the highest oxygen pressures shown in Fig. 1.28c, the cluster distribution has changed completely and new prominent mass signals appear indicating the formation of particular stable reaction products. The stoichiometry of the observed products and concurrent ab initio calculations suggest surprising structural arrangements for the stable antimony oxide clusters [163–167] eventually converging toward the chain-like modification of bulk Sb_2O_3 [23]. In turn, these cluster oxides and also the oxide clusters of the homologous element bismuth exhibit interesting reactive properties in the oxidation reactions of unsaturated small hydrocarbons, which have also been investigated in a flow tube reactor setup recently [168].

Collision Gas Cell

In the previous fast flow reactor setup, an unreactive buffer gas is often used for thermalization of the clusters before and in between the reactive collisions. In a different approach, *individual* collisions between clusters and reactive molecules are investigated to reveal the size-dependent cluster reactivity. This is achieved, e.g., in the collision gas cell experiment. In this case, a beam of neutral clusters passes through one or more cells with a low pressure of reactive gas only. Under such single-collision-like conditions, the determination of absolute numbers for the reaction probability in a collision is possible [3]. The schematic layout of this experiment is shown in Fig. 1.29. A beam of clusters is produced by a laser vaporization source, skimmed and passes two subsequent cells with reactive gas, in which the clusters experience one or a few collisions with the gas molecules. Product detection is accomplished by time-of-flight mass spectrometry after photoionization [3]. The depletion of the pure metal clusters and the appearance of products may be evaluated by employing statistical rate theory to yield the reaction probability. This method is claimed to be best suited for highly reactive systems, as the detection limit corresponds to a reaction probability of 0.01–0.05. Oxidation reactions of metal clusters, e.g., often proceed with high probability but are also highly exothermic. Thus, cluster fragmentation is likely for small clusters that cannot accommodate the excess energy liberated during the reaction causing the reaction product analysis ambiguous. For larger clusters, however the reaction probability determination is rather accurate and the method represents a very valuable complement to other cluster reactivity experiments operating at different pressure levels.

With this method, Andersson and Rosén [169] recently investigated the adsorption of hydrogen or deuterium and oxygen on neutral platinum clusters and discovered the catalytic water formation on the free clusters. Figure 1.30 displays mass spectra obtained with different partial pressures of hydrogen and oxygen in the separate collision gas cells. Panel a in Fig. 1.30 shows a mass spectrum of pure Pt_n clusters with no reactive gas in the collision cells. The mass spectrum in panel b was sampled after the cluster ions passed reaction cell 1 filled with 0.14 Pa of O_2. The additional peaks in the mass spectrum

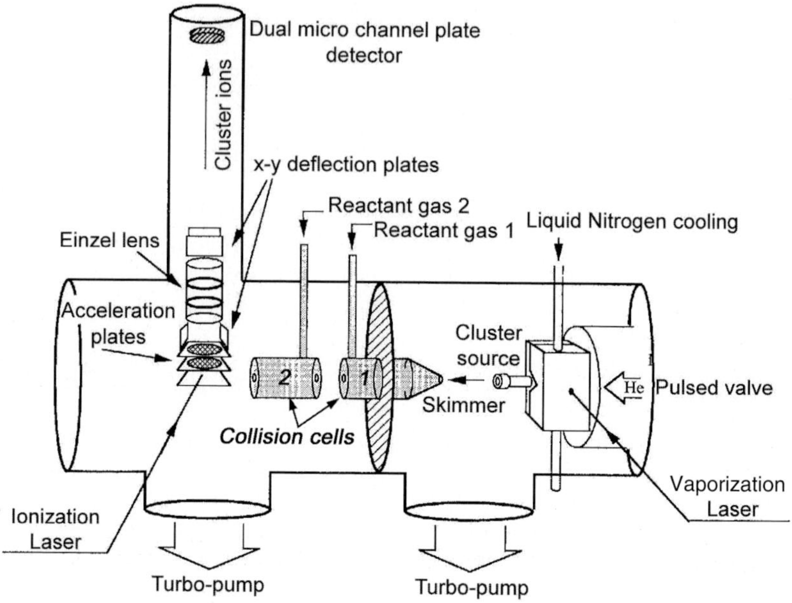

Fig. 1.29. Schematic sketch of the collision cell method for the study of metal cluster reactivity. The supersonic laser vaporization source is depicted on the right hand side. The clusters subsequently pass two collision cells in which reactions can take place. Finally, laser ionization mass spectrometry serves to detect the neutral reaction products [3]

correspond to clusters with one and, for $n > 12$, with two oxygen molecules adsorbed. If the oxygen pressure in reaction cell 1 is kept unchanged and deuterium is introduced in cell 2, the mass spectrum deviates significantly from a simple coadsorption spectrum in which both molecules would add to the mass of the corresponding platinum cluster. Instead, it is apparent from panel (c) in Fig. 1.30 that the abundance of clusters with adsorbed oxygen molecules decreases and that the amount of clusters without adsorbed molecules increases. This observation is explained by the reaction of oxygen and deuterium atoms on the clusters to form water, which subsequently desorbs [169]. The catalytic water-formation reaction is observed to proceed very efficiently on all investigated platinum cluster sizes with more than seven atoms displaying only moderate variations in the reaction probability with size.

Low-Energy Ion Guide

As discussed above, molecular beam experiments are often strongly handicapped by fragmentation phenomena, which commonly occur during reactive collisions, electronic transitions, neutralization or ionization processes. As a consequence, parent molecules and fragments can no longer be easily distinguished. In a typical low-energy guided ion beam experiment, this obstacle is

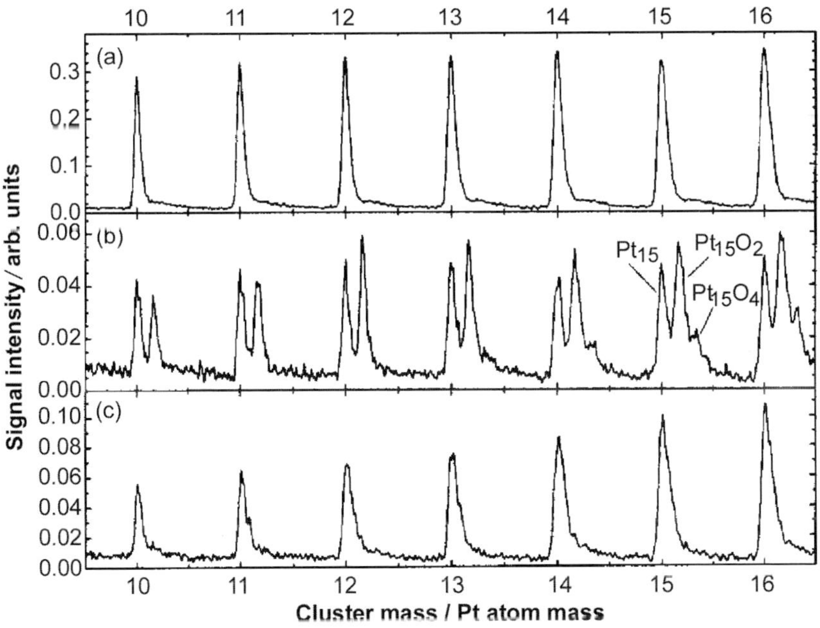

Fig. 1.30. Mass spectra of platinum clusters Pt_n before and after reaction with O_2 and D_2. In the upper mass spectrum (**a**) no reaction gas was added to the collision gas cells (cf. Fig. 1.29). Spectrum (**b**) was obtained with 0.14 Pa O_2 in reaction cell 1 and no gas in reaction cell 2. Spectrum (**c**) was sampled after the metal cluster beam passed cell 1 with 0.14 Pa O_2 and cell 2 with a deuterium pressure of 0.95 Pa. Note that the mass peaks in (**c**) are slightly broader and exhibit a small shift with respect to mass spectrum (**a**). This is due to multiple deuterium adsorption onto the platinum clusters which can, however, not be resolved in the mass spectrum [169]

circumvented by mass-selection of the cluster ion beam in advance of the reactive encounter [29, 30, 68, 70, 170–173]. The reactions commonly take place in an rf-ion guide drift tube as shown in Fig. 1.31 The cluster ions confined by the rf-field inside the ion guide have a defined low kinetic energy and experience multiple collisions with the reactive gas added to the ion guide. Mass-selected detection of the product ions is accomplished by a quadrupole mass filter followed by signal amplification with a secondary electron multiplier (SEM).

In the example depicted in Fig. 1.32, Ni_4^+ cluster ions are selected from the initial nickel cluster distribution. Treatment of this beam of tetranuclear clusters with a defined pressure of carbon monoxide and subsequent mass spectrometric analysis of the products reveals the formation of a series $Ni_4(CO)_k^+$ cluster ions. The mass spectrum in Fig. 1.32 was obtained by increasing the CO pressure gradually until no change in the product spectrum was observed anymore, i.e., when saturation of the cluster with carbon monoxide has occurred [31]. In this case, the highest molecular weight ion has the formula

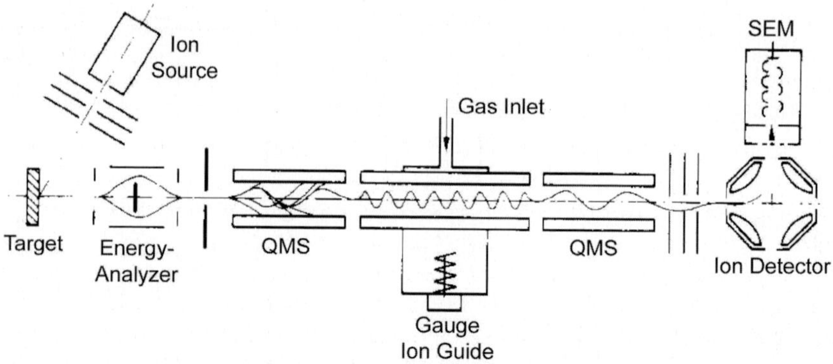

Fig. 1.31. Experimental setup employed by Wöste and coworkers to investigate the chemical reactivity of mass-selected low kinetic energy metal cluster ions with reactive gases like carbon monoxide. The metal clusters are generated by sputtering of a metal target with accelerated rare gas ions. The kinetic energy distribution of the resulting cluster ions is narrowed down during the passage through an energy analyzer. After mass-selection with a first quadrupole mass filter (QMS), the ions subsequently enter the ion guide drift tube where they are exposed to reactant molecules. Production analysis is again accomplished by a quadrupole mass filter followed by ion detection and signal amplification [31, 68]

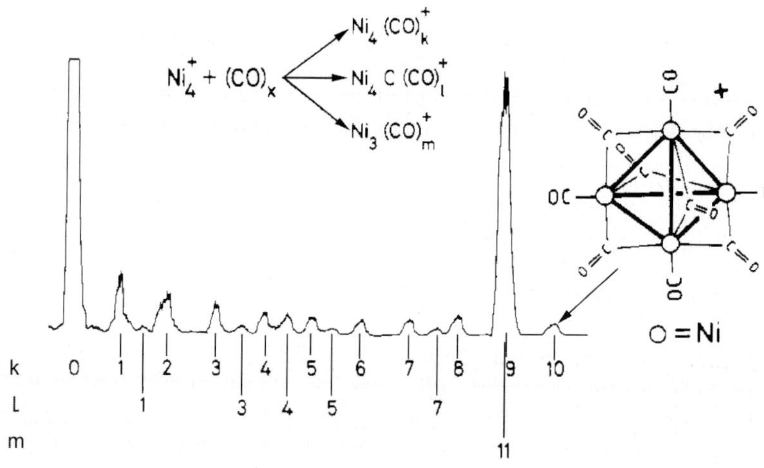

Fig. 1.32. Production mass spectrum obtained after reaction of a mass-selected Ni_4^+ cluster with approximately 3×10^{-3} mbar of CO. The geometric structure of the saturated carbonyl cluster proposed on the basis of simple electron count–structure correlations is also displayed [31]

$Ni_4(CO)_{10}^+$. According to simple electron counting rules originally introduced by Wade and by Mingos [174, 175] and extended later by Lauher to transition metal clusters [176], a four atom metal cluster will assume tetrahedral symmetry and be maximally stabilized when the total number of cluster valence electrons is 60 [31]. This total is made up of the metal valence electrons augmented by those supplied by the ligands (two from each carbon monoxide). Hence, the $Ni_4(CO)_{10}^+$ ion, produced by the reaction of excess CO with Ni_4^+, is suggested to possess a tetrahedral arrangement for which it is possible to draw a structure with four terminal and six bridging carbonyl ligands which exhibits full T_d symmetry and formally assigns to each nickel atom an 18-electron configuration. This structure is also depicted in Fig. 1.32. However, in the product ion mass spectrum in Fig. 1.32 also the development of two more series of cluster ions can be identified: these are the tetranickel carbido carbonyls $Ni_4C(CO)_l^+$ and the trinickel carbonyl clusters $Ni_3C(CO)_m^+$, which appear as a consequence of fragmentation processes [12, 31].

Ion Traps

Ion cyclotron resonance (ICR) mass spectrometry and also penning trap mass spectrometry have been demonstrated by several groups to be a powerful tool to investigate metal cluster reactivity in the gas phase under single collision conditions [24, 177–183]. From the schematics in Fig. 1.33, it can be seen that

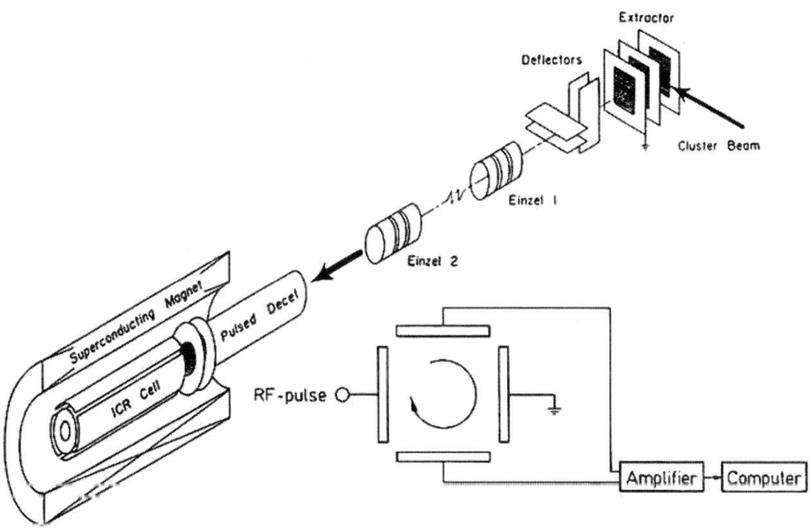

Fig. 1.33. Setup used by Smalley and coworkers to inject cluster ions into an ICR cell. The cluster beam is steered and focused by deflection plates and two Einzel lenses. Before entering the ICR cell, the cluster ions are decelerated. The inset shows the principle of an ICR mass spectrometer. Ions are excited by an rf-pulse to propagate in circular orbits. The image currents induced in the electrode plates are amplified and analyzed through computer-based Fourier transformation. The magnetic field in this scheme is perpendicular to the drawing plane. The schemes have been adopted from [71, 177, 178]

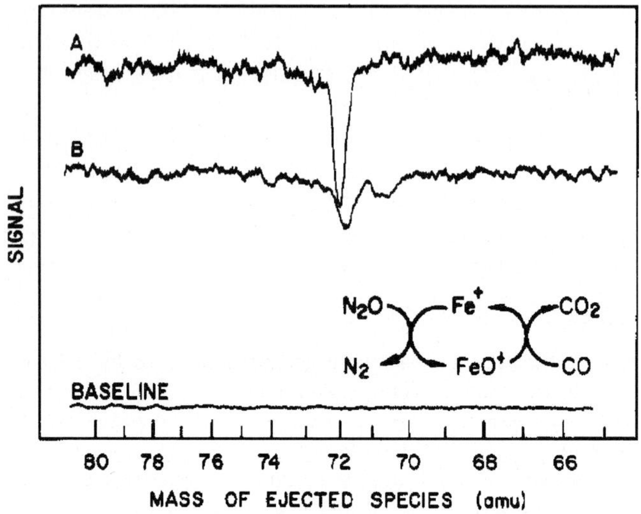

Fig. 1.34. Catalytic CO oxidation by N_2O in an ICR mass spectrometer [24]. The ICR signal shown obtained for Fe^+ while scanning the double-resonance oscillator to eject ions of a given mass from the cell. Trace B is obtained with only N_2O present in the trap. Trace A results when CO is added in addition to N_2O to the ion trap. The increase in Fe^+ signal after adding CO is caused by the regeneration of Fe^+ in the catalytic reaction cycle indicated in the inset. This is evidenced by the double resonance at 72 amu (FeO^+), which indicates that FeO^+ is reacting to Fe^+

the ICR cell is a small box or cylinder consisting of four isolated side plates and two isolated end plates. The cell is immersed in a homogeneous magnetic field. To investigate cluster reactions, the cluster ions are injected into the cell as depicted in the setup presented in Fig. 1.33. The cluster ion packets entering the cell are excited by an rf-pulse applied to one plate. Due to the homogeneous magnetic field traversing the ICR cell, the ions are thus lead to perform a circular motion in the cell. The cyclotron frequency of the circular motion depends on the ion mass and the magnetic field strength [71]. The image current induced on two opposite plates of the cell is amplified and analyzed by Fourier transformation. Each ion of different mass gives rise to a sine function. From the obtained superposition of sine waves, the Fourier analysis reveals the complete mass spectrum.

This method has a particular high mass resolution and sensitivity. The ICR instruments operate under UHV conditions and reactions are investigated under strict single collision conditions in this setup. In an early example presented in Fig. 1.34, ICR mass spectrometry was applied to reveal the catalytic oxidation of carbon monoxide by atomic iron ions [24]. The shown double-resonance spectra reveal that the relative increase in Fe^+ is due to a reaction by FeO^+ that occurs only when both, N_2O and CO, are present supporting the catalytic cycle depicted also in Fig. 1.34.

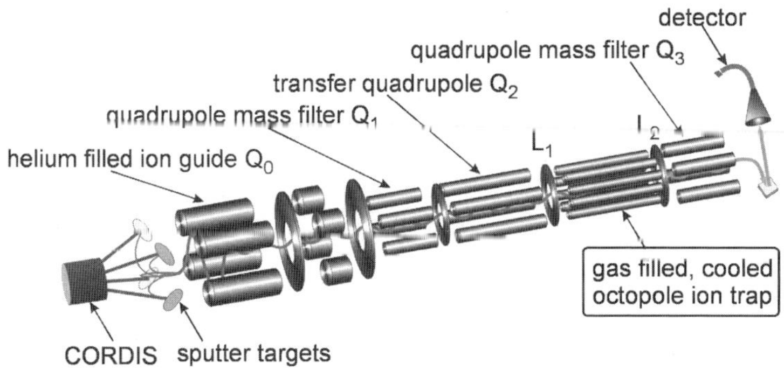

Fig. 1.35. Experimental setup for the investigation of gas-phase catalytic activity of mass-selected metal clusters. The cluster ions are sputtered from solid targets with a CORDIS, mass-selected (Q_1), and guided at low energies (Q_0 and Q_2) into the temperature controllable octopole ion trap. By means of appropriate switching of the lenses L_1 and L_2, the reaction products are extracted and subsequently mass-analyzed by another quadrupole mass filter (Q_3) [32, 186]

A rather new approach to investigate metal cluster reactivity and catalysis consists in storing the mass-selected cluster ions in a temperature variable rf-octopole ion trap filled with helium buffer gas and small partial pressures of reactant gases under multicollision conditions. The major advantages of this approach for the investigation of gas-phase reactions is the precise control over reaction time, reactant concentrations, and reaction temperature inside the ion trap. The trap acts like a test tube for gas-phase reaction kinetics studies. It is inserted into a standard guided ion beam apparatus [30, 68, 70]. The schematic layout of the experimental setup is displayed in Fig. 1.35. Metal clusters are prepared by a sputter source based on the cold reflex discharge ion source (CORDIS) [17] presented in Sect. 1.2.1. The charged clusters are steered into a first helium filled quadrupole Q_0 (cf. Fig. 1.35), which serves as "phase space compressor" to collimate and thermalize the cluster ions. The ion beam is further guided into a mass-selective quadrupole filter Q_1 to select one particular cluster size. Subsequently, the cluster ions are transferred with a third quadupole Q_2 into the home-built octopole ion trap [32, 184], which is filled with metal cluster ions up to the space charge limit (about 10^4 clusters mm^{-3}). The trap is prefilled with a helium partial pressure on the order of 1 Pa, and thermal equilibration of the clusters entering the trap is achieved within a few thousand collisions with the buffer gas, i.e., in a few milliseconds under our operating conditions. The cluster ions are stored in the trap for a considerably longer time period, typically for several seconds, without significant ion loss. A closed cycle helium cryostat attached to the trap allows temperature adjustment in the range between 20 and 350 K. Time-resolved kinetic measurements on the timescale of seconds are performed with small, well-defined partial pressures of the reactants O_2 and CO present in

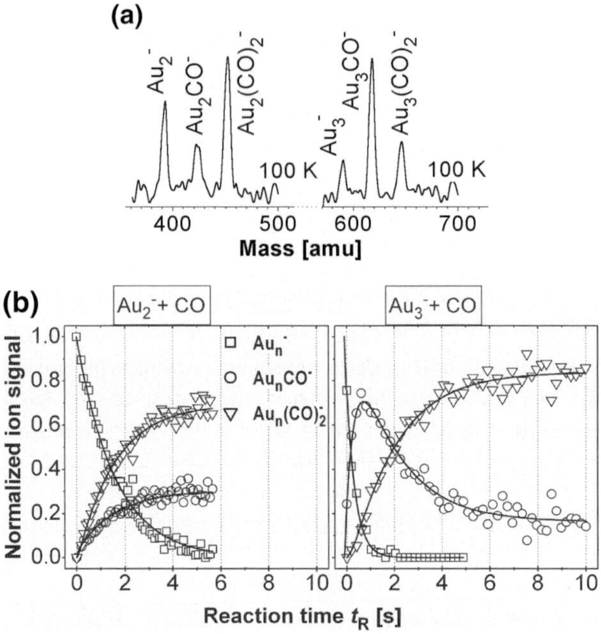

Fig. 1.36. (**a**) Production distributions analyzed after trapping Au_2^- and Au_3^-, respectively, inside the octopole trap filled with 1 Pa of helium and a small partial pressure of carbon monoxide. The mass spectra are obtained with the trap held at a temperture of 100 K. (**b**) Kinetic traces of the reaction of Au_2^- (*left*) and Au_3^- (*right*) with CO at a reaction temperature of 100 K. Plotted are the product ion concentrations as a function of the reaction time t_R (storage time in the ion trap). The open symbols represent the normalized experimental data. The *solid lines* are obtained by fitting the integrated rate equations of the reaction mechanism [(1.1) and (1.2)] to the experimental data. In the case of the Au_2^- reaction, the helium pressure in the trap is $p(\mathrm{He}) = 0.96$ Pa and the carbon monoxide partial pressure amounts to $p(\mathrm{CO}) =^- 0.24$ Pa. For the Au_3^- reaction $p(\mathrm{He}) = 1.08$ Pa and $p(\mathrm{CO}) = 0.02$ Pa

the trap. For this purpose, all ions are extracted from the trap after defined storage time by means of a pulsed electrostatic field (lenses L_1 and L_2 in Fig. 1.35) and are subsequently mass-analyzed by a final quadrupole mass filter (Q_3). By recording all ion concentrations as a function of the reaction time, the kinetics of the reaction may be obtained.

As an example, the reactions of the negatively charged gold dimer and trimer cluster ions with carbon monoxide inside the rf-ion trap are presented in Fig. 1.36 [185]. While no reaction products of Au_2^- and Au_3^- with carbon monoxide are detected at room temperature, cooling down leads first to the formation of mono-carbonyls and at the lowest temperatures around 100 K to a maximum adsorption of two CO molecules on Au_2^- and also on Au_3^- as can be seen from the mass spectra depicted in Fig. 1.36a. In order to deduce the reaction mechanism of the observed reactions, the reactant and product ion concentrations were recorded as a function of reaction time, i.e., the residence time of the ions inside the trap. The resulting kinetic traces for Au_2^- as well as for Au_3^- at 100 K reaction temperature are depicted in Fig. 1.36b. The single points represent the experimental data normalized to the total ion concentration in the trap during reaction. The kinetic traces of Au_2^- and Au_3^- have a strikingly different appearance. In both cases, the Au_n^- signal decreases exponentially, but the carbonyl product concentrations show very different evolution as a function of the reaction time. In the case of Au_3^-, the monocarbonyl can clearly be identified as an intermediate with decreasing concentration at longer reaction times, whereas for Au_2^-, the mono- and dicarbonyl concentrations rise simultaneously to reach an equilibrium.

The reaction mechanism for the observed kinetics is obtained by fitting the integrated rate equations of a proposed mechanism to the experimental data. This procedure is very sensitive to the type of mechanism, and it was possible to rule out all but one reaction mechanism. Most interestingly, the kinetics of both, Au_2^- as well as Au_3^-, are best fit by the same mechanism. This reaction mechanism is represented by the following equations ($n = 2, 3$):

$$\mathrm{Au}_n^- + \mathrm{CO} \rightarrow \mathrm{Au}_n\mathrm{CO}^- \tag{1.1}$$

$$\mathrm{Au}_n\mathrm{CO}^- + \mathrm{CO} \rightleftarrows \mathrm{Au}_n(\mathrm{CO})_2^- \tag{1.2}$$

The adsorption of CO occurs sequentially with Au_nCO^- as an intermediate product. As the carbon monoxide concentration in the trap is constant, all reaction steps are taken to be pseudo-first-order in the simulations. Purely consecutive reaction steps do not fit the experimental data, and it is therefore essential to introduce a final equilibrium (1.2). The fits of the integrated rate equations to the data are represented by the solid lines in Fig. 1.36b and are an excellent match to the experimental results.

The detailed kinetics and energetics of the reactions in the rf-ion trap can be understood by considering that the total pressure inside the ion trap is on the order of 1 Pa, which means that the experiment is operating in the kinetic low-pressure regime. Therefore, a Lindemann-type mechanism has to be considered for each reaction step, and the reaction rates depend on the buffer gas pressure [187, 188]. As a consequence, the obtained pseudo first order rate constant k contains the termolecular rate constant $k^{(3)}$ as well as the concentrations of the helium buffer gas and of the reactants; in the case of the adsorption reaction of the first CO molecule (1.1):

$$k = k^{(3)}[\mathrm{He}][\mathrm{CO}] \tag{1.3}$$

Gas-phase reactions in this pressure regime are described in more detail by the Lindemann energy transfer model for association reactions which is represented by the following equations [33, 187, 188]:

$$\mathrm{Au}_n^- + \mathrm{CO} \rightleftarrows \mathrm{Au}_n\mathrm{CO}^{-*} \tag{1.4}$$

$$\mathrm{Au}_n\mathrm{CO}^{-*} + \mathrm{He} \rightarrow \mathrm{Au}_n\mathrm{CO}^- + \mathrm{He}^* . \tag{1.5}$$

The reaction model includes the elementary steps of the initial formation of an energized complex Au_2CO^{-*} (rate constant k_a) and its possible unimolecular decomposition back to the reactants (k_d) in competition with a stabilizing energy transfer collision with helium buffer gas (k_s). Assuming all these elementary reaction steps to be again of pseudo-first-order and employing steady state assumption for the intermediate, the overall third-order rate expression is obtained to be [189]

$$k^{(3)} = \frac{k_a \cdot k_s}{k_d + k_s\,[\mathrm{He}]}. \tag{1.6}$$

As the experiment is operating in the kinetic low-pressure regime, the decomposition rate constant can consequently be considered to be much larger than the stabilization rate constant term: $k_d >> k_s[\mathrm{He}]$. This leads to a simplified expression for the termolecular rate constant, which can be applied to the experimental conditions present in the ion trap experiment

$$k^{(3)} = \frac{k_a \cdot k_s}{k_d}. \tag{1.7}$$

The ion–molecule association rate constant k_a as well as the final stabilization rate constant k_s are well represented by ion–molecule collision rate coefficients as specified by Langevin theory [189]. For each reaction step identified in the kinetic evaluation and fitting procedure, a Lindemann-type reaction scheme as described above has to be considered as elementary reaction scenario. One important consequence of this reaction scheme is that, according to this theory, ion–molecule reactions are basically charge-induced dipole interactions and exhibit no activation barrier, i.e., no temperature dependence. Thus, the only temperature dependent rate coefficient is the unimolecular decomposition coefficient k_d, but as this reaction leads to the reformation of the reactants, ion–molecule reactions generally exhibit an overall inverse temperature dependence. This has been confirmed experimentally [186]. For the case of the gold carbonyls, through statistical rate theory analysis [187] of the absolute termolecular rate coefficients, it was possible to estimate the binding energy of CO to Au_2^- and Au_3^- to be about 0.5 eV [185].

In addition, the apparent dissimilarity of the kinetic traces for Au_2^- and Au_3^- in Fig. 1.36b is found to originate from the different ratio of the adsorption and desorption rate coefficients of the second CO molecule for Au_2^- and Au_3^- (1.2). This kinetically favored formation of $Au_3(CO)_2^-$ points toward a significantly enhanced stability of this complex with respect to all the other investigated carbonyl compounds in this study [185].

1.2.4 Surface Analysis Techniques

Morphological Properties of Supported Clusters

The microscopic structure of small clusters on surfaces is of primordial interest as it influences most of their physical and chemical properties. Acquiring the exact arrangement of the atoms in a supported cluster is an enormous experimental challenge. There are experimental methods, most of them based on diffraction, which are perfectly adapted for obtaining the positions of the atoms in a material. However, diffraction methods were used only for obtaining structures of free clusters so far, as described in the previous section.

Local probes like scanning tunneling microscopy (STM) or atomic force microscopy (AFM), as well as optical spectroscopies in the visible or infrared (IRS) are other relevant experimental techniques to obtain information on the geometric structure of clusters on surfaces or at least part of it. Impressive STM images revealed recently the exact arrangement of all atoms in small linear gold clusters [190]. For 3D clusters, these methods are restricted to map the positions of the atoms on the cluster surface only, where the atomic arrangements on the lateral facets are in most cases difficult to obtain. Infrared spectroscopy can give in principle the complete cluster structure, if all normal modes are accessible. For metal clusters, the corresponding transitions lie, however, in the far infrared and conventional laboratory equipment can hardly reach this frequency range as of now. In this context, experiments on FELs may lead to important advancements as the important range of 40 to $-2,000\,\mathrm{cm}^{-1}$ can be covered. Nevertheless, infrared studies were carried out for measuring vibrational modes of ligand molecules adsorbed on metal clusters in laboratory experiments; these experiments indeed lead to some information on cluster's morphology and structure [191,192]. When operated in the local mode, infrared spectroscopy is of great promise and first results are obtained by using inelastic electron tunneling with STM (IETS-STM) [193]. Last but not least, highly sensitive surface photo-absorption spectroscopy in combination with ab initio calculations may also lead to cluster structures as absorption spectra are strongly dependent on the exact atomic arrangement [194]. In the following some illustrative examples are summarized.

Atomically Resolved STM Images of Clusters and Particles on Surfaces

A complete characterization of the morphology of a 27-atom palladium cluster supported on a cleaved MoS_2 single crystal was achieved by STM studies [195]. Information on the 3D shape, azimuthal orientation on the substrate, and arrangement of atoms on lateral facets was obtained (Fig. 1.37). The cluster clearly consists of two monolayers. The first layer is a regular hexagon with a three-atom wide side and an additional atom attached to the left hexagon side. This layer is composed of 20 atoms. The top layer has a regular hexagonal shape and the sides are composed of two atoms, resulting in a top layer of 7 atoms. This Pd_{27} cluster is schematically shown also in Fig. 1.37. The observed atomic arrangement is identical to the structure of bulk palladium with a (111) basal plane. From this atomically resolved STM image, in particular from the 2D representation, the relative orientation between cluster and the substrate is obtained in real space, showing that the azimuthal orientation of the particle's dense Pd atom rows are parallel to the rows of sulfur atoms of the MoS_2 surface.

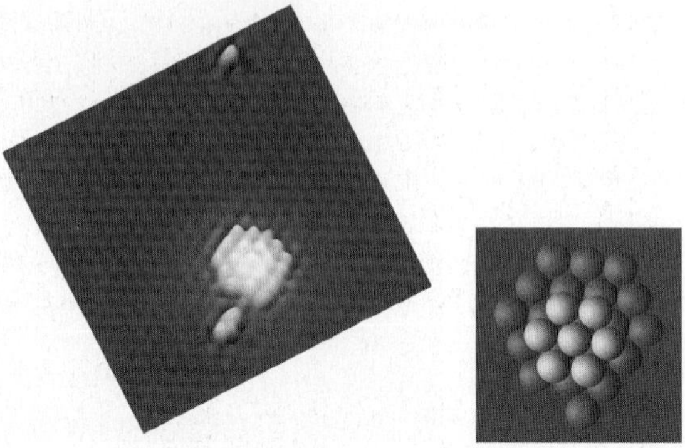

Fig. 1.37. 3D STM image of an isolated Pd cluster of ~15 Å in size supported on the (0001) surface of MoS_2. A single atom and a trimer are seen on the top and on the bottom of the figure, respectively. The dense rows of the Pd atoms in the cluster are in the [110] direction, which is parallel to the ([1120]) dense rows (sulfur atoms) of the MoS_2 substrate. A schematic representation of the cluster containing 27 atoms is depicted on the right [195]

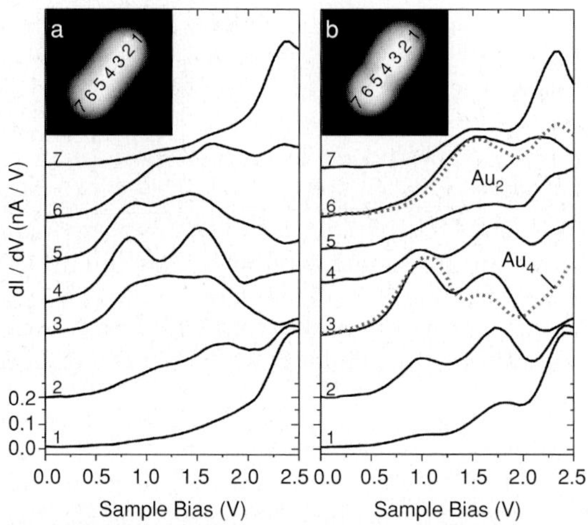

Fig. 1.38. dI/dV spectra of (**a**) a bare Au_7 chain and (**b**) an Au_7 chain with a single CO molecule adsorbed on the fifth atom, taken at positions indicated in the insets. The broken lines in (**b**) show spectra of a bare Au_2 and Au_4 chains for comparison [190]

In a recent study, small linear Au_n clusters on an NiAl(110) surface were assembled and imaged with atomic resolution [190]. An STM topographic image for a bare Au_7 chain is depicted in the inset of Fig. 1.38a. Interestingly and important for studying chemical properties of size-selected clusters on surfaces, even a single CO adsorbed on the linear gold cluster could be imaged and the adsorption site could be characterized in detail (inset Fig. 1.38b). In

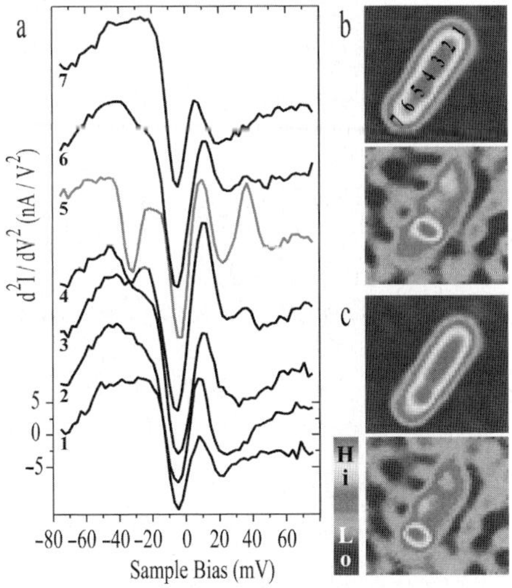

Fig. 1.39. (**a**) Vibrational spectra of an Au_7 chain with a CO molecule adsorbed on the fifth atom, taken at the positions indicated in (**b**). (**b**) Topographic (*top*) and vibrational (*below*, sample bias tuned to vibrational resonance shown in **a**) images of an Au_7 chain with a CO molecule adsorbed on the fifth atom and (**c**) same on the seventh atom [190]

these studies, information on the electronic structure (Fig. 1.38) of the bare Au_7 and CO-covered cluster were obtained, and vibrational frequencies of the adsorbed CO molecule could be measured (Fig. 1.39). From the change of the electronic structure of the cluster upon CO adsorption it is proposed that CO predominantly couples with the electronic states located on a single atom and not with the delocalized electronic entity of the cluster. The measured vibrational frequency at around $300\,cm^{-1}$ is attributed to the CO hindered rotation. This example impressively demonstrates the potential of local probes to obtain fundamental information on small clusters on surfaces.

Imaging large particles on oxide surfaces with atomic resolution was obtained for palladium nanoparticles on Al_2O_3 [196]. Figure 1.40a and b show STM images of a Pd particle with atomic resolution. They reveal a (111) layer for the top facet. The measured nearest neighbor distance is 2.76 ± 0.07 Å, indicating the absence of any strain in the cluster [$d = 2.75$ Å for Pd(111)]. In addition, these results show that for such large particles supported on oxide surfaces, it is possible to obtain atomic resolution across the entire cluster surface, although the tunneling conditions at the edge of the cluster change. As the tip apex moves away from the top layer of the particle when approaching the edge, tunneling occurs between the top layer of the particle and the atoms on the side of the tip. In some cases, it was even possible to obtain atomic resolution on the largest side facets of the particles, revealing (111) crystallographic orientation.

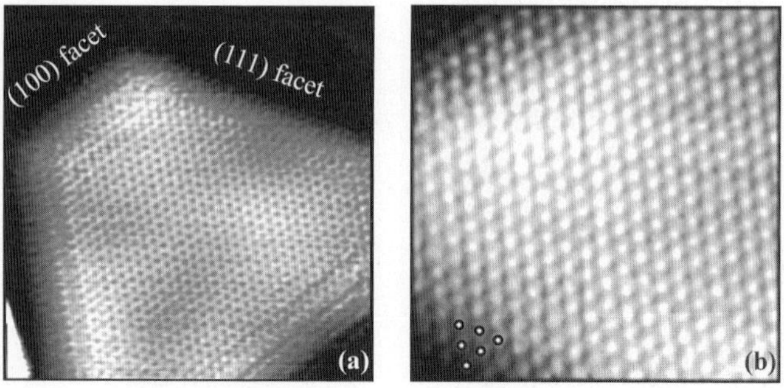

Fig. 1.40. Atomic-resolution images of crystalline nanosize Pd clusters. (**a**) $95 \times 95\,\text{Å}^2$, tunneling current $I_t = -0.8\,\text{nA}$, tunneling voltage $V_s = -5.0\,\text{mV}$. (**b**) $45 \times 45\,\text{Å}^2$. The resolution is also obtained a few layers down the sides, allowing identification of the side facets. The dots indicate atomic positions consistent with a (111) facet. $I_t = -1.8\,\text{nA}$, $V_s = -1.5\,\text{mV}$ [196]

In addition, the morphology of the particle was characterized by three parameters, which are the height of the particles, the width of the top facets, and the ratio between the side lengths of the top facets. The smallest particles of crystalline structure observed in these studies have a top facet of 20–30 Å width and a height of 5–10 Å corresponding to about 2–4 layers. Furthermore, these measurements revealed quantitative information about the work of adhesion (W_{adh}) by comparing the observed shapes of the particles with the one resulting from a Wulff construction based on calculated surface energies. For Pd on Al_2O_3, a value of $W_{adh} = 2.8 \pm 0.2\,\text{J}\,\text{m}^{-2}$ was reported.

These three examples clearly illustrate two cluster size domains. The Pd_{27} and the linear gold clusters are in a size regime where the physical and chemical properties are not scaleable from the bulk, and quantum size effects are important. This is the size regime most important for this discussion. The latter example shows a larger cluster consisting of thousands of atoms; the STM picture clearly reveals that in this size regime, the surface morphology can be understood with single crystal analogies. The chemical properties in this size regime are scalable from the bulk and they are discussed in depth in Chap. 3.

Counting Cluster Atoms by Using Rare Gas Decoration

An indirect method to image small clusters on surfaces was obtained for size-selected Ag_{19} clusters on a Pt(111) surface where the cluster was decorated with rare gas atoms [197].

This leads to a pronounced corrugation in the rare gas necklace around the cluster, making the rare gas atoms a sensitive probe to determine cluster size. Figure 1.41a and b show a high-resolution STM image and a linescan across the cluster of a gas-phase deposited Ag_{19} cluster on Pt(111) surrounded by 12 Kr atoms. This situation can easily be obtained as the binding energy of Kr to

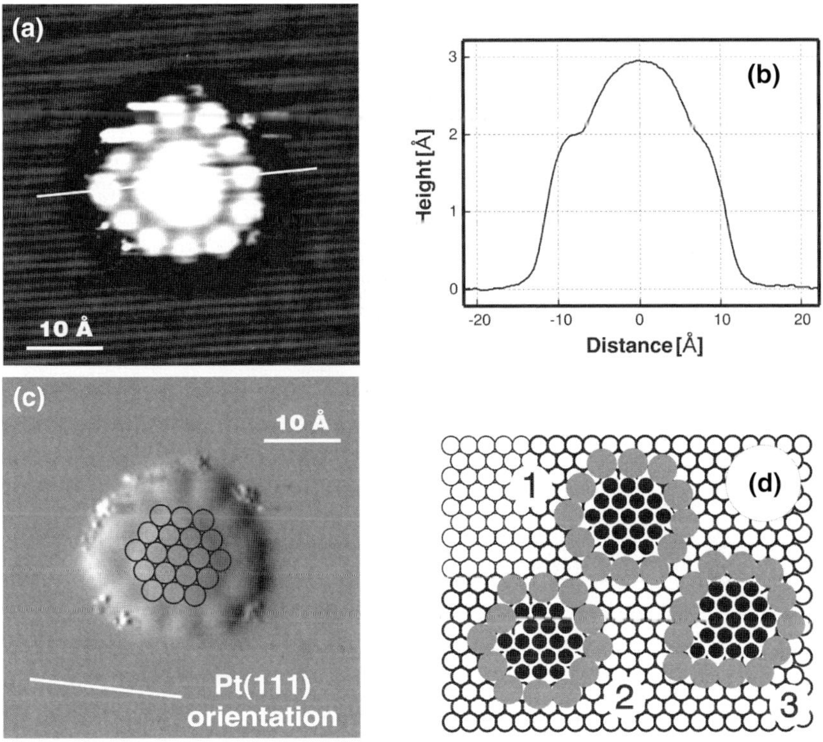

Fig. 1.41. Ag_{19} cluster on Pt(111) surrounded by 12 Kr atoms. (**a**) $I_t = 380\,pA$, $V_s = -200\,mV$. The image has been filtered to increase the contrast; (**b**) linescan across the cluster; (**c**) model superposed the image; (**d**) model [197]

step edges and around the cluster is enhanced. Based on the observed number of Kr atoms around the silver cluster and the shape of the ring, the cluster size is deduced from a simple geometric model. Assuming a silver atom diameter of 2.77 Å, knowing the Pt(111) lattice constant, and taking into account the height measurements of the cluster leading to only 1 ML structures, a regular hexagon containing 19 atoms fits perfectly the topographic image (Fig. 1.41c). Figure 1.41d shows that the Kr necklace is sensitive to the form of the cluster and the exact number of silver atoms in the cluster. This is the first experiment where a one-to-one correspondence after cluster deposition between cluster size on the surface and in the beam was obtained, and it directly shows that soft-landing small clusters onto solid surfaces is feasible.

Cluster-Based Materials

Few examples of materials made up of regular arrays of individual clusters have been reported so far. C_{60} and other fullerenes are the most prominent representatives in this respect. A new nanoscale cluster modification of antimony

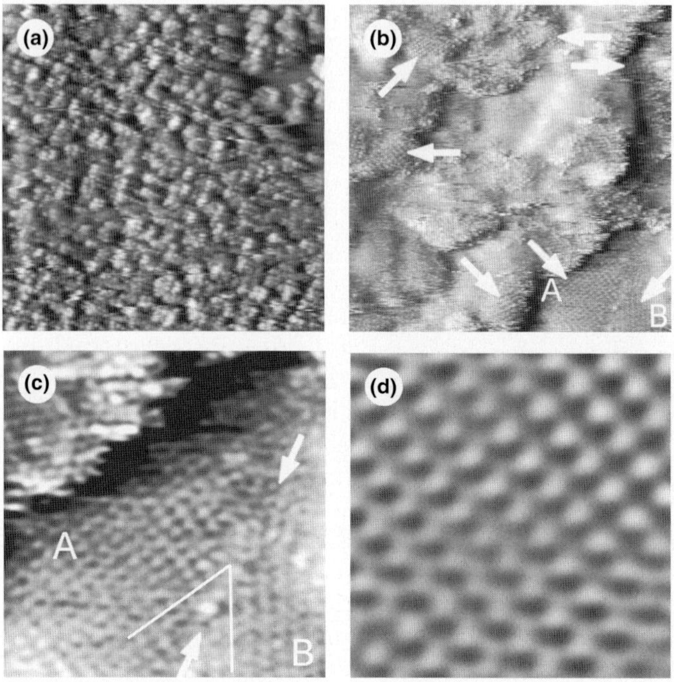

Fig. 1.42. STM data of a 2 ML antimony film on MoS_2 in different magnification. (**a**) $2,480 \times 2,480\,Å^2$ image showing the in general amorphous structure of the film. (**b**) The areas marked by arrows in the $310 \times 310\,Å^2$ image exhibit a well-ordered structure. (**c**) $100 \times 100\,Å^2$ magnification of the structures A and B in (**b**). Along the line marked by arrows, a grain boundary between two neighboring ordered areas is visible. (**d**) shows a $39 \times 39\,Å^2$ excerpt of area A. The image was high-frequency filtered to emphasize the well-ordered structure. Tunneling parameters of all images: 0.6 V, 1.0 nA [198]

consisting of Sb_4 tetrahedral units was recently discovered by STM in vapor deposited antimony thin films [198]. A solid state modification of antimony consisting of Sb_4 was hitherto unknown due to the instability of the Sb_4 clusters leading to polymerization and amorphous structures even when quenched with liquid nitrogen. On an $MoS_2(0001)$ surface, the thermal deposition of Sb_4 at a substrate temperature of $-190°C$ also results in macroscopically amorphous antimony films without apparent structural order. This can be seen from Fig. 1.42a, which displays a 2 ML thick film. If such a film is investigated at the highest resolution, local, well-ordered areas marked by arrows in Fig. 1.42b can be identified at coverages of 1–2 ML. These local structures always exhibit a simple, nearly cubic symmetry. The observed crystalline ordered antimony structures can only be explained, if each protrusion measured with STM (Fig. 1.42c and d) is assigned to one undissociated Sb_4 tetrahedron.

The STM data therefore indicate that a new allotropic modification of antimony composed of Sb_4 clusters exists on the nanometer scale.

Electronic Properties of Supported Clusters

Whereas the electronic structure of atoms, molecules, and extended solids is quite well explored from both experimental and theoretical viewpoint, little is known about the systematic development of the electronic structure as single atoms are combined to form clusters and eventually solids.

There are fundamental questions to be answered such as the onset of metallic behavior in small supported metal clusters, the influence of quantum size effects on the electronic structure of small particles, or the evolution of the electronic bands in matter. Therefore, it is of primordial interest to study individually, i.e., locally or in a size-selected manner, these objects of "embryonic" condensed matter [199]. In the context of chemical and catalytic properties of small clusters, the evolution of the electronic structure of small clusters is one of the primary keys for understanding the observed size-effects. The interaction of molecules with clusters and the potential energy surface of a reaction path are sensitively influenced by resonances of molecular states of the reactants and intermediates with the electronic structure of cluster. Subtle changes of the latter, induced by changing the cluster size even by a single atom or by changing the isomeric form of the cluster, can drastically alter the chemical behavior. The clusters' electronic structure can alternatively be used as characteristic fingerprints and thus stabilities and changes of clusters during reaction can be monitored. As for the geometric cluster structure, there are few experimental examples showing the complete picture of the evolution of the electronic structure with size for a variety of systems and the expected relation with the chemical properties. In the following, we present some experimental approaches presently in use together with obtained experimental results.

Electron Energy Loss Spectroscopy

The most widely used technique to get information on the electronic structure of clean surfaces, nanostructures on surfaces, or even molecules adsorbed on surfaces is ultraviolet photoelectron spectroscopy (UPS). The difficulty of this method, when applying it to clusters on surfaces, is to obtain sufficient spectral contrast between the low number of adsorbed clusters and the substrate [45]. Thus, electron energy loss spectroscopy (EELS) is more successfully used as a tool for the investigation of the electronic structure of supported clusters. An interesting test case for its suitability is the characterization of supported monomers, i.e., single Cu atoms on an MgO support material [200], as this system has been studied in detail before with various surface science techniques [201–204]. The adsorption site of Cu on MgO(100) is predicted

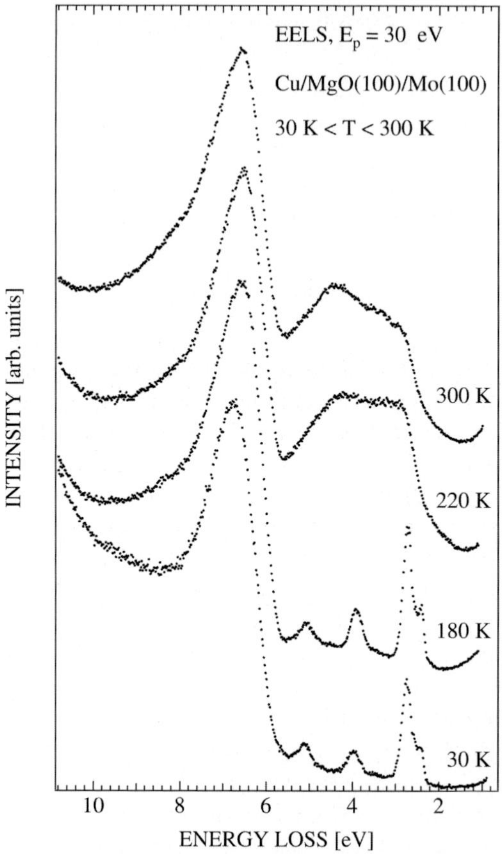

Fig. 1.43. EEL spectra of thermally evaporated Cu on an MgO(100) film. Note the difference in the spectra when going from 180 to 220 K [200]

to be on-top of an oxygen atom with calculated adsorption energies differing markedly in the literature (0.3–1.4 eV) [205–207]. Moreover, surface diffusion energy barriers of about 0.4 eV are expected for these systems [205, 207]. Figure 1.43 displays EEL spectra of Cu deposited on MgO(100), at a coverage of about 0.12 ML, at four different substrate temperatures. At low temperatures between 30 and 180 K, three main energy losses are discerned within the band gap of MgO at 2.7, 3.9, and 5.1 eV, respectively, which represent a fingerprint of single Cu atoms reflecting specific atomic transitions within neutral and ionic Cu [208]. Losses observed at 3.9 and 5.1 eV were assigned to transitions in neutral Cu. On the basis of recent reaction studies on supported Au clusters on MgO(100) [209] and ab initio modeling of metal adhesion on oxide surfaces with defects [210], the 2.7 eV loss was attributed to a transition due to the binding of the Cu atom to an $F^o{}_s$ center which in the case of Au clusters [209] and of Ag [210] is found to transfer charge of about 0.5e from the oxide surface toward the adsorbed metal atom (where e is the electron charge).

At $T = 220\,\mathrm{K}$, a drastic change is observed in the spectrum displaying now a broad emission band between 2.0 and 5.5 eV. From earlier experiments performed at room temperature [201–204,211,212], it is concluded that Cu atoms become mobile between 180 and 220 K and form islands (cf. also Sect. 1.2.2). If it is assumed that the temperature dependence of the diffusion is given by

$$T_\mathrm{d} = E_\mathrm{d}\left[k_\mathrm{B}\ln\left(\frac{4D_0}{\mu_j a^2}\right)\right]^{-1}, \tag{1.8}$$

where E_d is the energy barrier of the diffusion process, D_0 the prefactor in the diffusion coefficient, μ_j the hopping frequency, and $a \approx 3\,\mathrm{Å}$ the distance between neighboring adsorption sites [207,213]. An estimation of this barrier with typical values of $1\,\mathrm{hop\,s^{-1}}$ and a diffusion coefficient of $10^{-3}\,\mathrm{cm^2\,s^{-1}}$ leads to a value of $E_\mathrm{d} = 0.50(5)\,\mathrm{eV}$, in good agreement with the value of 0.4 eV [205] and 0.45 eV [207] predicted by first principle studies. This example shows how EELS can be used to characterize impurities on the oxide thin films and to get information on their stability and migration properties.

Furthermore, EEL spectra of small Ag_n ($n \leq 13$) and Cu_n ($n \leq 7$) clusters show clear evidence for a size effect in their electronic structure [214]. The clusters were generated by sputtering with an UHV-compatible Xe-ion gun [45]. After size-selection with a quadrupole mass spectrometer, they have been deposited in situ in submonolayer quantities on a magnesium oxide film. Figure 1.44 displays EEL spectra taken at $T = 45\,\mathrm{K}$ for 0.04 ML of Ag_n ($n \leq 13$) clusters, deposited at low kinetic energy ($E_\mathrm{k} = 3$–$6\,\mathrm{eV}$) to prevent their fragmentation [215]. Each deposition was made on a freshly prepared film to avoid creation of defects, which are known to act as pinning centers for deposited clusters [216,217].

Several distinct energy loss peaks appear within the MgO band gap (between 1 and 5.5 eV energy loss [218]) as a function of cluster size. These loss peaks cannot be assigned to low-lying transitions in the atom or in the ion [103,208,219,220]. EEL spectra of vapor deposited Ag, which forms islands and thin films via surface diffusion at sample temperatures between $T = 100$ and 500 K, have shown losses at 3.8 and 3.2 eV attributed to an Ag surface plasmon and to an Ag–MgO interface plasmon, respectively [218]. In contrast, the EEL spectra shown in Fig. 1.44 and recorded at $T = 45\,\mathrm{K}$ exhibit clearly a size dependence, which reflects the change in the electronic structure of the clusters. A similar behavior has been observed in optical absorption spectra of Ag_n ($n \leq 21$) clusters deposited in rare gas matrices [221], which has been interpreted as a manifestation of collective excitations (Mie plasmons) of the s electrons influenced by the ellipsoidal shape of the clusters. Some similarities but also some differences in the general trend with cluster size have been observed by comparing the optical absorption data shown in [221] with these EELS data [214]. In this context, it is important to note that EELS probes

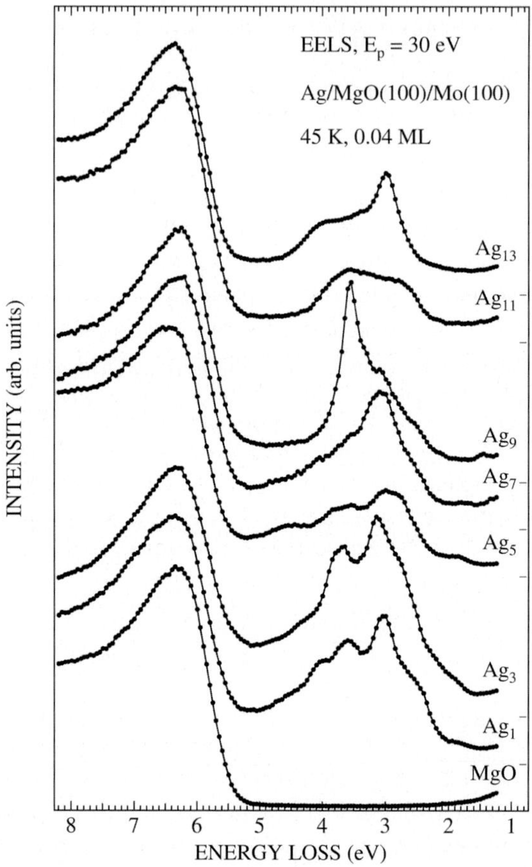

Fig. 1.44. EEL spectra of the bare MgO(100) film and of 0.04 ML of size-selected Ag_n $(n = 1-13)$ clusters deposited on identical oxide films. The spectra were taken at $T = 45\,K$. Note the distinct band gap of the MgO film and the size-dependent transitions within the band gap for the clusters [214]

electronic excitations and transitions which are dipole forbidden $(\Delta l \neq 1)$ in addition to those observed by absorption spectroscopy $(\Delta l = 1)$. The peaks shift toward lower energy loss $(\Delta E = 0.2 - 0.3\,\mathrm{eV})$ observed in EELS as compared to the optical absorption peaks as well as the different peak shapes in both techniques may be caused by the different environment (e.g., different dielectric constants: MgO support versus rare gas matrix), to the loss of central symmetry in the EELS experiment (surface versus matrix), and/or to different cross sections of both emission processes. Furthermore, charge transfer excitations from the MgO substrate to the cluster or vice versa are also present in the spectra, which have to be considered in a complete spectral analysis, a task that has not yet been achieved. Most importantly, these results illustrate that the characteristic size-dependence electronic structure of small Ag_n clusters is maintained upon deposition onto oxide surfaces and a strongly size-dependent chemistry is expected.

Near-Edge X-Ray Absorption Spectroscopy

Recently, small size-selected Cr_n clusters ($n = 1$–25), deposited onto a (1×1)O/Ru(001) single crystal surface, were investigated by near-edge X-ray absorption spectroscopy (NEXAFS) [47, 222]. These clusters were formed by sputtering of a chromium target and were mass separated in a magnetic dipole field. They were soft-landed by deposition into argon monolayers pread-sorbed on the substrate. The NEXA spectra show size-dependent shifts of the L_3 absorption edges (see Fig. 1.45), which display broad bulk-like line shapes. Furthermore, cluster–substrate interaction of the Cr clusters results in Cr_2O_3-like absorption signatures and inhibits diffusion and island formation even at elevated temperatures. In contrast, chromium atoms on Ru(001) have a high mobility, showing indications of coalescence and island formation. Sizeable shifts in the absorption energy after flash heating are attributed to diffusion or surface alloying. The difference between oxidation of the clusters on the oxide covered substrate and on the pure Ru metal is clearly visible in the absorption spectra, shown in Fig. 1.46.

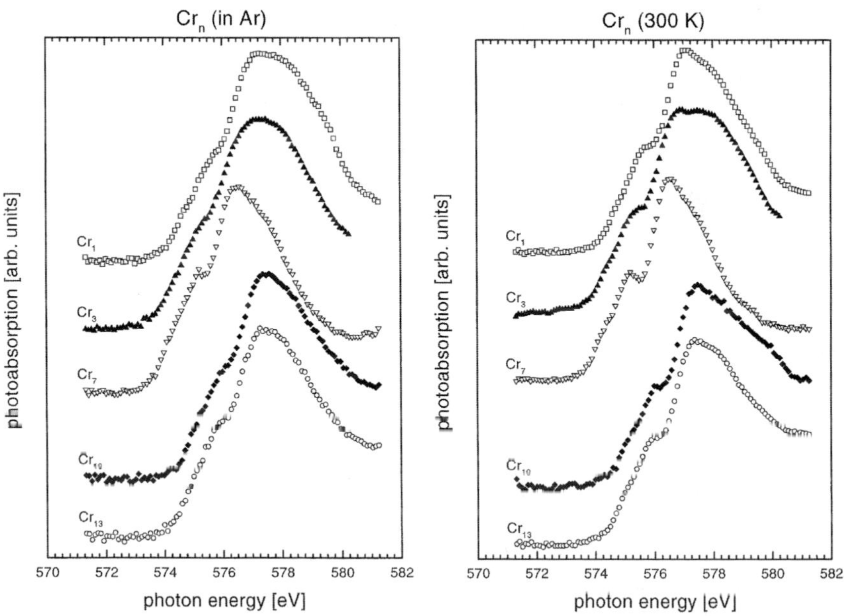

Fig. 1.45. L_3 photoabsorption spectra of size-selected Cr_n clusters as deposited and flash heated to 300 K. The spectra are normalized to the peak height of the main feature

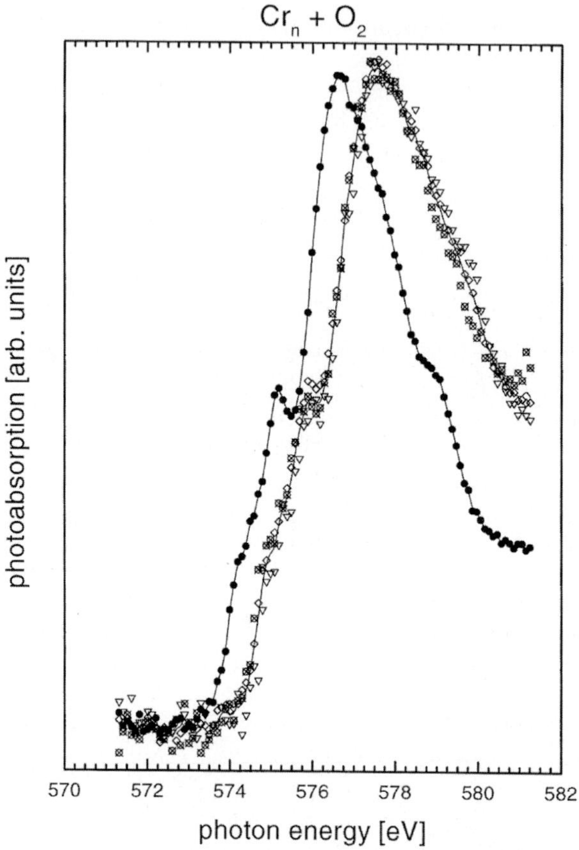

Fig. 1.46. L_3 photoabsorption spectra of $Cr_7/(1 \times 1)O/Ru(001)$ (*solid circle*) and $Cr_1/(1 \times 1)$ Ru(001), $Cr_{10}/(1 \times 1)$ Ru(001), $Cr_{13}/(1 \times 1)$ Ru(001) (*open symbols*) after oxygen treatment

Soft-landing of small Ag clusters ($n \leq 5$) at low surface concentration onto a well-characterized model oxide support (Al_2O_3) covered by Xe has been reported recently [46]. The clusters were produced by sputtering and mass-selected in a quadrupole mass spectrometer. Large core-level binding energy shifts up to 1 eV were observed in the photoelectron spectra, which were taken with monochromatized AlKα-radiation. These shifts were attributed to the different cluster sizes and to diffusion/agglomeration within the noble gas layer. Analysis of monomers and dimers deposited at low temperature in a thick Xe buffer layer indicated the mobility and aggregation of the clusters in such a film. A stable average cluster size was reached at 53 K that is likely to be the same for all initially deposited cluster sizes studied in the range from Ag_1 to Ag_5. Cluster aggregation progresses as the temperature is raised, tending toward the formation of metallic islands at high temperatures.

Cavity Ring-Down Spectroscopy

Optical absorption spectroscopy is a recent and very elegant method to characterize electronic transitions of small clusters on surfaces and gives not only insight into the electronic structures of clusters but also into the geometric structure, stability, and into the type of trapping centers, especially when compared to theoretical predictions. The application of optical absorption spectroscopy is manifold, as the electronic structure of clusters is extremely sensitive to the cluster's isomeric form and to the chemical environment. Although extended studies have been devoted to the absorption properties of size-selected clusters in the gas phase [157, 223] and in matrices [221], the low particle concentration of monodispersed materials is once again an experimental challenge, which has drastically refrained studies on supported species. In this respect, due to its extreme sensitivity, cavity ring-down spectroscopy (CRDS) is the method of choice to study such highly diluted systems. CRDS has been developed for gas-phase studies [224, 225], but recently it has also been applied to the liquid phases [226].

When applied to study the visible absorption response of monodispersed systems adsorbed on surfaces under UHV conditions, the method has proven to be extremely powerful in achieving an ultimate sensitivity to detect as little as 5×10^9 chromophors. In the example presented here, it has been applied to study the optical properties of highly diluted, monodispersed gold atoms and clusters adsorbed on an amorphous SiO_2 (α-SiO_2) surface [227].

In these experiments, the CRDS setup for studying clusters on surfaces consists of three 38 cm long optical resonators, which are mounted in parallel in a UHV chamber. A single resonator is defined by two high-reflectivity mirrors (reflectivity: ~99.995% at the optimal wavelength, bandwidth: ~100 nm). In order to increase the effective bandwidth of the optical setup, each resonator is built with a set of mirrors centered at a different wavelength. By this means, a wavelength range of about 300 nm can be covered without opening the UHV chamber. In principle, the bandwidth can even be enhanced by mounting a larger number of mirrors into the chamber. Tunable light pulses are coupled into one given resonator through the first mirror. The light leaking out the optical cavity through the second mirror is collected by a monochromator to filter out the amplified spontaneous emission (ASE) spectral contribution of the dye laser. The time evolution of the light intensity is then measured by a photomultiplier and sampled with a 500 MHz oscilloscope. The decay time of the light intensity, called ring-down time, can be extracted from the decay curve with a standard deviation of better than 1% by applying a nonlinear fitting procedure. This precision, combined with the low intrinsic loss of the optical cavity (<100 ppm per pass), puts the ultimate sensitivity of the optical setup below 1 ppm per pass.

For the actual measurements, the ringdown time of the optical cavity with the clean silica plate that presents the support material for cluster deposition is recorded before the deposition as a function of the photon energy.

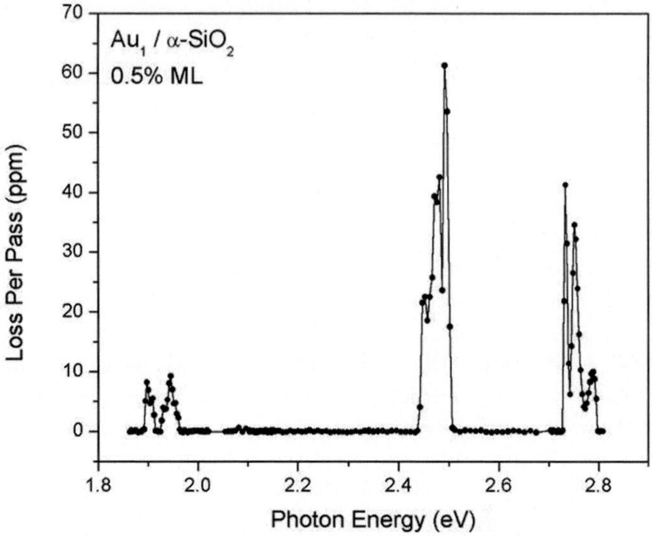

Fig. 1.47. Absorption spectrum of gold atoms deposited on an amorphous silica surface investigated by cavity ring-down spectroscopy

Subsequently, size-selected atoms or clusters are deposited on the plate. After deposition, ringdown times are measured for each photon energy at identical conditions as for the clean surface. Absorption losses

$$\Lambda(E_\gamma) = \left(\frac{L}{c}\right)\left(\tau^{-1} - \tau_0^{-1}\right) \tag{1.9}$$

as a function of the photon energy E_γ are then deduced from the ringdown times τ_0 and τ before and after the deposition, respectively, where L is the length of the cavity and c the velocity of light. From the absorption losses, absolute cross sections can be obtained if the cluster coverage is known:

$$\text{Cross section} = \Lambda \times (\text{surface coverage})^{-1} . \tag{1.10}$$

From this relation, we also can estimate the sensitivity of the method. As an example, for a typical limit of measurable absorption losses of 0.5×10^{-6} and a cross section of $0.1\,\text{Å}^2$ (typical for transitions in metal clusters) detectable cluster coverages are in the range of $5 \times 10^{10}\,\text{cm}^{-2}$ (~0.005% ML). This is indeed orders of magnitudes better than for conventional surface science techniques in laboratory conditions (XPS, UPS, FTIR, etc.).

To illustrate the power of this method, the absorption spectrum (between 1.85 and 2.82 eV) of gold atoms deposited on α-SiO_2 is depicted in Fig. 1.47. The spectrum clearly shows a first band of two well-resolved transitions at 1.90 and 1.95 eV. A second band is observed with three transitions at 2.46, 2.48, and 2.49 eV while a third band reveals transitions at 2.73, 2.75, and 2.79 eV.

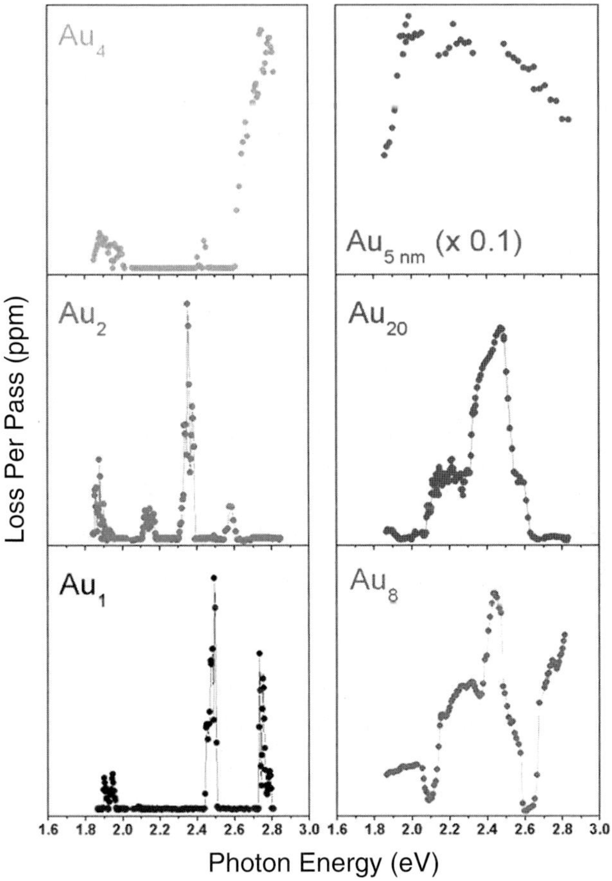

Fig. 1.48. Absorption spectra of small gold clusters Au_2, Au_4, Au_8, and Au_{20} in comparison to the spectra of gold atoms (lower left frame) and nanometer size gold islands (upper right frame). All spectra were obtained with the cavity ring-down method. The mass-selected clusters were soft-landed onto an α-SiO_2 substrate. The gold nanoparticles were grown on the same substrate by thermal evaporation

The three main bands at 1.90, 1.95, 2.48, and 2.75 eV can unambiguously be attributed to well-isolated Au atoms as there is no resemblance with the absorption spectra of Au_n/SiO_2 shown in Fig. 1.48. Particularly, the intense band at 2.35 eV for Au_2/SiO_2 is absent in the spectrum of Au_1/SiO_2. Thus, the formation of clusters can be excluded within the detection limit of this method. Furthermore, the weak transition at 2.49 eV in the Au_2 spectrum in Fig. 1.48 may be attributed to Au atoms, which are formed upon deposition of the dimer. With this interpretation, the fragmentation yield is at most about 5%, in fact this spectral feature may also be attributed to a characteristic transition of the dimer as predicted theoretically.

The optical transitions of the atoms and dimers on α-SiO_2 were intensively investigated by time-dependent density functional theory (TDDFT) [228]. The results show that Au atoms and dimers do not bind to the regular surface sites (Si–O–Si bridging oxygens), nor to the hydroxyl groups [isolated, $\equiv$ SiOH, or geminal, $=$ $Si(OH)_2$]. These interactions are very weak or of van der Waals type implying a facile diffusion and agglomeration of Au atoms and dimers at room temperature. The fact that cluster growth is not observed experimentally suggests a strong role of surface defects in stabilizing the deposited Au atoms and dimers. In fact, the optical transitions in the atoms and dimers can be understood if Si dangling bonds, $\equiv Si^{\bullet}$ (E′ center), nonbridging oxygen, $\equiv$ Si–$O^{\bullet}$ (NBO center), and charged NBO, $\equiv$ Si–O^- (NBO^- center) are considered. The binding energies of Au_1 and Au_2 to these defect centers range from 0.9 ($\equiv$ Si–O^-) to 3.2 eV ($\equiv Si^{\bullet}$), indicating stability at room temperature.

With these calculations, main transitions in the gold atom spectrum in Fig. 1.47 can be assigned in the following manner: The main two atomic lines, at 2.48 and 2.75 eV, are attributed to a $5d_{Au} \rightarrow \sigma^*(6s_{Au}\text{–}2p_O)$ transition [$\equiv$Si–O–Au] (transition energy $T_e = 2.59$ eV, oscillator strength $f = 0.037$), and and $6s_{Au} \rightarrow sp_{Si}$ transition $[\equiv\text{Si–O–Au}]^-$ species ($T_e = 2.86$ eV, $f = 0.034$). The weak bands around 1.90–1.95 eV are due to low-energy, low-intensity $5d_{Au} \rightarrow \sigma^*(6s_{Au}\text{–}2p_O)$ and $5d_{Au} \rightarrow 6s_{Au}$ transitions of [$\equiv$Si–O–Au] and $[\equiv\text{Si–O–Au}]^-$, respectively. No other trapping center considered can provide a similar agreement with the measured spectrum. Hence, it is concluded that Au atoms bind preferentially to $\equiv$ Si–$O^{\bullet}$ or $\equiv$ Si–O^- defect centers, and that these two types of centers are present in relatively high concentrations on the surface of the α-SiO_2 sample used. These conclusions are further corroborated by the results for the Au_2 clusters. When the dimer interacts with an NBO center, $\equiv$ Si–$O^{\bullet}$, an intense band at 2.38 eV, very close to that measured for Au_2/SiO_2, 2.35 eV is obtained and has a mixed O(2p)–Au(5d6s) character. The transitions for Au_2 bound to a charged $\equiv$ Si–O^- defect occur at energies larger than 3.2 eV and are outside the measured spectral range (1.85–2.82 eV).

In conclusion, CRDS can successfully be applied to highly diluted, monodispersed gold clusters on amorphous silica in order to measure the evolution of the optical properties with size, detect thermal stabilities of the cluster, characterize the trapping centers and in the future to get structural information on the deposited clusters.

Scanning Tunneling Spectroscopy

An ideal tool for the investigation of the electronic structure of metal clusters is offered by scanning probe methods which have been used in three pioneering studies of nanometer-size clusters of Au on GaAs(110) [229], Fe clusters on the same surface [230], and of size-selected Si_{10} clusters on a reconstructed Au(001) surface [231], which all have been published already in 1989. In the first investigation [229], a characteristic spectrum of band-gap states was observed for the Au particles grown on GaAs. Both donor and acceptor states have been observed and they have been identified with the first and second electron states of the Au–Ga bond, respectively. In the second study, the formation of a local Schottky-barrier has been realized by deposition and growth of Fe clusters (9–127 atoms per cluster) on GaAs surfaces in UHV. Scanning tunneling spectroscopy (STS) revealed the onset of the metallic character of the Fe clusters above about 35 atoms per cluster by the observation of a continuum of occupied cluster-induced states at the Fermi energy in the band gap of the semiconductor [230]. The third experiment [231] represents the first STM/STS study of supported size-selected clusters on a solid surface. The cluster images were found to depend on the sample bias voltage, and from STS measurements the clusters were found to exhibit a band gap of about 1 eV. A wide variety of cluster images were observed even though size-selected clusters were deposited. When Si atoms were deposited on the surface, flat islands were formed.

Surprisingly, in spite of these first promising results, up to the present time the reported experiments on size-selected supported clusters with local probes have been scarce. One interesting experiment used the fact that the tip-surface region of an STM emits light when the energy of the tunneling electrons is sufficient to excite luminescent processes. These processes provide access to dynamic aspects of the local electronic structure that are not directly amenable to conventional STM experiments. From monolayer islands of C_{60} buckminster fullerenes on gold (110) surfaces, intense emission of light was observed when the STM tip was placed above an individual molecule. The diameter of this emission spot associated with C_{60} is approximately 0.4 nm. These results demonstrated the highest spatial resolution of light emission to date with a scanning probe technique [232]. This technique has again been used recently to measure photon emission spectra of individual alumina-supported silver clusters obtained by evaporation and growth [233]. The light emission stimulated by electron injection from the tip of an STM has been assigned to a mode of the Mie-plasmon resonance in the small silver particles. As the cluster size decreases, the resonance shifts to higher energies and the linewidth increases. In the 1.5–12 nm size-range of the clusters studied, intrinsic size effects are proposed to be at the origin for the observed size dependence of the Mie resonance [233].

The energy gap of pristine silicon clusters supported on HOPG has been studied recently by STM and STS [234]. The clusters have been obtained by magnetron sputtering of an Si target and deposition on HOPG at room temperature, i.e., they were not size-selected but have been individually addressed by the tip of the STM. Clusters with sizes between 0.25 and 4 nm were studied and the size dependence of the band gap was determined. For clusters below 1.5 nm gaps up to 450 meV were found while for larger particles no gaps were recorded. The results were explained in terms of a transformation from diamond to a compact structure occurring at 1.5 nm (about 44 atoms per cluster). For clusters with diamond structure, the surface dangling bond density is high leading to electronic states filling the energy gap. On the other hand, the compact arrangement of the smaller clusters tends to eliminate dangling bonds. Therefore, finite gap values are observed for clusters with less than 44 Si-atoms [234].

Also, interesting STS results on individual non-size-selected Pt and Ag clusters produced by a PACIS source and deposited onto HOPG have been obtained recently [236]. The observed spectral structures for Pt have been interpreted in terms of a Fano resonance and those of Ag clusters in terms of 3D electron confinement. Effects of electron confinement have also been observed recently with STS for silver islands grown in nanopits on HOPG [237].

The power of STS to locally observe the electronic structure has again been nicely demonstrated with a study of a temperature dependent metallic to semiconducting phase transition of a metallic endohedral fullerene La@C_{60} supported on HOPG [235]. The question whether a metallic or a semiconducting behavior is observed depends on the type of the encapsulated atom. Ce@C_{60} shows a nonmetallic behavior with a band gap of about 0.5 eV, whereas La@C_{60} is found to be metallic at room temperature (Fig. 1.49). However, STS at low temperature reveals (Fig. 1.50) a reversible opening of a gap of about 40 meV below a temperature of 28 K. The spectra are interpreted in terms of free C_{60} including charge transfer and Jahn-Teller distortion [235].

Recently, STS and STM have been applied to study the onset of the catalytic activity of Au particles grown on titania [238], which appeared to be correlated with the layer thickness of the particles on the surface and, as shown above, the bonding of CO with small gold clusters could be characterized in STM/STS experiments (Fig. 1.39; see "Morphological Properties of Supported Clusters") [190]. The above presentation of the spectroscopic results on deposited, size-selected clusters clearly shows that valuable information on these nanosystems can only be obtained by the application of an arsenal of local and nonlocal surface science analysis methods. Therefore, in the near future a much more intense employment of scanning probe techniques such as STM, STS, AFM, and others will beyond any doubt improve considerably assembly, characterization, and functionalization of size-selected clusters on solid surfaces.

Chemical and Catalytic Properties of Supported Clusters

There are several experimental techniques available to study the chemical properties of clusters on surfaces and when used in combination, a detailed picture can be obtained. A classical approach is the use of temperature programmed desorption (TPD) and temperature programmed reaction (TPR) spectroscopy [39, 239, 240] in combination with temperature-dependent and eventually time-dependent Fourier transformed infrared spectroscopy (FTIR) [39, 192, 241–244]. This combination allows for obtaining information on estimates of binding energies of reactant molecules (TPD), for characterizing

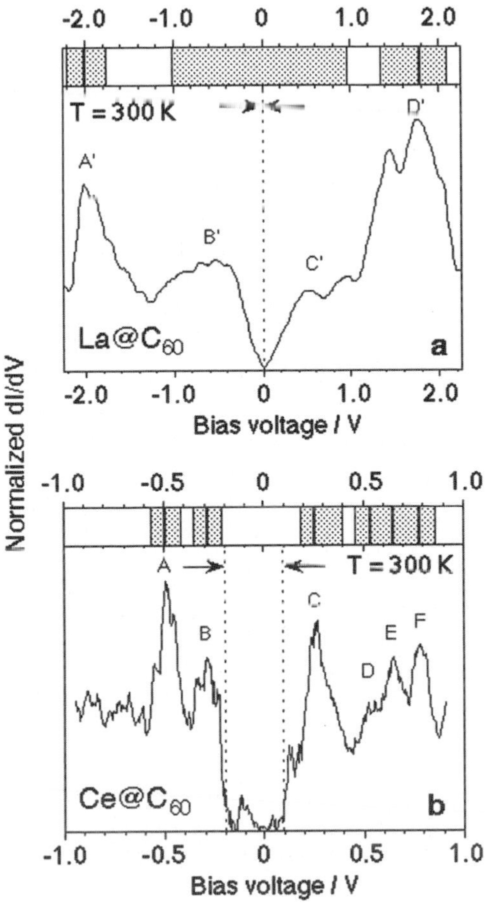

Fig. 1.49. Scanning tunneling spectra of endohedrally doped C_{60}. (**a**) La@C_{60}, (**b**) Ce@C_{60}. Note the band gap around 0 V bias voltage in the case of Ce@C_{60} (**b**) indicative of nonmetallic behavior [235]

vibrational frequencies, binding sites, and the activation of reactant molecules (FTIR), as well as for detecting product molecules of a chemical reaction on the clusters (TPR). In addition, when using FTIR as a function of temperature or in a time-resolved manner, important information on reaction mechanisms can be obtained [243, 244]. This combination can ideally be complemented by calorimetric studies, in order to get information on accurate binding energies and eventually on reaction heats [245–247]. At present, calorimetric studies have, however, not yet been applied to size-selected clusters. All these methods are not able to characterize a catalytic process on clusters or nanoparticles as, especially in a TPR experiment, only a single cycle of a catalytic process is monitored. In fact, a catalytic process comprises a large number of reaction

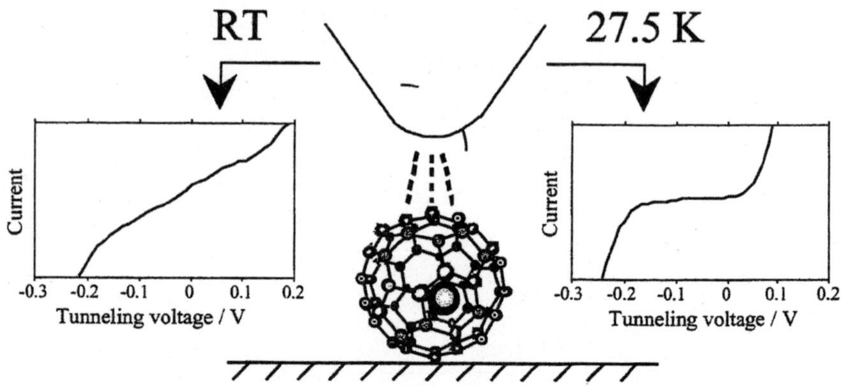

Fig. 1.50. Temperature dependent gap-opening in La@C_{60} [235]

cycles and is characterized by the turn over frequency (TOF), which is the number of product molecules per surface atom (or cluster) per second. Methods to estimate this quantity in UHV are molecular beam reactive scattering (MBRS) [243, 248, 249] or pulsed molecular beam reactive scattering (PMBRS) [250, 251]. In the following we describe to some detail these methods, specific examples are given in Sect. 1.5.

Thermal Desorption and Reaction Studies

Temperature-programmed desorption is the classical method to study chemical interactions of adsorbates with surfaces like the ones of amorphous materials, single crystals, or model catalysts [239, 240]. The desorption process or eventual chemical reactions can be described by kinetics and there are computer programs available for extracting the corresponding physical properties (binding energies, reaction orders, etc.). Basically, a molecule is adsorbed at low temperature on the material of interest (e.g., an oxide surface with a well-known density of size-selected clusters), the desorbing molecules are then detected by a differentially pumped mass spectrometer. When using an absolutely calibrated mass spectrometer, the number of desorbing molecules can be obtained. This calibration can, e.g., be done by studying the desorption from single crystals when the saturation coverage of an adsorbant is known [e.g., CO on Pt(111) with a saturation coverage of 1×10^{14} CO cm^{-2}). Applying this method to size-selected clusters on surfaces and knowing the cluster density, an average number of adsorbates per cluster can be obtained. By using differently labeled isotopes like, e.g., $^{13}C^{16}O$ and $^{12}C^{18}O$ one is able to distinguish between molecular and dissociative adsorption by monitoring the isotopic scrambling (formation of $^{13}C^{18}O$). An example for the molecular and dissociative adsorption of CO on supported Ni_{30} is depicted in Fig. 1.51.

Often TPD experiments are carried out as a function of coverage, and the obtained results directly lead to desorption energies when observing the

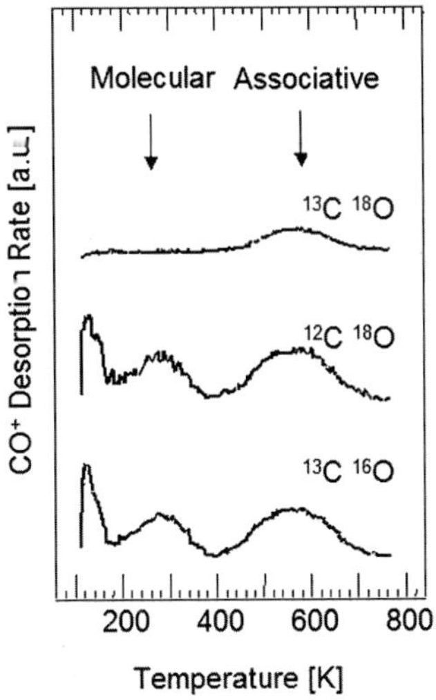

Fig. 1.51. Thermal desorption spectra of an isotopic mixture of $^{13}C^{16}O$ and $^{12}C^{18}O$ adsorbed on Ni_{30}/MgO/Mo. The cluster density is 0.14% of an ML (1 ML = 1.25 × 10^{15} cm^{-1}). Shown is the desorption of the three isotopes $^{13}C^{16}O$, $^{12}C^{18}O$, and $^{13}C^{18}O$. $^{13}C^{18}O$ can only be formed if carbon monoxide is dissociated by the cluster and subsequently is desorbed associatively. This experiment characterizes different binding sites on the cluster, showing that the low-temperature peaks at around 100 and 260 K originate from molecularly adsorbed CO and the high temperature peak from dissociated CO. $^{12}C^{16}O$ which is also formed is not shown due to bad signal-to-noise ratio. As the quadrupole mass spectrometer has been calibrated, the number of adsorbed CO molecules in the corresponding adsorption state could be obtained. For Ni_{30} about six CO molecules are bound molecularly whereas ~10 are dissociated upon adsorption

evolution of the maximal rate of desorption with coverage. In addition, several methods were developed to extract activation energies of desorption. For clusters on surfaces such experiments are often difficult to perform as cluster densities are kept low in order to prevent coalescence. Thus, the obtained signal-to-noise ratio is weak. Binding energies are normally extrapolated from the Redhead approximation at full coverage

$$\frac{E_{\mathrm{d}}}{RT_{\mathrm{max}}} = \ln\left(\frac{v_1 T_{\mathrm{max}}}{\beta}\right) \times 3.64 \tag{1.11}$$

with E_{d} the activation energy of desorption, T_{max} the maximum rate of desorption, v_1 preexponential factor (in this approximation 10^{13} s^{-1}), and β the heating rate of the experiment.

When the desorbing molecule is the product of a chemical reaction on the surface, the method is called TPR and in this mode it gives information on the distribution of product molecules of a surface- or cluster-catalyzed chemical reaction. As in the TPD mode, this method gives absolute numbers of product molecules per cluster when using a calibrated mass spectrometer. Thus, the evolution of the chemical reactivity with cluster size can be probed, and branching ratios for reactions with different product molecules can be extracted. In cases where the product molecule has a low interaction with the cluster or the reaction enthalpy is higher than the binding energies of the product molecule, the reaction energy can be estimated from the reaction temperature observed in the TPR experiment. One should be aware that this method only probes a single cycle of a chemical reaction and thus no information on the catalytic properties can be obtained. In addition, during the temperature scan, the surface coverage is changing and the TPR spectra are thus giving reactivities at varying coverage. In fact, at a given coverage only the reaction path with the lowest activation barrier is probed. In order to get detailed information on the catalytic properties other methods have to be applied. Later we will discuss the PMBRS.

Infrared Spectroscopy on Solid Surfaces

There exist two types of classical vibrational spectroscopies for studying the adsorption of small molecules on surfaces. These are the high-resolution electron energy loss spectroscopy (HREELS) and the FTIR. For both methods, there are excellent reviews [241] and only few details are given here, which are relevant to the study of adsorbates on deposited clusters. In HREELS electrons are scattered inelastically on solid surfaces and losses in the range of a few hundreds of meV are detected with a resolution of about $20\,\mathrm{cm}^{-1}$. In principle, it is possible to even detect vibrations of metal–metal bonds in small clusters. The loss process can be described mainly by dipole scattering and by resonance and impact scattering. This method is very surface sensitive (<1% ML). In the FTIR, dipole active adsorbed molecules are excited and the absorption of the IR light is measured (~1% ML). It has an excellent resolution, $1\ \mathrm{cm}^{-1}$; however, it is somewhat less sensitive than HREELS. In most surface experiments, FTIR is used in grazing angle reflection mode, where a metal single crystal acts as a mirror. In this mode, only molecules with a dynamic dipole moment perpendicular to the surface are detected as the electric field of the electromagnetic radiation perpendicular to the surface is amplified, whereas the component parallel to the surface disappears upon reflection. Most active vibrational modes of organic molecules can be detected; this method fails, however, when intramolecular bonds of clusters or bonds between the ligands and the cluster shall be characterized.

FTIR has been successfully applied for studying chemical properties of size-selected clusters on oxide surfaces [39,192,244,252]. In these experiments, the clusters were deposited onto thin oxide films grown on a metal single crystal. In this approach, the metal is used as the mirror in the optical setup; the oxide films ought to be thinner than the wavelength of the IR radiation. In this way, vibrations of adsorbates on size-selected clusters with densities as low as 0.2% ML could be detected. Mechanistic information of chemical reactions can be obtained when recording the infrared spectra as a function of reaction temperature. This leads to the evolution of the reactants and eventual product molecules along a thermally induced chemical reaction. In combination with TPR spectra, a detailed picture of the reactants and intermediates via infrared spectroscopy and the product molecules (TPR) is obtained, several examples are given later.

Molecular Beam Reactive Scattering Using Pulsed Molecular Beams

Size-selected clusters are deposited on solid surfaces under UHV conditions, and it is most convenient to study their catalytic properties in situ in a UHV chamber. To study a catalytic process, however one needs a high flux of reactants on the surface, which is not compatible with UHV conditions. One approach is to use continuous or pulsed molecular beams. Most often they are generated by expansion through a small orifice, and the molecular beam is then skimmed through several differentially pumped vacuum chambers to decrease the background pressure in the UHV chamber. When pulsed beams are required, the beam is chopped by a rotating, segmented disk (chopper). An alternative method is the use of very stable pulsed valves. A typical experimental setup is shown in Fig. 1.52. The characteristics of the used valves are summarized in the inset of Fig. 1.52; these are the pulse-to-pulse stability, as well as typical time and beam profiles. The valve is driven by a piezo-electric disk with a pulse-to-pulse stability of better than 1%, with variable pulse widths (minimal width 50 µs up to continuous opertation), and it allows to produce partial pressures of up to 100 Torr on the sample. With such valves, several experimental schemes are possible. At a constant, isotropic background pressure of one of the reactants, a defined pulse of the other reactant molecules hits the surface, and the product molecules are monitored by a differentially pumped mass spectrometer. These experiments lead to the following information. First, by studying the temporal evolution of the product pulses, the stability of the model catalysts can be investigated. If the mass spectrometer is absolutely calibrated, the integral of the product pulses will give the absolute number of produced product molecules. If normalized to the number of clusters on the surface and to the time, turn-over frequencies can be obtained as a function of temperature. They can then be compared for several sizes to obtain the evolution of the catalytic properties of clusters with size. Second, this experimental scheme allows to study the kinetics of a catalytic process when the product molecules and reactants are monitored as a function of time. Using kinetic simulation software [253] mechanistic information, rate constants and activation barriers of elementary reaction steps can be extracted. Third, several pulsed valves can be used to study catalytic reactions at various partial pressures of the reactants. In combination with TPR and FTIR, a detailed picture of chemical and catalytic properties of size-selected clusters can be obtained and results will be presented in Sect. 1.5.

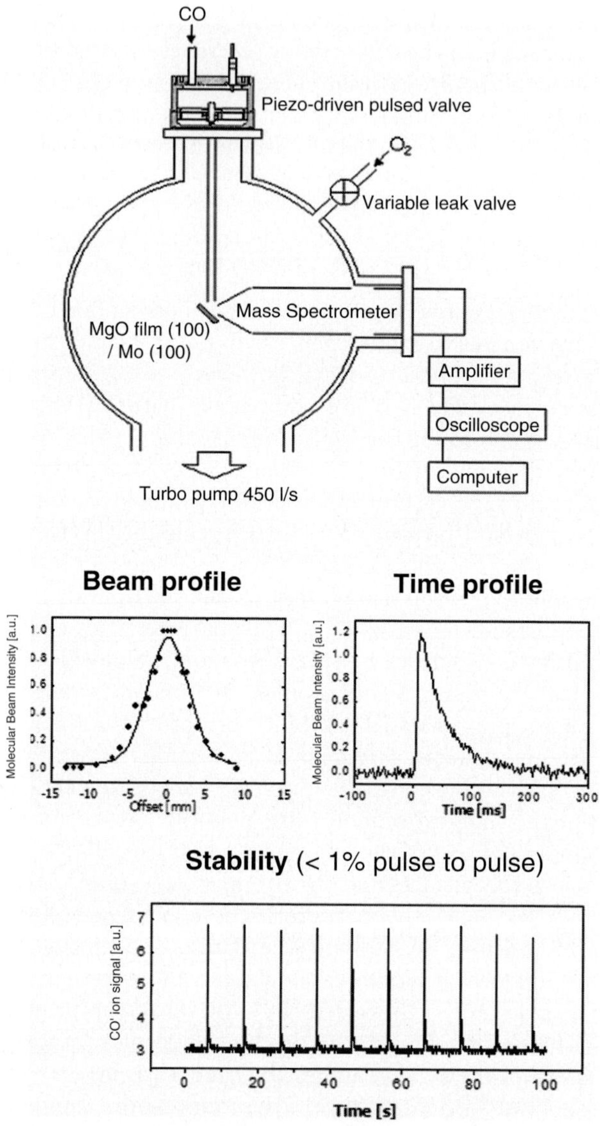

Fig. 1.52. Typical experimental setup for a pulsed molecular beam experiment for studying the catalytic properties of size-selected clusters on surfaces. It mainly consists of a pulsed valve for the generation of a pulsed molecular beam and a differentially pumped, absolutely calibrated quadrupole mass spectrometer. The length of the valve extension tube is adjusted to obtain a beam profile of similar dimensions as the sample under investigation. A typical time profile is also shown. It can be adjusted up to continuous operation. The pulse-to-pulse stability is better than 1%

1.3 Computational Techniques

Theoretical studies of the properties of the individual components of nanocatalytic systems (including metal nanoclusters, finite or extended supporting substrates, and molecular reactants and products), and of their assemblies (that is, a metal cluster anchored to the surface of a solid support material with molecular reactants adsorbed on either the cluster, the support surface, or both), employ an arsenal of diverse theoretical methodologies and techniques; for a recent perspective article about computations in materials science and condensed matter studies [254]. These theoretical tools include quantum mechanical electronic structure calculations coupled with structural optimizations (that is, determination of equilibrium, ground state nuclear configurations), searches for reaction pathways and microscopic reaction mechanisms, ab initio investigations of the dynamics of adsorption and reactive processes, statistical mechanical techniques (quantum, semiclassical, and classical) for determination of reaction rates, and evaluation of probabilities for reactive encounters between adsorbed reactants using kinetic equation for multiparticle adsorption, surface diffusion, and collisions between mobile adsorbed species, as well as explorations of spatiotemporal distributions of reactants and products.

The theoretical investigation described in the following sections of this chapter focus on studies aimed at understanding the correlations between the size-dependent properties of metal clusters of nanoscale dimensions, and their catalytic activity in the gas phase or when supported on a solid (metal-oxide) surface, on elucidation of the reaction mechanisms, and on the physical and chemical factors underlying the catalytic activity of nanoclusters. Consequently, we limit our discussion here to a brief review of the theoretical methods employed in investigations on which we focus in this chapter, with a particular emphasis on certain theoretical aspects pertaining to electronic structure calculations. In Sect. 1.3.1, we discuss fundamental aspects of the main method used in electronic structure calculations, that is, density functional theory (DFT); our presentation follows rather closely to that given in [255] to which the reader is directed for further details. This is followed in Sect. 1.3.2 by brief comments concerning theoretical methods used in investigations of nanocatalytic systems, including a discussion of first principles molecular dynamics simulations, and of methods employed for finding minimum energy paths of transitions (such as chemical reactions or structural transformations), and for estimation of reaction barriers – these methods are based on energy optimization coupled with constrained variation of a chosen reaction coordinate. This section is intended as an introduction to the principles of the above noted methods, rather then a comprehensive treatise. Readers who wish to learn about these methods in greater rigor and detail are directed to available reviews and to the original literature.

1.3.1 Electronic Structure Calculations Via Density Functional Theory

The Many-Body Problem

The basic building blocks of the materials world are electrons and nuclei interacting via electrostatic Coulomb interactions. Thus, the starting point of a theoretical treatment of the chemical and physical properties of materials is the Hamiltonian

$$\hat{H} = -\frac{\hbar^2}{2m_\mathrm{e}}\sum_i \nabla_i^2 + \sum_{i,k}\frac{Q_k q}{|\boldsymbol{r}_i - \boldsymbol{R}_k|} + \frac{1}{2}\sum_{i\neq j}\frac{q^2}{|\boldsymbol{r}_i - \boldsymbol{r}_j|}$$
$$-\sum_k \frac{\hbar^2}{2M_k}\nabla_k^2 + \frac{1}{2}\sum_{k\neq l}\frac{Q_k Q_l}{|\boldsymbol{R}_k - \boldsymbol{R}_l|}, \qquad (1.12)$$

where in terms of the elementary charge $e > 0$ and the atomic number is Z, the nuclear charge is $Q = Ze$, and the charge of the electron is $q = -e$. In the above Hamiltonian, the first and fourth terms are, respectively, the electrons and nuclei kinetic energy operators, the third and fifth terms represent the interelectron and internuclear Coulomb repulsion, respectively, and the second term describes the electrostatic interaction between the electrons and the nuclei. The Hamiltonian in (1.12) is a statement of the quantum many-body problem of coupled electrons and nuclei. For many purposes, it is desirable (for computational reasons), and it is a very good approximation (on physical grounds), to separate the electronic and nuclear degrees of freedom. To this end, we note that the only "small parameter" in the above Hamiltonian is the inverse of the nuclear mass ($1/M_k$ in the fourth term on the right), and neglect of the kinetic energy of the nuclei is found to be an excellent approximation for many cases. This is the essence of the Born–Oppenheimer (BO) or adiabatic (separation of time-scales) approximation [256], where the Schrödinger equation for the electrons is solved for a set of *fixed* nuclear positions which enter as parameters (and not as variables). The BO N-electron Schrödinger equation, with the (stationary) nuclear degrees of freedom entering only in the electron–nuclear interaction potential term $\hat{V}$, is given by

$$\left[\sum_i^N\left(-\frac{\hbar^2\nabla_i^2}{2m} + v(\boldsymbol{r}_i)\right) + \sum_{i<j} U(\boldsymbol{r}_i, \boldsymbol{r}_j)\right]$$
$$\times\, \Psi(\boldsymbol{r}_1, \boldsymbol{r}_2, \ldots, \boldsymbol{r}_N) = E\Psi(\boldsymbol{r}_1, \boldsymbol{r}_2, \ldots, \boldsymbol{r}_N), \qquad (1.13)$$

where $U(\boldsymbol{r}_i, \boldsymbol{r}_j)$ is the interelectron interaction potential

$$\hat{U} = \sum_{i<j} U(\boldsymbol{r}_i, \boldsymbol{r}_j) = \sum_{i<j}\frac{q^2}{|\boldsymbol{r}_i - \boldsymbol{r}_j|}, \qquad (1.14)$$

and the kinetic energy of the electrons is given by

$$\hat{T} = -\frac{\hbar^2}{2m}\sum_i \nabla_i^2, \tag{1.15}$$

where we dropped the subscript 'e' from in the symbol for the mass of the electron. The interaction between the electrons and the nuclei is represented by the multiplicative operator (which changes when the nuclear positions change, and in the following it will be called the "external" potential since it couples the electronic system to the nuclei)

$$\hat{V} = \sum_i v(\boldsymbol{r}_i) = \sum_{ik} \frac{Q_k q}{|\boldsymbol{r}_i - \boldsymbol{R}_k|}. \tag{1.16}$$

Other potentials representing (external) fields (e.g., electric or magnetic) that act on the electronic degrees of freedom may be included in the external potential. Note that for a single-body (that is, one electron) problem only the external potential (and the kinetic energy operator) act on the one-electron wave function.

For a given nuclear configuration [i.e., given external potential (1.16)], the solution of (1.13) is the many-body wave function from which observables can be calculated, including the particle density (which will play a key role in what follows)

$$n(\boldsymbol{r}) = N\int \mathrm{d}^3 r_2 \int \mathrm{d}^3 r_3 \ldots \int \mathrm{d}^3 r_N \Psi^*(\boldsymbol{r}_1, \boldsymbol{r}_2, \ldots, \boldsymbol{r}_N)\,\Psi(\boldsymbol{r}_1, \boldsymbol{r}_2, \ldots, \boldsymbol{r}_N). \tag{1.17}$$

The above scheme which constitutes the usual quantum-mechanical approach may be summarized schematically as follows

$$v(\boldsymbol{r}) \xrightarrow{\mathrm{SE}} \Psi(\boldsymbol{r}_1, \boldsymbol{r}_2, \ldots, \boldsymbol{r}_N) \xrightarrow{\langle\Psi|\ldots|\Psi\rangle} \text{observables}, \tag{1.18}$$

where SE above the left arrow indicates that knowledge of the potential in the Schrödinger equation leads to a solution for the wave function. In other words [257], if we consider a set of local one-body (external) potentials V and gather the nondegenerate ground state wave functions in the set Ψ, we may regard the solution of the corresponding Schrödinger equation [SE in the first step in (1.18)] as a map

$$C::ve \to \Psi. \tag{1.19}$$

This mapping is surjective; that is, Ψ contains no element which is not associated with some element of V.

The second step in (1.18), with the observable being the electron density (1.17) may be also regarded as a second surjective map

$$D:\Psi \to N, \tag{1.20}$$

where N is the set of ground state densities.

Many powerful methods have been developed for accurate treatment of many-body problems, ranging from configuration interaction (CI) techniques (employed mostly in quantum chemical calculations) based on systematic expansion in Slater determinants to diagrammatic perturbation theory based on Feynman diagrams and Greens function techniques (used mainly by physicists). However, common to this modes of attack are the limitations set by computational resources, with the largest many-atom system for which chemical properties have been computed (using first-principles methods, that is, not considering empirical techniques) falling short of a 100-atom molecule.

A simple estimate of the computational difficulties involved with the customary quantum mechanical approach to the many-electron problem illustrates vividly the point [255]. Consider a real-space representation of $\Psi(\boldsymbol{r}_1, \boldsymbol{r}_2, \ldots, \boldsymbol{r}_N)$ on a mesh in which each coordinate is discretized by using 20 mesh points (which is not very much). For N electrons, Ψ becomes a variable of $3N$ coordinates (ignoring spin), and 20^{3N} values are required to describe Ψ on the mesh. The density $n(\boldsymbol{r})$ is a function of three coordinates and requires only 20^3 values on the same mesh. CI and the Kohn–Sham formulation of DFT (see below) additionally employ sets of single-particle orbitals. N such orbitals, used to build the density, require 20^{3N} values on the same mesh. (A CI calculation employs in addition unoccupied orbitals and requires more values.) For $N = 10$ electrons, the many-body wave function thus requires $20^{30}/20^3 \approx 10^{35}$ times more storage space than the density and sets of single-particle orbitals $20^{30}/10 \times 20^3 \approx 10^{34}$ times more. Clever use of symmetries can reduce these ratios, but the full many-body wave function remains inaccessible for real systems with more than a few electrons.

The Density Functional Theory

In light of the above, it is natural that attempts have been made over the years to find alternative routes for quantitatively accurate and efficient calculations of the properties of many-electron systems. Indeed early heuristic attempts to express the ground state energy of a many-electron system solely in terms of the electron density have been considered by Thomas (1927) and Fermi (1927, 1928), with extensions made by Dirac (1930) and von Weizäcker (1935) [257]. However, the work of Hohenberg and Kohn (HK) (1964), and subsequently Kohn and Sham (KS) (1965) (for which Walter Kohn [258] shared the 1998 Nobel Prize in Chemistry with John Pople [259] who was instrumental in implementing DFT in computational chemistry), established DFT as the principal computational method for the description of the chemical and physical properties of materials in various phases (solid, liquid, gas), forms (crystalline, amorphous, ordered, disordered, neutral, or charged) and degrees of aggregation (extended solids, surfaces and interfaces, finite crystallites, clusters, molecules) [255, 257, 260–272].

The essence of the original basic theorem proved by HK [273] is that an exact representation of the ground state properties of a stationary, nonrelativistic, many-body system in terms of the ground state density alone is possible. The theorem ensures that the ground state can be calculated from a variational principle involving only the density, that is, without having to solve the Schrödinger equation. For a given system, e.g., particles interacting through Coulomb potentials and moving in an external potential, the exact ground state density and energy can be obtained via minimization of the energy expressed as a functional of the density (see below). In a schematic way the gist of the HK theorem may be summarized as

$$n(\boldsymbol{r}) \rightarrow \Psi\left(\boldsymbol{r}_1, \ldots, \boldsymbol{r}_N\right) \rightarrow v(\boldsymbol{r}), \tag{1.21}$$

that is, the "inverse" of the sequence (1.18).

While in its original version, the HK theorem stipulated a number of restrictions, such as the assumption of nondegenerate ground states, the rigorous foundation of DFT has been extended since to practically all circumstances of interest, including degenerate ground states, spin-polarized systems, thermodynamic systems, and relativistic systems. We note here that the HK theorem is not "constructive", in the sense that it does not provide a guide for the explicit construction of the functionals in question, and to date no exact functionals are known for any multiparticle system—indeed, this topic remains an active area of research and development.

The development that led to the emergence of DFT to its current status as the most widely used electronic structure computational method is that due to KS [274]. The ansatz used by KS replaces the interacting problem with an auxiliary independent particle one, with all the many-body effects included in an exchange-correlation functional. In practice, the KS scheme introduces an equivalent orbital picture (rigorously established), with the resulting KS equations solved self-consistently.

Prior to our following discussion of the HK theorem (in its original form), the KS scheme, and certain issues pertaining to the construction and use of various functionals, we comment first on the functional as a mathematical tool. Since the wave function is determined according to (1.21) by the density, it can be written as $\Psi_0 = \Psi[n_0]\left(\boldsymbol{r}_1, \boldsymbol{r}_2, \ldots, \boldsymbol{r}_N\right)$, indicating that Ψ_0 is a function of its N spatial variables but a functional of $n_0(\boldsymbol{r})$. More generally, a functional $F[n]$ can be defined (somewhat loosely mathematically) as a rule for going from a function to a number, just as a function $y = f(x)$ is a rule (f) for going from a number (x) to a number (y). A simple example of a functional is the particle number,

$$N = \int \mathrm{d}^3 r\; n(\boldsymbol{r}) = N[n], \tag{1.22}$$

which is a rule for obtaining the number N, given the function $n(\boldsymbol{r})$. Note that the name given to the argument of n is completely irrelevant, since the functional depends on the function itself, not on its variable. Another important case is that in which the functional depends on a parameter, such as in

$$v_{\mathrm{H}}[n](\boldsymbol{r}) = q^2 \int \mathrm{d}^3r' \frac{n(\boldsymbol{r}')}{|\boldsymbol{r}-\boldsymbol{r}'|}, \tag{1.23}$$

which is a rule that for any value of the parameter $\boldsymbol{r}$ associates a value $v_{\mathrm{H}}[n(\boldsymbol{r})])$ with the function $n(\boldsymbol{r}')$. This term is the so-called Hartree potential.

The Hohenberg–Kohn Theorem

To prove the HK theorem [274], it has to be proved that the maps $\mathcal{C}$ and $\mathcal{D}$, defined, respectively, by (1.19) and (1.20), are injective (one to one) and thus, in light of the surjectivity of these maps, they are also bijective (that is, fully invertible). The proof of injectivity is rather straightforward and can be found in Chap. 2 of [257].

As a consequence of the unique inversion capability of the map $\mathcal{D}$

$$\mathrm{D}^{-1} : n(r) \rightarrow |\Psi[n]\rangle \tag{1.24}$$

the first statement of the HK theorem can be restated: the ground state expectation value of any observable $\hat{O}$ is a unique functional of the exact ground state density

$$\langle\Psi[n]|\,\hat{O}\,|\Psi[n]\rangle = O[n]. \tag{1.25}$$

The full inverse map

$$(\mathrm{CD})^{-1} : n(r) \rightarrow v(r), \tag{1.26}$$

expresses the fact that knowledge of the ground state density determines the external potential of the system (to within a trivial constant) and thus, since the kinetic energy and the interparticle interaction are specified, the entire Hamiltonian.

The HK theorem establishes also the variational character of the energy functional

$$E_{v_0}[n] := \langle\Psi[n]|\,\hat{T}+\hat{U}+\hat{V}_0\,|\Psi[n]\rangle\,, \tag{1.27}$$

where $\hat{V}_0$ is the external potential of a specific system with ground state density $n_0(\boldsymbol{r})$ and ground state energy E_0. The states $|\Psi[n]\rangle$ are generated via $\mathcal{D}^{-1}$ from the elements of the set N. Due to the Rayleigh–Ritz variational principle $E_{v_0}[n]$ obeys the relations

$$E_0 < E_{v_0}[n] \quad \text{for} \quad n \neq n_0 \tag{1.28a}$$

and

$$E_0 = E_{v_0}[n_0]. \tag{1.28b}$$

Consequently, the exact ground state density can be determined by minimization of the functional $E_{v_0}[n_0]$. In summary,

$$E_0 = \min E_{v_0}[n] \quad \text{for} \quad n \subset N. \tag{1.29}$$

The map $\mathcal{D}^{-1}$ does not depend on the potential $\hat{V}_0$ of the particular system under consideration. Therfore, expressing the functional $E_{v_0}[n]$ as

$$E_{v_0}[n] = F_{\mathrm{HK}}[n] + \int \mathrm{d}^3 r \; v_0(\boldsymbol{r}) \; n(\boldsymbol{r}) \tag{1.30}$$

with

$$F_{\mathrm{HK}}[n] = \langle \Psi[n] | \, \hat{T} + \hat{U} \, | \Psi[n] \rangle \,, \tag{1.31}$$

allows us to conclude that the functional F_{HK} is *universal*, that is, it does not depend on the external potential $\hat{V}_0$; this means that F_{HK} is the same density functional for atoms, molecules, and solids, since in all these systems the electrons interact electrostatically via the Coulomb force.

The HK theorem tells us that it should be possible to calculate all observables using the density alone, since the HK theorem guarantees that they are all functionals of $n_0(\boldsymbol{r})$. In practice, however, one does not know how to do this explicitly. In particular, to perform such calculations one needs reliable approximations for $T[n]$ and $U[n]$ to begin with. Prior to discussing a practical way of attacking these problems, we recall an older, but still occasionally useful, alternative: the Thomas–Fermi (TF) approximation [257].

In this method, one approximates the full interparticle interaction energy by the Hartree energy

$$U[n] \approx U_{\mathrm{H}}[n] = \frac{q^2}{2} \int \mathrm{d}^3 r \int \mathrm{d}^3 r' \frac{n(\boldsymbol{r}) n(\boldsymbol{r}')}{|\boldsymbol{r} - \boldsymbol{r}'|} . \tag{1.32}$$

Furthermore one sets for the kinetic energy functional

$$T[n] \approx T^{\mathrm{LDA}}[n] = \int \mathrm{d}^3 r \; t^{\mathrm{hom}} \left(n(\boldsymbol{r}) \right) , \tag{1.33}$$

where $t^{\mathrm{hom}}(n)$ is the kinetic-energy density (that is, per unit volume) of a homogeneous interacting system with (constant) density n. Since it refers to interacting electrons, $t^{\mathrm{hom}}(n)$ is not known explicitly and further approximations are needed. As it stands, however, this formula is already a first example of a local-density approximation (LDA). In this type of approximation one imagines the real inhomogeneous system [with density $n(\boldsymbol{r})$ in potential $v(\boldsymbol{r})$] to be partitioned into small cells in each of which $n(\boldsymbol{r})$ and $v(\boldsymbol{r})$ are approximately constant. In each cell (i.e., locally), one can then use the per volume energy of a homogeneous system to approximate the contribution of the cell to the real inhomogeneous one. Making the cells infinitesimally small and summing over all of them yields (1.33).

For a noninteracting system (specified by subscript S, for "single-particle") the function $t^{\mathrm{hom}}(n)$ is known explicitly, $t_{\mathrm{S}}{}^{\mathrm{hom}}(n) = 3\hbar^2(3\pi^2)^{2/3} n^{5/3}/(10m)$. This is used to further approximate the kinetic energy functional

$$T[n] \approx T^{\mathrm{LDA}}[n] \approx T_{\mathrm{S}}^{\mathrm{LDA}}[n] = \int \mathrm{d}^3 r \; t_{\mathrm{S}}^{\mathrm{hom}} \left(n(\boldsymbol{r}) \right) , \tag{1.34}$$

where $T_S^{LDA}[n]$ is the local-density approximation to $T_S[n]$, the kinetic energy of noninteracting electrons of density n. Equivalently, it may be considered as the noninteracting version of $T^{LDA}[n]$. The TF approximation consists of combining (1.32) and (1.34) to express the energy functional as

$$E[n] = T[n] + U[n] + V[n] \approx E^{TF}[n] = T_S^{LDA}[n] + U_H[n] + V[n]. \quad (1.35)$$

A major defect of the TF approximation is that within it molecules are unstable: the energy of a set of isolated atoms is lower than that of the bound molecule.

This fundamental deficiency, and the lack of accuracy resulting from neglect of correlations in (1.32) and from using the local approximation (1.34) for the kinetic energy, limit the practical use of the TF approximation. However, it is found as a most useful starting point for a large body of work on improved approximations in chemistry and physics [266,275]. More recent approximations for $T[n]$ can be found, e.g., in [276–278]. The extension of the local-density concept to the exchange–correlation energy is at the heart of many modern density functionals.

The Kohn–Sham Equations

In light of the severe limitations of the TF approximation, more accurate schemes for treating the kinetic energy functional of the interacting electrons, $T[n]$, and for the inclusion of exchange and correlation effects have been explored. First, $T[n]$ is decomposed into a part that represents the kinetic energy of noninteracting particles, $T_S[n]$, and another that represents the remainder, $T_C[n]$ (the subscripts S and C stand for "single-particle" and "correlation," respectively),

$$T[n] = T_S[n] + T_C[n]. \quad (1.36)$$

$T_S[n]$ is defined as the expectation value of the kinetic energy operator with the Slater determinant arising from the density n, i.e., $T_S[n] = \langle \Phi[n] | \hat{T} | \Phi[n] \rangle$, and the full kinetic energy is defined as $T[n] = \langle \Psi[n] | \hat{T} | \Psi[n] \rangle$.

$T_S[n]$ is not known exactly as a functional of n [and using the LDA to approximate it leads one back to the TF approximation (1.34)], but it is easily expressed in terms of the single-particle orbitals $\phi_i(\boldsymbol{r})$ of a noninteracting system with density n, as

$$T_S[n] = -\frac{\hbar^2}{2m} \sum_i^N \int \mathrm{d}^3 r \phi_i^*(\boldsymbol{r}) \nabla^2 \phi_i(\boldsymbol{r}), \quad (1.37)$$

because for noninteracting particles the total kinetic energy is just the sum of the individual kinetic energies. Since all $\phi_i(\boldsymbol{r})$ are functionals of n, this expression for T_S is an explicit orbital functional but an implicit density functional, $T_S[n] = T_S[\{\phi_i[n]\}]$.

We now rewrite the exact energy functional as

$$E[n] = T[n] + U[n] + V[n] = T_{\mathrm{S}}\left[\{\phi_i[n]\}\right] + U_{\mathrm{H}}[n] + E_{\mathrm{XC}}[n] + V[n], \quad (1.38)$$

where by definition E_{XC} contains the differences $T - T_S$ (i.e., T_{C}) and $U - U_{\mathrm{H}}$. Unlike (1.35), (1.38) is formally exact but of course E_{XC} is unknown—although the HK theorem guarantees that it is a density functional. This functional, $E_{\mathrm{XC}}[n]$, is called the exchange–correlation (XC) energy. It is often decomposed as $E_{\mathrm{XC}} = E_{\mathrm{X}} + E_{\mathrm{C}}$, where $\mathrm{E_X}$ is due to the Pauli principle (exchange energy) and E_{C} is due to correlations. (T_{C} is then a part of E_{C}.) The exchange energy can be written explicitly in terms of the single particle orbitals

$$E_{\mathrm{X}}\left[\{\phi_i[n]\}\right] = -\frac{q^2}{2}\sum_{jk}\int \mathrm{d}^3 r \int \mathrm{d}^3 r' \frac{\phi_j^*(\boldsymbol{r})\phi_k^*(\boldsymbol{r}')\,\phi_j(\boldsymbol{r}')\,\phi_k(\boldsymbol{r})}{|\boldsymbol{r}-\boldsymbol{r}'|}, \quad (1.39)$$

but no exact expression in terms of the density is known.

A simple way to understand the origin of correlations is to recall that the Hartree energy is obtained in a variational calculation in which the many-body wave function is approximated as a product of single-particle orbitals. Use of an antisymmetrized product (a Slater determinant) results in the Hartree and exchange energy. The correlation energy is then defined as the difference between the full ground-state energy (obtained with the correct many-body wave function) and the one obtained from the (Hartree–Fock or Kohn–Sham) Slater determinant. Note, that since the correlation energy arises from the use of a more general trial wave function (that is, one with more variational freedom) than a single Slater determinant, correlations cannot raise the total energy, i.e., $E_{\mathrm{C}}[n] \leq 0$. Since a Slater determinant is itself more general than a simple product, we also have $E_{\mathrm{X}} \leq 0$, and thus the upper bound $E_{\mathrm{XC}}[n] \leq 0$. A lower bound is provided by the Lieb–Oxford formula [279, 280].

Recalling the quantum-mechanical interpretation of the wave function as a probability-amplitude, we see that a product form of the many-body wave function corresponds to treating the probability amplitude of the many-electron system as a product of the probability amplitudes of individual electrons (the orbitals). Mathematically, the probability of a composed event is only equal to the probability of the individual events if the individual events are independent (i.e., uncorrelated). Physically, this means that the electrons described by the product wave function are independent. Such wave functions thus neglect the fact that, as a consequence of the Coulomb interaction, the electrons try to avoid each other. The correlation energy is the additional energy lowering obtained in a real system due to the mutual avoidance of the interacting electrons.

Expressing the single particle kinetic energy T_{S} as an orbital functional (1.37) prevents direct minimization of the energy functional (1.38) with respect to n. Instead, one commonly employs a scheme suggested by Kohn and Sham [274], which starts by writing

$$0 = \frac{\delta E[n]}{\delta n(\boldsymbol{r})} = \frac{\delta T_{\mathrm{S}}[n]}{\delta n(\boldsymbol{r})} + \frac{\delta V[n]}{\delta n(\boldsymbol{r})} + \frac{\delta U_{\mathrm{H}}[n]}{\delta n(\boldsymbol{r})} + \frac{\delta E_{\mathrm{XC}}[n]}{\delta n(\boldsymbol{r})}$$
$$= \frac{\delta T_{\mathrm{S}}[n]}{\delta n(\boldsymbol{r})} + v(\boldsymbol{r}) + v_{\mathrm{H}}(\boldsymbol{r}) + v_{\mathrm{XC}}(\boldsymbol{r}). \quad (1.40)$$

Since $V[n] = \int \mathrm{d}^3 r\, n(\boldsymbol{r}) v(\boldsymbol{r})$ [see, e.g., (1.30)], $\delta V/\delta n = v(\boldsymbol{r})$, the external potential acting on the electrons. The term $\delta U_{\mathrm{H}}/\delta n$ yields the Hartree potential, introduced in (1.33). For the term $\delta E_{\mathrm{XC}}/\delta n$, which can only be calculated explicitly once an approximation for E_{XC} has been chosen, one commonly writes v_{XC}.

Consider now a system of noninteracting particles moving in the potential $v_{\mathrm{S}}(\boldsymbol{r})$. For this system, the minimization condition is simply

$$0 = \frac{\delta E_{\mathrm{S}}[n]}{\delta n(\boldsymbol{r})} = \frac{\delta T_{\mathrm{S}}[n]}{\delta n(\boldsymbol{r})} + \frac{\delta V_{\mathrm{S}}[n]}{\delta n(\boldsymbol{r})} = \frac{\delta T_{\mathrm{S}}[n]}{\delta n(\boldsymbol{r})} + v_{\mathrm{S}}(\boldsymbol{r}), \quad (1.41)$$

since there are no Hartree and XC terms in the absence of interactions. The density solving this Euler equation is $n_{\mathrm{S}}(\boldsymbol{r})$. Comparing this with (1.40), we find that both minimizations have the same solution $n_S(\boldsymbol{r}) \equiv n(\boldsymbol{r})$, if v_{S} is chosen as

$$v_{\mathrm{S}}(\boldsymbol{r}) = v(\boldsymbol{r}) + v_{\mathrm{H}}(\boldsymbol{r}) + v_{\mathrm{XC}}(\boldsymbol{r}). \quad (1.42)$$

Consequently, one can calculate the density of the interacting (many-body) system, acted upon by the external potential $v(\boldsymbol{r})$, described by a many-body Schrödinger equation of the form (1.13), by solving the equations of a noninteracting (single-body) system with the external potential $v_{\mathrm{S}}(\boldsymbol{r})$; the question whether such a potential $v_{\mathrm{S}}(\boldsymbol{r})$ always exists in the mathematical sense is called the noninteracting v-representability problem. For lattice systems, a complete solution is available [281, 282]. The Schrödinger equation of this auxiliary (noninteracting) system

$$\left[-\frac{\hbar^2 \nabla^2}{2m} + v_{\mathrm{S}}(\boldsymbol{r})\right] \phi_i(\boldsymbol{r}) = \varepsilon_i \phi_i(\boldsymbol{r}), \quad (1.43)$$

yields orbitals that reproduce the density $n(r)$ of the original system [these are the same orbitals employed in (1.37)],

$$n(\boldsymbol{r}) \equiv n_{\mathrm{S}}(\boldsymbol{r}) = \sum_i^N f_i \left|\phi_i(\boldsymbol{r})\right|^2, \quad (1.44)$$

where f_i is the occupation of the ith orbital. Equations (1.42–1.44) are known as the Kohn–Sham equations. They replace the problem of minimizing the functional $E[n]$ by that of solving a noninteracting Schrödinger equation. (Recall that the minimization of $E[n]$ originally replaced the problem of solving the many-body Schrödinger equation.) Since both v_{H} and v_{XC} depend on n, which depends on the ϕ_i, which in turn depend on v_{s}, the problem of solving the KS equations is a nonlinear one. Usually, one starts with an initial guess for $n(\boldsymbol{r})$, calculates the corresponding $v_S(\boldsymbol{r})$, and then solves the differential equation (1.43) for ϕ_i. From these, one calculates a new density using (1.44) and starts again. The process is repeated until it converges. The technical name for this procedure is "self-consistency cycle." Different convergence criteria (such as convergence in the energy, the density, or some observable calculated from these) and various convergence-accelerating algorithms (such as mixing of old and new effective potentials) are in common use.

Once one has a converged solution n_0, one can calculate the total energy from (1.38) or, equivalently and more conveniently, from

$$E_0 = \sum_i^N \varepsilon_i - \frac{q^2}{2}\int \mathrm{d}^3r \int \mathrm{d}^3r' \frac{n_0(\boldsymbol{r})n_0(r')}{|\boldsymbol{r}-\boldsymbol{r}'|} - \int \mathrm{d}^3r\; v_{\mathrm{XC}}(\boldsymbol{r})n_0(\boldsymbol{r}) + E_{\mathrm{XC}}\left[n_0\right]. \tag{1.45}$$

Equation (1.45) follows from writing $V[n]$ in (1.38) by means of (1.42) as

$$V[n] = \int \mathrm{d}^3r\; v(\boldsymbol{r})n(\boldsymbol{r}) = \int \mathrm{d}^3r\left[v_S(\boldsymbol{r}) - v_H(\boldsymbol{r}) - v_{\mathrm{XC}}(\boldsymbol{r})\right]n(\boldsymbol{r}) \tag{1.46}$$

$$= V_S[n] - \int \mathrm{d}^3r\left[v_H(\boldsymbol{r}) + v_{\mathrm{XC}}(\boldsymbol{r})\right]n(\boldsymbol{r}) \tag{1.47}$$

and identifying the energy of the noninteracting (Kohn–Sham) system as $E_S = \sum_1^N \varepsilon_i = T_S + V_S$. Equation (1.45) shows that E_0 is not simply the sum of all ε_i; indeed the difference between E_0 and $\sum_1^N \varepsilon_i$ is due to electron–electron interactions. In fact, it should be clear from our derivation of (1.43) that the ε_i are introduced as artificial objects; they are the eigenvalues of an auxiliary single-body Schrödinger equation whose eigenfunctions (orbitals) yield the correct density. It is only this density that has strict physical meaning in the KS equations. The KS eigenvalues, on the other hand, bear only a semiquantitative resemblance to the true energy spectrum [284] but are not to be trusted quantitatively.

An exception to the above is the highest occupied KS eigenvalue (sometimes called highest occupied KS orbital, or HOKSO). Denoting by ε_N the Nth eigenvalue, one can show rigorously that $\varepsilon_N = -I$, the negative of the first ionization energy of the N-electron system, and $\varepsilon_{N+1}(N+1)$, the negative of the electron affinity of the same N-electron system. These relations hold *only* for the exact functional (they are similar to the Koopmans' theorem proved in the context of the Hartree–Fock method [285, 286]). When calculated with an approximate functional of the LDA or GGA type (see below), the highest eigenvalues usually do not provide good approximations to the experimental I and A. Better results for these observables are obtained by calculating them as total-energy differences, according to $I = E_0(N-1) - E_0(N)$ and $A = E_0(N) - E_0(N+1)$, where $E_0(N)$ is the ground-state energy of the N-body system. Alternatively, self-interaction corrections (SIC) can be used to obtain improved ionization energies and electron affinities from Kohn–Sham eigenvalues [287].

Practical Considerations and Further Developments: Functionals

The main issue involved in using DFT and the KS scheme pertains to construction of expressions for the XC functional, $E_{XC}[n]$, containing the many-body aspects of the problems (1.38). The main approaches to this issue are: (a) local functionals; the Thomas Fermi (TF) and LDA, (b)semilocal or gradient-dependent functionals; the gradient-expansion approximation (GEA) and generalized gradient approximation (GGA), and (c) nonlocal functionals; hybrids, orbital functionals, and SIC. For detailed discussions the reader is referred to the reviews [257, 260–272].

Local Functionals: LDA

We have already introduced the concept of a local-density approximation in the context of the TF approximation (1.34). Using a similar approach in the KS scheme, one employs a local-density approximation for the exchange energy density

$$e_{\mathrm{X}}^{\mathrm{hom}}(n) = -\frac{3q^2}{4}\left(\frac{3}{\pi}\right)^{1/3} n^{4/3}, \tag{1.48}$$

so that

$$E_{\mathrm{X}}^{\mathrm{LDA}}[n] = -\frac{3q^2}{4}\left(\frac{3}{\pi}\right)^{1/3} \int \mathrm{d}^3 r\, n(\boldsymbol{r})^{4/3}. \tag{1.49}$$

If one adds this expression to the TF expression (1.35) one obtains [257] the Thomas–Fermi Dirac approximation to $E[n]$. Because of problems associated with the local-density approximation to the kinetic energy (used in the TF approach), it is advantageous to maintain the orbital description of the kinetic energy and use the above local approximation for the exchange energy in the KS equations (1.43). The situation is more complicated for the exact correlation energy, $E_{\mathrm{C}}(n)$, since $e_{\mathrm{C}}{}^{\mathrm{hom}}(n)$ is not known. Modern expressions for $e_{\mathrm{C}}{}^{\mathrm{hom}}(n)$ [288–290] are parametrizations of highly accurate Quantum Monte Carlo (QMC) calculations for the electron liquid [291]. These expressions are implemented in most standard DFT program packages and in typical applications give almost identical results.

Independent of the parametrization, the LDA for the $E_{\mathrm{XC}}[n]$ functional is formally expressed as

$$E_{\mathrm{XC}}[n] \approx E_{\mathrm{XC}}^{\mathrm{LDA}}[n] = \int \mathrm{d}^3 r \left[e_{\mathrm{X}}^{\mathrm{hom}}(n(\boldsymbol{r})) + e_{\mathrm{C}}^{\mathrm{hom}}(n(\boldsymbol{r}))\right] = \int \mathrm{d}^3 r\, e_{\mathrm{XC}}^{\mathrm{hom}}(n(\boldsymbol{r})). \tag{1.50}$$

This approximation for $E_{\mathrm{XC}}[n]$ has proved to be remarkably successful, even when applied to systems that are quite different from the electron liquid that forms the reference system for the LDA. For many decades the LDA has been applied in, e.g., calculations of band structures and total energies in solid-state physics. In quantum chemistry, it is much less popular because it fails to provide results that are accurate enough to permit a quantitative discussion of the chemical bond in molecules; so-called "chemical accuracy" requires calculations with an error of not more than about 1 kcal mol^{-1} = 0.0434 eV per particle.

Semilocal Functionals. GEA and GGA

The particular way in which the inhomogeneous many-body problem is decomposed and the various possible improvements on the LDA are responsible for the success of DFT in practical applications of quantum mechanics to real materials. In the LDA, one exploits knowledge of the density at point $\boldsymbol{r}$. Any real system is spatially inhomogeneous, i.e., it has a spatially varying density $n(\boldsymbol{r})$, and it would clearly be useful to also include information on the rate of this variation in the functional. A first attempt at doing this were the so-called "gradient-expansion approximations" (GEA). In this class of approximation, one tries to systematically calculate gradient corrections of the form $|\nabla n(\boldsymbol{r})|, |\nabla n(\boldsymbol{r})|^2, \nabla^2 n(\boldsymbol{r})$, etc., to the LDA. A famous example is the lowest-order gradient correction to the Thomas–Fermi approximation for $T_{\mathrm{S}}[n]$

$$T_{\mathrm{S}}[n] \approx T_{\mathrm{S}}^{\mathrm{W}}[n] = T_{\mathrm{S}}^{\mathrm{LDA}}[n] + \frac{\hbar^2}{8m}\int \mathrm{d}^3r \frac{|\nabla n(\boldsymbol{r})|^2}{n(\boldsymbol{r})}. \quad (1.51)$$

A major breakthrough occurred when it was realized, in the early eigthies, that instead of power-series-like systematic gradient expansions one could experiment with more general functions of $n(\boldsymbol{r})$ and $\nabla n(\boldsymbol{r})$, which need not proceed order by order. Such functionals, of the general form

$$E_{\mathrm{XC}}^{\mathrm{GGA}}[n] = \int \mathrm{d}^3r f\left(n(\boldsymbol{r}), \nabla n(\boldsymbol{r})\right), \quad (1.52)$$

have come to be known as generalized-gradient approximations (GGAs) [292, 293]. Different GGAs differ in the choice of the function $f(n, \nabla n)$. In particular, GGAs used in quantum chemistry are typically developed by fitting parameters to test sets of selected molecules. On the other hand, GGAs used in physics tend to emphasize exact constraints. Nowadays the most popular (and most reliable) GGAs are PBE (denoting the functional proposed in 1996 by Perdew et al. [292, 293]) in physics, and BLYP (denoting the combination of Becke's 1988 exchange functional [294] with the 1988 correlation functional of Lee Yang and Parr [295]) in chemistry. Many other GGA-type functionals are also available and new ones continue to appear.

Quite generally, current GGAs seem to give reliable results for all main types of chemical bonds (covalent, ionic, metallic, and hydrogen bond). Both in physics and chemistry, the widespread use of GGAs has lead to major improvements as compared to LDA. "Chemical accuracy," as defined above, has not yet been attained but is not too far away either. A useful collection of explicit expressions for some GGAs can be found in the appendix of [296], and more detailed discussion of some selected GGAs and their performance is given in [297] and in the chapter Kurth and Perdew in [269, 270].

Nonlocal and Orbital Functionals

In spite of these advances, the quest for more accurate functionals continues, and both in chemistry and physics various formulations that attempt to go beyond the GGA functionals have appeared. Perhaps the most popular functional in quantum chemistry is B3LYP. This is a combination of the LYP GGA for correlation [295] with Becke's three-parameter hybrid functional B3 for exchange [298]. Common hybrid functionals, such as B3, mix a fraction of Hartree–Fock exchange into the DFT exchange functional (other mixtures are also possible). This mixing involves a certain amount of empiricism (the weight factors given to the HF and DFT exchange terms are adjustable) and optimization for selected classes of molecules. Formally, this might be considered a drawback but in practice B3 has proven to be the most successful exchange functional for chemical applications, in particular, when combined with the LYP GGA functional for E_C. More extreme examples of this semi-empirical mode of construction of functionals are Becke's 1997 hybrid functional [299,300], which contains 10 adjustable parameters, and the functionals of [301,302], each of which contains 21 parameters.

Another recent development that aims at going beyond the GGA is the emergence of so-called Meta-GGAs, which depend, in addition to the density and its derivatives, also on the Kohn–Sham kinetic energy density [303–308]. In several recent tests [283,304–307,309–311], Meta-GGAs have given favorable results even when compared to the best GGAs, but the full potential of this type of approximation is only beginning to be explored systematically.

As we have discussed already, the kinetic energy, T_S, can be readily represented as a functional written in terms of single-particle orbitals (1.37), whereas expressing it in terms of the density directly is much harder. Such functionals are termed orbital functionals. Another important example of an orbital-dependent functional is the exchange energy (1.39). The Meta-GGAs and hybrid functionals mentioned above are also orbital functionals.

Still another type of orbital functional is the self-interaction correction SIC. Self-interaction refers to the fact that in DFT the Hartree contribution to the energy (1.32) includes an unphysical interaction of an electron with itself, that is, while in the Hartree–Fock method the ith electron interacts with the full electron density minus the density of the ith electron, this subtraction is not made in DFT. However, in exact treatments of the exchange, such self-interaction in the Hartree term is cancelled (as in the HF method), but this is not the case for approximate treatments of the exchange–correlation energy, and the errors can be large since they are Columbic in nature. The self-interaction problem can be particularly severe for systems in which electrons tend to localize and strongly interact such as transition metal oxides and rare earth elements and compound. Other systems that may be particularly susceptible to self-interaction errors are negatively charged ones (i.e., anionic species).

The earliest SIC method predates DFT, and is due to Hartree in 1928 [312–314], in his calculations of atoms. He subtracted a self-term due to the charge density of the ith occupied state, thus defining the ith-state potential; that is, each occupied state was associated with a different potential. This method applies easily to finite systems, but the SIC vanishes for an extended system.

Most current implementations of SIC make use of the expressions proposed in [289]. This correction can be applied to any density functional, and it ensures that the resulting corrected functional satisfies $E_C[n] \equiv 0$ and $E_X[n] \equiv -E_H[n]$ for a one-electron system. This SIC method has been shown to greatly improve the uncorrected LDA [315–317] for systems such as transition metal oxides, but for thermochemistry it does not seem to be significant [318]. Unfortunately this approach does not lead to Kohn–Sham equations of the usual form, and the number of equations that need to be solved scales as the number of electrons in the system. As a consequence, a number of specialized algorithms for minimizing the SIC energy functional have been developed. For more details on these algorithms and some interesting applications in solid-state physics see [315–317, 319] and the contribution of Temmerman et al. in [271].

Spin-Density Functional Theory

Till now we have discussed DFT in terms of the charge (or particle) density $n(r)$ as the fundamental variable. In order to reproduce the correct charge density of the interacting system in the noninteracting (Kohn–Sham) system, one must apply to the latter the effective KS potential $v_S = v + v_H + v_{XC}$ in which the last two terms represent the effect of electron–electron interactions on the charge density. This is the originally proposed form of DFT [273]. In a formulation called spin-density functional theory (SDFT) [320, 321], $n_\uparrow(\boldsymbol{r})$ and $n_\downarrow(\boldsymbol{r})$ (the densities of the up and down spins, respectively) are treated as the fundamental variables, and they can be used to calculate the charge density $n(\boldsymbol{r})$ and the spin-magnetization density $m(\boldsymbol{r})$ from

$$n(\boldsymbol{r}) = n_\uparrow(\boldsymbol{r}) + n_\downarrow(\boldsymbol{r}), \tag{1.53}$$

$$m(\boldsymbol{r}) = \mu_0 \left(n_\uparrow(\boldsymbol{r}) - n_\downarrow(\boldsymbol{r}) \right), \tag{1.54}$$

where μ_0 is the Bohr magneton. This formulation can be expressed as associating with the density for each spin-direction (or with each density-like quantity), one effective potential in the KS system.

More generally, the Hohenberg–Kohn theorem of SDFT states that in the presence of a magnetic field $B(\boldsymbol{r})$ that couples only to the electron spin (via the familiar Zeeman term), the ground-state wave function and all ground-state observables are unique functionals of n and m or, equivalently, of $n_\uparrow$ and $n_\downarrow$. In the particular field-free case, the SDFT HK theorem still holds and continues to be useful, e.g., for systems with spontaneous polarization. Almost the entire further development of the HK theorem and the KS equations follows and can be rephrased in a rather straight forward manner for the SDFT, just by attaching a suitable spin index to the densities.

In order to construct an exchange energy functional for the SDFT, one uses the relation [322]

$$E_{\mathrm{X}}^{\mathrm{SDFT}}\left[n_{\uparrow}, n_{\downarrow}\right]=\frac{1}{2}\left(E_{\mathrm{X}}^{\mathrm{DFT}}\left[2 n_{\uparrow}\right]+E_{\mathrm{X}}^{\mathrm{DFT}}\left[2 n_{\downarrow}\right]\right), \tag{1.55}$$

which can be used to construct an SDFT exchange functional from a given DFT exchange functional. In the context of the LSDA, von Barth and Hedin [320] wrote the exchange functional in terms of an interpolation between the unpolarized and fully polarized electron gas, which by construction satisfies (1.55). Alternative interpolation procedures can be found in [288]. GGA exchange functionals also satisfy (1.55) by construction. For the correlation energy no scaling relation of the type (1.55) holds, so that in practice correlation functionals are either directly constructed in terms of the spin densities or written by using, without formal justification, the same interpolation already used in the exchange functional. In the case of the LSDA, this latter procedure was introduced in [320] and further analyzed and improved in [288].

The Kohn–Sham equations of SDFT are

$$\left[-\frac{\hbar^{2} \nabla^{2}}{2 m}+v_{s \sigma}(\boldsymbol{r})\right] \phi_{i \sigma}(\boldsymbol{r})=\varepsilon_{i \sigma} \phi_{i \sigma}(\boldsymbol{r}), \tag{1.56}$$

where $v_{\mathrm{S}\sigma}(\boldsymbol{r})=v_{\sigma}(\boldsymbol{r})+v_{\mathrm{H}}(\boldsymbol{r})+v_{\mathrm{XC},\sigma}(\boldsymbol{r})$, and

$$v_{\mathrm{XC}, \sigma}(\boldsymbol{r})=\frac{\delta E_{\mathrm{XC}}^{\mathrm{SDFT}}\left[n_{\uparrow}, n_{\downarrow}\right]}{\delta n_{\sigma}(\boldsymbol{r})}. \tag{1.57}$$

SDFT is the most widely used form of DFT. Recent work on SDFT is described in [323]. Detailed discussions can be found in [257, 321, 324], and a particularly clear exposition of the construction of XC functionals for SDFT is the contribution of Kurth and Perdew in [269, 270].

Accuracy

The above discussion about the continuing efforts aimed at the development of functionals to be employed in practical applications of DFT, highlighted the particular importance of developing improved approximations for the exchange-correlation functionals. Underlying these efforts is the realization that while E_{xc} is a small fraction of the total energy of an atom, molecule or a solid, it may be the dominant contribution to the chemical bonding or atomization energy (the energy needed to break up the system to individual atoms).

Typical relative errors found from self-consistent calculations within the local spin-density (LSD) or GGA approximations are given in Table 1.1.

Table 1.1. Typical errors for atoms, molecules, and solids from self-consistent Kohn–Sham calculations within the LSD and GGA approximations

Property	LSD	GGA
E_X	5% (not negative enough)	0.5%
E_C	100% (too negative)	5%
Bond length	1% (too short)	1% (too long)
Structure	Overly favors close packing	More correct
Energy barrier	100% (too low)	30% (too low)

Note that there is typically some cancellation of errors between the exchange and correlation contributions to E_{XC}. The "energy barrier" is the barrier to a chemical reaction. After Kurth and Perdew in [270]

Table 1.2. The mean absolute error of the atomization energies for 20 molecules, evaluated by various approximations

Approximation	Mean absolute error (eV)
Unrestricted Hartree–Fock	3.1 (underbinding)
LSD	1.3 (overbinding)
GGA	0.3 (mostly overbinding)
Desired "chemical accuracy"	0.05

After Kurth and Perdew in [270]

The mean absolute error of the atomization energies for 20 molecules, evaluated by various approximations is given in Table 1.2.

We note that in the Hartree–Fock method exchange is treated exactly but correlations are neglected completely. The Hartree–Fock total energy is an upper bound to the true ground-state total energy, while the LSD and GGA energies are not.

1.3.2 Computational Methods and Techniques

First-Principles Molecular Dynamics

The progress achieved in electronic structure calculations has made it possible to perform molecular dynamics (MD) simulations with the forces acting on the nuclei derived directly through the solution of the quantum mechanical equations for the electrons for given spatial configurations of the nuclei [325–327]. This is to be contrasted with "classical" MD simulations where the forces on the aroms are given by interatomic interaction potentials (which can express pair-wise or many-particle interactions), derived through semiempirical fitting of chosen functional forms (e.g., Lennard-Jones, Morse potentials, etc.), or through parametrization of separate quantum mechanical studies for various spatial arrangements of the atoms.

In the simulations that are referred to as first-principles molecular dynamics (FPMD), or ab initio MD, simulations, the motions of the nuclei are obtained from a solution of Newton's equations of motion with the force on the Ith ion given by

$$\boldsymbol{F}_I = -\frac{\partial E}{\partial \boldsymbol{R}_I}, \tag{1.58}$$

where the energy is written (using DFT) as

$$E = \frac{\langle \Psi|\hat{H}|\Psi\rangle}{\langle \Psi|\Psi\rangle} \equiv \left\langle \hat{H} \right\rangle = \left\langle \hat{T} \right\rangle + \left\langle \hat{V}_{\text{int}} \right\rangle + \int \mathrm{d}^3 r\, V_{\text{ext}}(\boldsymbol{r}) n(\boldsymbol{r}) + E_{II} \tag{1.59}$$

with the last term on the right-hand side representing the direct ion–ion Coulomb repulsion. From the above we obtain

$$\boldsymbol{F}_I = -\frac{\partial E}{\partial \boldsymbol{R}_I} = -\int \mathrm{d}^3 r n(\boldsymbol{r}) \frac{\partial V_{\text{ext}}(\boldsymbol{r})}{\partial \boldsymbol{R}_I} - \frac{\partial E_{II}}{\partial \boldsymbol{R}_I}, \tag{1.60}$$

where $n(\boldsymbol{r})$ is the unperturbed electron density (obtained trough an electronic structure calculation, e.g., solution of the KS DFT equations for the instantaneous configuration of the nuclei), and the partial derivative indicates that all the nuclei (except the Ith one) are held fixed. Note that even though the kinetic energy and interelectron interaction do change as the nuclei move, these terms cancel and only the external potential (and interion interactions) enter the expression for the force. Equation (1.60) expresses the Hellmann–Feynman theorem [328,329] when the energy is expressed in the form given by (1.59). In the case of nonlocal potentials (for example pseudopotentials), the force cannot be expressed only in terms of the electron density as in (1.60). Nevertheless, the general expression

$$-\frac{\partial E}{\partial \boldsymbol{R}_I} = -\left\langle \Psi \left| \frac{\partial \hat{H}}{\partial \boldsymbol{R}_I} \right| \Psi \right\rangle - \frac{\partial E_{II}}{\partial \boldsymbol{R}_I} \tag{1.61}$$

still holds true.

Often the FPMD schemes are referred to as Born–Oppenheimer (BO) molecular dynamics (BOMD) simulations, since the most often used electronic structure calculation methods employ the separation of time-scales between the nuclear and electronic motions, introduced by Born and Oppenheimer (see discussion at the beginning of section 1.3.1, "The Many-Body Problem"). Such BO FPMD simulations have been implemented in several ways [330–334].

In the investigation of nanocatalytic systems described in this chapter, the method used mostly is the BO-SDFT-MD method [333] with GGA PBE [292, 293] corrections. This method was designed specifically for treatment of charged systems, as well as situations where charge separation or multipolar polarization develop. Such circumstances present problems when periodic conditions are used, since the interaction of a charged system (or a system exhibiting charge separation or polarization) represented in the primary (calculational) cell with its periodically replicated images are long ranged. To circumvent the influence of such spurious contributions to the energy and dynamics of the system, the method used here does not employ periodic boundary conditions or a supercell. Thus, in the BO-SDFT-MD method [333], the KS SDFT equations are solved using a plane-wave basis (using a 3D periodically replicated grid and fast-Fourier transforms), and employing norm-conserving, soft, nonlocal pseudopotentials [324] (in the case of heavy metals, particularly gold, scalar relativistic ones [119,335], but without periodic replication of the ionic system. Instead, the system is defined on a spatial grid whose dimensions are chosen such that the wave functions vanish on the grid boundaries.

Reaction Paths and Barriers

Determination of microscopic, atomic scale, reaction mechanisms and reliable estimations of reaction barriers are among the main challenges of theoretical research in catalysis, nanocatalysis in particular. The problem may be formulated as a search for the lowest (minimum) energy (free energy) path (MEP) connecting a stable configuration of the reactive system prior to the reactive event (i.e., on the "reactant" side) with another stable configuration after reaction (i.e., on the "product" side). Often one uses the concept of a "reaction coordinate" for the transition between the reactant and product spaces [336]; for example, when the reaction involves a dissociation event, the distance between the dissociating fragments may serve as the reaction coordinate. The energy maximum along the MEP is the saddle point energy, giving the activation energy barrier, which plays a central role in estimating the reaction rate within the (harmonic) transition state theory [337]. Many methods have been developed for finding reaction paths and saddle points (see [338,339] for reviews).

In one of the oldest and simplest ways for determination of the potential energy barrier for reaction, one starts from a well-defined stable state on the reactant side, and for each state of the reactants along the prescribed reaction coordinate (for example, for each value of the distance between the dissociation fragments) one relaxes fully all the configurational (atomic positions) degrees of freedom of the system, except the one lying along the reaction coordinate, which is constrained to take prescribed values (connecting the reactant and product states). A plot of the total (potential) energy of the system versus the reaction coordinate gives an energy profile of the reaction, and the maximum of the curve is the (potential) energy barrier (i.e., no entropic contributions are included).

To determine the free-energy profile (rather then the potential energy profile as above), one performs a (typically short) constant temperature (canonical ensemble) FPMD simulation of the system (while constraining the reaction coordinate to the prescribed value) and records the mean force of constraint along the chosen reaction coordinate. This process is then repeated for a number of points along the reaction coordinate connecting the reactant and product states. An integral of the force times the distance along the reaction coordinate, evaluated between a lower limit associated with the initial stable reactant state and a point along the reaction coordinate, gives the reversible work of taking the system between these points. In the canonical ensemble, the reversible work equals the Helmholtz free-energy difference between the two points. A plot of these free energies evaluated for successive points along the reaction coordinate yields the free-energy profile of the reaction, and the maximum value of this profile is the reaction free-energy barrier. This method is close to the thermodynamic integration method employed in investigation of physical and chemical processes in solution, condensed phases, and biophysical studies [340].

For cases where the free energy barriers are energy dominated, a method that gained popularity is the so-called "nudged elastic band" (NEB) method [341]. The NEB method assumes that both the reactant and product states are known. A number of replicas of the original system are prepared, initially located equidistantly along a "linear" path between the reactant and product states. In the next step, the position of the path is relaxed to find a new reaction path. This relaxation is achieved through connecting the replicas along the path by harmonic springs that tend to keep the replicas equidistant to each other—this is the "elastic band" (EB). In addition to these fictitious springs, the replicas are acted upon by the intermolecular forces (expressed as the gradients of the intermolecular potential that tend to drive them to a minimum of the potential energy—FPMD based on DFT may be used to obtain the corresponding values of the intermolecular potential and its gradients). Only the intermolecular force projections perpendicular to the local tangent to the path are allowed to act on the replicas, while the EB forces are allowed to act only along the local tangent to the path. This process is repeated till the transverse intermolecular force projection vanishes, indicating that convergence to the MEP has been achieved. All along, the elastic band forces prevent all the replicas from collapsing into the reactant or product state. For further details and refinements see [342, 343]. The NEB algorithm is ideally suited for computations using parallel computers, allowing time-efficient reaction path investigations.

1.4 Concepts for Understanding Chemical Reactions and Catalytic Properties of Finite Systems

1.4.1 Basic Mechanisms of Catalytic Reactions

Langmuir–Hinshelwood Reaction Mechanism

According to the IUPAC, the Langmuir–Hinshelwood mechanism is defined as a mechanism for surface catalysis in which the reaction occurs between species that are adsorbed on the surface. This mechanism is expected to exhibit a second order kinetics with respect to the surface coverage of the two reactants.

As an example, we briefly discuss the oxidation of carbon monoxide by molecular oxygen, which is considered a model surface reaction in heterogeneous catalysis [240, 344]. The CO combustion reaction,

$$2CO + O_2 \rightarrow 2CO_2 \tag{1.62}$$

although exothermic by 283 kJ mol^{-1} [345], does not proceed in the gas phase under ambient conditions due to a high-activation energy barrier, which mainly results from the necessity to break the oxygen molecular bond. Thus, the activation or cleavage of the O–O bond is the major step to facilitate the reaction. This is readily achieved by transition metal catalysts, in particular by the platinum-group metals [240]. The reaction on the metal catalyst surface proceeds via the Langmuir–Hinshelwood mechanism [240], i.e., CO and O are first adsorbed on the surface, whereby the spontaneous dissociation of the diatomic oxygen molecule is presumed. The reactants subsequently diffuse on the surface before forming a reactive surface complex which leads to the CO_2 formation. Coadsorption studies reveal that repulsive interactions due to the necessity of electronic charge restructuring around the surface complex dominate the encounter between neighboring adsorbed CO and O particles [347] and lead to a reaction–activation barrier. Furthermore, the shape of the product (CO_2) molecule usually differs from the shape of the surface intermediate, i.e., the transition state or activated complex, which is assumed to be bent in this case. The respective reorientation of atomic or molecular orbitals will also contribute to an activation energy barrier. Figure 1.53 presents a 2D representation of the potential energy surface, where the adsorption bond length for the CO_2 product molecule is plotted on the x-axis and the mutual distance R(O–CO) between the adsorbed reactants O_{ad} and CO_{ad} is plotted on the y-axis. The graphic is based on the report by Ertl on CO oxidation on a platinum (111) surface [346]. All energies are referred to the gaseous reactants (CO + ½ O_2) in molar concentrations. Starting from the upper left, the adsorption of 1 mol of CO and O is accompanied by a net energy gain of 259 kJ. Adsorbed O and CO reside in the deep potential energy minimum on the upper left side of the diagram. As the reaction proceeds, both species must move uphill along the broken line to reach the transition state (denoted by the symbol ‡) before CO_2 can actually be formed. This passage requires the aforementioned activation energy ΔE_{LH} of about 100 kJ. The CO_2-surface complex is very weakly adsorbed in a shallow potential well from which it can easily escape and leave the surface (adsorption energy of ~20 kJ mol^{-1}). The gaseous CO_2 then exhibits the well-known thermodynamic reaction enthalpy of 283 kJ mol^{-1} as compared to the reactants.

Eley–Rideal Mechanism

According to IUPAC, a mechanism for surface catalysis in which the reaction occurs between a reactant molecule in the gas phase and one that

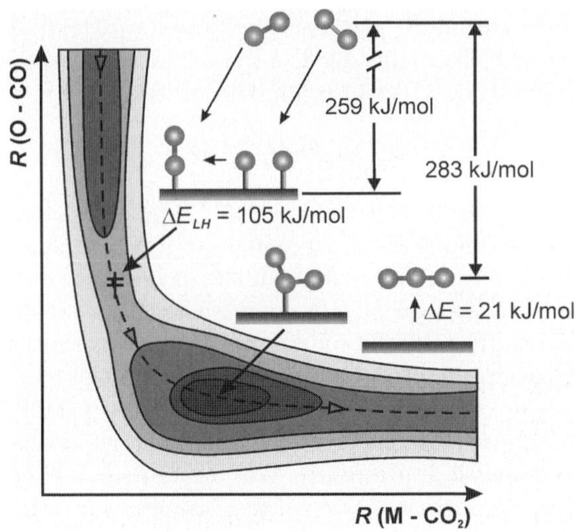

Fig. 1.53. 2D potential energy diagram of the CO oxidation reaction on a platinum(111) surface with the involved energies indicated. The mutual O–CO distance of the unreacted system R(O–CO) is plotted versus the CO_2–Pt distance R(M–CO_2). The reaction proceeds (at low coverages) as a Langmuir–Hinshelwood mechanism. The reaction coordinate is marked by the broken line (adapted from [240,346]). The blue *spheres* represent carbon atoms, red *spheres* oxygen atoms

is adsorbed on the surface is termed Eley–Rideal mechanism. A gas-phase molecule collides according to this mechanism with another molecule adsorbed on the surface for the reaction to proceed. The rate of product formation is in this case proportional to the partial pressure of the gas-phase reactant molecule above the surface and to the surface coverage of the adsorbed reactant. In the high-pressure regime, where the surface is saturated with reactant molecules, the rate law for this mechanism becomes first-order depending only on the partial pressure of the reactant colliding from the gas phase.

For the case of supported model catalyst under surface science conditions an Eley–Rideal mechanism has so far not been unambiguously identified. However, for gas-phase clusters, an Eley–Rideal type mechanism, in which the reactant has to collide at the right position with the cluster-complex to initiate the reaction, is more likely to proceed due to the reduced size of the catalytically active unsupported cluster.

Pressure and Material Gap

Molecular insight into catalytic reaction mechanisms is mostly based on studies employing well-defined model catalysts under ultrahigh vacuum conditions. The obtained information, however, can often not be directly related to real

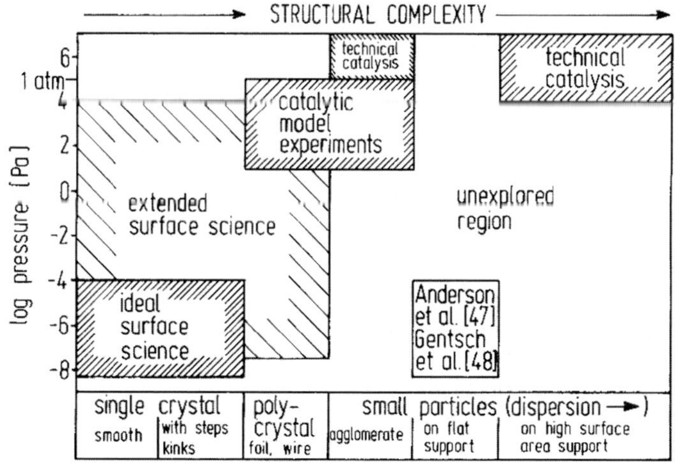

Fig. 1.54. Schematic diagram illustrating the pressure gap (ordinate) and the material gap between surface science and practical technical catalysis. "Structural complexity" plotted on the abscissa reflects the material gap in the surface morphology of the catalytic materials and ranges from small single-crystalline areas ($\sim 1\,cm^2$) on the left-hand side to technical catalysts with areas of several hundred square meters per gram. The diagram is taken from Christmann [240]

catalysis under industrial conditions because of the so-called pressure and material gap (Fig. 1.54). In fact, real catalysts usually consist of ill-defined systems operating at elevated pressures and temperatures which prevent the application of standard surface science analysis techniques. The pressure gap occurs because of the different pressure regimes under which model and real catalytic systems are investigated. Accordingly, the material gap represents the discrepancy between the effort to design idealized and well-defined model systems for mechanistic investigations and the structural complexity of real catalyst systems.

Therefore, the commonly applied procedure relies on the extrapolation of catalytic mechanisms identified under low pressure and idealized material properties.

In recent years, considerable efforts have been made in a variety of research groups to bridge the pressure and material gap and to experimentally as well as theoretically establish the new methods required to continuously vary the conditions starting form the model catalyst side (bottom up approach) or form the real catalyst side (top down approach). First examples support the importance of a detailed understanding of environmental and material complexity effects on catalytic reactions. The common distinction of pressure gap and material gap as depicted in Fig. 1.54 might, however, also be oversimplified as has been demonstrated in a recent contribution on the oxidation of CO on a ruthenium surface. The first step in this reaction consists in an oxidative transformation of the ruthenium surface. The catalytic activity of Ru in high-pressure oxidation reactions is thus likely due to nanometer thin but nevertheless bulk-like oxide patches of Ru which evades the assignment of the catalytic effect to the pressure or the material gap (see also discussion by Reuter in Chap. 5).

Spillover, Reverse Spillover, and Adlineation

Already in 1929 it was proposed by Schwab and Pietsch that the catalytic reaction on supported metal catalysts often takes place at the metal–oxide interface. This effect is known as adlineation, however, up to the present there is only little direct experimental evidence. In one example, the oxidation of CO on nanoscale gold, it is presently discussed whether the catalytic action takes place at the particle–support interface. Adlineation is strongly related to the effect of reverse spillover, where the effective pressure of the reactants in a catalytic process is enhanced by adsorption on the oxide material within the so-called collection zone and diffusion to the active metal particle (see Fig. 1.55 and also The Reactivity of Deposited Pd Clusters). The area of the collection zone and thus the reverse spillover are dependent on temperature, on the adsorption and diffusion properties of the reactants on the oxide material, as well as on the cluster density.

The spillover, in contrast to the reverse spillover, is a migration of intermediates, which are formed on the catalyst particle, onto the support material where it further reacts for instance on defect sites to the product molecules (Fig. 1.56).

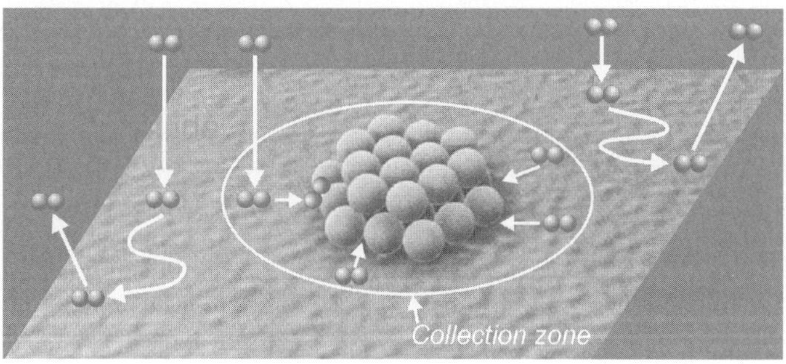

Fig. 1.55. Schematic representation of collection zone, adlineation, and reverse spillover for the case of a gold cluster supported on a metal–oxide surface. The reactants (red *spheres*) might either adsorb from the gas phase in the vicinity of the cluster, within the so-called collection zone, and be directly attracted toward the catalytically active cluster. Or the adsorption might be followed by random diffusion and eventually lead to desorption back to the gas phase, if the primary adsorption places are outside the collection zone of the cluster (graphics adapted from [348])

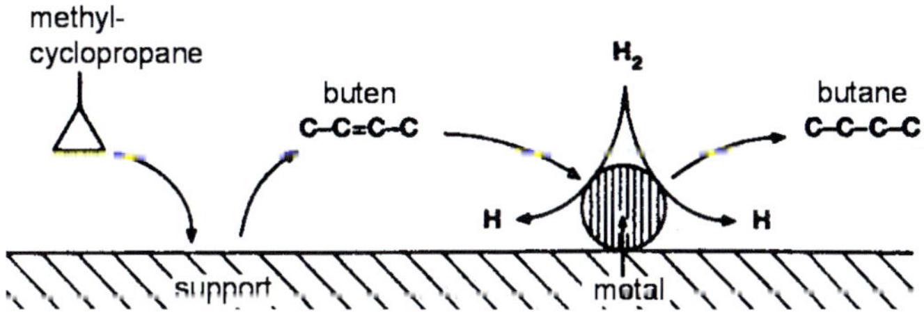

Fig. 1.56. In the example depicted in this figure, a bifunctional catalyst like platinum on silicon oxide enables the isomerization of methyl-cyclopropane to buten-2 as well as its hydrogenation to butane [349]. The isomerization proceeds on the oxide support. Hydrogen dissociates on the metal particle and the H-atoms "spillover" onto the oxide support material where they are mobile to react with butene

1.4.2 Cluster-Specific Mechanisms

Electronic Size Effects

Small metal clusters show strong variations in their electronic structure as a function of the cluster size. These quantum-size effects are particularly pronounced for clusters of the alkali and coinage metals with singly occupied s-orbitals. The apparent consequence of the resulting quantized electronic cluster structure is a distinct odd–even alternation in the electron binding energy. Measured vertical electron detachment energies (VDEs) for negatively charged gold clusters are depicted in Fig. 1.57. The alternating closed and open shell valence electron configuration in these clusters has direct implications on the

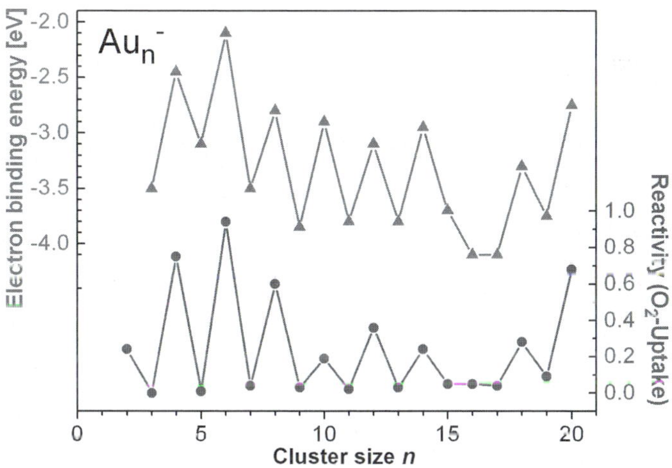

Fig. 1.57. Electron binding energies measured as VDEs of gas phase negatively charged gold cluster ions in comparison to the reactivity of the same cluster sizes toward molecular oxygen (adapted from [160])

chemical reactivity toward adsorbate molecules that involve the corresponding electronic levels. For instance, molecular oxygen binds to these clusters as an electron acceptor. The metal s-valence orbitals resonate with the $O_2\pi^*$ antibonding orbitals resulting in a charge transfer. This charge transfer is facilitated for clusters with low-electron binding energies, i.e., low VDEs as can be seen in the corresponding reaction extent displayed also in Fig. 1.57 [160].

Analogous effects have also been discussed for the reactivity of different bulk metal surfaces, where the center of gravity of the d-band determines the bonding [350]. In this case, the effects on the reactivity are considerably smaller than in the case of clusters in the nonscalable size regime and are size independent. Additional factors that change the d-band structure for a particular element are strain effects and the crystallographic orientation of the respective surface plane (see Chap. 3).

For supported clusters, electronic size effects are not only relevant for the adsorption of the reactants and the catalytic activity but also for the cluster–support interaction. As for the adsorbates, the bonding to the substrate is strongly influenced by the electronic structure of the clusters. This may lead to strongly size-dependent binding energies, size-dependent charge transfer from or to the clusters, as well as size-dependent change of the geometric cluster structure upon adsorption on the surface. Besides these static factors, also the dynamics of the clusters on surfaces is influenced by the size-dependent electronic structure of the cluster. Particularly important in nanocatalysis of small clusters is also the size-dependent migration and coalescence of the clusters at a given temperature, which is of course connected to the relative binding energies of the clusters to the various geometric sites on the substrate along the migration path.

Cooperative Adsorption

Coadsorption phenomena in heterogeneous catalysis and surface chemistry quite commonly consider competitive effects between two reactants on a metal surface [240, 344]. Also cooperative mutual interaction in the adsorption behavior of two molecules has been reported [240]. Recently, this latter phenomenon was found to be very pronounced on small gas-phase metal cluster ions too [351–354]. This is mainly due to the fact that the metal cluster reactivity is often strongly charge state dependent and that an adsorbed molecule can effectively influence the metal cluster electronic structure by, e.g., charge transfer effects. This changed electronic complex structure in turn might foster (or also inhibit) adsorption and reaction of further reactant molecules that would otherwise not be possible. An example of cooperative adsorption effects on small free silver cluster ions identified in an ion trap experiment will be presented in the following.

In this example, the observed adsorption of multiple oxygen molecules onto negatively charged odd size Ag_n^- clusters will be related to cooperative adsorbate effects. The reaction behavior of free Ag_n^- with O_2 is in marked contrast to the reaction behavior of free Au_n^- clusters with oxygen as presented in the previous section. Whereas Ag_n^- with even n adsorb only one O_2, similar to even n Au_n^-, silver clusters with odd number n of silver atoms adsorb at maximum two oxygen molecules, in contrast to odd size Au_n^- which do not react with O_2 at all [34, 160, 355] (Fig. 1.57). Ag_4^- even first forms $Ag_4O_2^-$ but then reacts with additional two O_2 molecules at long reaction time and low temperature. No products with an odd number of oxygen atoms are detected. The corresponding adsorption kinetics data are best fitted by a sequential adsorption reaction mechanism with the reaction rate constants k_1 and k_2:

$$Ag_n^- + 2O_2 \xrightarrow{k_1} Ag_nO_2^- + O_2 \xrightarrow{k_2} Ag_nO_4^- \tag{1.63}$$

Adsorption of two O_2 molecules as observed in this experiment for Ag_n^- had been predicted before for the gold cluster anions [356]; however, it was never observed experimentally which lead to further discussion in the literature [357, 358]. In combination with a systematic theoretical study performed by the group of Bonačić-Koutecký [351] an understanding of the measured rate constant evolution with cluster size depicted in Fig. 1.58a emerges and reasons for the distinct behavior of silver cluster anions can be given [351].

The obtained rate constants are termolecular under the low-pressure multicollision conditions of the experiment (Lindemann model for gas-phase association reactions [33, 188]). They include the elementary steps of the initial formation of an energized complex (rate constant k_a) and its possible unimolecular decomposition back to the educts (k_d) in competition with a stabilizing energy transfer collision with helium buffer gas (k_s) [33, 188]. As the ion–molecule reaction rate constants k_a ($\approx 5.5 \times 10^{-10}\,cm^3\,s^{-1}$) and k_s ($\approx 5.3 \times 10^{-10}\,cm^3\,s^{-1}$) show no significant cluster size dependence [359], the observed strong size effects must be contained in k_d (see below). The experimentally obtained reaction rate constants (Fig. 1.58a) compare particularly well with the trend in the calculated O_2 binding energies E_b (Fig. 1.58b) because, according to statistical rate theory [188], E_b largely determines the rate of unimolecular decomposition of the initially formed energized complex (k_d).

The results concerning multiple adsorption of O_2 onto the anionic silver clusters according to (1.63) can be assessed by qualitative frontier orbital considerations for the binding of molecular oxygen. The oxygen molecule, as an one-electron acceptor, binds strongly to the anionic silver clusters with odd number of electrons (even n) and low VDE values leading to a small k_d and hence to a fast reaction rate k_1. In contrast, it binds only weakly to silver clusters with even number of electrons (odd n) and closed shell electronic structure resulting in low reaction rates k_1 in these cases as can be seen from Fig. 1.58a. However, recent work on reactivity of hydrated anionic gold clusters with molecular oxygen has shown that this behavior can be inverted by binding of a strong electron acceptor such as the OH group [360]. Due to the electron withdrawing effect, electron transfer from the cluster occurs leaving an unpaired electron on the clusters with even number of electrons and

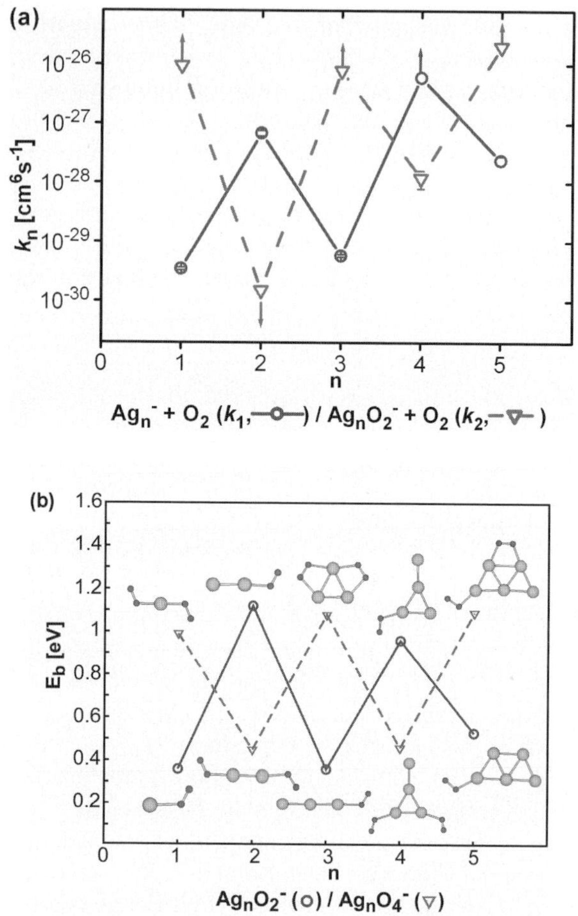

Fig. 1.58. (**a**) Experimental rate constants k_n obtained for the adsorption of the first and second O_2 molecule onto gas-phase Ag_n^- at 300 K reaction temperature. (**b**) Theoretical binding energies E_b and lowest energy structures (gray *spheres*, silver atoms; dark *spheres*, oxygen atoms). Note that the lowest energy structures of AgO_2^- and $Ag_3O_4^-$ contain dissociated oxygen (not shown), which however requires considerable activation energy not available under the experimental conditions [351]

inducing subsequent stronger binding of the molecular oxygen to hydrated clusters.

An analogous mechanism is proposed for the activation of molecular oxygen and the cooperative binding of two oxygen molecules on the anionic silver clusters in which the first adsorbed O_2 serves as an activator [351]. Since anionic silver clusters have generally lower VDE values than gold clusters, weaker electron acceptors such as O_2 can already induce electron transfer and activate them, which is not possible in the case of Au_n^-. The first oxygen molecule is bound to the silver clusters with even number of electrons by 0.36, 0.36, and 0.53 eV for Ag^-, $Ag_3{}^-$, and $Ag_5{}^-$, respectively [351]. The mechanism of the bonding involves the electron transfer from the metal cluster into the π^*-MO of O_2. The binding of the first O_2 molecule changes the electronic structure of the cluster and induces a stronger cooperative binding with the second O_2. This results in the case of $Ag_3{}^-$ and $Ag_5{}^-$ in new oxide species with doubly bound, superoxo-like O_2 subunits as can be seen from the calculated structures in Fig. 1.58b. The cooperative effect is reflected in larger binding energies as shown also in Fig. 1.58b. Moreover, the binding of the second molecular oxygen shows a reversed pattern being stronger for clusters with odd number of atoms and weaker for the ones with even number of atoms. The latter ones with one unpaired electron bind strongly only one oxygen molecule since the electron transfer leaves them with a closed shell electronic structure. The above mechanism can be extended to the absorption of three oxygen molecules which should qualitatively exhibit the similar behavior as the adsorption of one O_2. Adsorption of three O_2 molecules has indeed been detected experimentally for $Ag_4{}^-$.

Hence, experimental rate constant measurements in combination with theoretical simulations show a pronounced size and structure selective activity of anionic silver clusters toward molecular oxygen due to cooperative effects. In particular, for $Ag_n{}^-$ clusters with odd n, a weakly bound first O_2 promotes the adsorption of a second O_2 molecule which is then (for $n = 3, 5$) differently bound with the O_2 bond elongated to 1.32 Å and thus potentially activated for further oxidation reactions such as CO combustion, which have indeed been observed for larger cluster sizes [361].

Structural Dynamic Fluxionality

One of the hallmarks of atomic clusters is the presence of various isomeric structures at finite temperatures. This means that at a given temperature, a Boltzmann distribution determines the population of the different isomers. Often these isomers have very close lying ground state energies, which on the one hand present complications for a discrete structure determination but on the other hand are fundamental to the catalytic activity as each isomer has its distinct chemical properties. In the course of a chemical reaction, the system has the possibility to find the energetically optimal reaction path by choosing the most suited isomeric structures. This is possible because of the structural dynamic fluxionality, which is the interconversion between different isomeric arrangements along the reaction coordinate at finite temperatures.

As an example, for the oxidation of CO on Au_8 supported on magnesia surface, three different isomers are relevant. Those structures are quasiplanar, bicapped pyramidal (structures a and b in Fig. 1.59.), and bicapped octahedral. The quasiplanar structure is the lowest energy isomer. The bicapped pyramidal and the bicapped octahedral structures are higher in energy by 0.2 and 0.3 eV, respectively. Upon adsorption of O_2, the energetic order of the

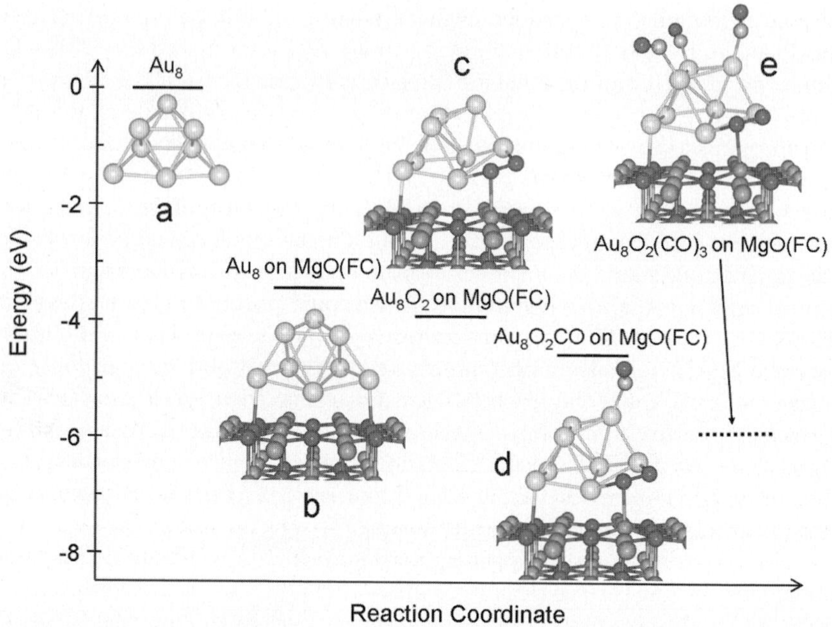

Fig. 1.59. Dynamic structural fluxionality of Au_8 upon deposition, adsorption of reactants (O_2, CO) and oxidation reaction. Note that the most stable geometric structure for Au_8 in the gas phase is planar. On an F center of the MgO(100) surface, Au_8 shows two energetically very close lying isomers, a quasi-planar structure (not shown here, as this isomer reveals unfavorable energetics in the reaction path of the oxidation reaction) and the bicapped octahedron shown in this figure as structure a and b. Upon adsorption of oxygen, the bicapped octahedral structure is transformed into a bicapped bipyramid (structures c, d, and e). Interstingly, when freezing the structure of the bicapped octahedron no adsorption of O_2 is possible. Thus, the reaction takes place only when the bicapped bipyramidal structure is populated. Upon adsorption of CO only minor structural changes are observed

structures is reversed and the bicapped bipyramid now presents the ground state isomer (structure c). Interestingly, the bicapped octahedron is inert for the adsorption of O_2. The CO oxidation reaction proceeds most favorably via the bicapped pyramidal complexes d and e with three CO molecules adsorbed on the top facet of the gold cluster and the oxygen molecule bound to the interface gold atoms. After the final reaction step, which is the production of CO_2, this structure becomes again the least stable one.

1.5 Specific Examples

Intrinsic size effects in the chemical properties of small clusters are most purely observed in the gas phase in absence of any environment. In addition, their chemical properties can be studied for different charge states (cations, neutrals, anions). Thus, guide lines for understanding the size evolution of chemical and catalytic properties of clusters based on these intrinsic properties can established. Up to now there are only few experiments demonstrating catalytic reactions on free clusters. The most intriguing prospect of research of free clusters might, however, consist in a direct comparison with identical supported cluster systems in order to reveal the importance and dominance of the concepts presented in Sect. 1.4. In addition, the interpretation of catalytic reaction mechanisms on surfaces is very complex and results from the gas phase might contribute to the identification of the reaction mechanism. Therefore, in the following, we present results from the gas phase and clusters on surface together for selected systems and reactions.

Support materials most adapted for stabilizing clusters on surfaces are oxide materials. They reveal a relatively large band gap and thus the characteristic, discrete electronic levels of the clusters are maintained to a certain extent. Furthermore, the interaction of the clusters with oxide surfaces, especially with their defects may be substantial and trapping of the clusters is feasible. Oxide supported metal particles are also relevant in industrial catalysis.

It has to be pointed out, however, that upon deposition, the cluster's geometric and electronic structure is altered in comparison to the gas phase, and this will change their chemical properties drastically in most cases. Nevertheless, we feel that comparing the chemical and catalytic properties of free and supported clusters is illustrative and relevant also for understanding the cluster support interaction. There are, however, effects which dominate chemical and catalytic properties of supported clusters and which are solely present on the surface. These are for instance the spillover and reverse spillover. In addition, when studying the chemical and catalytic properties of clusters on surfaces and extracting information on the intrinsic cluster properties, one has to be aware that also the oxide surface has its own chemical and catalytic properties. Thus the investigation of the pure oxide surface itself is also relevant for these studies.

Although these systems are already quite complex, state-of-the-art ab initio simulations are able to describe catalytic reactions on small clusters in great detail for both the free and supported clusters. In many examples, it has been shown that theory not only contributes to a fundamental understanding of nanocatalysis but also has a predictive potential.

We open the Sect. 1.5.1 by illustrating the chemistry of magnesium oxide and their defects. This is also the support material of choice of the cluster deposition experiments presented here. These comprise three catalytic reactions, the oxidation of CO, the reduction of NO by CO, and the polymerization of acetylene. These studies are reviewed in Sects. 1.5.2–1.5.5.

1.5.1 Chemical Reactions on Point Defects of Oxide Surfaces

Chapter 3 presents a detailed description of oxide surfaces and a classification of possible surface defects of oxide materials [362]. At least four major kinds of irregularities are described: low-coordinated sites, divacancies, impurity atoms, and surface vacancies. The last category includes cation vacancies, usually called V centers, as well as oxygen vacancies, which are called color centers or F centers (from the German word for "color"—"Farbe"). As shown below, F centers will play an important role in the discussion of the chemical and catalytic properties of size-selected clusters on MgO. In a broader sense, an F center is defined as an electron trapped in an anion vacancy and was first introduced to explain chemical- and radiation-induced coloration of alkali halides. In bivalent oxides, like MgO and CaO, two types of F centers exist. The vacancy containing two electrons is called F center and the one containing only one electron is labeled as F^+ center. In both vacancies, the electrons are confined by the Madelung potential of the crystal. Thus, in a first approach the F/F^+ centers can be described as two/one electrons moving in a quantum box of nanometer length scale; i.e., F/F^+ centers are systems with highly confined electrons and can therefore even act by themselves as nanocatalysts. The unique chemical activity of F centers has been indeed recently demonstrated.

In a combined theoretical and experimental study, it was shown that the heterolytic breaking of methanol on MgO thin films is catalyzed by surface oxygen vacancies [363]. The films were epitaxially grown onto Mo(100) single crystals by evaporation of Mg in an oxygen background. The oxygen vacancies are generated by changing the preparation method [209], e.g., the Mg evaporation rate and the oxygen background pressure. In this way, two kinds of films are prepared: defect-poor (Mg evaporation rate: 0.3–0.5 ML min^{-1}, O_2 background: 5×10^{-7} Torr) and defect-rich (Mg evaporation rate: 2–5 ML min^{-1}, O_2 background: 55×10^{-7} Torr) films. Both kinds of films are annealed to 840 K during 10 min. Auger electron spectroscopy (AES) measurements show a one-to-one stoichiometry for magnesium and oxygen and the absence of any impurity [39]. Typical thicknesses are about 10 ML, as determined by AES peak intensities [39] and by X-ray photoemission (XPS), using the intensity attenuation of the Mo 3d core level with increasing film coverage [73].

Both films have also been studied by electron energy loss spectroscopy [73]. In contrast to defect-poor films, which are characterized by a loss at about 6 eV in the EEL spectra (Fig. 1.60a, in good agreement with previous studies on MgO(100) single crystals [201]), the EEL spectra of defect-rich films exhibit characteristic losses between 1 and 4 eV, lying within the MgO band gap (Fig. 1.58a). Similar loss structures have been observed before [364], and according to first-principle calculations using large cluster models, they have been attributed to transitions, characteristic for neutral surface F centers in various coordinations on flat terraces, at steps and at kinks [365]. The density of these oxygen vacancies is estimated to be larger than $5 \times 10^{13}\,cm^{-2}$. The interaction of methanol with the defect-poor and defect-rich films (Fig. 1.60b) was studied using thermal desorption spectroscopy (TDS) (Fig. 1.60c). For both films desorption of physisorbed methanol at around 180 K is most dominant. On the defect-poor films small amounts of chemisorbed methanol desorb up to around 350 K. On defect-rich films desorption of chemisorbed methanol evolves in three distinct peaks at 200, 260, and 340 K. A small reproducible feature is observed at around 500 K. Most importantly, H_2 desorbs at 580 K only on defect-rich films. The corresponding infrared spectra taken at 90 K (insets of Fig. 1.60c) confirm the presence of mainly physisorbed CH_3OH with the typical vibrational band for the OH-group at 3,285 cm^{-1}, bands due to the symmetric C–H stretch (2,930 cm^{-1}/2,828 cm^{-1}) and the C–H bending (1,475 cm^{-1}) modes of the CH_3 group, as well as the C–O vibrational mode at 1,080 cm^{-1}/1,050 cm^{-1}. For defect-poor films most of the intensity of the vibrational bands disappear between 180 and 200 K, consistent with desorption observed at 180 K in the TDS. Up to temperatures of about 400 K, small bands are observed for the CH_3– and C–O vibrations, which decrease in intensity with temperature. These bands are attributed to chemisorbed methanol. As for the defect-rich films, the evolution of the IR-spectra with temperature is consistent with the corresponding TDS. Physisorbed methanol desorbs before 200 K (as shown by the disappearance of the OH-band), and the more intense vibrational features of chemisorbed methanol are unambiguously detected up to 360 K. At higher temperatures, a clear peak is observed at 1,070–1,085 cm^{-1} [363]. It is interesting to note that the disappearance of this band correlates with the desorption of H_2 from the surface at around 580 K and therefore this peak is attributed, in accordance to the theoretical studies, to a proton trapped in the cavity of an F center. The calculations show that F centers easily dissociate methanol giving a CH_3O group and an H^+ ion adsorbed into the F center (Fig. 1.60c). This adsorbed proton is strongly bound as the desorption of a neutral H atom, $F/H^+ \rightarrow F^+ + H^\bullet$, costs 4.1 eV. This represents a crude estimate of the barrier to overcome in order to observe H_2 desorption from the surface. Once the H atom is detached from the F center, it will rapidly diffuse on the surface. In fact, the binding of an H atom to an oxygen ion on a terrace is of about 0.5 eV. Diffusion will eventually lead to recombination with a second H to form H_2; H_2 is weakly bound to the MgO terrace sites and at 580 K will immediately desorb. The adsorbed hydrogen gives rise to vibrations of strong intensity at 830–950 cm^{-1} when isolated and to an intense band at 1,030 cm^{-1} in the chemisorbed complex shown in Fig. 1.60b. Therefore, the stable species is assigned to H atoms incorporated into oxygen vacancies.

In conclusion, the combined experimental and theoretical study of methanol adsorbed on MgO films with different defect densities allows for a better identification of the surface sites responsible for the MgO reactivity. On the inert terrace sites only physisorption is observed. Molecular chemisorption, activation, and heterolytic dissociation occur on irregular sites. The low-coordinated Mg–O pairs of ions located at edges and steps can lead to strongly activated and even dissociated methanol molecules. Adsorption of CH_3O^- and H^+ fragments seems to be preferred over dissociation into $CH_3{}^+$ and OH^- units. All these species are stable on the surface for temperatures up to 350 K and account for the TDS spectra of the defect-poor films. On defect-rich films (F centers) the O–H bond is selectively dissociated, resulting in the observed

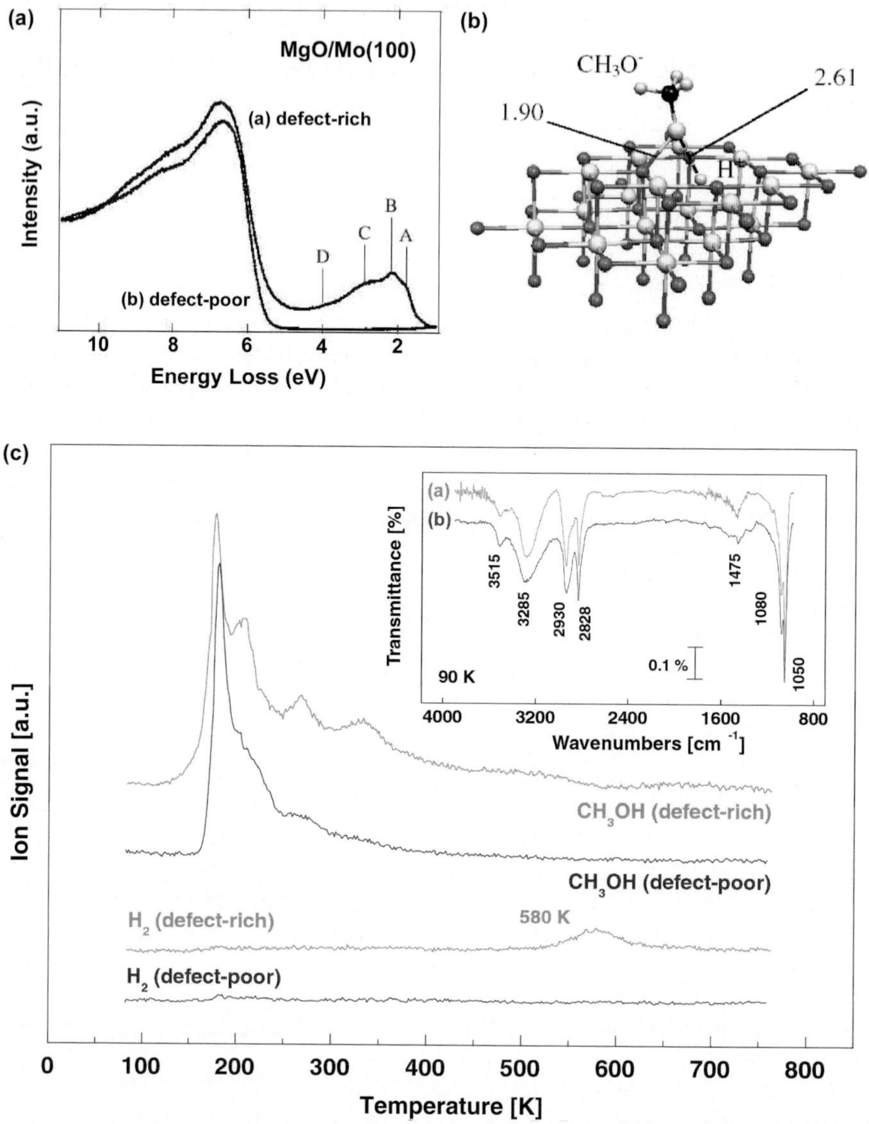

Fig. 1.60. (**a**) EEL spectra of thin defect-poor and defect-rich MgO(100) films grown on Mo(100) at different experimental conditions. A–D are losses which are attributed theoretically to transitions characteristic of neutral F centers on MgO. (**b**) Model of an oxygen vacancy at a terrace of an MgO(100) surface with chemisorbed CH_3–OH (CH_3OH). (**c**) Thermal desorption spectra of CH_3OH and H_2 on defect-poor and defect-rich MgO(100) films. Note the desorption of H_2 at 580 K for defect-rich films. The insets show FTIR spectra recorded at 90 K for adsorbed CH_3OH on both defect-poor (**a**) and defect-rich films (**b**)

desorption of H_2 at high temperature. Thus, these oxygen vacancy centers, called F centers, act as nanocatalysts for certain reactions. But these two kinds of MgO supports with different types of defects are also relevant for studying the cluster–support interaction and its influence on the catalysis of small clusters. We will present in the following sections several examples where small clusters change their chemical and catalytic properties when deposited either on defect-poor or defect-rich MgO(100) films.

1.5.2 The Oxidation of CO on Small Gold Clusters

For long time, gold has been considered as not interesting for catalysis as low adsorption energies and high activation barriers for dissociating small molecules on extended gold surfaces are responsible for the nobleness of gold as bulk material [350]. Nanosize particles of gold, however, show high and rather unusual catalytic reactivity [366,367]. While the novel heterogeneous catalysis by nanosize gold aggregates supported on oxides is of great significance, the processes underlying the catalytic activity of gold in reduced dimensions and the reaction mechanisms are not yet understood. In the following, these issues are addressed for the nonscalable size regime through investigations of the low-temperature combustion of CO on gas-phase gold cluster anions, as well as on size-selected gold clusters, Au_n ($n \leq 20$), and strontium-doped gold clusters, $SrAu_n$ ($n \leq 20$), supported on defect-poor and defect-rich MgO (100) films [33, 209, 368, 369].

The Reactivity of Free Gold Clusters

Positively charged free gold clusters in the size range up to about 20 atoms per cluster are essentially unreactive toward molecular oxygen [354, 370, 371]. The same result was found for neutral gold clusters. In contrast, gold cluster anions show a pronounced odd–even alternation with cluster size in their reactivity toward O_2 as has been indicated already in Electronic Size Effects. This remarkable size and charge state dependence of the reactivity of free gold clusters toward small molecules was first recognized more than a decade ago [355, 370] and confirmed by several other groups later on [34, 160, 372]. Figure 1.61a presents a compilation of the results on the reactivity of Au_n^- toward O_2. Since gold is an s^1 valence electron metal, its physical properties, such as, e.g., ionization potential [373] and electron affinity [103, 104, 374], also alternate with cluster size in the small size regime, and the close relation of gold cluster electron affinity and reactivity toward molecular oxygen has been discussed by several groups [34, 160, 370, 372]. The reactions of gold cluster ions, positively as well as negatively charged, with carbon monoxide have also been reported [352, 372, 375–378]. However, although strong size effects in reactivity are apparent, no odd–even alternations are observed in this case as can be seen for Au_n from Fig. 1.61b, which is based on the results of Lee and Ervin [372], as well as Wallace and Whetten [352]. The first step toward

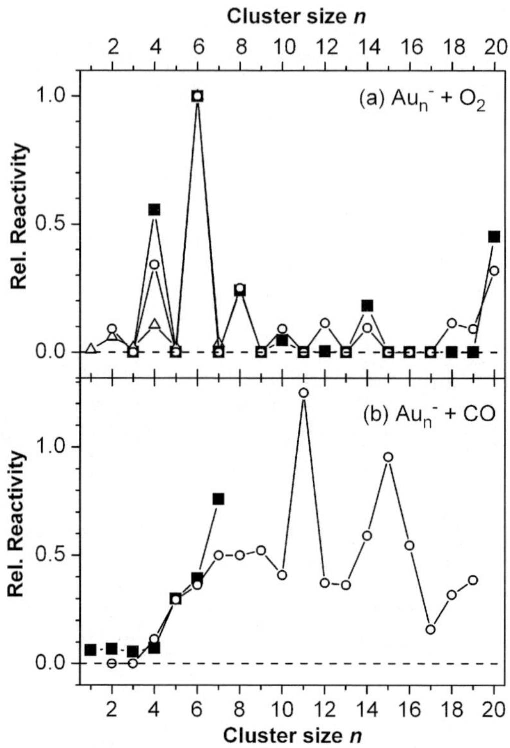

Fig. 1.61. Compilation of experimental results from the literature on the relative reactivity of gold cluster anions in the adsorption reaction of one O_2 or one CO molecule, respectively, as a function of the cluster size n. (**a**) Reactions of Au_n^- with O_2: (*filled squares*) data from [370], (*open triangles*) data from [372], (*open circles*) data from [160]. (**b**) Reactions of Au_n^- with CO: (*filled squares*) data from [372], (*open circles*) data from [352]. For comparison, all data are normalized to the reactivity of Au_6^- toward O_2

a catalytic reaction on gas-phase cluster ions consists in the simultaneous adsorption of the reactants, O_2 and CO, on the metal clusters which will be discussed in the following section with an emphasis on cluster specific cooperative effects.

Cooperative Coadsorption Effects on Small Gold Clusters. Two examples of cooperative adsorption effects on small gold cluster anions identified in temperature dependent rf-ion trap experiments (see Chemical and Catalytic Properties of Gas-Phase Clusters for experimental details) will be presented in the following. Au_3^- does not react with O_2 in the ion trap experiment at any reaction temperature [34]. It, however, adsorbs a maximum of two CO molecules at reaction temperatures below 250 K [185]. If the gold trimer is exposed simultaneously to CO and O_2 inside the octopole ion trap, still no reaction products are observed at reaction temperatures above 250 K as can be seen

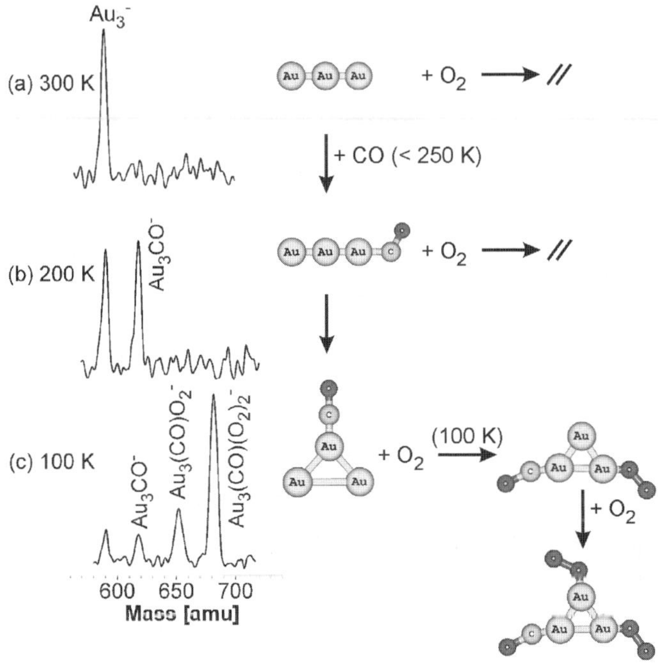

Fig. 1.62. Temperature dependent mass spectra of the reaction of Au_3^- with O_2 and CO. The production distributions are analyzed after trapping Au_3^- for 500 ms inside the octopole ion trap filled with 0.02 Pa O_2, 0.05 Pa CO, and 1.23 Pa He. Reaction temperatures: (**a**) 300 K, (**b**) 200 K, (**c**) 100 K [34, 371]. The proposed cooperative adsorption mechanism is indicated schematically. Large *spheres* indicate gold atoms, small, bright *spheres* carbon atoms, and small, dark *spheres* oxygen atoms

from Fig. 1.62a. Figure 1.62b shows that at 200 K, one CO molecule adsorbs onto Au_3^-. Further cooling, however, does not lead to adsorption of a second CO molecule as has been seen when only carbon monoxide is in the trap, but instead results in the additional adsorption of up to two O_2 molecules under formation of the coadsorption products $Au_3(CO)O_2^-$ and $Au_3(CO)(O_2)_2^-$ (Fig. 1.62c) [34]. Hence, the adsorption of CO onto Au_3^- seems to change the cluster complex electronic structure in a way that it is now able to react with oxygen. In other words, CO preadsorption conditions the gold cluster to enable O_2 coadsorption. This temperature dependent product formation demonstrates that first one CO has to be adsorbed onto the gold cluster before subsequent O_2 adsorption becomes possible.

The following idea for a possible molecular mechanism of this unexpected cooperative action of two adsorbate molecules on the small Au_3^- cluster is based on recent ab initio simulations of CO adsorption and CO/O_2 coadsorption energetics and structures [379] as well as on previous experiments on the femtosecond dynamics of noble metal clusters initiated by a change of the charge state [49, 380, 381]. At temperatures below 250 K, Au_3^- adsorbs CO under our experimental conditions. The interaction of the first CO molecule proceeds mainly through π back-bonding leading to a charge transfer from the metal cluster d-orbitals into the 2pπ* antibonding orbitals of the CO molecule. From femtosecond laser spectroscopic investigations, it is known that the initially linear noble metal trimer anions undergo bending motion to reach a triangular geometry, if the additional electron is detached and thus the charge reduced [49]. Therefore, it is likely to assume that the CO adsorption might lead to a change in the cluster complex geometry as depicted schematically in Fig. 1.62. According to the theoretical simulations, the linear Au_3CO^- has a high VDE of 3.37 eV and does not form stable complexes with O_2 [379]. The triangular Au_3CO^- complex, however, exhibits a considerably lower VDE of only 2.83 eV [379] and is therefore predicted to react with oxygen under formation of the coadsorption complex $Au_3(CO)O_2^-$ which has indeed been observed experimentally. Also the formation of $Au_3(CO)(O_2)_2^-$ is confirmed by this calculations [379] (cf. Fig. 1.62). Hence, the observed cooperative adsorption on Au_3^- might be attributed to an adsorbate-induced geometry change of the metal cluster, which in turn results in a reduced HOMO energy (lower VDE) enabling the subsequent O_2 adsorption.

The gold anion dimer, Au_2^-, reacts more than one order of magnitude faster with O_2 than with CO [34]. Accordingly, when the octopole ion trap is filled with similar partial pressures of both reactive gases, O_2 adsorption will most likely precede CO adsorption [33]. Production distributions at two different temperatures for the case when O_2 and CO are present in the trap are depicted in Fig. 1.63. As can be seen from these mass spectra, the dimer Au_2^- forms the dioxide at 300 K, but a new, additional peak at the mass of $Au_2(CO)O_2^-$ appears at a temperature of 100 K (hatched peak in Fig. 1.63b). This is the final, major reaction product which shows that Au_2^- clearly favors the simultaneous coadsorption of both, oxygen and carbon monoxide, over adsorption of just one sort of reactive molecule [34]. This kind of cooperative effect was also found for larger gold cluster anions in a flow-reactor study [352]. Furthermore, under the conditions of the ion trap experiment, small amounts of $Au_2O_2^-$ are present too (Fig. 1.63b). But most interestingly, no Au_2CO^- or $Au_2(CO)_2^-$ carbonyls are observed at any temperature or reaction time, although these complexes are stable products at cryogenic temperatures. The absence of carbonyl complexes can be rationalized by the above mentioned observation that the reaction of Au_2^- with oxygen appears to be considerably faster than with carbon monoxide.

Catalytic CO Oxidation by Free Au_2. The potential catalytic activity of Au_2^- in the CO combustion reaction was first predicted by Häkkinen and Landman [382]. The subsequent experimental investigation employing an rf-ion trap indeed revealed the catalytic reaction of the gold dimer and, in conjunction with theory, a detailed reaction cycle could be formulated [33]. Also for particular larger gold cluster anions evidence for catalytic CO_2 formation has

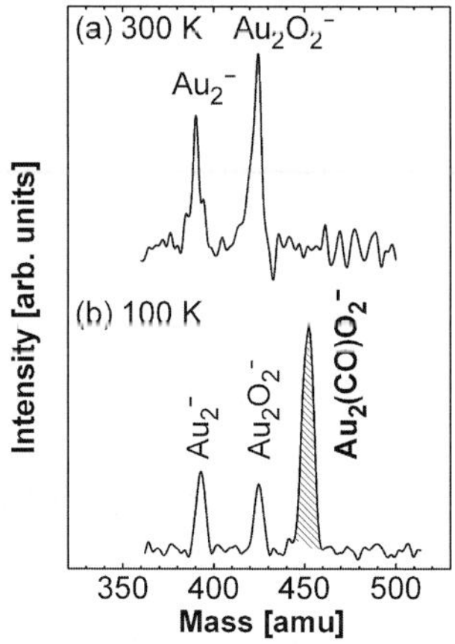

Fig. 1.63. Temperature dependent mass spectra of the reaction of Au_2^- with O_2 and CO. The production distributions are analyzed after trapping Au_2^- for 500 ms inside the octopole ion trap filled with 0.02 Pa O_2, 0.05 Pa CO, and 1.23 Pa He. (**a**) At a reaction temperature of 300 K only Au_2^- and $Au_2O_2^-$ are detected. No further ion signals are observed at temperatures above 200 K. Cooling down further reveals an additional ion signal appearing at the mass of $Au_2(CO)O_2^-$(hatched peak). Mass spectrum (**b**) shows the ion distribution at 100 K [33]

been reported [352]. In the following, the experimental reaction mechanism and first principles simulation will be presented and discussed in detail.

Experimental Reaction Mechanism. As discussed above, Au_2^- reacts under conditions when solely O_2 is present in the rf-ion trap under formation of the single production $Au_2O_2^-$. Analysis of the kinetics of this reaction reveals a straight forward association reaction mechanism with Au_2^- completely reacting to yield the oxide product within several seconds at room temperature. If CO is added to the ion trap, no new reaction products besides Au_2^- and $Au_2O_2^-$ are formed at $T = 300$ K as can be seen from the mass spectrum in Fig. 1.63a. However, surprisingly enough, Au_2^- is no longer completely transformed into oxide, but an offset appears in the gold cluster concentration at long reaction times which is apparent from the corresponding kinetics depicted in Fig. 1.64a. The most simple reaction mechanism that fits the data is the equilibrium reaction in which oxide is formed but bare gold dimer is reformed to a certain extend.

$$Au_2^- + O_2 \rightleftharpoons Au_2O_2^- \tag{1.64}$$

The solid lines in Fig. 1.64a are obtained by fitting this equilibrium mechanism to the experimental data. However, the extend in which the gold dimer is reformed increases with increasing CO concentration. Thus, the reaction mechanism must involve more intermediate steps representing the influence of CO.

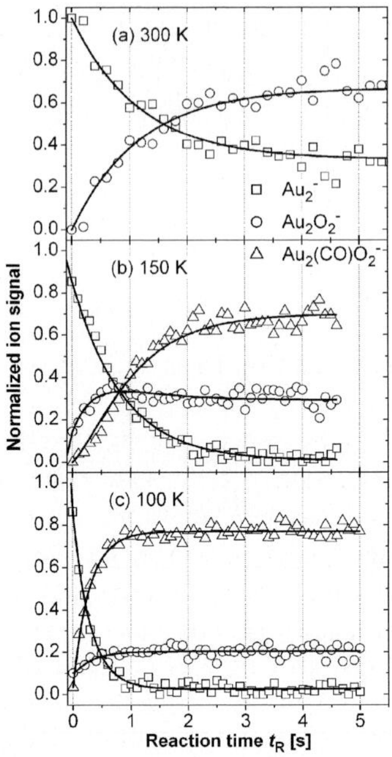

Fig. 1.64. Production concentrations as a function of the reaction time for three different reaction temperatures and different reactant gas concentrations. (**a**) T = 300 K; $p(O_2)$ = 0.12 Pa; p(CO) = 0.12 Pa; p(He) = 1.2 Pa. (**b**) T = 150 K; $p(O_2)$ = 0.04 Pa; p(CO) = 0.04 Pa; p(He) = 1.0 Pa. (**c**) T = 100 K; $p(O_2)$ = 0.02 Pa; p(CO) = 0.03 Pa; p(He) = 1.0 Pa. Open symbols represent the normalized experimental data ((*squares*), Au_2^-; (*circles*), $Au_2O_2^-$; (*triangles*), $Au_2(CO)O_2^-$). The *solid lines* are obtained by fitting the integrated rate equations of the catalytic reaction cycle (1.65) to the experimental data

To reveal the complete reaction mechanism, the reaction was investigated at lower temperatures. The product ion mass spectrum recorded at 100 K with O_2 and CO in the ion trap (Fig. 1.63b) shows the appearance of the coadsorption complex $Au_2(CO)O_2^-$ discussed above. This complex represents a key intermediate in the reaction mechanism of the catalytic oxidation of CO to CO_2 as has been predicted in the earlier theoretical study [382]. The experimental evidence obtained so far demonstrates that O_2 adsorption is likely to be the first step in the observed reaction mechanism. Subsequent CO coadsorption yields the observed intermediate (Fig. 1.63b) and finally the bare gold dimer ion must be reformed. The further strategy to reveal the full reaction mechanism consists in varying the available experimental parameters, i.e., reaction temperature and reactant partial pressures. This procedure leads to a series of kinetic traces similar to the one shown in Fig. 1.64b and c [33]. The goal then is to find one reaction mechanism that is able to fit all experimental kinetic data obtained under the various reaction conditions. This kinetic evaluation method results in the most simple mechanism which is able to fit the experimental data [33]. In this way it is possible to rule out all but one possible reaction mechanism which is represented by the following reaction equations:

$$Au_2^- + O_2 \rightarrow Au_2O_2^- \quad (1.65a)$$

$$Au_2O_2^- + CO \rightleftarrows Au_2(CO)O_2^- \quad (1.65b)$$

$$Au_2(CO)O_2^- + CO \rightarrow Au_2^- + 2CO_2 \quad (1.65c)$$

The solid lines in Fig. 1.64b and c represent the fit of this mechanism to the experimental data. It equally well fits all other obtained kinetic data [33]. In this catalytic reaction cycle, $Au_2O_2{}^-$ reacts with CO to form $Au_2(CO)O_2{}^-$, which will either redissociate to the oxide or further react with a second CO molecule to reform $Au_2{}^-$ while liberating two CO_2 molecules. It should be noted that the quality of the fit is very sensitive to the postulated reaction steps and that the kinetic evaluation procedure that was employed is clearly able to discriminate against alternative mechanisms, as has been demonstrated before [32,187,188]. The replacement of the equilibrium in reaction (1.65b), e.g., by a simple forward reaction will lead to a mechanism that yields an inadequate fit to the experimental data. The $Au_2O_2{}^-$ signal will then disappear at long reaction times, which is not the case as can be seen from Fig. 1.64c.

Adsorption Sequence. The mechanism reveals that O_2 adsorption precedes CO adsorption in the catalytic reaction. This is further supported by the fact that no signal for the ion Au_2CO^- is observed and by the reaction kinetics of $Au_2{}^-$ when only O_2 or only CO are present in the trap; because the adsorption of O_2 is by about an order of magnitude faster than the adsorption of CO molecules [185]. Further insight into the catalytic reaction mechanism can be obtained from partial pressure dependent measurements. The dependence of a pseudo-first-order rate constant of the mechanism (1.65) on the concentration of a reactant (O_2 or CO) demonstrates the involvement of the reactant in this particular reaction step. The variation of the oxygen partial pressure inside the trap, e.g., solely affects the rate constant of reaction (1.65a) and all other rate constants remain unaffected [33]. However, most interestingly, it appears that at very high CO partial pressures and low temperatures, CO adsorption starts to compete with O_2 adsorption in the first reaction step. This is reflected in a systematic increase of the rate constant of reaction (1.65a) with increasing p(CO) at lower temperatures. From this side reaction, a new product $Au_2(CO)O_2{}^-$ with different structure than the discussed intermediate in (1.65) will form competing with the catalytic cycle [33].

As discussed above, the postulation of an equilibrium in step (1.65b) is essential to the reaction mechanism. This in turn has implications on the possible structure of the intermediate species $Au_2(CO)O_2{}^-$ in the catalytic cycle (1.65) which might be a simple coadsorption of the two molecules, e.g., on different sides of $Au_2{}^-$ or already a reacted carbonate CO_3-like species adsorbed onto the gold dimer. Apparently, the observed intermediate can

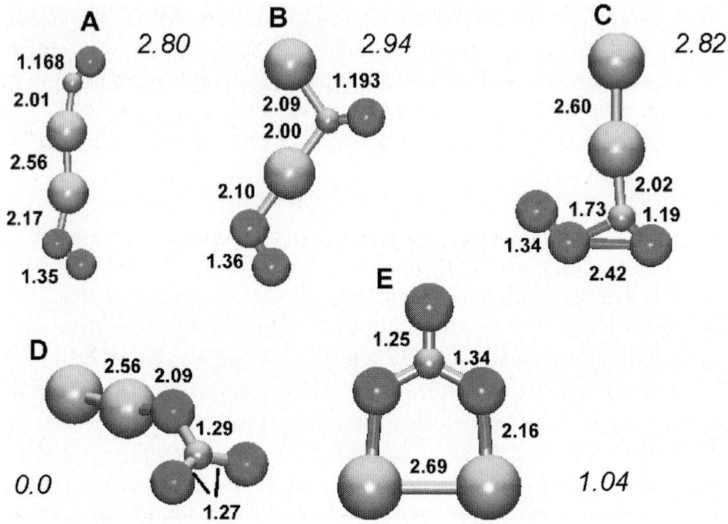

Fig. 1.65. Five optimized structures A–E of $Au_2(CO)O_2^-$, with bond lengths in Å. The relative stability of these structures is indicated by the numbers in italics (in eV). A, B, C, and E are planar, and the two carbonate species (D and E) have C_{2v} symmetry. Structures C and D are the ones pertinent to the reactions discussed in the text. Au atoms are depicted by large grey *spheres*, a small gray *sphere* corresponds to the carbon atom, and the oxygen atoms are depicted by dark *spheres* [33]

have different isomers and one of them is formed by molecular coadsorption of CO and O_2 because otherwise the possible CO loss required by (1.65b) would not be feasible. Therefore, several possible structures of $Au_2(CO)O_2^-$ have been calculated by Häkkinen and Landman, which will be presented in the following.

Five structures corresponding to the mass of the complex $Au_2(CO)O_2^-$ were studied theoretically, and the structural and energetic information is given in Fig. 1.65 [33]. The formation and stability of these structures is discussed with respect to preformed $Au_2O_2^-$. This analysis leads to the conclusion that structures C and D (Fig. 1.65) are the ones pertinent to the observed reaction steps (1.65b) (equilibrium between CO association to $Au_2O_2^-$ to form $Au_2(CO)O_2^-$ and dissociation of the latter to form $Au_2O_2^-$) and (1.65c) (formation of CO_2).

Structures A and B correspond to molecular coadsorption of CO and O_2 to Au_2^-. From the two molecularly coadsorbed species, CO can readily (without barrier) bind to the end of the Au–Au axis (structure A) whereas a barrier of 0.2 eV was found for CO association from the gas phase to the Au–Au bridging site of structure B, where the Au-Au bond is significantly elongated to 3.34 Å. The barrier for forming B from A via displacement of CO from the end of the complex to the Au–Au bridge is rather high (on the order of 0.9 eV). In both structures A and B, the O–O bond is activated to a value typical to a superoxo-species (about 1.35 Å). The stability of structure C is close to that of A. It contains a reacted O–O–C–O group that is attached through the carbon atom to the gold dimer anion. The O–O bond is activated to a superoxo-state, and this species bears some resemblance to the gold–peroxyformate complex identified in the early experiments of gold atoms in cryogenic CO/O_2 matrices [383].

By far, the most stable structures corresponding to the mass of $Au_2CO_3^-$ are the two carbonate species D and E [33]. Both structures were proposed by Häkkinen and Landman [382] before. Structure E requires a preformed $Au_2O_2^-$, where the molecular axes of Au_2 and O_2 lie parallel to each other. Since this structure of $Au_2O_2^-$ is 1 eV less stable than the ground state discussed above (the binding energy of oxygen in this configuration is 0.39 eV versus the optimal binding energy of 1.39 eV), it is unlikely to be formed and consequently structure E is not expected to play a relevant role in the catalytic cycle. On the other hand, formation of the most stable structure D by insertion of CO(g) into the O–O bond in $Au_2O_2^-$ (where the O_2 molecule is end-on bonded to the gold dimer anion) requires a barrier of only 0.3 eV, which is easily overcome under our experimental conditions.

Activation Barriers: ER Versus LH Mechanism

In Langmuir–Hinshelwood Reaction Mechanism, the general reaction mechanism of the catalytic CO oxidation reaction was presented. On an extended catalyst surface, the reaction proceeds via a Langmuir–Hinshelwood(LH)-type mechanism relying fundamentally on the diffusion of the reactants on the catalyst surface. From the theoretical simulations presented above, however, it becomes apparent that a diffusion-type motion of the coadsorbed reactants on the cluster surface is not to be expected. The formation of structure C in Fig. 1.65, e.g., via a sequence of structures like A and B by CO motion is energetically impossible under the reaction conditions in the ion trap. Consequently, Eley–Rideal (ER)-type reaction scenarios are suggested for the formation of the proposed intermediate structures C and D (Fig. 1.65). Most notably, the adsorption of O_2 as well as the coadsorption of CO leading to the peroxyformate structure C are barrierless according to the calculations. The formation of the carbonate structure D requires only a small energy barrier as discussed above. This is in excellent agreement with the experimental observation that all but one steps of the reaction cycle show a negative dependence of the corresponding termolecular rate constants on the reaction temperature [33]. Within the Lindemann model for low pressure gas-phase kinetics, this negative temperature dependence is indicative of a barrierless reaction.

Only the last reaction step (1.65c) representing the reformation of Au_2^- and the liberation of CO_2 displays a positive temperature dependence of the rate constant [33]. This demonstrates the presence of an activation barrier in this particular reaction step. This experimental information on the reaction

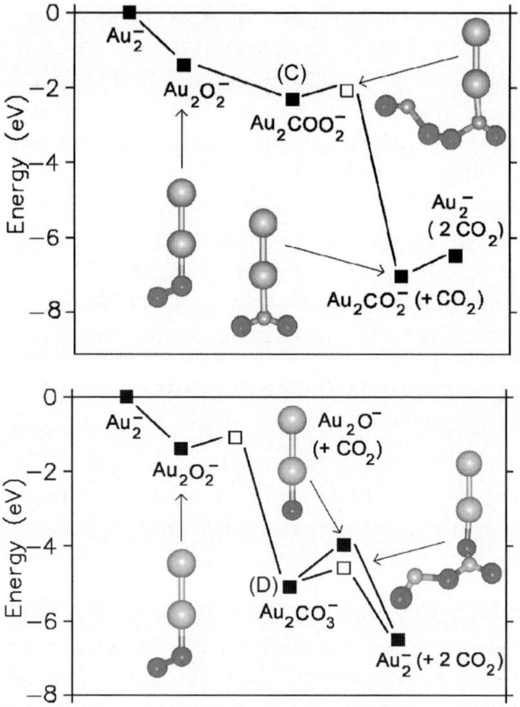

Fig. 1.66. The energetics of the ER mechanisms of the catalytic reaction. In (**a**) the peroxyformate-like species $Au_2COO_2^-$ (configuration C in Fig. 1.65) is the metastable intermediate state. The *open square* denotes the reaction barrier connecting the peroxyformate-like state with the $Au_2CO_2^- + CO_2$ product, and the corresponding transition state configuration is shown at the top right. The last step of the reaction is the desorption of CO_2. The initial energy level at zero corresponds to the sum of the total energies of all the reactants ($Au_2^- + O_2 + 2\,CO$). In (**b**) the carbonate species $Au_2CO_3^-$ (configuration D in Fig. 1.65) is the metastable intermediate state. The *open squares* again denote the reaction barriers. The first reaction barrier is associated with the insertion of CO into the O–O bond of $Au_2O_2^-$ (see structure on the left) leading to formation of $Au_2CO_3^-$. Subsequently, two reaction paths are shown. One path involves thermal dissociation of the carbonate to produce Au_2O^- (see the structure shown at the top), which then reacts with CO(g) releasing another CO_2 molecule. The other path proceeds through an ER reaction of the carbonate with CO(g) and it results in the formation of two CO_2 molecules. The latter path involves a barrier of 0.5 eV, and the corresponding transition-state configuration is shown on the right [33]

energetics is again in accord with theoretical modeling of the reaction pathway. Figure 1.66 shows the calculated energetics along the catalytic reaction path involving the intermediate structures C (Fig. 1.66a) and D (Fig. 1.66b), respectively. In both cases, the mechanism proceeds via an ER reaction of

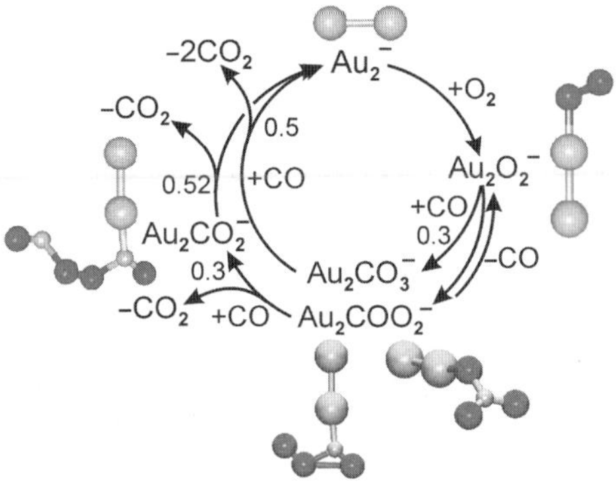

Fig. 1.67. Schematic representation of the gas-phase catalytic cycle for oxidation of carbon monoxide by gold dimer anions based on the reaction mechanism determined by kinetic measurements in conjunction with first-principles simulations. The numbers denote calculated energy barriers in eV. Also displayed are geometric structures of reactants and intermediate products according to the calculations (large, grey *spheres*, Au; small gray *spheres*, C; dark *spheres*, O) [33]

CO(g) with the intermediate complex structures releasing two CO_2 molecules. The formation of CO_2 from the reaction between CO(g) and structure C (Fig. 1.66a) involves a low barrier of 0.3 eV (see the transition state configuration in Fig. 1.66a), resulting in the formation of a metastable $Au_2CO_2^-$ complex, where CO_2 is bound to Au_2 via the carbon atom (see the structure of this complex in Fig. 1.66a). However, the heat of reaction (4.75 eV) evolving from the formation of the first CO_2 molecule is large enough in order to overcome the binding energy (0.52 eV) of the remaining CO_2 to Au_2^-, thus facilitating its desorption from the metal cluster.

The second scenario involves two branches (Fig. 1.66b) [382]. In the first one, thermal dissociation of CO_2 from the carbonate D (which is endothermic by 1.12 eV) produces a highly reactive species, Au_2O^-, which reacts spontaneously (i.e., without an activation barrier) with CO(g) to produce CO_2. The production Au_2O^- has, however, never been observed in the experiment. Therefore, the second branch is favored consisting of an ER reaction of CO(g) with species D to produce CO_2. While this step involves a modest barrier of 0.5 eV (denoted by an open square, with the corresponding transition-state configuration shown on the right in Fig. 1.66b), it releases readily *two* CO_2 molecules, since the remaining $Au_2CO_2^-$ species, where CO_2 is bound to Au_2 via one of the oxygen atoms, is unstable under our experimental conditions (100–300 K) [33].

The detailed catalytic reaction cycle emerging from experimental and theoretical evidence for the CO oxidation by gas-phase Au_2^- clusters is depicted in Fig. 1.67. Also included are the calculated energy barriers for the different reaction steps and the simulated intermediate structures. In addition to a comprehensive molecular mechanistic understanding based on experiment

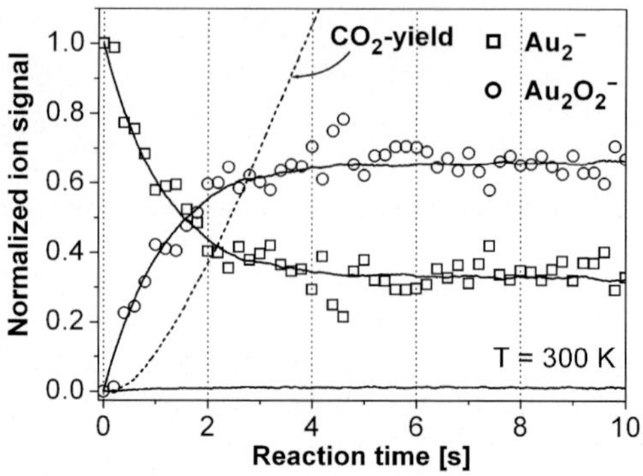

Fig. 1.68. Simulated CO_2 formation (*dashed line*) for the experimental conditions of Fig. 1.64a. The CO_2 concentration increases exponentially with time until the Au_2^- and $Au_2O_2^-$ signals reach equilibrium. The rise then remains linear and from the corresponding slope the TOF is estimated [384]

and theory, the efficiency of the catalytic reaction can be estimated from the experimental kinetic data as presented in the following section.

Catalytic Turn-Over-Frequency. From the kinetic fit, it is possible to simulate the CO_2 formation rate and thus to obtain the turn-over-frequency (TOF) of the catalytic reaction. The simulated CO_2-yield under the reaction conditions of Fig. 1.64a is displayed in Fig. 1.68. The corresponding TOF amounts to 0.4 CO_2 molecules per gold cluster per second. For the conditions of Fig. 1.64c, a TOF of 0.3 CO_2 molecules per gold cluster per second was estimated. These values are in the same order of magnitude as the catalytic activity of oxide supported gold cluster particles with a size of a few nanometers, which ranges between $0.2\,s^{-1}$ per Au atom (∼2 nm diameter particles at 273 K) and $4\,s^{-1}$ per Au atom (3.5 nm particles at 350 K) [238, 366, 367].

In order to approach an understanding of the deduced TOFs in the gas-phase reactor experiment, a rough estimation of the number of collisions of a metal cluster with the required reactant molecules (O_2 and 2 CO) in the trap will be given [384]. The collision frequency in the trap is about $10^5\,s^{-1}$ [32]. The reactant gas concentrations are about 10% or less of the helium concentration. Hence, ∼100 collisions s^{-1} of an Au_2^- ion might potentially lead to the catalytic formation of one CO_2 molecule. Considering the measured TOFs, a reaction efficiency (successful versus total number of collisions) of 0.3–0.5% depending on the exact reaction conditions can be estimated. Baring in mind that the gas-phase catalytic reaction proceeds via an Eley–Rideal mechanism in which the reactant molecule must collide in exactly the right location with exactly the correct orientation in order for the reaction to proceed, an efficiency of about 1% of all collisions leading to CO_2 formation can be considered surprisingly large. This indicates that the TOF in the ion trap reactor experiment is mainly limited by the collision frequency, i.e., the reactant partial pressures. These considerations also confirm the high efficiency of catalytic gas-phase clusters ion reactions previously observed by Ervin and coworkers [7].

The Reactivity of Pure and Mixed Gold Clusters on Surfaces

The mechanistic details for the combustion of CO on *supported* gold clusters are discussed next. Small gold clusters, Au_n ($n \leq 20$) were deposited after size-selection from the gas phase onto defect-poor and defect-rich MgO(100) films. As described in Sect. 1.5.1, defect-rich films are characterized by a given density ($\sim 5\%$ ML) of extended defects and point defects (F centers), whereas for defect-poor films the density of F-centers is negligible. The CO-oxidation was studied by combined temperature programmed reaction (TPR) and Fourier transform infrared spectroscopy and the obtained results were compared to extensive ab initio calculations [209, 368, 369].

Experimental Observation of the Size-Dependent Evolution. Subsequent to the verification that oxidation of CO does not occur on bare MgO substrate films (with and without defects), the catalyzed combustion of CO by nanosize gold clusters adsorbed on the two different MgO films was investigated. In these experiments, isotopically labeled $^{18}O_2$ and ^{13}CO were used to disentangle the CO_2 production on the cluster from an eventual catalytic oxidation of CO involving oxygen atoms from the MgO substrate. Indeed, the ^{13}CO molecule is exclusively oxidized by $^{18}O_2$ since only the $^{13}C^{16}O^{18}O$ isotopomer is detected. The experiments on defect-rich films revealed that the pure gold clusters up to the heptamer are inert for the oxidation of CO and that Au_8 is the smallest size for which the low-temperature ($T < 250$ K) combustion of CO occurs. For larger sizes, a distinct size dependency has been observed (Fig. 1.69, left). Gas-phase reactivities for the whole CO combustion cycle on gold clusters, except Au_2^-, do not exist, thus the results for the supported clusters are compared here to the O_2 adsorption on the corresponding, negatively charged free gold clusters. As discussed above, only negatively charged gold clusters are able to adsorb a single O_2 molecule, and as shown in Fig. 1.69, this adsorption is related to the electron binding energies of the clusters (see above, cf. also Figs. 1.57 and 1.61). This comparison is interesting as the adsorption and activation of O_2 is indeed the first step in the CO-combustion, as has also been observed for free Au_2^- catalyzing the CO oxidation. For cluster sizes smaller than $n = 12$, the evolution of the CO combustion and O_2 adsorption is completely different, whereas larger clusters reveal a surprisingly similar size-dependent behavior. Furthermore, experiments on the influence of the type of defects present on the MgO surface reveal a strong and size-dependent behavior. Whereas Au_8, the smallest active cluster size, turns inert when deposited on defect-poor films (Fig. 1.70), other cluster sizes remain active, however, at different temperatures. This clearly demonstrates that substrate defects are important for understanding size-dependent reactivities of small, supported gold clusters.

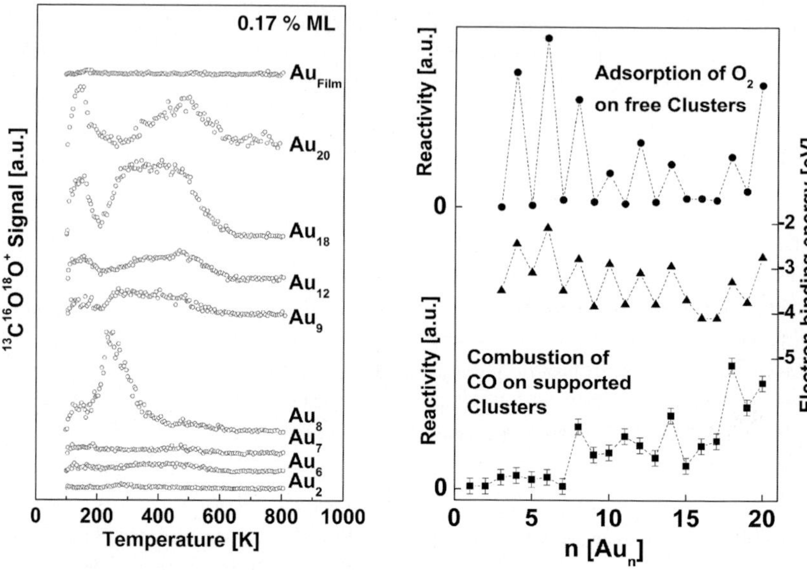

Fig. 1.69. Left: TPR experiments for the CO oxidation on selected Au_n clusters on defect-rich MgO(100) films. The model catalysts were saturated at 90 K with ^{13}CO and $^{18}O_2$, and the isotopomer $^{13}C^{18}O^{16}O$ was detected with a mass spectrometer as a function of temperature. Right: The reactivities for Au_n expressed as the number of formed CO_2 per cluster (lower trace) are compared to the O_2-adsorption properties and electron binding energies of the corresponding free Au_n clusters (upper traces)

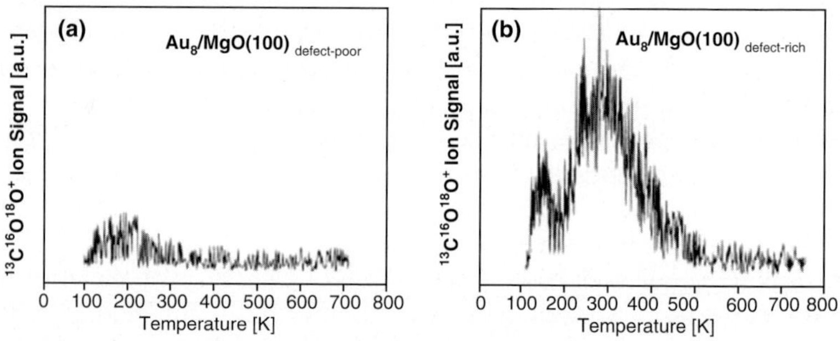

Fig. 1.70. TPR experiments for the CO oxidation on Au_8 clusters on defect-poor (**a**) and defect-rich (**b**) MgO(100) films. The model catalysts were saturated at 90 K with ^{13}CO and $^{18}O_2$, and the isotopomer $^{13}C^{18}O^{16}O$ was detected with a mass spectrometer as a function of temperature

Detailed Discussion of the Reactivity on Au_8. When deposited on MgO(100) films with a vanishing density of F centers, Au_8 is inert (Fig. 1.70a). Surprisingly, Au_8 turns active on defect-rich MgO(100) films and forms CO_2 at 140 and 280 K (Fig. 1.70b). Evaluating the area of the TPR signal yields

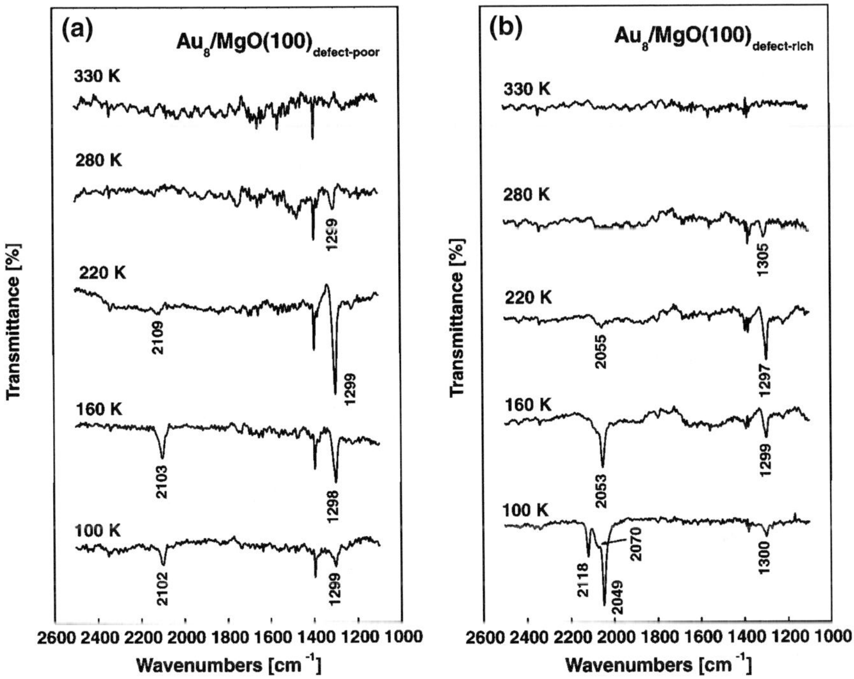

Fig. 1.71. TIR spectra taken at 90 K after exposing the model catalysts (**a**) $Au_8/MgO(100)_{defect-poor}$ and (**b**) $Au_8/MgO(100)_{defect-rich}$ to ^{13}CO and O_2 and annealing the sample to the indicated temperature. The frequencies between 2,050 and 2,100 cm^{-1} are attributed to adsorbed CO, whereas the band at 1,300 cm^{-1} may originate from activated O_2 molecules. Note the redshift of the CO stretch by 50 cm^{-1} on defect-rich films. This shift is consistent with the predicted charging of the cluster when deposited on an F center (**c**), where the difference of the charge densities of the isolated (Au_8, MgO_{5Fc}) and the model system (Au_8/MgO_{5Fc}) was calculated. The charging was estimated to be $0.5e$

in the production of about one CO_2 molecule per Au_8 cluster. Information about the origin of this observation is obtained from the FTIR spectra of the two reactants during CO combustion (Fig. 1.71). Both reactants, CO and O_2, are adsorbed molecularly at 90 K, as inferred from the CO stretch frequencies at 2,102 cm^{-1} (Au_8 on defect-poor films) and 2,077/2,049 cm^{-1} (Au_8 on defect-rich films), respectively, as well as from the frequency at 1,300 cm^{-1}, which is tentatively attributed to highly activated O_2. For the reactive model catalyst (Au_8 on defect-rich films), these vibrational bands disappear in concert with the formation of CO_2 (Fig. 1.69a), thus both of these molecules are involved in the reaction. In the other case, the two reactants simply desorb from the sample as monitored by mass spectrometry. The adsorption of the two reactants on Au_8 was investigated theoretically

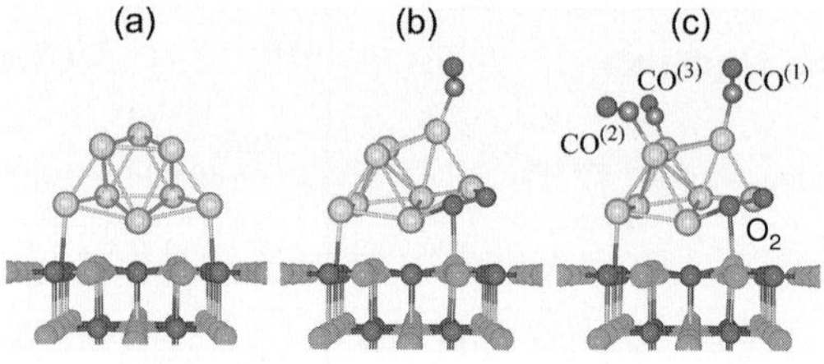

Fig. 1.72. Optimized configurations of: (**a**) a bare gold octamer (light grey *spheres*) adsorbed on an F center of an MgO (001) surface (oxygen atoms in black and magnesium in grey); (**b**) a surface-supported gold octamer with one adsorbed CO molecule and with an oxygen molecule adsorbed at the interface between the cluster and the magnesia surface. Note the significant change in the geometry of the cluster compared to the one shown in (**a**); (**c**) the gold octamer on the magnesia surface (MgO(FC)) with three CO molecules carbon in dark grey and oxygen in black adsorbed on the top facet of the cluster and an oxygen molecule preadsorbed at the interface between the cluster and the magnesia surface. Note that the molecule marked $CO^{(2)}$ lies parallel to the surface and is thus not infrared active in the experimental configuration employed here. The C–O bond length $d(CO^{(i)})$, the charge transferred to the $CO^{(i)}$ molecule $\Delta Q^{(i)}$, and the calculated C–O vibrational frequency $\nu^{(i)}$, $i = 1$, 2, and 3, as well as the corresponding values for the O_2 molecule, are: $d(CO\ ^{(i)})$ [Å] = 1.151, 1.158, 1.153, $d(O_2) = 1.422$; $\Delta Q^{(i)}[e] = 0.306$, 0.346, 0.319, $\Delta Q(O_2) = 1.125$; $\nu^{(i)}$ [cm^{-1}] = 1,993; 1,896; 1,979, $\nu(O_2) = 895$

Three energy-optimized deposited cluster configurations pertinent to the experimental work are displayed in Fig. 1.72 [369]. First we note that the bare adsorbed Au_8 cluster (Fig. 1.72a) exhibits only a slight distortion from the structure of the corresponding gas-phase neutral cluster. It consists mainly of a displacement of the gold cluster atom that is closest to the surface oxygen vacancy. The cluster is found to be anchored strongly to the defective MgO surface (binding energy of 3.445 eV) compared to a significantly weaker binding to the defect-free surface (1.225 eV). Optimal geometries with a single adsorbed CO molecule, $Au_8/O_2/CO/MgO(FC)$, and at saturation coverage of the cluster [i.e., with three adsorbed CO molecules, $Au_8/O_2/(CO)_3/MgO(FC)$] are shown in Fig. 1.72b and c, respectively. Comparison between geometry of the bare cluster in Fig. 1.72a with those shown in Fig. 1.72b and c reveals a significant change in the geometry of the gold cluster caused by the adsorption of the reactant molecules—this configurational change, which is a manifestation of the propensity of catalytic clusters to exhibit "structural fluxtionality," is caused largely by the peripherially adsorbed oxygen molecule. The system displayed in Fig. 1.72c possesses two infrared-active ^{13}CO molecules ($CO^{(1)}$, $CO^{(3)}$), as only these modes have a dynamic dipole moment perpendicular to the surface and obey the surface selection rule of IR spectroscopy in the reflective mode. This is consistent with the experimental observation of two absorption bands with frequencies corresponding to internal stretch vibrations of the molecule. Other configurations were tested too. Most important for the

Table 1.3. Vibrational frequencies of the $Au_8O_2{}^{13}CO$ complex (gas phase) as function of charging: Results for free ($Au_8O_2{}^{13}CO$) complexes as a function of excess electron charge ΔQ

ΔQ ($Au_8O_2{}^{13}CO$)	Spin	ν(cm^{-1})	BE(O_2+^{13}CO) (eV)	$d(^{13}CO)$ (Å)	$\Delta Q(^{13}CO)$	$d(O_2)$(Å)	$\Delta Q(O_2)$
0	1	2,005	1.060	1.149	0.29	1.336	0.71
0.25	0.875	1,987		1.150	0.32	1.344	0.75
0.5	0.75	1,968		1.154	0.35	1.350	0.77
1.0	0.5	1,926		1.160	0.43	1.364	0.86
0	0	2,009	0.753	1.148	0.28	1.378	0.88
0.25	0	1,990		1.150	0.31	1.381	0.89
0.5	0	1,975		1.153	0.34	1.385	0.92
1.0	0	1,920		1.158	0.41	1.398	1.00

Results are shown for various values of the spin; for the neutral cluster we show triplet ($S = 1$) and singlet ($S = 0$) states. The quantities that we display are: ν, the CO stretch frequency; BE(O_2+^{13}CO), the CO binding energy to the gold cluster with a preadsorbed oxygen molecule; the bond distances, $d(^{13}CO)$, $d(O_2)$, and excess electronic charges, $\Delta Q(^{13}CO)$, $\Delta Q(O_2)$, of the two adsorbed reactants CO and O_2. For reference we remark that the calculated vibrational frequency of gas-phase ^{13}CO is 2,070 cm^{-1} which is 25 cm^{-1} smaller than the experimental value of $\nu(^{13}CO) = 2,095$ cm^{-1}

discussion on possible reaction mechanisms is the possibility to adsorb O_2 on the top facet of the cluster (see below).

Experimental Evidence of Charge Transfer. Now let us turn our attention to the redshift in the CO stretch frequency by 25–50 cm^{-1} when the molecule is adsorbed on Au_8 supported on defect-rich MgO surfaces instead of defect-poor films. This shift is a key for elucidating the change in the charge state of the gold octamer bound to defective magnesia surface. It is well established that the absorption frequency of CO adsorbed on metal surfaces depends strongly on the population of the antibonding $2\pi^*$ orbital. Thus, extensive ab initio calculations were carried out for the isolated $Au_8/O_2/{}^{13}CO$ complex and the results are summarized in Table 1.3. The free complex was studied for this purpose in order to allow a clear distinction between charging and other support-related effects. These calculations reveal that the ^{13}CO stretch frequency shifts as a function of the charge state of the complex. In detail, the shift is correlated with variations in the population of the antibonding state. Indeed, for neutral complexes [ΔQ ($Au_8/O_2/{}^{13}CO$) $= 0$ in Table 1.3] with zero spin, a vibrational frequency of 2,009 cm^{-1} is obtained for the adsorbed ^{13}CO. The calculated decrease in frequency (61 cm^{-1}) in comparison to the value calculated for the free molecule is attributed to a net excess charge (0.285e$^-$) on the adsorbed molecule resulting from backdonation into the CO($2\pi^*$) antibonding state. As expected, such donation of charge from occupied orbitals of the metal to unoccupied states of the adsorbed molecule is even more pronounced (0.878e) for the more electronegative di-oxygen molecule. Upon charging the complex with up to $0.5e$, the net excess charge on the ^{13}CO molecule is increased to $0.341e$ and it is reflected in a redshift of the CO stretch frequency ($1{,}975\,cm^{-1}$) by $34\,cm^{-1}$ with respect to the neutral cluster case ($2{,}009\,cm^{-1}$). The clearly observed correlation between the increased amount of backdonation and the increased degree of charging of the metal cluster (ΔQ in Table 1.3) is also portrayed in the concomitant increase in the C–O bond length that is found to vary from 1.148 Å for $\Delta Q = 0$ to 1.153 Å for $\Delta Q = 0.5e$. A similar redshift of the adsorbed carbon monoxide stretch frequency and increase in bond length is also found when starting from the triplet state of the neutral complex (Table 1.3). It is interesting to note that the calculated redshift is comparable to the experimentally observed decrease in the frequency of the C–O stretch mode between the case of a molecule adsorbed on a gold octamer bound to a surface F center and that when the molecule binds to a gold cluster deposited on a defect-poor surface. Consequently, the degree of charging of the surface-supported gold cluster anchored to an F center site is higher (by several tenth of the electron elementary charge e) than that of the octamer deposited on a defect-poor surface. In fact, this charge transfer can also be deduced directly from ab initio calculations, where the charge density of the $Au_8/MgO(F_{5c})$ system is compared with the isolated system (Au_8, $MgO(F_{5c})$) [209, 368]. The difference of the two charge density plots predicts a charge transfer of $0.5e$ into the cluster (see insets of Fig. 1.73).

Bonding and Activation of the Reactant Molecules on Au_8. Having established the importance of charging for the activation of Au_8 for the CO combustion, we discuss in the following its consequences for the binding and activation of the two reactants. In the case of O_2, this is done by the analysis of the spectra of the local density of electronic states (LDOS) projected on the oxygen molecule and on the metal cluster. A similar analysis is often employed in the context of the interaction of adsorbates with extended surfaces [385]. Figure 1.73a shows the LDOS projected on the O_2 molecule which is adsorbed at the peripheral site of the most reactive isomer of the $Au_8/MgO(F_{5c})$ model catalyst. All the prominent peaks of the LDOS spectrum in Fig. 1.73a can be unambiguously assigned to orbitals of (free) molecular oxygen origins. In addition, these states overlap with the entire d-band of the Au_8 cluster (shown in Fig. 1.73b) in the range of $-7\,eV \leq E \leq E_F$, where E_F is the Fermi energy. Bonding of the oxygen molecule to the gold octamer involves mainly hybridization of the 5σ, $1\pi_{||}$, and $1\pi_{\perp}$oxygen states with the gold d-band (Fig. 1.73b) Most importantly, the full spin-manifold of the antibonding $2\pi^*$ states of O_2 is located below E_F, resulting in strong activation of the O_2 molecule through occupation of the antibonding orbitals. In the case of the defective surface, this charge transfer is directed from the oxygen vacancy into the deposited gold cluster and adsorbed molecule. This leads to weakening of the O–O bond that is reflected in a significant increase of its length (1.43 Å) compared to that of the free molecule (1.25 Å). Accompanying the activation process is a change in the spin state of the molecule from a triplet state in the gas phase to a peroxo-like adsorbed state with a zero net spin. We observe

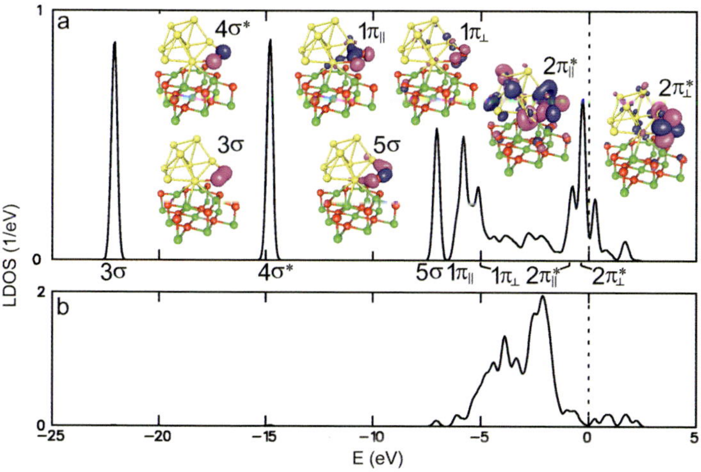

Fig. 1.73. LDOS of $Au_8/O_2/MgO(100)(FC)$. For these studies, the O_2 molecule was adsorbed on the periphery and no CO molecules were coadsorbed in order to separate the interaction of O_2 with the cluster from coupling effects between the reactants. The prominent peaks of the oxygen LDOS are labeled following the conventional nomenclature for the molecular orbitals of the gas-phase O_2 molecule, with $\perp$ and $||$ meaning perpendicular and parallel to the MgO surface, respectively. The Fermi energy E_F is at 0 eV

that O_2 binding to the cluster is stronger in the presence of the F center (0.469 and 0.329 eV for the spin 0 and 1 states, respectively) in comparison to the case without F centers (0.329 and 0.299 eV for the spin 0 and 1 states, respectively). Thus, in contrast to solid gold, Au_8 adsorbed on an F center of an MgO(100) surface not only binds di-oxygen but also highly activates the molecule. The bonding of CO to the $Au_8/O_2/MgO(FC)$ complex can be understood by analyzing the correlation diagram shown in Fig. 1.74. It depicts the local density of states (LDOS) for the two reactants, CO (Fig. 1.74, left) and the $Au_8/O_2/MgO(FC)$ complex (Fig. 1.74, right), as well as for the product complex $Au_8/O_2/CO/MgO(FC)$ with the CO adsorbed on the deposited cluster (Fig. 1.74, middle). As expected, the 3σ and the nonbonding 4σ (oxygen lone pair) molecular orbitals of the isolated CO are not involved in the bonding to the cluster complex as their energies are not changed upon adsorption. On the other hand, the nonbonding 5σ (carbon lone pair) orbital is stabilized by the interaction (by about 3 eV) and thus it contributes to the chemisorption of carbon monoxide to the surface-supported complex through hybridization with the metal cluster orbitals. The consequent depletion of the 5σ frontier orbital is often referred to in surface science studies as *donation* from CO to the metal. The 1π level is slightly pushed upward in energy (<1 eV).

The largest contribution in the bonding of CO to the complex occurs via hybridization of the antibonding $2\pi^*$ levels of CO and the occupied frontier

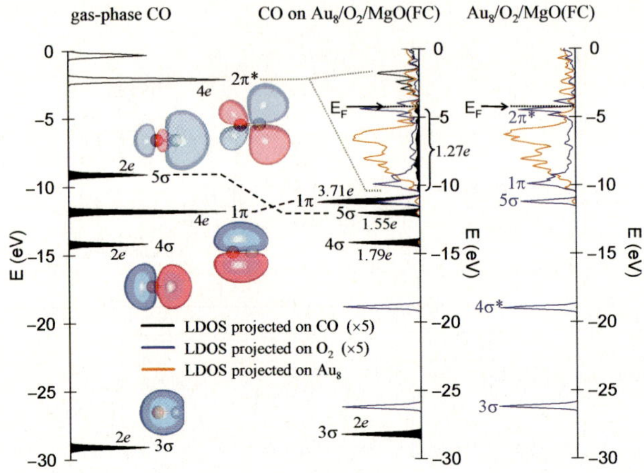

Fig. 1.74. LDOS correlation diagram of free CO (*left*) and Au_8/O_2/MgO(FC) (*right*) resulting upon adsorption of the CO molecule in the complex adsorbed on Au_8/O_2/CO/MgO(FC) (*middle*). The color assignment in the LDOS diagrams are given in the figure. The electron populations of the various levels are given as 2*e*, 4*e*, etc., and iso-surface images of the orbitals of the free CO molecules are also shown (*left*). Dashed lines indicate orbital shifts and redistribution caused by the adsorption of the CO molecule

electronic states of the cluster complex. The LDOS of the complex projected on the adsorbed CO clearly reveals contributions of the $2\pi^*$ orbitals that are spread over the entire energy range of the cluster's electronically occupied spectrum (see in the middle panel of Fig. 1.74, the black shaded LDOS for energies smaller in absolute value than 10.5 eV). The $2\pi^*$ orbitals are therefore pushed below the Fermi level and this mechanism of (partially) populating the molecular state represents the origin of the so-called backdonation (i.e., hybridization and population of the initially unoccupied antibonding orbitals of the CO molecule through interaction with occupied surface orbitals). All these features are also present in the correlation diagram of the cluster complex bound to the defect-free MgO surface, where backdonation, however, is less pronounced.

The amount of backdonated electronic charge has been quantified by integration over the squared amplitude of the $2\pi^*$ orbital located below the Fermi energy. For the Au_8/O_2/CO bound to the F center of the MgO support, it was found that 1.27*e* are backdonated into the $2\pi^*$ orbital, whereas a smaller degree of backdonation (1.18*e*) is estimated for the complex bound to the undefective surface. This difference in the degree of backdonation is manifested, as noted above, in a variation of the stretch frequencies of the adsorbed CO molecules, and these correlate with the aforementioned variation in the degrees of substrate-induced charging of a gold octamer deposited on magnesia surfaces with or without F center defects. Indeed, ν(CO) for the complex bound

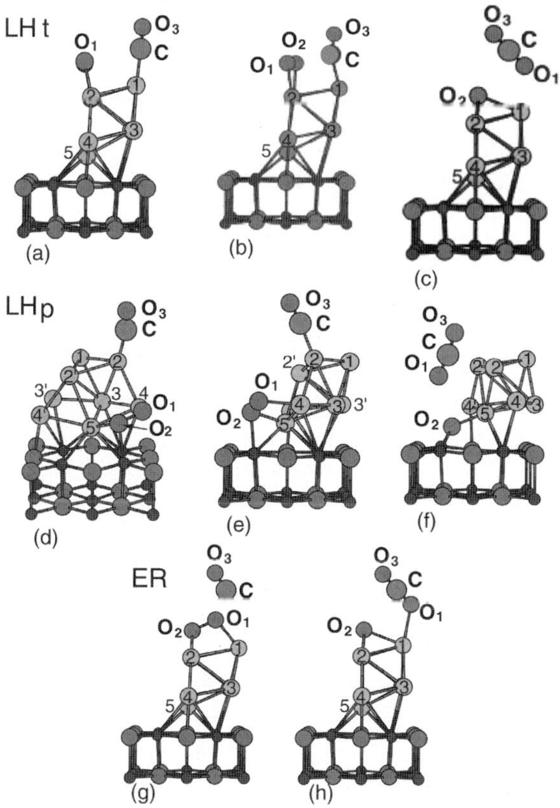

Fig. 1.75. Reaction mechanisms of the CO oxidation on supported Au_8 [209]

to a defect-rich surface is measured to be red shifted by 25–53 cm^{-1} with respect to the frequency of a CO molecule bonded to the complex deposited on a defect-poor support. This compares favorably with the corresponding calculated red-shift of 34 cm^{-1}.

Reaction Mechanism of the CO-Combustion on Au_8. Next we turn our attention to the two reaction pathways, which may mainly contribute to the observed oxidation of CO at 140 and 280 K (Fig. 1.70) and which can be characterized as Langmuir–Hinshelwood (LH) types of reaction mechanisms [209]. For the formation of CO_2 at 140 K, the two reactants are initially coadsorbed on the triangular top facet of the Au_8 (Fig. 1.75). The distance between the carbon atom and one of the peroxo-oxygens in this local minimum state is equal to 3.11 Å ($d(CO_1)$, Fig. 1.75a). In the following, this LH top-facet reaction mechanism is labeled LHt. Through mapping of the potential energy surface along the $C–O_1$ reaction coordinate a rather low-energy barrier $\Delta E_{b(LHt)} = 0.1$ eV occurring at $d(CO_1) \approx 2.0$ Å is determined for the LHt oxidation channel with the end product of a weakly adsorbed carbon dioxide molecule (∼0.2 eV). The total exothermicity of the reaction $Mg(F_{5c})/Au_8 + O_2(g) + CO(g) \rightarrow MgO(F_{5c})/Au_8O + CO_2(a)$ is ∼4.8 eV and it is ∼5 eV when the CO_2 product is desorbed, here (g) and (a) denotes gas phase and adsorbed molecules, respectively. This LHt low-temperature oxidation mechanism was found by these theoretical studies (with similar energetics) for reactions on the gold cluster deposited on either a defect-free MgO(100) surface or one containing an F_{5c}, and both are expected to be relatively insensitive to the Au_n cluster size, correlating with the experimental results (Figs. 1.69 and 1.70). As aforementioned, the higher temperature channel is strongly enhanced for Au_8 supported on a defect-rich MgO (100) support. This trend correlates with the simulations of an LH-periphery (LHp) reaction mechanism, starting with the CO adsorbed on the top-facet of the Au_8 cluster and the peroxo $O_2{}^*$ molecule bonded to the periphery of the interfacial layer of the cluster, where the distance between the C atom and the oxygen marked O_1 is $d(CO_1) = 4.49$ Å. Indeed, mapping of the potential energy surface along the C–O_1 reaction coordinate revealed for $MgO(F_{5c})/Au_8$, a rather broad reaction barrier $\Delta E_b(\mathrm{LHp}) \sim 0.5$ eV at $d(CO_1) \sim 2.0$ Å (Fig. 1.75), while for the defect-free substrate, a significantly higher barrier was found, $\Delta E_b(\mathrm{LHp}) \sim 0.8$ eV. The reaction product is shown in Fig. 1.75f for the $MgO(F_{5c})/Au_8$ catalyst. This change in activation energy is at the origin of the observed change in reactivity, as for the defect-free substrate, the CO molecule desorbs prior to reaction. The question may arise why the two reaction channels are observed experimentally, as the low-temperature mechanism should always be dominant as its reaction rate is orders of magnitude higher. In fact, theoretical studies revealed that only one oxygen molecule adsorbs on the clusters. This observation is also confirmed by gas-phase studies where it is known that small negatively charged gold clusters adsorb just a single oxygen molecule. Thus, there exists an ensemble of Au_8 with an oxygen molecule adsorbed on the periphery and another ensemble with the oxygen molecule bound to the top facet of the cluster. In fact, the ratio of the number of CO_2 formed at low and high temperature is in the same range as the ratio of O_2, which directly adsorbs on Au_8 (leading to O_2 adsorbed on the top facet of Au_8) and which adsorbs via reverse spillover from the MgO substrate (leading to O_2 adsorbed on the periphery of Au_8), see "Spillover, Reverse Spillover, and Adlineation". A third reaction path was theoretically found starting from the optimal configuration of $O_2{}^*$ adsorbed on the $MgO(F_{5c})/Au_8$ catalyst. A gas-phase CO molecule brought to the vicinity of the peroxo molecule reacts spontaneously (without an energy barrier) forming a carbon dioxide molecule weakly bound to the catalyst. This Eley–Rideal (ER)-type reaction mechanism (Fig. 1.75g and h) corresponds to a low-temperature generation of (weakly adsorbed) CO_2 through direct reaction of gaseous CO with a preadsorbed peroxo $O_2{}^*$ molecule. This reaction channel can occur even at 90 K, that is, during the initial dosing stage in the experiment.

The Reactivity of Pure Au_n *and Mixed* Au_nSr *Clusters.* Motivated by the result that charging is essential for the activation of small Au_8 clusters,

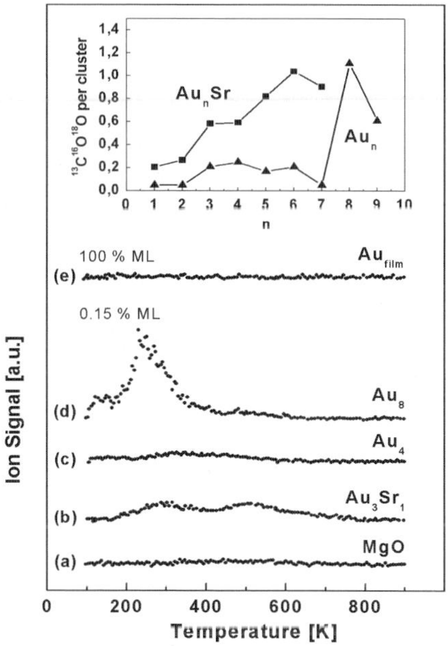

Fig. 1.76. TPR spectra of CO_2 formation on: (**a**) an MgO(100) film, (**b**) Au_3Sr/MgO(FC), (**c**) Au_4/MgO(FC), (**d**) Au_8/MgO(FC), and (**e**) a thick gold film grown on MgO(100). The Au_4/MgO(FC), pure MgO film, and the thick Au film are catalytically inert. The inset compares the chemical reactivity R of pure Au_n and doped Au_nSr clusters, with $1 \leq n \leq 9$, expressed by the number of product CO_2 molecules per deposited cluster. The TPR spectra are recorded after exposure of the model catalyst to isotopically labeled $^{18}O_2$ and $^{13}C^{16}O$ at 90 K. Only the production of the $^{13}C^{16}O^{18}O$ isotopomer is detected by measuring its ion current (I_s), which indicates that the oxidation of CO occurs only on the cluster and that no oxygen from the MgO substrate is involved

experiments on mixed Au_nSr clusters were performed [368]. It is well-known that gold is strongly electronegative and that mixing with alkaline- or alkaline earth metals can even lead to ionic alloys. The best known example is cesiumaurid [386]. Experimental considerations determined the choice of Au_nSr clusters for these studies. For this purpose Au Sr target disks with various compositions were produced by mixing the two metals and annealing the mixture to high temperatures in vacuum environment.

The formation of CO_2 is then measured for the mixed cluster samples and compared to the formation on pure Au_n clusters. Fig. 1.76 shows the one cycle TPR spectra for the clean MgO and gold films as well as for selected cluster sizes, which are discussed in detail below for understanding the impurity doping and the size effects. Whereas the pure gold tetramer is inert, the strontium-doped Au_3Sr cluster produces CO_2 at 250 and 500 K. These

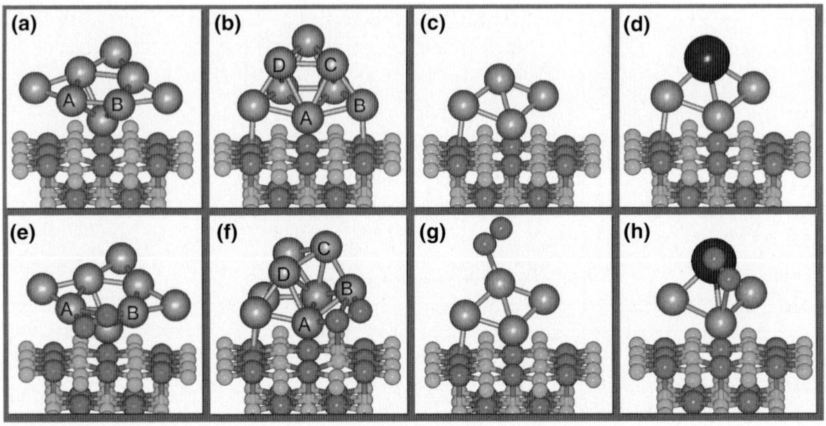

Fig. 1.77. The optimized atomic structures of model catalysts comprising Au_8 (**a** and **b**), Au_4 (**c**), and Au_3Sr (**d**) clusters adsorbed at an F center of an MgO(100) surface. For Au_8, two relevant isomers are depicted, separated by an energy difference of 0.29 eV with the quasi-planar geometry being the ground state. The optimal geometries for the adsorption of the O_2 molecule on these model catalysts are shown in panels **e**–**h**

reaction temperatures are higher than those observed for Au_8. The evolution of the size-dependent reactivity, shown in the inset of Fig. 1.76, clearly reveals that the onset for the CO combustion is shifted to lower sizes (Au_3Sr) in comparison to the pure clusters (Au_8). As already shown above, the clean MgO(100) surface as well as multilayer gold films grown on MgO(100) are catalytically inert, the latter reflecting the noble character of bulk gold.

To understand the origin of this size-dependent and element-specific behavior, the atomic structure and the electronic spectra of these model catalysts were studied by extensive first-principles simulations, and the optimized structures of Au_8(with two relevant isomers), Au_4, and Au_3Sr adsorbed on MgO(F_{5c}) are shown in Fig. 1.77, before (a–d) and after (e–h) O_2 adsorption. As shown above, these clusters bind to the MgO(F_{5c}) surface quite strongly (calculated binding energy of 2.65–4.06 eV), and their binding is significantly enhanced (typically by about 2 eV) due to the F center defects. These high binding energies and the "anchoring" effect by the F centers correlate well with the observed thermal stability of the supported clusters in the TPR experiments. As mentioned above and supported by the experimental results, another important finding pertains to the charge state of the adsorbed clusters—in particular, the calculations predict that the interaction with the surface is accompanied by charge transfer [209] of 0.5e, 0.3e, and 0.3e to the adsorbed Au_8, Au_4, and Au_3Sr clusters, respectively. Binding and activation of molecular O_2 by the model catalyst are necessary elementary steps in the CO oxidation process [387, 388]. Thus, understanding the structural, dynamical, electronic, and compositional factors that govern these processes are a key to the elucidation of the observed size-dependent activity of the $Au_n/MgO(F_{5c})$ model catalysts. Indeed, these first-principles simulations show that the binding energy of O_2 to the supported Au_n clusters and the degree of O–O bond activation are strongly dependent on the cluster size as discussed previously (see also below). In contrast, the adsorption energy of CO to the supported clusters is relatively insensitive to the cluster size, and it is higher than that of oxygen (0.7–1.0 eV per CO molecule, depending on coverage). Consequently, in the following, the adsorption and activation of O_2 by the supported gold and gold-strontium nanoclusters is studied in more detail to illustrate "nanocatalytic" factors which are the dynamic structural fluxionality as well as electronic size and impurity-doping effects.

Structural Dynamical Fluxionality. The capability of small clusters to exhibit several structural forms (isomers) of comparable energies, and to interconvert between such isomers at finite temperature, is one of the hallmarks of cluster science (see also "Structural Dynamic Fluxionality"). This unique structural variability may influence the chemical reactivity of nanocatalytic systems in two main ways. First, at finite temperature, the model catalyst (and in particular the cluster component) will form an equilibrium ensemble of coexisting structural configurations, with various isomers exhibiting different chemical reactivities. This is illustrated here via the two structural isomers of the Au_8 cluster with the two-layered one (Fig. 1.77b), which is thermodynamically less stable (by 0.29 eV) than the quasi-planar isomer (Fig. 1.77a), showing a higher energy gain upon oxygen adsorption (0.47 eV compared to 0.28 eV for the latter isomer). Second, and most importantly, is the structural dynamic fluxionality of clusters in the course of reactions that expresses itself in the ability of a given structural isomer to dynamically adapt its structure such as to allow the reaction to evolve on the most favorable free-energy path. Such fluxional propensity is illustrated in Fig. 1.77b and f, where the more reactive two-layered Au_8 cluster is shown to undergo a large structural transformation upon adsorption of molecular oxygen at the "peripheral site"; i.e., the approximate bicapped octahedral geometry (Fig. 1.77b) is transformed to a bicapped trigonal prism (Fig. 1.77f). Note that the structural fluxionality is essential for the reaction to occur, since it was found that constraining the cluster to maintain its original geometry (Fig. 1.77b) prevents the adsorption and activation of O_2.

Electronic Size Effects. Understanding the size-dependent electronic structure of the $Au_n/MgO(F_{5c})$ model catalysts, which is fundamental for elucidation of their atom-by-atom controlled reactivity, is facilitated by analysis of the spectra of the LDOS projected on the oxygen molecule and on the metal cluster (see also "Electronic Size Effects"). Figure 1.78a shows the LDOS projected on the O_2 molecule which is adsorbed at the peripheral site (Fig. 1.77f) of the more reactive isomer of the $Au_8/MgO(F_{5c})$ model catalyst. As shown above, bonding and activation of O_2 on the octamer is enabled by resonances formed between the cluster's electronic states and the $2\pi^*$ molecular states

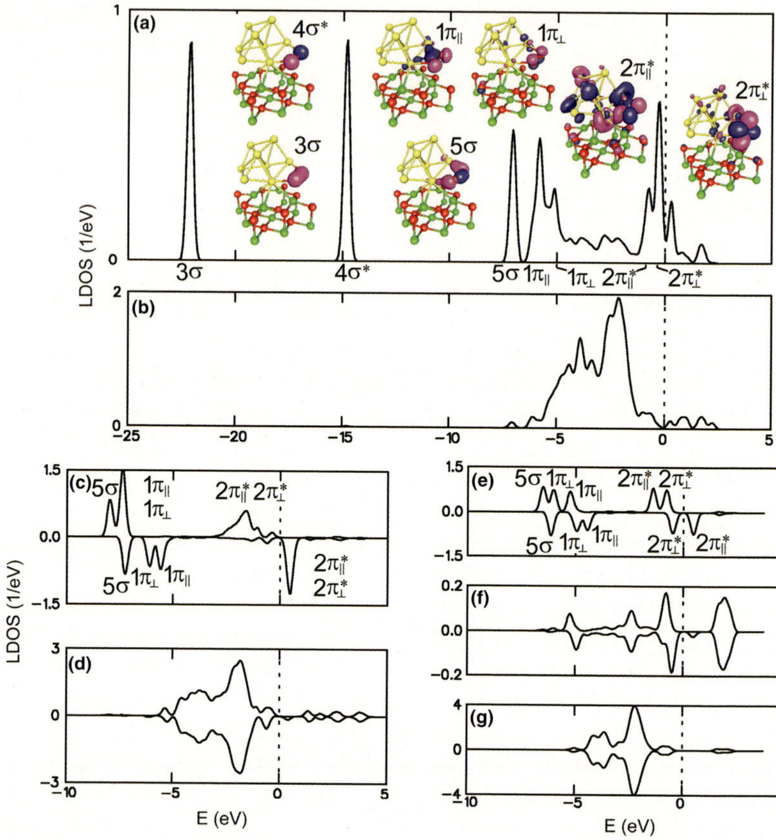

Fig. 1.78. Local density of spin-up and spin-down electronic states (LDOS) of the model catalysts shown in Fig. 1.77f–h for oxygen (**a, c, e**) and the metal part (**b, d, f, g**). The prominent peaks of the oxygen LDOS are labeled following the conventional nomenclature for the molecular orbitals of the gas-phase O_2 molecule, with $\perp$ and $||$ meaning perpendicular and parallel to the MgO surface, respectively. The Fermi energy E_F is at 0 eV

of oxygen as these antibonding states are pulled below the Fermi level of the system upon interaction with the cluster. The population of the antibonding states is accompanied by a change in the spin state of the molecule from a triplet state in the gas phase to a peroxo-like adsorbed state with a zero net spin. A drastically different scenario is found for the interaction of O_2 with the smaller gold cluster, $Au_4/MgO(F_{5c})$, where molecular oxygen adsorbs in an "on-top" configuration, with one of the oxygen atoms binding to a single gold atom (Fig. 1.77g). This system exhibits rather weak binding of the molecule to the metal cluster (0.18 eV), an almost unperturbed O–O bond length (1.28 Å), and a spin splitting of the oxygen-projected LDOS spectrum (see Fig. 1.78c). The weak binding is attributed to the narrower d-band of the adsorbed Au_4 cluster compared to that of Au_8 (see Fig. 1.78b and d), with a consequent lack of overlap between the states at the bottom of the d-band of the gold cluster (Fig. 1.78d) and the molecular oxygen states with energies $E < -5$ eV. Moreover, the spin-down antibonding $2\pi^*{}_\perp$ and $2\pi^*{}_{||}$ orbitals of the adsorbed oxygen molecule are located now above E_F (unlike the case of the larger cluster, compare Fig. 1.78c and a), which results in no activation of the molecule by the adsorbed Au_4 cluster.

Impurity-Doping Effect. Finally, we address the possibility of enhancing the catalytic activity of a nanocluster by designed incorporation (doping) of an impurity, demonstrated here by the catalyzed oxidation of CO on Au_3Sr/MgO (F_{5c}). The LDOS spectra projected onto the oxygen molecule, the Sr atom, and the Au_3 part of the metal cluster are displayed in Fig. 1.78e, f, and g, respectively. Doping by a single impurity atom changes significantly the bonding and activation of O_2 compared to the pure gold tetramer, Au_4, case. The bonding of O_2 is mainly to the strontium atom of the Au_3Sr cluster (Fig. 1.77h), and it is characterized by a substantially higher adsorption energy (1.94 eV compared to 0.18 eV for the configuration shown in Fig. 1.77g) and a significant activation of the O–O bond that is reflected in an increased bond length of 1.37 Å. This activation is due to occupation of the spin-down $2\pi^*{}_\perp$ oxygen orbital (compare Fig. 1.78e and c, where in the latter, which corresponds to the pure cluster, this state contributes to the peak just above E_F), resulting in a superoxo-like state of the adsorbate. Bonding of the oxygen molecule to Au_3Sr/MgO(F_{5c}) occurs via resonances formed between the Sr states in the energy intervals 5–6 and 0–1 eV below E_F, with the spin-up $1\pi_{||}$ and $2\pi^*{}_\perp$ states, as well as with the spin-down $1\pi_\perp$ and $2\pi^*{}_\perp$ orbitals, of the adsorbed activated O_2.

These studies established the importance of charging the cluster upon interacting with the surface. This is fully supported by the gas-phase results, showing that only negatively charged Au_n clusters are observed to be catalytically reactive. Interestingly, the gold dimer is inert on the MgO support, whereas it catalyzes the reaction in the gas phase. The comparison of the two studies suggests that on the surface, the charge transfer is only partial not leading to enough activation. In addition, considering the proposed mechanism in Fig. 1.75, it becomes clear that steric effects might prevent the coadsorption of the reactants and the formation of the intermediate on the surface.

The Effect of Water on the Catalytic Activity of Gold Nanoclusters

Evidence for an enhancement by moisture of the catalytic activity of small gold particles has been provided already early on [389]. This finding and other experimental studies that followed [390–395] are particularly interesting, both intrinsically as a property of gold nanocatalysts as well as in comparison with common catalysts where the presence of moisture is found to be detrimental or nonbeneficial to their activity [389]. Additional interest in the effect of moisture on gold catalysis derives from expected applications of these catalysts at ambient conditions, that is, moderate temperatures and humid atmospheres.

In light of the above, and findings that for certain supports moisture is able to increase the activity of gold catalysts by about two orders of magnitude [389, 392], correlating with the amount of moisture adsorbed on the catalyst [390], it is rather surprising that to date only a few recent experimental studies discuss the mechanisms of water enhancement of the catalytic activity of gold nanocatalysts ([389–394] and citations in [392]). Furthermore, with the exception of a recent experimental report [392] and a most recent theoretical study [396], only effects occurring at high-moisture levels have been examined.

The experimental investigations noted above [392] addressed the effect of moisture on the low-temperature oxidation of CO by gold particles with diameters of 3.0, 3.9, and 8.2 nm supported, respectively, on TiO_2, Al_2O_3, and SiO_2 surfaces. The H_2O concentration has been varied over a wide range, from about 0.1–6,000 ppm. The degree of enhancement of the CO oxidation rate showed a dependence on the type of support, varying from high for insulating alumina and silica to moderate for titania; at 3,000 ppm H_2O the activity of Au/TiO_2 reaches full conversion of CO (100%). Interestingly, despite the large effect of moisture, the apparent activation energies for the titania and alumina supports were found to be essentially unchanged, indicating that the reaction mechanisms over these catalytic systems are unaffected by moisture. It has been also concluded that the key factor governing the moisture effect on the CO oxidation reaction is the amount of H_2O on the catalyst surface. Furthermore, the effect of moisture was found to be essentially reversible, and no significant change in the gold particles size was detected by TEM measurements after the reaction.

The above study [392] concludes with a discussion about several scenarios pertaining to the origin of the moisture effect. First, the possibility of the water–gas shift (WGS) reaction was ruled out since the reaction temperatures used in this study were well below the WGS reaction temperature (~573 K) over Au/TiO_2. Moreover, the H_2O concentrations employed were far below that required by stoichiometry for the reaction with CO (10,000 ppm). From the results of the experiments in [392] and those of previous experimental investigations, it has been concluded that moisture may influence the CO oxidation reaction through two main routes: (a) activation of oxygen, and (b) decomposition of carbonate. Reaction schemes pertaining to these mechanisms can be found in [392].

To focus on the influence of water on the chemical activity of gold in nanocatalytic reactions, quantum mechanical ab initio calculations were performed [396] for model systems that do not exhibit charging effects originating from excess electrons bound to free Au clusters (i.e., gold cluster anions) [33, 382, 397] or from defects at the metal-oxide surface [209, 368] that have been shown, as described in the previous subsections, to promote the

Table 1.4. Energies (in eV) for the adsorption and coadsorption of O_2 and H_2O on free (Au_8 and Au_{30}) clusters and on a gold octamer supported on MgO(100), that is, Au_8/MgO

	O_2	H_2O	H_2O–O_2
Au_8	Unbound	~ 0.3	0.4–0.9
Au_{30}	≤ 0.4	0.3–0.6	0.7–0.9
Au_8/MgO–T	≤ 0.1	0.2–0.3	0.5–1.2
Au_8/MgO–P	0.3–0.8	0.4–0.6	1.3–2.1

In the case of the Au_8/MgO system, results are given for both the adsorption on the top facet of the gold cluster (–T) and at the peripheral interface of the cluster with the substrate (–P)

catalytic activity of gold nanoclusters. Accordingly, two free neutral Au clusters containing 8 (Au_8) and 30 (Au_{30}) atoms were considered, and in the case of a surface-supported cluster a gold octamer, Au_8, adsorbed on a defect-free MgO surface described by a two-layer MgO(100)-(3×3) slab was examined. We comment here that the characteristics of the perfect MgO(100) surface and of the corresponding oxygen vacancy were found to be well reproduced by a two-layer film [398].

O_2 *Adsorption on Free and Supported Gold Clusters.* While molecular oxygen does not adsorb on clean Au(110) and (111) extended surfaces [399, 400], it has been suggested that the adsorption propensity of O_2 to finite size Au particles is increased, particularly, at low-coordinated sites [356, 369, 401] as also discussed above. For each of the relaxed configurations of O_2 at the various adsorption sites on the Au_8 and Au_{30} clusters, the adsorption energy, defined as the difference between the optimized energy of the combined system, $Au_n + O_2$, and that of the separated relaxed components, Au_n and O_2, were calculated. These calculations showed that O_2 does not bind to Au_8 and for the larger cluster the adsorption energies at the various sites range only up to 0.4 eV (Table 1.4).

Next, the properties of the gold octamer adsorbed on a defect-free MgO(100) surface were examined. The presence of the support causes broadening of the O_2 adsorption energies distribution. Two spatial regions may be identified: (a) the top facet of the cluster, where O_2 adsorbs weakly (adsorption energies up to 0.1 eV), in agreement with other first-principles calculations [209, 401, 402], and (b) peripheral sites (at the Au_8/MgO interface), where O_2 adsorbs with energies larger than 0.3 and up to 0.8 eV (Table 1.4), with the O–O bond extended to 1.37–1.49 Å (i.e., in the range typical of superoxo, or peroxo, states).

Coadsorption of O_2 *and* H_2O *on Free and Supported Gold Clusters.* It has been established experimentally that extended clean gold surfaces are hydrophilic [403]. The calculations showed that water adsorbs (relatively weakly) on free and supported Au clusters (Table 1.4), with adsorption energies that

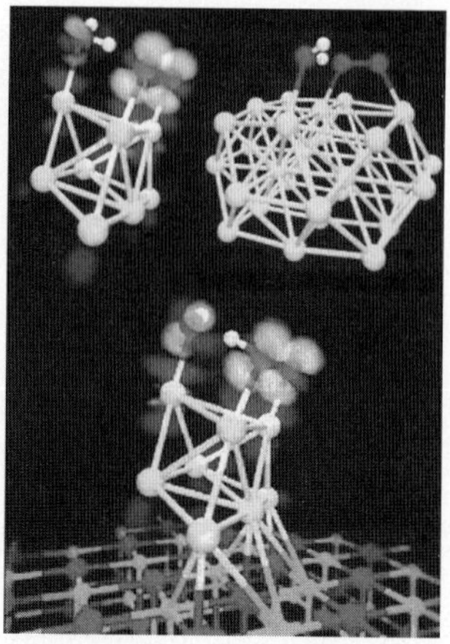

Fig. 1.79. Relaxed atomic configurations of H_2O and O_2 coadsorbed on free Au_8 and Au_{30} clusters (top) and on Au_8/MgO(100) (bottom). For the free and supported Au_8 cases, the difference-charge-density between the complete adsorption system and the separated Au_8/MgO(100) and O_2–H_2O complex are displayed, superimposed on the atomic configuration. Charge depletion is shown in dark grey and charge accumulation in gray-white. Light grey, dark, and white *spheres* correspond to Au, O, and H atoms, respectively

vary from 0.2 to 0.6 eV, and without apparent correlation between the adsorption strength and the coordination of the adsorption site. It was also found that an adsorbed H_2O is an "attractor" for molecular O_2. Indeed, in the presence of an adsorbed water molecule, O_2 preferentially adsorbs (in a singlet spin state) at a neighboring site (Fig. 1.79, top). Moreover, the coadsorbed O_2 does not show preference to particular sites on the gold cluster, and it may bind even at sites where the adsorption of O_2 (without coadsorbed H_2O) is energetically unfavorable. In the case of free Au_8 and Au_{30} clusters, the calculations gave a range of coadsorption energies of 0.4–0.9 eV (Table 1.4). The observation that these values are larger then the sum of the adsorption energies of the two separated species on the Au clusters indicates that the coadsorption of H_2O and O_2 exhibit a synergetic effect and implies possible formation of a stable complex of the two adsorbed molecules.

In the coadsorbed state on the free clusters, the O_2 molecule shows peroxo-like characteristics, with a bond length of about 1.45 Å (18% larger than the gas-phase value). Comparison with the properties of O_2 adsorbed under dry condition highlights the enhanced O–O bond-activation induced by the H_2O molecule. In the coadsorbed configurations (Fig. 1.79, top), a proton is partially shared between the H_2O and one of the oxygens of the O_2 molecule, with the distance between the two oxygens that share the proton being close to 2.5 Å. To estimate possible electronic charge rearrangement and charging effects in the coadsorbed system, the specific adsorption complex shown for

Au_8 was considered (see atomic configuration at the top left of Fig. 1.79). For this particular set of nuclear positions, it is useful to analyze the difference between the electronic charge density of the full system and the sum of the densities of the separated components (the Au_8 cluster and the $O_2 \cdots H_2O$ species, respectively, keeping the geometries as found for the adsorption system). This procedure allows one to assess the charge redistribution occurring by joining the Au cluster and the coadsorbed species. According to this analysis, it was found that electronic charge that is depleted in the region of the Au_8 cluster is accumulated at the location of the adsorbed O_2 molecule (in particular in the originally empty $2\pi^*$ orbital, see gray-white lobes in Fig. 1.79, top left). Integrating over the semispaces defined by the normal plane bisecting the Au–O bonds yields an estimated charge transfer of about $0.27e$ (where e is the electron charge).

In general, the optimized configurations of H_2O and O_2 coadsorbed on the top facet of Au_8/MgO are similar to those found for the unsupported clusters. However, in some occasions the proton shared by the H_2O and O_2 molecules may actually be considered as transferred to the O_2 species (see, e.g., bottom of Fig. 1.79). In such instances, coadsorption leads to the formation of an OH and a hydroperoxyl-like (HO_2) group. The distance between the O atoms sharing the proton takes a value of about 2.49 Å, while the O–O hydroperoxyl bond length reaches a value of about 1.48 Å (that is 21% larger than in a free molecule), reflecting a very high degree of bond activation. The electronic charge density of the combined system, referenced to that of the separated Au_8/MgO and $OH \cdots HO_2$ components (Fig. 1.79, bottom), exhibits a charge redistribution pattern similar to that described above for the free Au cluster (Fig. 1.79, top left). In particular, a charge analysis show that an amount of approximately $0.31e$ is transferred from the Au_8/MgO system to the coadsorbed species; similar analysis applied to the bare Au_8/MgO complex led to an estimate that upon adsorption an electronic charge of about $0.4e$ is transferred from the MgO substrate to the adsorbed Au_8 cluster.

These results suggest that the coadsorption of H_2O and O_2 stabilizes partially charged, highly activated states of the adsorbed oxygen molecule (with the extra electronic charge donated by the underlying supported gold cluster). Such activated states include the hydroperoxyl-like intermediate shown at the bottom of Fig. 1.79, as well as a partially negatively charged oxygen molecule in a superoxo or peroxo state (with an O–O bond elongated by 0.1–0.2 Å, respectively, with reference to the free molecule), stabilized through a proton-sharing mechanism. No such bond activation is found for the adsorption of O_2 on the top facet of the supported gold octamer without water coadsorption (recall the small adsorption energies given for the Au_8–MgO–T configuration in Table 1.4).

MgO surfaces are hydrophilic, and H_2O molecules adsorb with energies of about 0.4 eV. In the vicinity of the peripheral interfacial sites of the Au cluster, the calculations showed that the adsorption energy of H_2O increases by 0.1–0.2 eV (depending on the particular site and adsorption configuration). Thus,

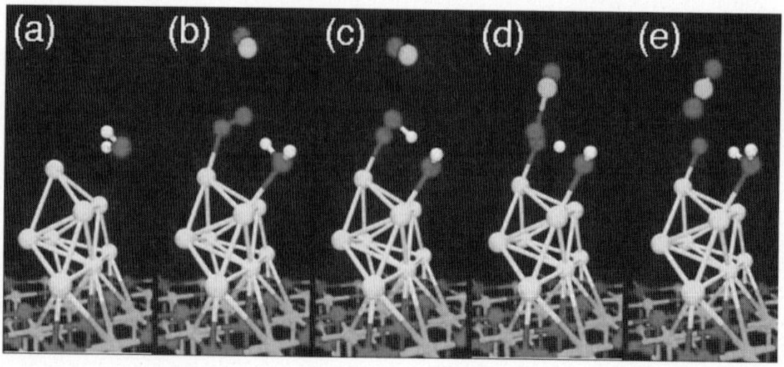

Fig. 1.80. Relaxed atomic configurations displaying several stages in the simulation of the coadsorption of H_2O and O_2 on the top facet of an Au8 cluster supported on MgO(100) and the subsequent reaction with gaseous CO to form CO_2. (**a**) The approach of an H_2O molecule to the cluster. (**b**) Coadsorbed H_2O (right) and O_2 (left) with an approaching CO. Note the preferential orientation of the H_2O and partial proton sharing. (**c**) CO-induced proton transfer resulting in formation of a hydroperoxyl-like group (left) and a hydroxyl (right). (**d**) Transition state configuration with the CO binding to the activated species. The proton is about midway between one of the oxygens of the transition-state complex (left) and the hydroxyl (right). (**e**) The proton shuttles back to reform an adsorbed H_2O, and the CO_2 product desorbs from the surface, leaving an adsorbed oxygen atom that reacts in the next step with a CO molecule to yield a second CO_2. Light grey, dark grey, white, and grey *spheres* correspond to Au, O, H, and carbon atoms, respectively

the gold cluster acts as an attractor for adsorbed water (reverse spillover, see "Spillover, Reverse Spillover, and Adlineation"). Hence, at the interface between the Au cluster and the MgO surface, peripheral sites show a high propensity to bind both H_2O and O_2 (Table 1.4). The markedly larger binding energies of the coadsorbed complex (compared to the individual adsorbates) reflect a synergetic effect, expressed through the occurrence of the aforementioned proton sharing and proton transfer processes.

To address the reactivity of O_2 coadsorbed with H_2O on the top facet of the adsorbed gold cluster, a sequence of adsorption and reaction steps that result in the catalytic oxidation of CO is displayed in Fig. 1.80. Starting from the bare Au_8/MgO system an H_2O molecule adsorbs first (Fig. 1.80a). Subsequently, an O_2 molecule is coadsorbed and the system is exposed to incident CO (see the proton-sharing configuration in Fig. 1.80b). In Fig. 1.80c, a proton transfer process, induced by the incoming gaseous CO molecule, is shown, leading to formation of a hydroperoxyl-like complex. Upon reaction of the CO molecule with the complex, the proton shuttles back toward the hydroxyl group (Fig. 1.80d), with the process culminating in the desorption of a CO_2 molecule and reformation of an adsorbed H_2O molecule that is preferentially oriented with respect to the remaining adsorbed oxygen atom (Fig. 1.80e). The above Eley-Rideal like reaction mechanism involves relatively low barriers, which were estimated from a constrained molecular dynamic approach. It was found that formation of the transition state (shown in Fig. 1.80d) entails a readily accessible energy barrier of ~0.5 eV. An added CO molecule reacts readily (a barrier of 0.1 eV) with the single adsorbed oxygen atom, and the (barrierless) desorption of the CO_2 product closes the catalytic cycle. For a peripherally adsorbed O_2 reacting with a CO molecule adsorbed in its vicinity, a Langmuir–Hinshelwood reaction barrier of 0.4 eV was found (with or without a neighboring coadsorbed H_2O molecule). The barrier for desorption of the CO_2 product is 0.6 eV under dry conditions and 0.3 eV with coadsorbed H_2O.

1.5.3 The Oxidation of CO on Small Platinum and Palladium Clusters

Platinum, palladium, and rhodium are the major materials used industrially for the catalytic oxidation of CO. This initiated a variety of studies on single crystals, on supported particles, as well as on real catalysts in order to obtain important details about this catalytic process. In this section, an overview over the reactive properties of free Pt and Pd clusters will be presented and the catalytic CO combustion reaction on clusters of these metals will be discussed. Subsequently, the activity of deposited mass-selected Pt and Pd clusters in this reaction will be analyzed in order gain insight into the details of the catalytic reaction mechanisms on these materials.

The Reactivity of Free Pt and Pd Clusters

Platinum and palladium were among the first metals that were investigated in the molecular surface chemistry approach employing free mass-selected metal clusters [159]. The clusters were generated with a laser vaporization source and reacted in a pulsed fast flow reactor [18] or were prepared by a cold cathode discharge and reacted in the flowing afterglow reactor [404] under low-pressure multicollision reaction conditions. These early measurements include the detection of reaction products and the determination of reaction rates for CO adsorption and oxidation reactions. Later, anion photoelectron spectroscopic data of cluster carbonyls became available [405, 406] and vibrational spectroscopy of metal carbonyls in matrices was extensively performed [407]. Finally, only recently, the full catalytic cycles for the CO oxidation reaction with N_2O and O_2 on free clusters of Pt and Pd were discovered and analyzed [7, 408].

Pt and Pd Carbonyl Complexes. From reactivity measurement with carbon monoxide, Cox et al. [18] determined the relative CO bonding strength to transition metal clusters:

$$\text{Pd, Pt, Co} > \text{V, Ni, Nb, Ru, Ir, W} > \text{Mo, Fe} > \text{Cu, Al}$$

This relative scale demonstrates the exceptional reactivity of clusters of platinum and palladium. The cluster–surface analogy for CO adsorption is persued in theoretical contributions, e.g., for small Pd clusters in [409]. Experimentally, Ervin and coworkers [404, 410] investigated the reactions of negatively charged Pd and Pt clusters with CO in the flow tube reactor. For clusters with more than four atoms, the measured bimolecular rate coefficients for CO adsorption were near the ion-dipole collision rate constant. Saturation coverages were determined showing that the CO coverages of the Pd clusters are, similar to the homologous Ni, typical for an 18-electron transition metal [12,31,411], whereas the saturation limits for platinum are lower, reflecting the electron deficient structures observed in condensed phase chemistry. Collision-induced dissociation measurements of CO-binding energies on Pt and Pd clusters lead to the identification of specific CO-binding sites on the clusters associated with particular carbonyl complex structures. $Pt_3(CO)_6^-$, e.g., has not only the same stoichiometry but also the same structure as the condensed phase species $Pt_3(CO)_3(\mu\text{-}CO)_3^{2-}$. Thus, the site specifity of the individual binding energies allows the assignment of distinct binding configurations to the different CO adsorbates [22, 412–414].

The electronic structure of free carbonyls of Pd_n and Pt_n have been investigated by mass-resolved anion photoelectron spectroscopy. From these measurements, the vibrational frequencies of the C–O stretch mode of the electronic ground state could be obtained [6, 405]. The C–O stretch vibration frequency found for several saturated species are reported to agree roughly with the corresponding surface data [405]. However, the stretch frequencies appear to be strongly coverage dependent, as it is the case on extended surfaces. Photoelectron spectroscopic data obtained for the monocarbonyl complexes indicate the particular importance of backdonation of d_π-electron density into the π^* orbital of CO [406, 415]. A strong π-backdonation was suggested to be one key information to understand the catalytic properties of small transition metal clusters [405]. IR-spectroscopic measurements and ab initio calculations of the CO vibrational frequencies of the Pd and Pt carbonyl complexes in rare gas matrices support the gas-phase observations [407]. The authors of this contribution [407] conclude that the frequencies for CO attached to metal centers in complexes and supported catalyst systems can be used as a probe for the local charge at the metal (M) center using neon matrix-isolated MCO^+, MCO, and MCO^- frequencies as a scale.

Oxidation of Free Pt and Pd Clusters

In their flow tube reactor studies, Hintz and Ervin found that Pt_n^- and Pd_n^- cluster anions react with molecular oxygen under addition of multiple O_2 units [416]. Whereas no monoxide formation is detected, increasing palladium cluster fragmentation with cluster size is observed upon reaction with O_2 (loss of metal atoms for Pt_3^- and Pt_4^-). The observation of cluster fragmentation induced by the O_2 reactions supports a mechanism in which the oxygen molecule dissociates exothermically on the cluster surface, releasing more energy than can be efficiently thermalized by collisions with the buffer gas [416]. Computational studies of the interaction of oxygen with neutral platinum clusters indicate that a superoxo state is first formed when the O_2 molecule approaches the metal cluster. With the dioxygen molecule moving closer to the cluster surface, the O–O bond is further stretched, and the adsorbed molecule evolves into a peroxo state, which will eventually dissociate into two adsorbed atomic states [417]. The reaction of positively charged platinum clusters with molecular oxygen was found to lead to cluster degradation in an FT-ICR experiment under single collision conditions [183, 354]. Interestingly, as a result of the fragmentation, the release of the neutral product PtO_2 was postulated for clusters with two to five Pt atoms:

$$Pt_n^+ + O_2 \rightarrow Pt_{n-1}^+ + PtO_2. \quad (1.66)$$

The oxidation reactions with nitrous oxide N_2O are even more appealing because direct oxygen-atom transfer is expected, releasing the particularly stable N_2 molecule. This process is indeed observed for the negatively charged Pd and Pt clusters [416]. As the N_2O concentration increases, additional oxygen atoms sequentially bind to the cluster to form $M_nO_m{}^-$ products. Oxygen-atom transfer from N_2O is also the sole reaction observed for positively charged free platinum clusters [183, 418]. However, the reaction rates measured for $Pt_n{}^+$ in this case are surprisingly low and for $n = 2$ and 4 the occurrence of the reaction could not be established at all [183]. An explanation for this striking observation is put forward by Schwarz and coworkers, as well as Armentrout et al. [183, 419, 420]. These authors note that the electronic ground states of cationic oxides MO^+ correlate with atomic $O(^3P)$ whereas adiabatic oxygen release from N_2O yields $O(^1D)$, meaning that the overall reaction leading to PtO^+ in the electronic ground state is formally spin forbidden. A similar kind of reaction barrier is expected to be present for the clusters $Pt_n{}^+$ as well, although the potential energy surfaces are certainly much more complicated in this case [183].

The comparison with the reactivity of platinum surfaces toward N_2O also bears some appealing similarities: early studies found adsorption to be inefficient at low temperatures, accompanied by a minor extent of decomposition [421, 422]. The application of higher N_2O pressures at 363 K, however, leads to the formation of a complete oxygen monolayer on the surface of a supported Pt/SiO_2 catalyst [423]. Schwarz and coworkers [183] note that the fact that N_2O is not dissociating on bulk platinum at low temperatures, whereas oxygen-atom transfer is effective at elevated temperatures, point to barriers being operative that are probably also important in the corresponding reactions of $Pt_n{}^+$.

CO Oxidation Catalysis on Free Pt and Pd Clusters. Platinum cluster anions $Pt_n{}^-$ with $n = 3$–6 [7], the atomic cation Pt^+ [408] and selected larger cluster

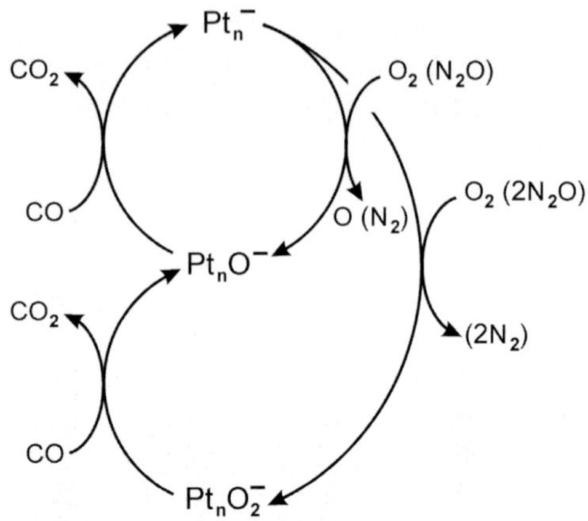

Fig. 1.81. Reaction cycles for the oxidation of CO catalyzed by gas-phase platinum cluster anions as formulated by Shi and Ervin [7]

cations $Pt_n{}^+$ (n = 5–8) [424] are found to actively catalyze the gas-phase oxidation of carbon monoxide. The previously discussed oxide complexes Pt_nO^- and $Pt_nO_2{}^-$, as well as Pt_nO^+ and $Pt_nO_2{}^+$, have been identified as intermediates in the catalytic reaction cycles. In the following, details of the corresponding catalytic reaction mechanisms, obtained by mass-spectrometric measurements, will be outlined.

The first catalytic cluster ion reaction cycle under *thermal* conditions was observed by Ervin and coworkers. This group demonstrated that the gas-phase cluster anions $Pt_n{}^-$ ($n = 3$–7) efficiently catalyze the oxidation of CO to CO_2 by N_2O or O_2 near room temperature in a combined flow tube and gas cell reactor instrument under multicollision conditions [7]. At the end of the process, the intact cluster is regenerated and each step is exothermic and occurs rapidly at thermal energies. The observed catalytic reaction cycles are summarized in Fig. 1.81.

In these experiments, the produced $Pt_n{}^-$ clusters are first thermalized by collisions with a buffer gas. When either O_2 or N_2O is introduced downstream in the flow tube, Pt_nO^- and $Pt_nO_2{}^-$ ions are formed in rapid exothermic reactions [416]. With the introduction of CO into the reaction gas cell of the instrument, the $Pt_n{}^-$ clusters reappear as observed with mass spectrometry. It was concluded that neutral CO_2 is stoichiometrically formed on the clusters and that negligible fragmentation of the metal clusters for $n \geq 4$ occurs. Collisional stabilization of the cluster oxides in the presence of buffer gas was pointed out as an important factor for the observation of catalytic activity because under single-collision conditions in the guided beam, extensive fragmentation would prohibit the reformation of the initial metal clusters. Further experiments revealed that at low energies, the cross section approaches the calculated collision limit meaning that the reactions are quite efficient. When

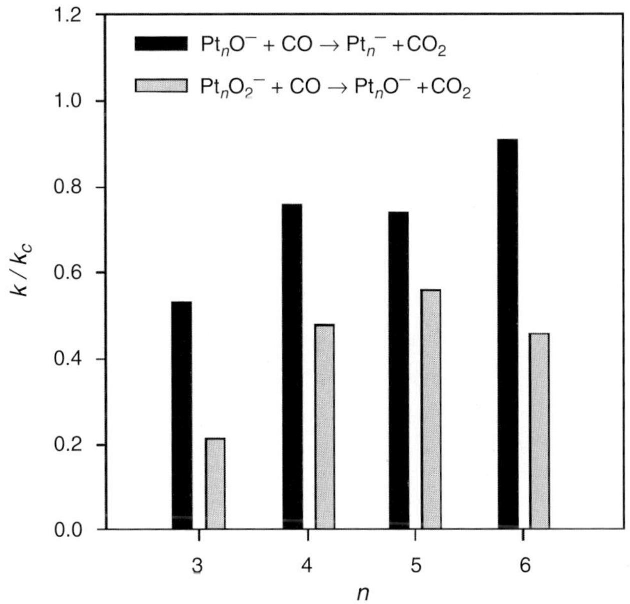

Fig. 1.82. Relative reaction efficiencies k/k_c for the conversion of CO to CO_2 by $Pt_nO_m{}^-$ cluster of various compositions (k is the measured reaction rate coefficient and k_c represents the Langevin ion–molecule collision rate coefficient) [22]

$Pt_nO_2^-$ ions are selected initially, sequential loss of oxygen atoms is observed to form two CO_2 products, as shown by CO pressure dependence studies (cf. Fig. 1.81). This observation implies that in this case, O_2 is dissociatively adsorbed on the $Pt_n{}^-$ clusters as oxygen atoms rather than chemisorbed or physisorbed as molecular O_2, as discussed for $Au_n{}^-$ in the previous section. Two other observations support this conclusion. (a) $Pt_nO_2{}^-$ ions produced by reaction of the bare cluster with either O_2 or N_2O show the same reactivity and (b) Collision-induced dissociation of $Pt_nO_2{}^-$ shows loss of oxygen atoms with no O_2 loss. The regeneration of $Pt_n{}^-$ ions at low energies proves that a full catalytic oxidation cycle can be completed at near room temperature in either a single-step or a two-step process as indicated in Fig. 1.81. Figure 1.82 shows the measured reaction efficiencies for the oxidation of the first CO molecule by Pt_nO^- or $Pt_nO_2{}^-$ ($n = 3$–6). Apparently, the presence of the second oxygen atom decreases the reactivity. Possible reasons for the lower reactivity of $Pt_nO_2^-$ might include the displacement of the more favorable binding sites for CO or the thermodynamical stabilization of the cluster oxide by the second oxygen atom. However, the reaction efficiencies for a single CO reaction are definitely larger than 40% for $n \geq 4$; so only a few collisions with CO reactants would be required for complete conversion. From this it was concluded that at near room temperatures, the gas-phase metal cluster anions are better catalysts than the supported catalysts used in current technology

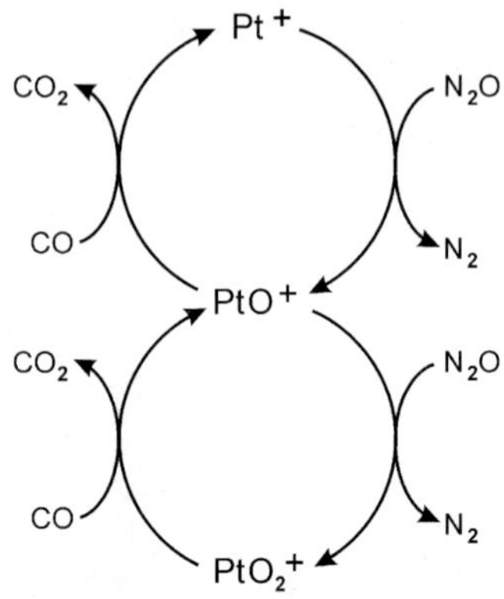

Fig. 1.83. Observed reaction cycles for the oxidation of carbon monoxide by nitrous oxide catalyzed by gas-phase atomic platinum cations [408]

for automotive catalytic converters. These are needed to be heated to high temperatures on platinum surfaces; temperatures of 400–500 K are typically required for oxidation of CO [425].

In further work of the group of Ervin, the activity of palladium cluster anions in the CO oxidation catalysis was examined and it was found that Pd_n^- also efficiently catalyzes the CO combustion reaction. The palladium clusters, however, are reported to exhibit more fragmentation than the platinum clusters, consistent with the weaker meal-metal bond strength in palladium relative to platinum [22].

The activity of Pt^+ ions in the catalytic conversion of CO to CO_2 has been investigated by Schwarz and coworkers in an FT-ICR mass spectrometer [408]. Atomic platinum cations do not react with molecular oxygen [354]. Therefore the oxygen-atom transfer reaction with nitrous oxide is used for the generation of the active intermediates PtO^+ and ${PtO_2}^+$ (Fig. 1.83). Isotopic exchange reactions with $H_2{}^{18}O$ and ${}^{18}O_2$ suggest that the structure of ${PtO_2}^+$ is rather described as a dioxide ($OPtO^+$) than as a peroxide ($PtOO^+$) or as a dioxygen complex ($(Pt(O_2))^+$). The absence of an O–O bond in this ion is confirmed by calculations using correlated ab initio methods at the CASPT2 level of theory [408].

In the presence of carbon monoxide, both oxide ions readily react, releasing carbon dioxide. Upon treating ${PtO_2}^+$ with a mixture of CO and N_2O, quasi-stationary concentrations of Pt^+, PtO^+, and ${PtO_2}^+$ are obtained in the FT-ICR experiment. Identical stationary concentrations of these ions evolve when starting from Pt^+ or PtO^+, respectively. Thus, reaction cycles as depicted in Fig. 1.83 can be formulated, describing the effective catalysis of the conversion $CO+N_2O \rightarrow CO_2+N_2$ by the gaseous platinum species. The turnover number was found to be basically limited by side reactions with background impurities. In different runs turnover numbers ranging from 80 to 300 were obtained [408]. The catalytic cycles in Fig. 1.83 are very similar to the cycles depicted in

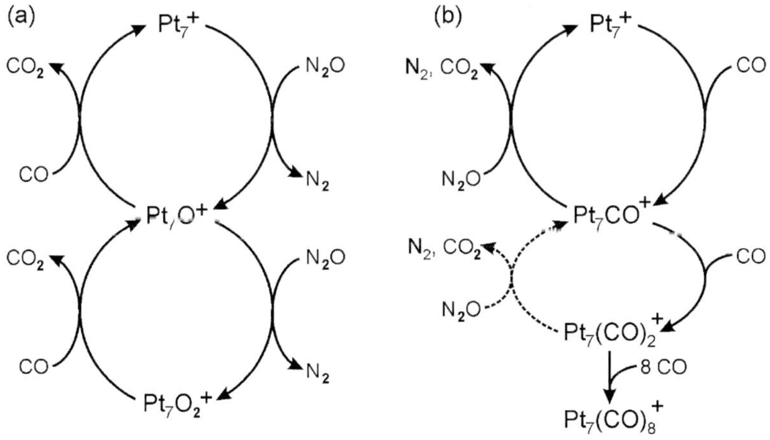

Fig. 1.84. Catalytic cycles of the Pt_7^+ ion in a 1:6 mixture of CO and N_2O. The experiment was carried out under single collision conditions in an FT-ICR instrument. (**a**) The catalytic cycles involving Pt_7O^+ and $Pt_7O_2^+$ are identical to the Pt^+ cycles in Fig. 1.83. (**b**) In addition, a cycle involving the initial formation of Pt_7CO^+ is observed. Additional CO molecules, however, increasingly poison the cluster. The conversion of $Pt_7(CO)_2^+$ back to Pt_7CO^+ might contribute to a minor extent to the catalytic activity [424]

Fig. 1.81 for the negatively charged platinum clusters. The difference results from the fact that, in contrast to Pt^+, Pt_n^- clusters react with O_2 leading to a direct formation of the intermediate dioxide $Pt_nO_2^-$ without previous formation of the monoxide Pt_nO^-, as it is the case for the oxidation reaction involving nitrous oxide.

A gas-phase catalytic reaction similar to the one described for Pt^+ in Fig. 1.83 was recently observed also for small Pt_n^+ clusters in reaction with CO and N_2O in an FT-ICR experiment [424]. Catalytic activity was observed for platinum cluster cations with 5–8 atoms, whereas Pt_9^+ and particularly Pt_{10}^+ were found to predominantly merely adsorb carbon monoxide. Detailed investigations of the Pt_7^+ reaction revealed not only the prevalence of the cycles depicted in Fig. 1.84a, but also pointed toward the role of CO adsorption in competition with the oxygen-transfer reaction with N_2O. Evidently, by addition of multiple CO molecules, some bare Pt_7^+ clusters were removed from the catalytic cycle. At high CO partial pressures in the ICR cell, the previously mass-selected Pt_7CO^+ was observed to subsequently react via two equally abundant mechanisms (cf. Fig. 1.84b): (a) Oxidation of the Pt_7CO^+ complex with N_2O back to the bare Pt_7^+ under liberation of N_2 and CO_2 or (b) adsorption of a second CO molecule to form $Pt_7(CO)_2^+$. $Pt_7(CO)_2^+$, in turn, was found to eagerly adsorb further CO molecules and only a very minor part could be oxidized back to the bare metal cluster. In this way, platinum clusters are increasingly removed from the catalytic reaction cycles at enhanced CO partial pressures. With regard to the other cluster sizes, it was observed that the catalytic activity with respect to CO oxidation by N_2O is largely determined by the reactivity of the bare cluster with N_2O: A high reactivity toward N_2O apparently shifts the cluster intensities in favor of the oxide species, effectively hindering the poisoning reaction by sequential CO adsorption on the bare cluster [424].

The Reactivity of Deposited Pt Clusters

The oxidation of CO on platinum was extensively studied by using single crystals. On highly coordinated Pt atoms, like on a Pt(111) single crystal, carbon monoxide can be oxidized at low temperature (160 K) [387, 426, 427]. In this case, oxygen atoms that are produced during the dissociation process of O_2 on the surface oxidize carbon monoxide. In such a mechanism oxygen atoms react with CO before they reach the energy minimum of the oxygen–surface interaction potential and therefore no additional activation for the actual oxidation step is needed [387]. A second reaction channel on Pt(111) involves chemisorbed oxygen atoms, which oxidize adsorbed CO at temperatures higher than 200 K. This mechanism can also be observed on low-coordinated platinum sites, e.g., the ones on a Pt(355) surface [428, 429], and the oxidation temperature of this mechanism is sensitively dependent on the character of the reactive site [428]. Real catalysts for the oxidation of CO consist, however, of highly dispersed metal particles on refractory metal oxides [430–432], and the findings from single crystals can be used to understand partly the catalysis of nanoparticles, a few nanometers in size, exposing different crystalline facets [248, 433]. For small platinum clusters, the size-dependent reactivity is illustrated in the following by presenting TPR and FTIR results for the oxidation of CO on Pt_n $(1 \leq n \leq 20)$ deposited on thin (defect-rich) MgO films.

Figure 1.85 shows the TPR spectra for the CO oxidation on supported $Pt_n (1 \leq n \leq 20)$ clusters. Since only CO_2 molecules containing one isotopically labeled ^{18}O atom are detected, oxygen is dissociated prior or during the catalytic oxidation of CO. Thus already these small Pt clusters are catalytically active. Remarkably, each cluster size reveals different oxidation temperatures, labeled α, β_1, and β_2, and a different catalytic activity as reflected in the different ion signal intensities. The main CO_2 production, labeled β_1, occurs at temperatures between 200 and 400 K. For Pt_{15}–Pt_{20} the CO_2 production is also detected at temperatures around 150 K, labeled α, while Pt_{20} clusters display an additional reaction at 460 K, labeled β_2. The different oxidation temperatures (peaks α, β_1, β_2 in Fig. 1.85) indicate different catalytic processes occurring on different active sites on the Pt clusters. Results of FTIR experiments, displayed in the insets of Fig. 1.85, support this assumption, showing only one CO absorption frequency (2,065 cm^{-1}) for Pt_8 but two absorption frequencies (2,045 and 1,805 cm^{-1}) for Pt_{20}.

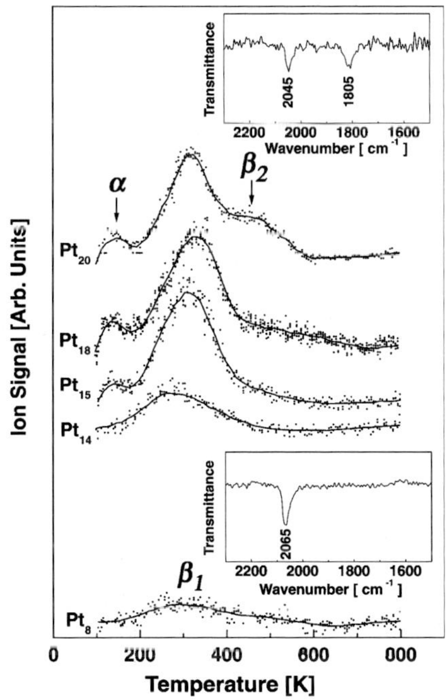

Fig. 1.85. Shown is the formation of CO_2 on selected cluster sizes after saturating the samples at 90 K with ^{13}CO and O_2 and detecting the product molecule by a mass spectrometer as function of temperature. Different CO_2 formation mechanisms (α, β_1, β_2) are labeled according to single crystal studies. The insets show measured vibrational frequencies of CO adsorbed on deposited Pd_{20} and Pd_8

Expressing the catalytic activity of the monodispersed Pt clusters as the number of catalyzed CO_2 molecules per cluster (Fig. 1.86a) and per atom (Fig. 1.86b), it is recognized that the CO_2 production increases nonlinearly with cluster size. Up to Pt_8 less than one CO_2 molecule per cluster is formed (Fig. 1.86a). The reactivity increases abruptly when going from Pt_8 to Pt_{15}, on which about six CO molecules are oxidized. For the oxidation of CO per atom (Fig. 1.86b) a maximum of 0.4 and minimum of 0.1 are observed for Pt_{15} and Pt_8, respectively. Considerable catalytic activity is also observed for Pt_1 (not shown). Whether the Pt atom is stable during the catalytic reaction is, however, on open question. The reactivity shows a local decrease for Pt_{13} clusters.

In lack of detailed theoretical investigations, these observations are first compared to the catalytic oxidation of CO on Pt single crystal surfaces. Two main mechanisms have been observed. On Pt(111) terrace sites the reaction α:

$$O_2^- + 2CO_{(\text{terrace})} \rightarrow 2CO_2 \tag{1.67}$$

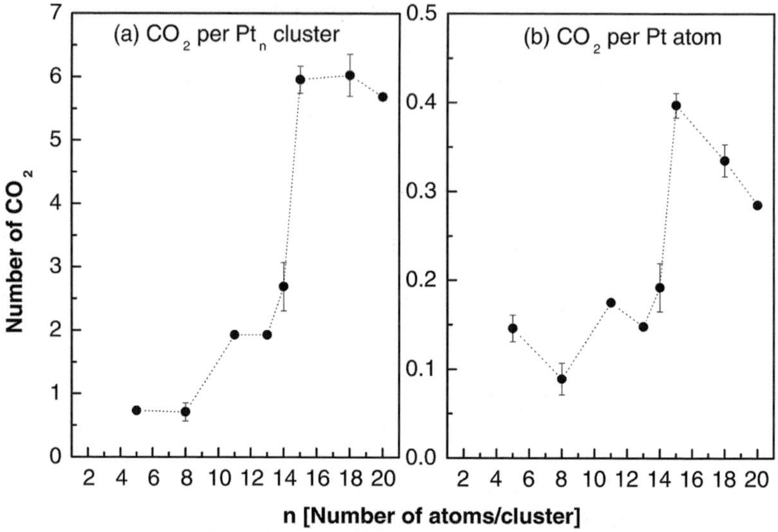

Fig. 1.86. (**a**) Total number of formed CO_2molecules per cluster as a function of cluster size and (**b**) total number of formed CO_2 per atom as a function of cluster size

has been found at 160 K involving hot oxygen atoms [387, 434]. On a stepped Pt(355) surface the reactions β:

$$\begin{array}{ll} O_{(terrace)} + CO_{(terrace)} \rightarrow CO_2 & \beta_1 \\ O_{(step)} + CO_{(step)} \rightarrow CO_2 & \beta_2 \\ O_{(step)} + CO_{(terrace)} \rightarrow CO_2 & \beta_3 \end{array} \tag{1.68}$$

at 290, 350, and 200 K, respectively, have been identified [428]. Since the oxidation temperatures on size-selected Pt clusters are in the similar range, the reaction sites are identified accordingly in this case. The throughout presence of the β mechanism indicates dissociation of O_2 on all cluster sizes. The α mechanism is only observed for larger clusters, Pt_{15}–Pt_{20}, implying that oxygen is adsorbed molecularly in an ionic state only on these species. Thus, the presence of higher coordinated and therefore less reactive atoms on the cluster (as on single crystal terraces) is suggested. This is not unexpected as the average coordination number of atoms in clusters is increasing with size. If we now compare these reactivities per atom with the ones obtained from Pt single crystal surfaces, one has to consider that the reactivity converges to zero for the latter case because of their small surface-to-bulk atom ratio. Nevertheless, on Pt(112), a reactivity per Pt surface atom of < 0.2 has been determined [429], which is more than a factor of 2 smaller than the one found for Pt_{15} (see Fig. 1.86b).

Second the results on supported clusters are compared to the above discussion of the CO oxidation on free Pt clusters. The α mechanism is also consistent with the gas-phase studies, where a relatively low reactivity toward molecular oxygen was observed and explained theoretically by the presence of a peroxo-like complex previous to the O_2 dissociation on the clusters. The main formation of CO_2 on Pt-clusters in the gas-phase studies, however, is in accordance with the β mechanism on supported Pt clusters proceeding via atomic oxygen. The sudden increase of the reactivity of supported clusters with a size larger than 8 atoms (Fig. 1.86a) may be explained by the observation that free Pt clusters of that size increasingly adsorb CO molecules (cf. Fig. 1.84).

Extensive theoretical investigations on the catalytic properties of small Pt_n clusters are missing at present and therefore a detailed understanding as in the case of small gold clusters is not available. One approach to rationalize the size-dependent observations is to relate the evolution of the reactivity with the geometric structures of the corresponding clusters. Open structures with low-coordinated atoms may be considered to be more reactive than structures with highly coordinated atoms. Local density functional (LDF) calculations (in a non-self-consistent Harris functional version) showed that free clusters up to Pt_6 are planar [435] and it is likely that they remain planar on MgO as additional energy is gained by the cluster substrate interaction for 2D structures in comparison to 3D structures. The optimized geometry of Pt_{10} with C_s point group symmetry has ten atoms arranged into three layers resembling (111) planes, as obtained by using DFT calculations including relativistic effects [436]. Self-consistent LDF [437] and embedded atom method (EAM) [438] calculations predict 3D structures for Pt_{13} clusters with cuboctahedron symmetry (O_h) and a highly asymmetric, irregular, and defected structure, respectively. The larger clusters Pt_n ($n > 13$) are considered to be 3D [438]. The change in the measured reactivity, expressed in the number of produced CO_2 per cluster (Fig. 1.86a), thus also correlates with the predicted morphological transition from 2D to 3D structures for Pt cluster sizes between Pt_6 and Pt_{10}. It is interesting to note that Pt_{13} shows a local minimum in the reactivity per atom (Fig. 1.86b), with Pt_{13} being the first atomic-shell closing of both icosahedral and cuboctahedral structures. This type of rationalization is often used in surface chemistry for explaining reactivities of different facets of single crystals. For understanding cluster reactivities, however, such an argumentation may often fail as electronic quantum size effects are dominant in the nonscalable size regime and clusters often reveal several energetically close-lying isomers being the origin of the above mentioned dynamic structural fluxionality.

The Reactivity of Deposited Pd Clusters

In the examples discussed so far, important information on the catalytic properties of small clusters were obtained by comparing the chemical reactivity of cluster model catalysts obtained by one cycle experiments with ab initio calculations. That such small clusters are indeed catalytically active during many reaction cycles cannot be deduced from such experiments. In the following, the catalytic reaction of cluster model catalysts was studied by isothermal, pulsed molecular beams in order to extract the most important catalytic quantity, which is the turn-over frequency (TOF).

After exposing the model catalysts to CO molecules by molecular beam pulses with typical widths of 40 ms (resulting in maximal effective pressures of 2×10^{-4} mbar) and at a constant isotropic O_2 pressure of 5×10^{-7} mbar, the resulting product molecules (CO_2) were detected within isothermal conditions by an absolutely calibrated quadrupole mass spectrometer. From the integral of the detected CO_2 pulses, the absolute number of formed CO_2 per cluster could be obtained. Dividing this number by the reaction time (40 ms) and the number of atoms in the clusters results in the TOFs of this reaction for various cluster sizes with an estimated accuracy of 10%. From these results, a reaction probability can then be defined by dividing the TOF by the total flux of CO (composed of the direct adsorption or adsorption through reverse spillover) on the cluster. With these two numbers, a catalytic process is defined and allows the cluster's catalytic properties to be compared to other model systems.

Size Effects

The catalytic formation of CO_2 on Pd_{30} is shown in Fig. 1.87b expressed as the number of produced CO_2 per deposited Pd-atom per second for selected temperatures and for a constant partial pressure of O_2 of 5×10^{-7} mbar and a CO-effective pressure of 1×10^{-4} mbar. Up to about 300 K no catalytic reaction occurs. Between 300 and 375 K a drastic increase in reactivity is observed, and the maximal reactivity at 375 K is about 29 CO_2/(Pd-atom s). The reactivity is then decreasing with temperature and reaches a value of ~2 CO_2/(Pd-atom s) at around 700 K. A small delayed contribution to the CO_2 formation is observed at 351 K at around 1s; the delay time is decreasing with temperature. At 450 K, it merges the main peak. The origin of this delayed contribution is the competitive adsorption of the two reactants and it was also observed for larger nanoparticles under similar experimental conditions [243, 439–442]. In fact, it disappears at temperatures (>450 K) where CO readily desorbs from the palladium clusters. The evolution of the reactivity of Pd_8 is similar to the one of Pd_{30}, although the increase is less steep and occurs between 300 and 425 K (Fig. 1.88). In addition, the maximal reactivity is with 39 CO_2/ (Pd-atom s), distinctly different.

Furthermore, the temperature of maximal reactivity is shifted by 50 K to higher temperatures. The small delayed CO_2 formation is also observed for Pd_8 at the same experimental conditions. For Pd atoms, the evolution of the TOFs as function of temperature is different with the onset of the reactivity setting in at 375 K only and reaching a maximal value of 21 CO_2/(Pd-atom s) at 510 K (Fig. 1.89). As for the other cluster sizes, the reactivity is decreasing at higher temperatures. Although Pd-atoms form CO_2 in a one-heating cycle TPR experiment at around 280 K as depicted in the inset of Fig. 1.89b, this

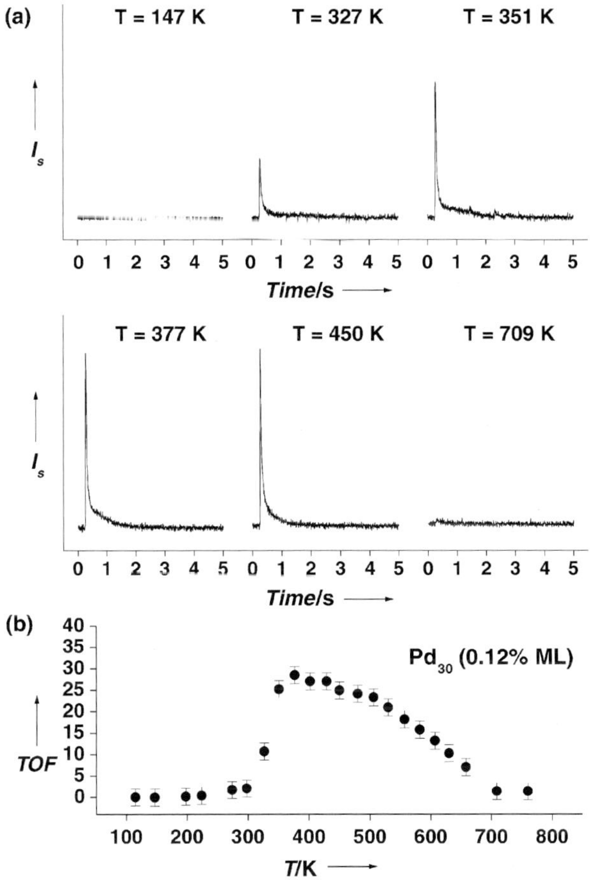

Fig. 1.87. (**a**) Shown are CO_2 transients for representative temperatures obtained for Pd_{30}/MgO(100) model catalysts at an isotropic pressure of O_2 of 5×10^{-7} mbar and an effective pressure of CO of 1×10^{-4} mbar obtained by a CO-pulse of 40 ms duration. (**b**) Shown is the evolution of the absolute TOFs (obtained from the integrals of the transients) as a function of temperature

model catalyst is not stable at 300 K after CO_2 formation and forms larger clusters. This is inferred from FTIR studies where the absorption bands of the probe molecule CO adsorbed on Pd-atoms were monitored during the CO oxidation, e.g., with increasing temperature (Fig. 1.90) [443,444]. At 90 K and at saturation coverage a main absorption band between 2,045 and 2,005 cm^{-1} (Fig. 1.90a) is attributed to two on-top bound CO molecules on the Pd atoms (Fig. 1.90b).

In the presence of O_2, the high frequency CO is efficiently oxidized as observed by the decrease of the corresponding band in the temperature range of 180–250 K. At a temperature of 300 K, a band at 1,830 cm^{-1} grows in

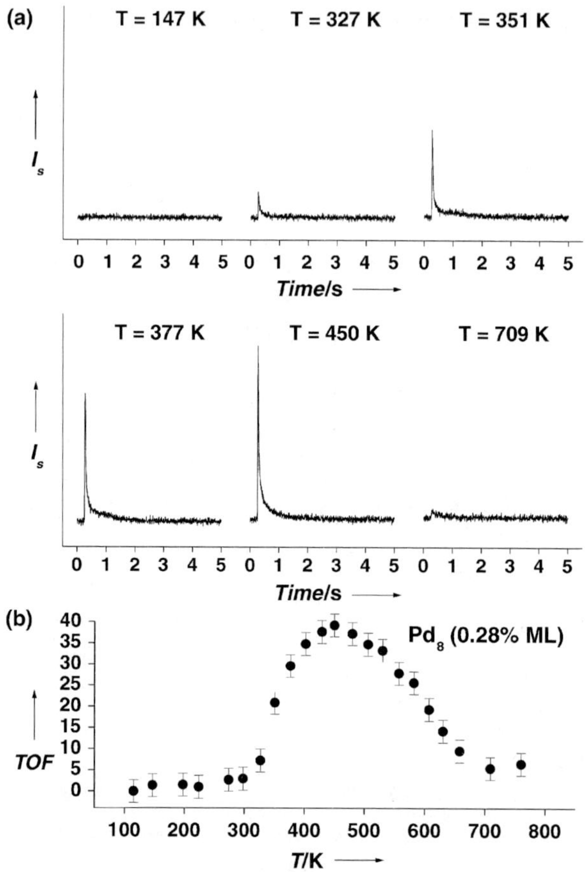

Fig. 1.88. (**a**) Shown are CO_2 transients for representative temperatures obtained for Pd_8/MgO(100) model catalysts at an isotropic pressure of O_2 of 5×10^{-7} mbar and an effective pressure of CO of 1×10^{-4} mbar obtained by a CO pulse of 40-ms duration. (**b**) Shown is the evolution of the absolute turn-over frequencies as a function of temperature

(Fig. 1.90a) and can be attributed to bridge bonded CO. This binding mode is only possible when larger clusters are formed via coalescence. In fact, the whole process has been described with ab initio calculations [444]: the low-temperature mechanism observed in the one-heating cycle experiment involves the perpendicular bonded CO and a molecular bonded, highly activated O_2 on the atom–substrate interface (Fig. 1.90b). Once the temperature of 300 K is reached and a first reaction cycle took place, the trapped Pd atoms start to migrate and coalesce to larger clusters. This sudden migration is caused by the filling of the trapping center of the atom (oxygen vacancy) by the remaining oxygen atom after the formation of one CO_2 molecule. This strongly

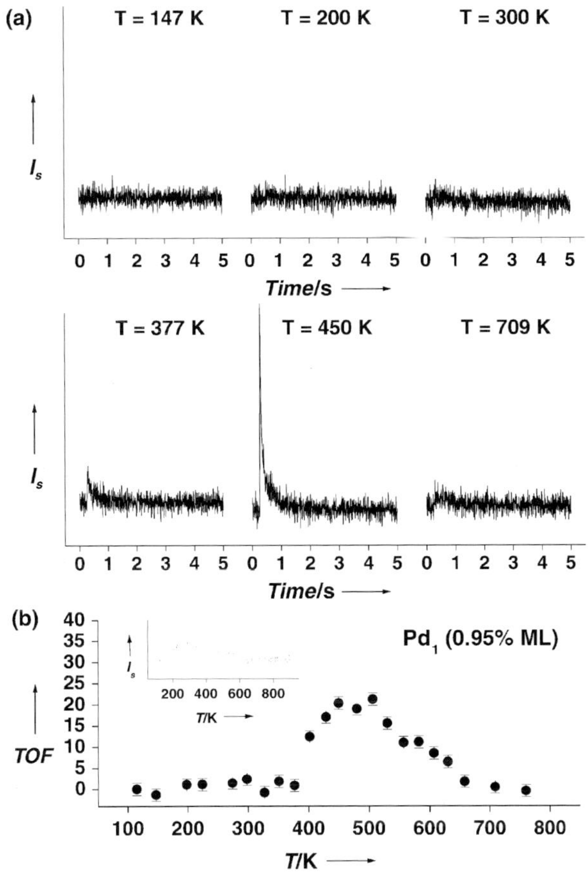

Fig. 1.89. (**a**) Shown are CO_2 transients for representative temperatures obtained for $Pd_1/MgO(100)$ model catalysts at an isotropic pressure of O_2 of 5×10^{-7} mbar and an effective pressure of CO of 1×10^{-4} mbar obtained by a CO pulse of 40-ms duration. (**b**) Shown is the evolution of the absolute TOFs as a function of temperature. The inset shows the CO_2 formation obtained by a one-heating cycle experiment

reduces the binding energy of the Pd atom with the substrate. Accepting this mechanism, it becomes evident that the catalytic formation at temperatures larger than 400 K and measured by pulsed molecular beams originates form larger Pd clusters. Thus, single Pd atoms do not withstand a first reaction cycle. These examples show that small clusters indeed show size-dependent catalytic properties, which may on the one hand be induced by the changing intrinsic chemical properties of small clusters with size but on the other hand also by the different stabilities of the clusters on the support material.

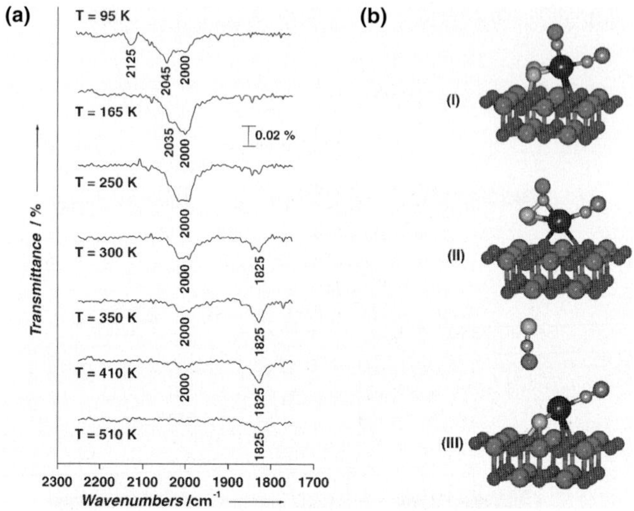

Fig. 1.90. (**a**) Shown are the CO absorption bands during the CO oxidation on Pd_1/MgO(100) model catalysts. In these experiments, the samples were annealed to the indicated temperatures and the IR-spectra were recorded at 90 K. (**b**) Shown is the reaction mechanism of the CO oxidation on a single Pd-atom adsorbed on an F-center, obtained by intensive ab initio calculations

Influence of the Substrate. The results discussed so far were obtained by depositing the clusters on defect-rich surfaces. By depositing the size-selected clusters on defect-poor MgO films, the reactivity is changing as shown in Fig. 1.91. For Pd atoms, the reactivity of the high-temperature formation of CO_2 is decreased by about a third. Surprisingly, the onset of the high-temperature formation is constant, although ab initio calculations revealed the binding energy of Pd atoms to F centers to be a factor of 2.5 larger than the one of Pd-atoms bound to terrace sites [444]. This can only be understood when accepting the mechanism described above that after the fist reaction cycle the oxygen vacancy is filled with an oxygen atom and the trapped Pd atom becomes mobile from the beginning like on defect-poor MgO surfaces. For clusters, the influence of the F-center density on the reactivity is most pronounced for Pd_8, where the reactivity is decreasing from 39 CO_2/(Pd-atom s) to 26 CO_2/(Pd-atom s). A similar mechanism as described for the Pd atoms cannot account for the observed reactivity change, as in this case the reactivity should be identical over the whole temperature range as the oxygen vacancy is filled already after the first reaction cycle. Thus, the F centers stay intact and the different reactivities are due to a change of the intrinsic cluster properties when depositing on various MgO substrates. Similar to small Au_n clusters, there might be a substantial charge transfer from the F centers to the Pd_n clusters. This would increase the charge density on the cluster and strengthen the bonding of the reactant molecules and thus

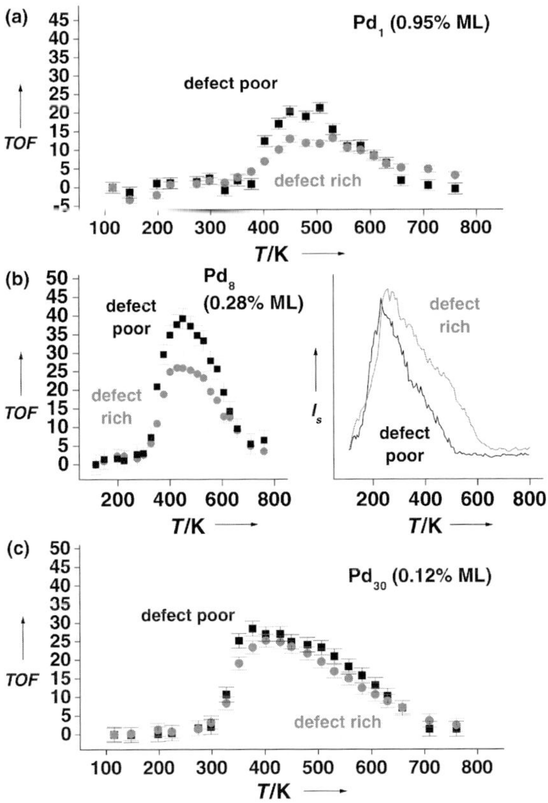

Fig. 1.91. Shown are the evolutions of the absolute turn-over frequencies as a function of temperature for (**a**) Pd-atoms, (**b**) Pd_8, and (**c**) Pd_{30} deposited on defect-poor and defect-rich films, respectively. In (**b**) the one-heating cycle experiments for Pd_8 are also shown. Note that for Pd_8 on defect-rich films, the contribution of CO_2 formed at high temperatures is increased

augment the activation energy of the catalytic process. This is confirmed by temperature programmed reaction experiments, where the formation of CO_2 is studied for the two cases. Indeed, after depositing the Pd_8 clusters on oxygen vacancies, the contribution of CO_2 formed at high temperature is clearly higher (Fig. 1.91b) [445]. For Pd_{30}, the influence of the defect sites is almost absent and the reactivity remains constant. This difference shows that a possible additional charge density is highly localized in small clusters and changes the reaction sites whereas on larger clusters, the charge density is more delocalized or extended over a few atoms close to the trapping center and therefore not accessible for the reactants. In other words, for larger clusters, the density of states at the Fermi energy is so high that charging plays a dominant role. These results suggest that the support may change the intrinsic properties of small clusters more efficiently than it does for larger clusters.

Spillover and Adlineation. In this section, we show that the reverse spillover significantly influences the reactivity of the Pd_n clusters. According to the collection zone model, each cluster exhibits a circular diffusion area, a so-called collection zone, A_c. If for instance a CO molecule adsorbs within the collection zone, it will reach the cluster by surface diffusion and the effective CO flux is increased. This effect is also called reverse spillover (see "Spillover, Reverse Spillover, and Adlineation"). The collection zone is dependent of the cluster radius, r_c, and the mean diffusion length, X_s. As the diffusion length is temperature dependent, the collection zone also depends on temperature and decreases with increasing temperature. It is also important to note that, the collection zone may be limited by the cluster coverage. As an example, for coverages of 0.28% ML (Pd_8) and 0.12% ML (Pd_{30}) and assuming closed packed structures as well as uniform cluster distributions, the uncovered MgO surface area per cluster (e.g., largest possible collection zone without overlap) is approximately 40 nm^2 for both sizes. When increasing cluster coverage to, e.g., 0.65% ML (Pd_8) and 0.28% ML (Pd_{30}), this area is reduced to about 17 nm^2.

With this model, we can distinguish between a direct flux (F_{direct}, the reactant impinges the cluster directly from the gas phase) and an indirect flux coming from the collection zone, F_{diff} (or flux from the reverse spillover). Assuming the overlapping of the adjacent collection zones as the important boundary condition (it is fulfilled at low surface temperatures and relatively high cluster coverages) the following equations can be used for calculating the ratio $F_{\text{diff}}/F_{\text{direct}}$.

$$\frac{F_{\text{diff}}}{F_{\text{direct}}} = \frac{\alpha_g}{A_C - 1}$$
$$\alpha_g = n_c \pi r_c \left(r_c + 2\alpha X_s P\right), \tag{1.69}$$

with α_g being the global sticking coefficient and n_c representing the number density of clusters. α is the sticking probability of CO, and P represents a solution of the differential diffusion equation. By multiplying (1.69) with the direct flux (1.70), F_{diff} and F_{total} are obtained as function of the substrate temperature and of the cluster size.

$$F_{\text{direct}} = \pi r_c^2 J_i \tag{1.70}$$

The ratio, $F_{\text{diff}}/F_{\text{direct}}$, for CO and the clusters Pd_8 and Pd_{30} is plotted for two cluster coverages in the inset of Fig. 1.93. For low temperatures, the ratio is almost constant as F_{diff} is constant (limited by the cluster coverage). For higher temperatures, F_{diff} changes as it is not cluster coverage limited and as X_s decreases with increasing temperature. Note that the direct flux onto a cluster is independent of the substrate temperature.

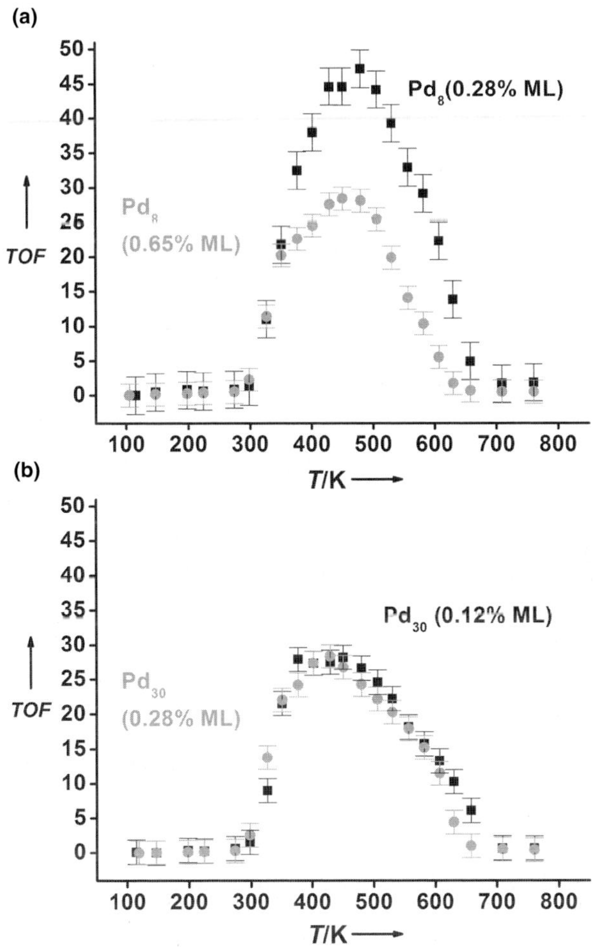

Fig. 1.92. Shown are the evolutions of the TOFs as a function of temperature for different surface coverages of the Pd_8 (**a**) and Pd_{30} (**b**) clusters. Note that the reactivity is notably changing for the Pd octamer

To study the effect of the reverse spillover, the reactivity for the CO oxidation was measured for Pd_8 and Pd_{30} at two different coverages. The results of the TOF measurements (Fig. 1.92) exhibit distinct differences for the two cluster sizes. For Pd_8, the evolutions of the measured TOFs are different for the two coverages with the low-coverage sample being more reactive. For Pd_{30}, the reactivities are similar for both coverages. For Pd_8, this reflects the influence of the reverse spillover with the higher reactivity for low coverages due to the enhanced indirect flux. This effect should be corrected if the TOFs are normalized to the total flux, which is estimated in the upper paragraph. The obtained reaction probabilities (Fig. 1.93) should then be coverage independent. This is to a rough estimation true for Pd_8. For Pd_{30}, the evolution of the

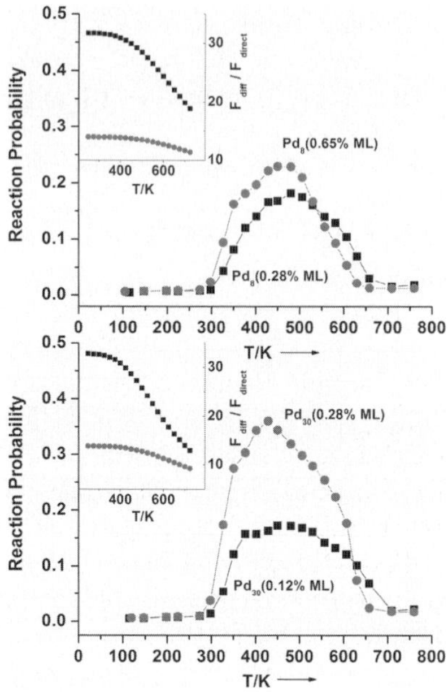

Fig. 1.93. Shown are the reaction probabilities as a function of temperature for different surface coverages of the Pd_8 and Pd_{30} clusters. The insets depict the calculated ratio of $F_{\mathrm{diff}}/F_{\mathrm{direct}}$ as a function of temperature

reaction probabilities are, however, strongly dependent of the cluster coverage, showing higher reaction probabilities for higher coverages. This observation can be understood when assuming F_{direct} and F_{diff} not having the same effect on the reaction. In fact, the increased probability for high coverages indicates F_{diff} to be less important.

1.5.4 The Reduction of NO by CO on Pd Clusters: Cooler Cluster Catalysis

During the last decades, the NO+CO reaction has been studied in great detail, resulting in a tremendous number of studies, both experimental and theoretical (see [446–449] and references therein). The reason for this interest is mainly the aim to reduce air pollution. Nitrous oxides NO_x (NO, NO_2, etc.) are among other molecules like O_2, CO, H_2, and hydrocarbons present in the exhaust gases from internal combustion engines. The NO_x species need to be removed because they undergo photochemical reactions with hydrocarbons and oxygen and produce ozone in the atmosphere. To reduce the toxic properties from car exhausts, the first catalysts where manufactured at the beginning of the 1970s and the first three-way catalysts (TWC) where built at the end of that decade. The idea of the TWC is to remove CO and NO_x and to oxidize unburnt hydrocarbons [446–448]. Nowadays, the commercial catalysts contain rhodium as active part for the reduction of NO to N_2 ($2NO + 2CO \rightarrow N_2 + 2CO_2$). Rh reduces NO with high efficiency and is so far the most active catalyst for the reaction. A number of experimental and theoretical studies revealed a detailed picture of the reaction mechanism [449]. With this knowledge, one aims to find cheaper and more abundant metals instead of rhodium. Palladium is considered to be the most promising replacement. It is more durable at higher temperatures and thus catalysts which contain Pd instead of Rh could be positioned closer to the engine.

The CO + NO reaction on Pd single crystals and supported particles has been investigated in detail too [441, 442, 450–461]. In short, the reaction is structure sensitive and the activity is increasing from more open surfaces to closed packed ones [Pd(100) < Pd(110) < Pd(111)] [455, 459, 460]. For supported Pd particles, the efficiency of the CO_2 formation changes with clusters size, and clusters with the highest density of closed packed facets are most active [441, 442, 457, 459]. It was found that adsorbed NO dissociates on the surface, resulting in two different types of adsorbed N_a, with one kind stronger bound to the surface than the other one. The more strongly bound N_a is poisoning the surface. To understand the structure sensitivity, it is important to know that the relative concentration of poisoning N_a is decreasing from Pd(100) to Pd(111) [459]. All these studies dealt with nanometer-sized particles consisting of hundreds to thousands of atoms.

In the following experimental results and the reaction mechanisms of the CO + NO reaction on small, size-selected supported Pd_n clusters with $n \leq 30$ will be summarized. Before, however, the reactions of free Pd clusters will be presented briefly and a recent experimental survey will be reviewed which presents the first example of the homogeneous catalysis of the reaction of NO with CO in the gas phase with atomic transition metal cations serving as catalysts [462].

The Reactivity of Free Palladium Clusters and Atomic Metal Ions

Although nothing is known so far about the reduction of NO by CO on free palladium clusters in the nonscalable size regime, a series of interesting results with gas-phase palladium clusters have been reported [463]. Reactions with several molecules, such as H_2, D_2, N_2, CH_4, CD_4, C_2H_4, C_2H_6, have demonstrated strongly size-dependent reactivity patterns. Small Pd clusters can bind up to three deuterium atoms per Pd atom in the clusters. Strong discontinuities in D_2 uptake may be an indication of cluster structure differences. The observed general reactivity trends were H_2, $D_2 > N_2 > C_2H_4 > CH_4$, CD_4, C_2H_6. In several theoretical studies, the adsorption of NO on clusters of palladium was investigated [464–467]. Atomic palladium cations were observed to react with NO by second order ionization according to the mechanism [462]:

$$Pd^+ + NO + NO \rightarrow NO^+ + PdNO. \quad (1.71)$$

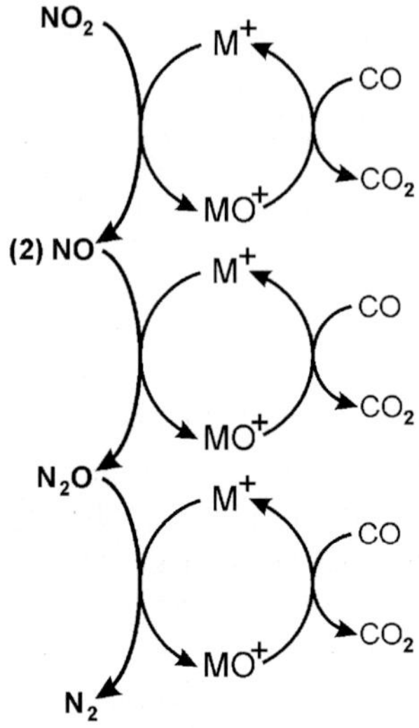

Fig. 1.94. Catalytic cycles for the homogeneous reduction of nitrogen oxides by carbon monoxide mediated by various atomic transition metal cations M^+. The ions Fe^+, Os^+, and Ir^+ were found to be most effectively performing the outlined reaction cycles out of 29 investigated transition metal elements. Adapted from [462]

This reaction is driven by the formation of the stable metal nitrosyl molecule PdNO. Pd^+, as well as Cu^+, Rh^+, Ag^+, Cd^+, and Pt^+ that also induce NO ionization, thus do not catalyze the reduction of nitric oxide.

Catalytic activity in the homogeneous gas-phase conversion of nitrogen oxides and carbon monoxide to nitrogen and carbon dioxide was observed to be most effective for the atomic ions Fe^+, Os^+, and Ir^+ out of the investigation of 29 different transition metal cations M^+ [462]. The overall catalytic scheme that was established in this study consists of the three catalytic cycles shown in Fig. 1.94. The catalysis occurs in two steps in which NO is first reduced to N_2O. An analogous three-step catalytic reduction of NO_2, in which NO_2 is first reduced to NO, was also discovered. The three cycles in Fig. 1.94 were characterized with laboratory measurements of reactions of each of the three nitrogen oxides NO_2, NO, and N_2O with the different transition metal ions in an inductively coupled plasma/selected-ion flow tube tandem mass spectrometer [462].

The Reactivity of Deposited Pd Clusters

The reduction of NO by CO was studied on small supported palladium clusters with sizes up to 30 atoms [468]. Combined TDS and pulsed molecular beam studies showed small clusters to stay indeed active during a catalytic process and to be catalytically active at temperatures 100 K below those observed for larger palladium particles [441, 442, 454] and bulk systems [459]. In

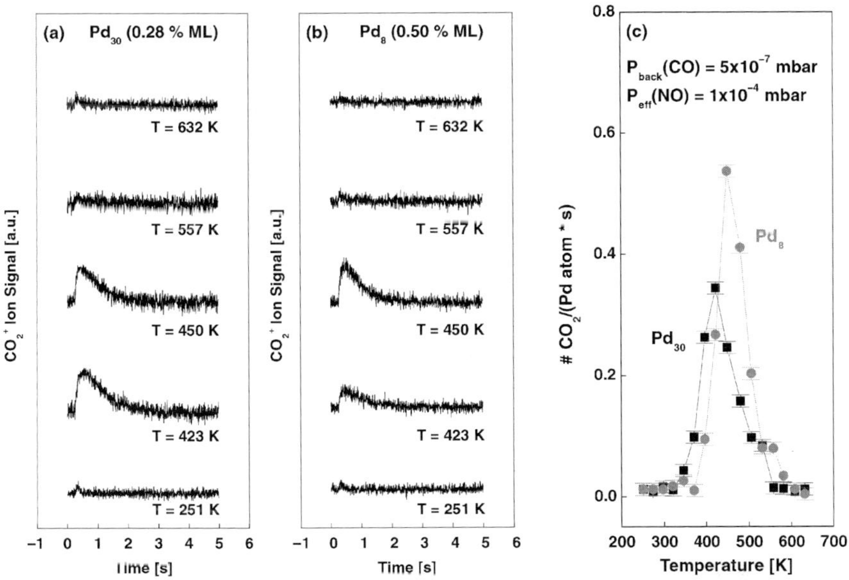

Fig. 1.95. CO_2 formation measured for Pd_{30} (**a**) and Pd_8 (**b**) at several temperatures. These CO_2 transients were measured at an isotropic CO pressure of 5×10^{-7} mbar CO and an NO pulse with a width of 100 ms, resulting in an effective pressure of 1×10^{-4} mbar. From the calibrated integral of the CO_2 peak and by dividing by the number of deposited atoms and the pulse width, the turn-over frequencies (TOFs) at various temperatures are obtained. The evolution of the absolute TOFs as a function of temperature are shown in (**c**). Pd_8 is more reactive, the temperature of maximal reactivity, however, is 30 K higher [468]

combination with FTIR studies information on the reaction mechanisms was obtained.

Catalytic Reactivity. The size-dependent cluster catalysis was first studied by pulsed molecular beams. In these experiments, a nitric oxide molecular pulse is injected onto the cluster catalysts and the product molecules CO_2 and N_2 are quantitatively detected by a mass spectrometer as a function of cluster size, temperature, and CO background pressure [468]. The catalytic formation of CO_2 on Pd_{30} and Pd_8 is shown in Fig. 1.95a and b for selected temperatures and for a constant CO partial pressure of 5×10^{-7} mbar and an NO effective pressure of 1×10^{-4} mbar. The width of the NO pulse was 100 ms. Pd_8 and Pd_{30} show almost no catalytic reactivity up to about 390 K. For Pd_{30}, maximal reactivity is observed at 420 K whereas Pd_8 is most reactive at 450 K. At higher temperatures, the formation of CO_2 decreases. The CO_2 formation on both cluster sizes at temperature of maximal reactivity is stable even after hundreds of NO pulses.

Since the mass spectrometer is calibrated absolutely, from these data, absolute turn-over frequencies as a function of cluster size and temperature

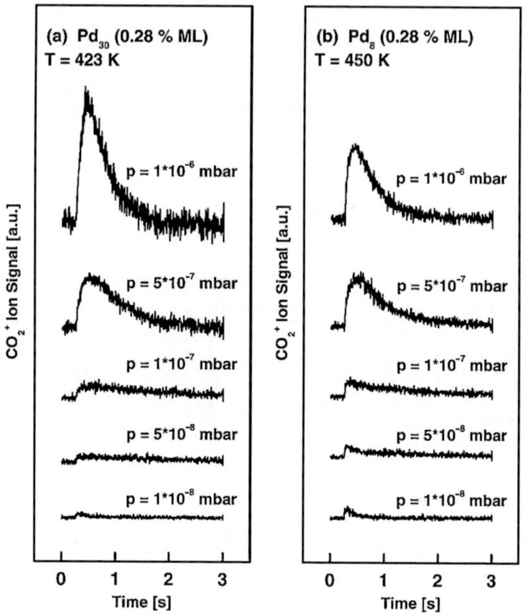

Fig. 1.96. CO_2 formation measured for Pd_{30} (**a**) and Pd_8 (**b**) at 423 K and 450 K, respectively. These CO_2 transients were measured at varying isotropic CO pressures (1×10^{-8} to 1×10^{-6} mbar) and an NO pulse with a width of 100 ms, resulting in an effective pressure of 1×10^{-4} mbar. As expected, the reactivity is increasing with CO pressure and saturation is not yet attained at these experimental conditions

were extracted by integrating the CO_2 peaks and dividing the obtained numbers by the number of deposited atoms and the duration of the NO pulse. The evolution of the TOF as a function of temperature is shown in Fig. 1.95c. These small clusters exhibit a size-dependent catalytic reactivity with Pd_8 ($0.54\,CO_2$ atom^{-1} s^{-1}) being significantly more reactive than Pd_{30} ($0.34\,CO_2$ atom^{-1} s^{-1}) whereas the latter reveal maximal reactivity at lower temperatures.

Interestingly, Pd_4 shows no measurable catalytic reactivity under any of these experimental conditions. Before attributing these differences to intrinsic cluster properties (quantum size effects), one has to assure that they cannot be explained by purely geometric effects related to the increased proportion of surface atoms with decreasing cluster size. As an example, a truncated octahedron of order 1 contains 38 atoms where the surface is composed of 32 atoms, representing 84% of the total number of atoms. In contrast, for Pd_8, all atoms are on the surface. Thus, using these geometric arguments, an increase in reactivity of only ~19% is expected for Pd_8, which is, however, significantly smaller than the one observed experimentally (~59%). For both cluster sizes, the catalytic reactivity was measured with increasing CO background pressure, and as expected the reactivity is increasing (Fig. 1.96).

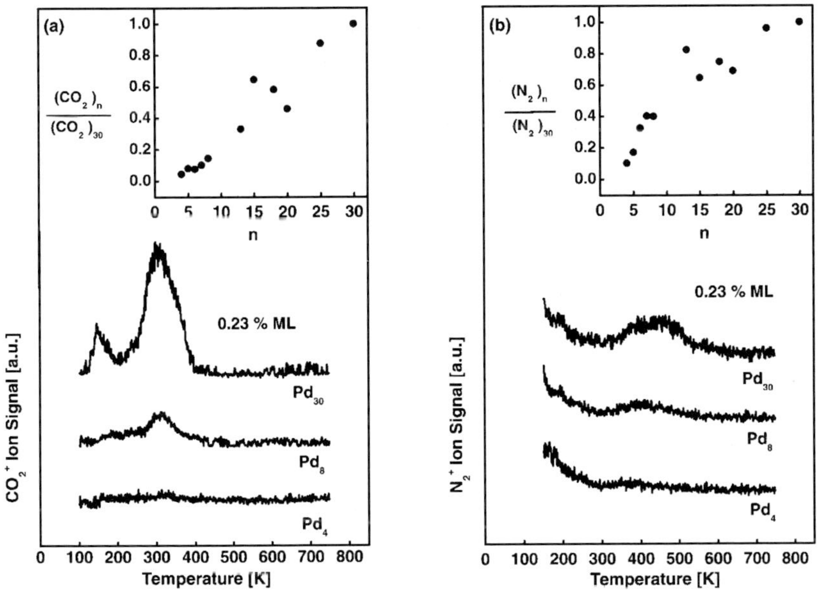

Fig. 1.97. Formation of CO_2 (**a**) and N_2 (**b**) for Pd_4, Pd_8, and Pd_{30} measured by a temperature programmed reaction (TPR) experiment. In these experiments, the cluster samples were first dosed at 90 K to CO (1 Langmuir) and subsequently to NO (1 Langmuir). The reactivities for various clusters, expressed as the normalized number of product molecules per cluster, are shown in the insets

In fact, it is known that a shallow reactivity maximum is reached at a mole fraction of the two reactants of 0.5 [469]. This shows that these experimental conditions are far from the optimum. In addition, for Pd_8 (Fig. 1.96a) and Pd_{30} (Fig. 1.96b), the peak width of the CO_2 transients is decreasing with increasing isotropic CO pressure. This can be understood by the competitive adsorption of the two reactants. When increasing the pressure of the less abundant reactant, in this case CO, the NO molecule is replaced more efficiently and the reaction can take place. Thus the reaction probability is increased leading to narrower transients.

Reaction Mechanism. To understand the size-dependent reactivity of palladium clusters on MgO surfaces in more detail, combined Fourier transform infrared (FTIR) and thermal desorption (TDS) studies were performed. The cluster model catalysts were first exposed to 1 Langmuir of ^{13}CO at 90 K and subsequently to the same amount of ^{15}NO. Upon linearly heating the model catalysts, the product molecules $^{13}CO_2$ and $^{15}N_2$ were detected by mass spectrometry as a function of the cluster size for Pd_n with $n \leq 30$. While for Pd_4, the formation of $^{13}CO_2$ is negligible, Pd_8 and Pd_{30} form $^{13}CO_2$ at 305 K or 145 K and 300 K, respectively (Fig. 1.97).

The reactivities expressed as the number of $^{13}CO_2$ per deposited cluster are shown in the insert of Fig. 1.97a. Clusters with up to four atoms are inert, whereas for larger clusters, the number of produced CO_2 increases non-monotonically with cluster size. Pd_8 and Pd_{30} form $^{15}N_2$ between 300 and 500 K (Fig. 1.97a). For Pd_4, traces of $^{15}N_2$ may be present below 400 K. These results also show that for this reaction ($2NO + 2CO \rightarrow 2CO_2 + N_2$), a minimum cluster size is indeed required and that the critical sizes are between Pd_5 and Pd_8. Furthermore, by combining FTIR and TDS results, the reaction mechanisms at 145 and 300 K were obtained [244] (Fig. 1.98). The main reaction mechanism, which is observed under CO-rich conditions for all reactive cluster sizes and occurring at around 300 K involves molecularly adsorbed CO and an adsorbed oxygen atom resulting from the dissociation of NO. This Langmuir–Hinshelwood type mechanism is also indicated by the presence of a strong CO absorption band at 2,055 cm^{-1} (Fig. 1.98a, inset) and the fact that NO is dissociated on the clusters (formation of N_2). The low-temperature mechanism at 145 K is observed for Pd_{20}–Pd_{30} only, and can be ascribed to the direct reaction of molecularly bonded CO with an absorption band at 2,055 cm^{-1} and a threefold bound NO [244] with an absorption band at 1,550 cm^{-1} (Fig. 1.98a, inset). Interestingly, under NO-rich conditions, the formation of CO_2 is largely inhibited (Fig. 1.98b), as the cluster is poisoned mainly by molecularly adsorbed NO with an absorption band at 1,720 cm^{-1} (Fig. 1.98b, inset). This observation is consistent with the results obtained by the pulsed molecular beam experiments. The clean MgO films are completely inert for this reaction (Fig. 1.98c).

Cluster Catalysts, Nanoparticles, and Single Crystals. In the following, the catalytic reactivity of the clusters is compared with published data on Pd nanoparticles [441, 442, 457, 469, 470] and single crystals [453, 457]. Most importantly, the model catalysts consisting of small palladium clusters reach maximal reactivities at temperatures which are 100 K lower than those for palladium particles with sizes of 2.5 nm and larger [441, 442, 470] (Table 1.5). Although the reaction temperatures depend on the partial pressures of the reactants in the experiments [469], this interesting trend is also observed in independent TPR experiments for the different systems. The reaction temperatures for the various systems and experimental conditions are summarized in Table 1.5. Before comparing the reactivities of the different systems, we note that the absolute TOFs are strongly dependent on the experimental conditions. As the experiments reported here were carried out at NO pressures (10^{-4} mbar) between those of nanometer-sized Pd particles on MgO(100) [441] (6×10^{-8} mbar) and those of nanometer-sized Pd particles on alumina [454] or Pd single crystals [455] (6 mbar) an accurate comparison is not possible. If the catalytic reactivity is expressed as the reaction probability (i.e., the conversion factor) with respect to the flux of NO, this comparison is more meaningful. Here the probabilities are calculated by taking the ratio of the number of CO_2 molecules produced in a single pulse per Pd atom and the number of NO in a beam pulse reaching a single Pd atom. The obtained values for the cluster

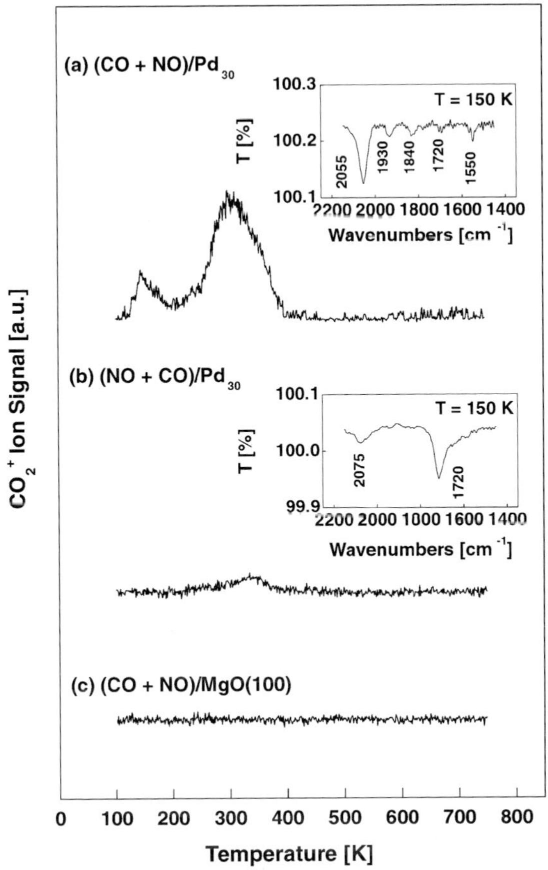

Fig. 1.98. Shown is the formation of CO_2 on Pd_{30} obtained by a TPR experiment for CO-rich conditions (CO is dosed prior to NO) with the formation of CO_2 at 145 and 300 K (**a**). The same for NO-rich conditions (NO is dosed prior to CO) with a small formation of CO_2 at around 350 K (**b**). Note that a clean MgO film is completely inert (**c**). The insets show the FTIR spectra for the corresponding conditions taken at 150 K

model catalysts are in the range of 10^{-4}. This reaction probability is about four orders of magnitude smaller than the one for nanoparticles on MgO(100) but one order of magnitude larger than for nanoparticles on alumina or for Pd single crystals.

Low-Temperature Catalysis. Next, the results for Pd_{30} and those for single crystals and Pd nanoparticles are compared in detail in order to understand the low-temperature catalysis on clusters. On Pd_{30}, dinitrogen is produced as a broad peak centered at around 450 K, and CO_2 gives rise to two sharper peaks at 300 and 145 K. On Pd nanoparticles supported on alumina or silica [457], nitrogen recombination is observed at 500 and 680 K, which is 50–230 K

Table 1.5. Reaction temperatures for maximal formation of CO_2 and N_2 measured in single cycle TPR experiments and measured under steady-state conditions or using pulsed molecular beams

	T_{max} (TPR): CO_2	T_{max} (TPR): N_2	T_{max} (TOF): CO_2
Pd_8/MgO(100)/ Mo(100)	305 K	∼400 K	450 K
Pd_{30}/MgO(100)/ Mo(100)	145/300 K	∼450 K	420 K
Pd/MgO(100) Pd particles 2.8 nm			520 K [441, 442]
Pd/Al_2O_3/Ta(110) Pd particles 2.5 nm		500 + 650 K [454]	> 650 K [454]
Pd(111)		450 + 560 K [454]	> 650 K [454]
Pd(100)	480 K [450, 451]	500 + 700 K [450, 451] 480 + 580 K	> 650 K

Note 1. Note that for both experimental schemes, the cluster catalysts are most reactive at temperatures at least 100 K smaller than for single crystals or Pd nanoparticles

higher than for the cluster model catalysts. This behavior is similar to the one for open Pd surfaces like Pd(100) [455] or Pd(320) [453] where a loosely bound nitrogen adspecies is involved for the formation of N_2 at 500 K, whereas the N_2 formation occurs via a strongly bound nitrogen adspecies at 600–700 K. Interestingly, the second mechanism takes place at temperatures much higher than the CO_2 production (500 K) [453]. This is the reason why at steady state conditions the strongly adsorbed nitrogen poisons the catalyst. In contrast, for Pd(111), the high-temperature peak is almost absent and even disappears when the reactants are dosed at low temperature (as in the present experiments). The very small amount of the strongly bound nitrogen on Pd(111) is at the origin of the higher reactivity of this surface [454]. Interestingly, the behavior of Pd_{30} clusters is much more similar to that of a Pd(111) surface than it is to Pd particles in the nanometer-size range. The similarity of the character of adsorption sites on these tiny clusters to those of Pd(111) is also manifested by the IR spectra of adsorbed NO. NO adsorbed on Pd (111) shows two bands near 1,560 (main peak above RT) and 1,740 cm^{-1} (main peak below RT) corresponding to threefold and bridge sites, respectively [457]. On Pd_{30}, a strong absorption is observed at 1,725 cm^{-1} at low temperature while above RT an absorption band appears near 1,590 cm^{-1}. On Pd(100), NO gives only one absorption band that shifts from 1,630 to 1,672 cm^{-1} when the temperature decreases from 500 to 100 K. When comparing the adsorption behavior and reactivity of Pd_{30} with Pd(111) further, distinct differences are, however, observed. First a very low-temperature CO_2 peak at 145 K is observed on Pd_{30} but not on Pd(111). As shown before, the corresponding reaction mechanism involves molecularly bound CO and NO. This mechanism is not relevant for the catalytic action observed by the pulsed molecular beam experiments as at such low temperature no CO_2 is observed. Furthermore, desorption of CO and NO from the small clusters is observed at much lower temperatures in comparison to the Pd(111) surface (Pd_{30}: ~300 K for CO and 370 K for NO). This may either indicate smaller intrinsic binding energies of these molecules with the cluster or higher local densities of the adsorbates where the larger repulsion leads to lower desorption temperatures. In addition, the energy of formation of N_2 on the cluster is distinctly smaller than on extended surfaces and nanoparticles with crystalline facets. This may be due to a decreased interaction of atomic nitrogen atoms with the clusters and/or an increased dissociation probability of NO resulting in high local N_{ad} densities and a concomitant strong repulsion favoring N_2 formation. Thus, for small clusters poisoning through N_{ad} is neutralized at distinctly lower temperatures in comparison to surfaces of bulk palladium.

This example shows that clusters consisting of only a few atoms are catalytically active at relatively low temperature and stay active during the process. A minimum cluster size of 5 atoms is needed for the catalytic reaction to take place. The catalytic properties of these small clusters are different from that of nanoparticles or extended surfaces. With the mechanistic details obtained for the CO + NO reaction on small Pd_n ($n \leq 30$), the catalytic process and the measured TOF can be described as follows. Below 300 K, very small amounts of CO_2 are formed for Pd_{20}–Pd_{30} only and this formation is described by a mechanism in which CO is directly oxidized by molecularly bonded NO. This mechanism is not efficient in the catalytic process. The formation of CO_2 above 400 K is orders of magnitude larger and the corresponding reaction mechanism is described by the oxidation of CO by adsorbed oxygen atoms, O_{ad}, resulting from the dissociation of NO. This is the mechanism relevant in the catalytic process. The much reduced reaction temperatures for cluster materials in comparison to larger particles or single crystals may on the one hand be explained by the fact that clusters consist almost exclusively of surface atoms and show structural fluxionality. This means that oxygen cannot be accommodated in the bulk, which may largely reduce the activation energy for the formation of CO_2. On the other hand and more important, the formation of N_2 on the cluster materials at low temperatures prevents poisoning of the model catalysts. Only when N_2 is formed can the reaction cycles take place.

1.5.5 The Polymerization of Acetylene on Pd Clusters

The Reactivity of Free Clusters

Atomic metal ions are known to activate hydrocarbons as described, e.g., in the reviews by Schwarz and Schröder [471, 472]. Cox and coworkers have also studied palladium and platinum clusters interacting with a series of alkanes and aromatics [463, 473, 474]. In the case of CH_4 on Pt_n with up to $n = 24$ atoms, activation was realized for the first time on an unsupported metal cluster, and the reaction had a distinct cluster size dependence, with Pt_2–Pt_5 being most reactive. Charged ${Pt_n}^{+/-}$ clusters ($n = 1$–9) react with methane under single collision conditions through activation of the C–H bonds leading to the subsequent elimination of molecular hydrogen H_2, to form the final metal carbene complex ${Pt_n}^{+/-}$ CH_2 [475]. The cation cluster reactions were found in general to proceed more than one order of magnitude faster than the anionic clusters. The platinum tetramer anion is unique in this respect, reacting more efficiently than the corresponding cation. Indications for a correlation of reactivity with the availability of low coordination metal atoms were discussed by Trevor et al. [474]. Thus, it was speculated that low coordination metal atoms activate CH_4 more readily than closed packed metal surface atoms [476]. The propensity of small Pd clusters to activate methane in a similar manner was also confirmed theoretically [477–479].

The CH_4 reactions with gas-phase platinum clusters yielded $Pt_nC_{1,2}H_y$ species but larger clusters retained more hydrogen [474]. However, platinum clusters reacting with larger alkanes caused extensive dehydrogenation down to a C/H ratio of nearly 1, indicating that aromatization of certain alkanes was facile and cluster size dependent. Mono- and diadducts of alkanes with Pt_2–Pt_4 were extensively dehydrogenated. Similar, strongly size-dependent dehydrogenation patterns were observed for the reactions of CH_4, CD_4, C_2H_4, and C_2H_6 with Pd clusters [463].

For Pt_2–Pt_8 clusters reacting with *n*-hexane and 2,3-dimethylbutane, the degree of dehydrogenation increased with cluster size. In the case of cyclohexane, dehydrogenation to benzene occurred. Benzene itself was dehydrogenated on Pt_3 and larger clusters [473]. Qualitative comparisons of rates indicated that these clusters have at least the same reactivity for reversible C–H bond breaking in benzene as a clean surface of Pt(111). However, the analogy with crystal surfaces broke down when considering the seemingly excessive number of adsorbed aromatic moieties, more than could be accommodated considering steric factors.

Dehydrogenation reaction steps are characteristic for metal ion- or metal cluster-mediated cyclization reactions in the gas phase, leading eventually to the formation of stable aromatic complexes. The most general example is the gas-phase trimerization of ethylene catalyzed by ${Fe_4}^+$, W^+, and U^+, which proceeds in a consecutive manner [25, 481, 482]. Fig. 1.99 displays the corresponding general catalytic reaction cycle. The sequence starts with the formation of a cationic metal–acetylene complex via dehydrogenation of ethylene by M^+ [480]. In the next step, which is often rate-determining, the acetylene complex undergoes dehydrogenation of a further ethylene molecule to yield $MC_4{H_4}^+$. With the third addition of ethylene, benzene complexes are formed via the loss of H_2. However, although the latter reaction step is very exothermic, the heat of reaction liberated is usually not enough to overcome the large

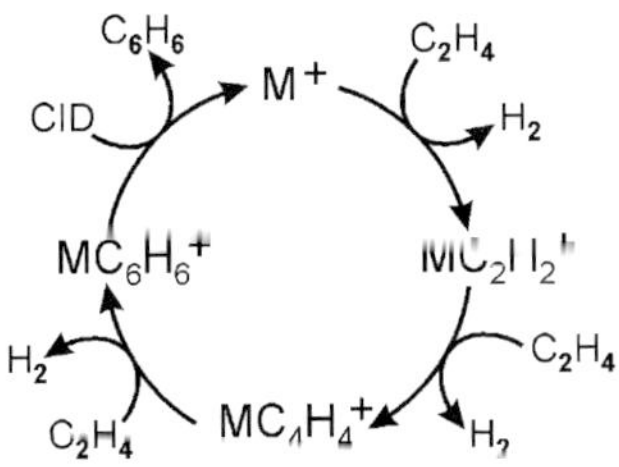

Fig. 1.99. General schema for the trimerization reaction of ethylene to yield benzene in a final collisionally activated reaction step (CID). This reaction has been shown to proceed with W^+ and U^+ cations or the iron tetramer cluster $Fe_4{}^+$ as catalysts [480]

bond dissociation energy of M^+-(benzene) [480]. Therefore, the regeneration of the bare metal ion or cluster accompanied by the liberation of a benzene molecule is not observed. This dehydrogenative cyclization of hydrocarbons may be driven in a catalytic manner by detaching the newly formed benzene ligand from the metal by collision-induced dissociation (CID), thus regenerating the bare metal cation or cluster to start a new catalytic cycle [25]. In an ideal catalytic system, however, the catalytically active species should be regenerated in the reaction system without additional supply of energy. This is conveniently achieved by employing "high-energy" reactants as, e.g., acetylene instead of ethylene. With acetylene as reactant, the exothermicity of the final catalytic reaction step, i.e., the reaction $MC_4H_4{}^+ + C_2H_2 \rightarrow MC_6H_6{}^+$ is increased approximately by the heat of dehydrogenation of ethylene to acetylene [480]. This additional reaction energy is stored completely in the reactive complex. Consequently, the final detachment of the benzene ligand is eased and, thus, the catalyst M^+ becomes available to start a further catalytic cycle.

This concept was proven valid for Ru^+ and Rh^+ cations in an FT-ICR mass spectrometry experiment [480]. A catalytic cycle for the formation of benzene from butadiene and acetlyene, mediated by these metal cations with a turnover number of 6 for the rhodium and 4 for the ruthenium system, has been realized. In the first step, dehydrogenation of butadiene by M^+ generates $MC_4H_4{}^+$ ($M = Ru, Rh$). This key intermediate was found to react with acetlyene to yield C_6H_6 and bare M^+. CID experiments suggest a metallacyclopentadiene strucure for the intermediate $RuC_4H_4{}^+$ and the formation of a cyclobutadiene ligand in the case of $RhC_4H_4{}^+$ as the most probable structures. Also C_4H_4 complexes of all other group 8 transition metal cations have been generated and reacted with acetylene. Only $FeC_4H_4{}^+$ was observed to produce benzene in these investigations, while the analogous complexes of Co^+ and Ni^+ are unreactive toward acetlyene [480].

The Reactivity of Supported Clusters

Although a lot of information on the stoichiometry of the intermediates has been obtained in the gas phase, the unambiguous assignment of neutral reaction products has not been realized directly so far. Thus the investigation of the selectivity of a catalytic process on free clusters is difficult. On supported clusters, the detection of neutral product molecules is straight forward by applying TPR. In the following example, we illustrate such a size-dependent selectivity with the polymerization of acetylene on palladium nanocatalysts [483]. This reaction over supported Pd particles reveals a direct correspondence between reactivities observed on model systems and the behavior of industrial catalysts under working conditions [484]. In ultrahigh vacuum (UHV) [485] as well as under high pressure, large palladium particles of typically thousands of atoms show an increased selectivity for the formation of benzene with increasing particle size [484]. In contrast, small palladium particles of typically hundreds of atoms are less selective for the cyclotrimerization and catalyze butadiene and butene as additional products [484].

The Trimerization Reaction on Single Atoms. First, we discuss the cyclotrimerization on single Ag, Pd and Rh atoms. The nanocatalysts were exposed at 90 K, using a calibrated molecular beam doser, to about 1 Langmuir of acetylene. In a TPR study, catalytically formed benzene (C_6H_6), butadiene (C_4H_6), and butene (C_4H_8) were detected by a mass spectrometer and monitored as a function of temperature. Surprisingly and in contrast to single crystal studies, where an ensemble of seven atoms is needed in the case of Pd(111) [486] and Rh bulk systems are inert, already single Pd and Rh atoms catalyze the cyclotrimerization reaction and the benzene molecule (C_6H_6) is desorbing at 300 K (220 K) and 430 K, respectively (Fig. 1.100). No other product molecules of the polymerization reaction (e.g., C_4H_6, C_4H_8) are observed. Thus these model catalysts are highly selective for the polymerization reaction. As shown above, the formation of benzene on Rh^+ has also been suggested in the gas phase. In contrast, on single Ag atoms no benzene is detected (Fig. 1.100). We note that on Ag and clean MgO(100) surfaces none of the products are formed.

In the following, the key factors governing the observed element-specific cyclotrimerization of acetylene on Ag, Pd, and Rh atoms on MgO surfaces are discussed. The influence of the *electronic structure* is best analyzed on free atoms with the help of ab initio calculations [487]. Free Ag atoms do not adsorb C_2H_2 due to the presence of the partially filled 5s level. This orbital does not hybridize with the filled 4d shell and, as it is spatially expanded, gives rise to a strong Pauli repulsion with the electronic states of the reactant in the entrance channel so that the reaction is not possible. A free Pd atom ($4d^{10}$) readily adsorbs two C_2H_2 molecules by an average binding energy of 1.35 eV per molecule and transforms them into the C_4H_4 intermediate with an energy gain of 3.9 eV. A third acetylene molecule, however, is only weakly bound and practically not activated. Thus, also on a free Pd atom, the cyclotrimerization does not occur. This changes for Rh atoms with their open d-shell configuration. The first two acetylene molecules adsorb with an energy gain of 1.57 and 1.23 eV, respectively, and transform with a very small barrier (0.04 eV) and

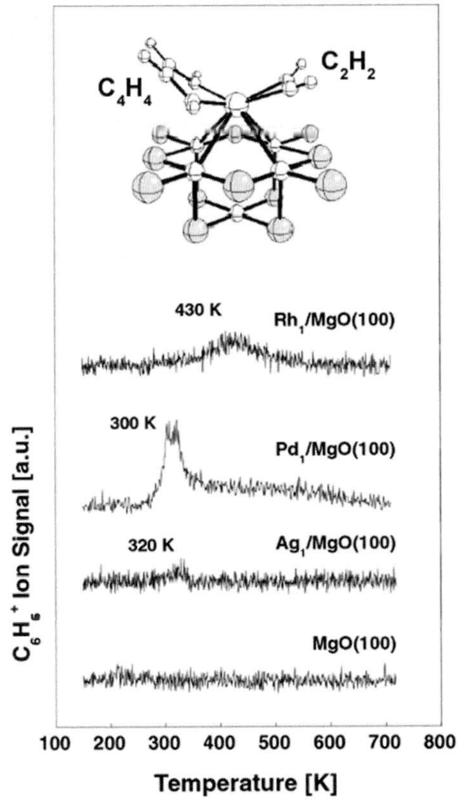

Fig. 1.100. TPR spectra of C_6H_6 formed on Ag, Pd, and Rh atoms deposited on defect-rich MgO thin films grown on Mo(100) surfaces. For comparison, the same experiment was performed on a clean defect-rich MgO film. Shown is also the calculated $(C_4H_4)(C_2H_2)/Pd_1/F_{5c}$ intermediate of the cyclotrimerization reaction on Pd atoms adsorbed on an F center of the MgO(100) surface. For Pd atoms, the formation of benzene was also observed at 220 K

an energy gain of 0.86 eV into the metallopentacycle, $Rh(C_4H_4)$. The third acetylene molecule is readily adsorbed (0.76 eV) and activated as shown by a change from sp to partly sp^2 hybridization. Finally, benzene is formed with a low barrier (0.24 eV) and an energy release of 2.1 eV. Thus, free Rh atoms are not only predicted to be reactive but, due to the very low barriers along the reaction path, they should form benzene at very low temperature.

We now consider the influence of the support material on the reactivity of the three atoms. The experiments show that the coinage metal atom silver remains inert when deposited on the MgO surface and exposing this model system to 1 Langmuir of acetylene at 90 K and detecting benzene as function of temperature (Fig. 1.100). Thus, the electronic structure of Ag is not substantially changed even when trapped at an F center. Calculations confirm this observation and show no ability of the Ag/MgO(F) complex to bind acetylene. In fact, activation of Ag atoms requires depopulation of the 5s orbital, which is not possible on basic oxides like MgO. In contrast to the inert character of an Ag atom, a single Pd atom is turned into an active model catalyst when deposited on MgO films, and benzene is detected in the TDS spectra as a sharp peak at 215 or 300 K, respectively. Such a change in reaction temperature is only observed for Pd and is related to the thickness of the MgO film. This surprising result can be rationalized by studying theoretically a Pd atom adsorbed on different MgO sites represented by a cluster of ions embedded in an array of point charges. First a Pd atom was adsorbed on a five-coordinated oxygen ion on the MgO(001) terrace, O_{5c}, the binding energy is about 1 eV. It was found that indeed the $Pd(C_4H_4)$ complex is formed but the third acetylene molecule is not bound to the complex and therefore this configuration is catalytically inactive. Other bonding sites for the adsorption of Pd atoms are considered. On four-coordinated step or three-coordinated corner oxygen sites, O_{4c} and O_{3c}, respectively, the Pd atom binds slightly stronger with an energy of 1.2–1.5 eV, in addition, the atom is more reactive. However, on both O_{3c} and O_{4c} sites, the third C_2H_2 molecule is only weakly bound or even unbound to the $Pd(C_4H_4)$ surface complex, with the binding energy smaller than the activation energy of the formation of C_6H_6. Thus, Pd atoms adsorbed on O_{nc} sites cannot explain the observed activity. This is consistent with the results of the study on the adsorption properties of CO on Pd_1/MgO [443], as the experimental results have been rationalized in terms of Pd atoms, which are stabilized on oxygen vacancies (F centers) in neutral or charged states, F or F^+, respectively. The interaction of a Pd atom with the F center is much stronger, 3.4 eV, which makes these centers good candidates for Pd binding. On F^+ centers, binding energies of about 2 eV have been computed [443]. The presence of trapped electrons at the defect site results in a more efficient activation of the supported Pd atom. In fact, the complex $(C_4H_4)(C_2H_2)/Pd_1/F_{5c}$, Fig. 1.100, shows a large distortion and a strong interaction of the third C_2H_2 molecule. These results indicate that F and F^+ centers can act as basic sites on the MgO surface and turn the inactive Pd atom into an active catalyst. Notice that the supported Pd atoms on defect sites not only activate the cyclization reaction but also favor benzene desorption, as shown by the very small $(C_6H_6)/Pd_1/F_{5c}$ adsorption energy. The complete reaction path for this specific nanocatalyst has been calculated [488] and is shown in Fig. 1.101. The first barrier of the reaction path is the one for the formation of the intermediate $Pd(C_4H_4)$ and is 0.48 eV only. The formation of the C_4H_4 intermediate is thermodynamically favorable by 0.82 eV. On $(C_4H_4)/Pd_1/F_{5c}$, the addition of the third acetylene molecule is exothermic by 1.17 eV, leading to a very stable $(C_4H_4)(C_2H_2)/Pd_1/F_{5c}$ intermediate (Fig. 1.101). To transform this intermediate into benzene, one has to overcome a barrier of 0.98 eV. The corresponding energy gain is very large, 3.99 eV, and mainly related to the aromaticity of the benzene ring. Once formed, C_6H_6 is so weakly bound to the supported Pd atom that it immediately desorbs.

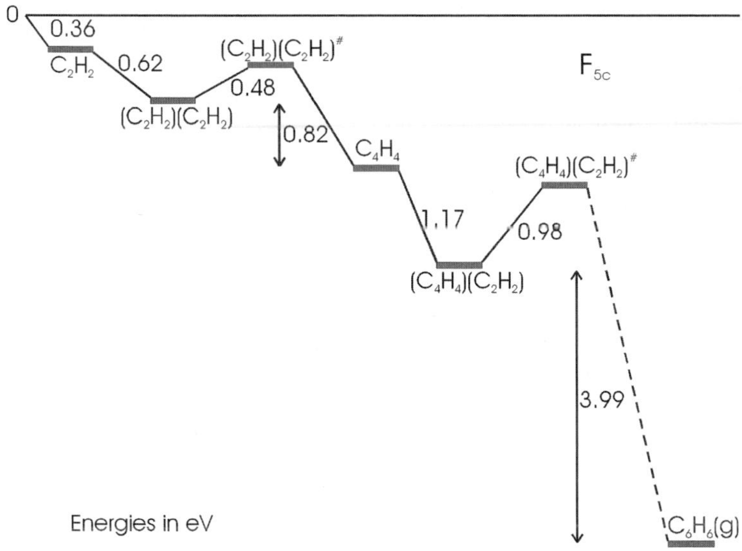

Fig. 1.101. Computed reaction path for the formation of benzene starting from acetylene promoted by a Pd atom supported on a neutral oxygen vacancy at an MgO terrace, F_{5c} (DFT BP results)

Thus, the reaction on Pd/F_{5c} is rate limited in the last step, the conversion of $(C_4H_4)(C_2H_2)$ into C_6H_6. This is different from the Pd(111) surface where the rate determining step for the reaction is benzene desorption. The calculations are consistent with the experimental data. In fact, on Pd_1/F_{5c}, the computed barrier of 0.98 eV corresponds to a desorption temperature of about 300 K, as experimentally observed (Fig. 1.100). On Pd (111) surfaces, the bonding of benzene is estimated to be ≈1.9 eV. This binding is consistent with a desorption temperature of 500 K as observed for a low coverage of C_6H_6 on Pd(111) [489].

Thus, this atom is activated on specific sites on the MgO surface. From FTIR studies, it is known that Pd atoms already diffuse upon deposition to F centers and it is this defect site, which activates Pd by a substantial *charge transfer* [490]. This increases the possibility for binding and activating the third acetylene molecule. Theory predicts Pd atoms to be also activated at MgO(O_{lc}) low coordinated sites [490], but the energetics of the reaction on these sites is totally incompatible with the desorption peak of the product molecule at 215 and 300 K (Fig. 1.100). This further supports the presence of a single active center, namely Pd/MgO(F).

A change in reactivity upon deposition is also observed for Rh. In this case, however, Rh atoms are *deactivated*, as the reaction occurs at higher temperature than predicted for the gas-phase by DFT calculations. The product molecule, benzene, is desorbing as a broad peak between 350 and 500 K

(Fig. 1.100). The identification of the involved reaction mechanisms is more complex, as Rh atoms are stabilized at two different trapping centers after deposition and as migration on the MgO surface occurs upon heating. After deposition and prior to acetylene exposure, the majority of the Rh atoms bind to step edges. The calculations show that these Rh atoms strongly interact with a first acetylene molecule by 1.76 eV and the reactant is highly activated as indicated by the long C–C distance (d(C–C) = 1.315 Å), and by the small HCC angle [(H–C–C): 140.2°]. The $Rh(C_2H_2)$ complex is strongly bound (2.5 eV) to the surface where Rh is simultaneously interacting with the O_{4c} atom at the step and with the O_{5c} anion of the basal plane. The large activation of the acetylene molecule reflects the interaction of the complex with the two surface basic sites. The high stability of the $Rh(C_2H_2)$ complex together with steric effects prevent adsorption of a second and third acetylene molecule to form the $Rh(C_2H_2)_2$ and $Rh(C_2H_2)_3$ complexes essential for promoting the reaction. As comparison, for free Rh atoms, the two acetylene molecules are bound at opposite sites, a configuration which is not possible when Rh is bound to step edges. Thus, in this configuration Rh atoms do not contribute to the observed formation of benzene. Heating up the model catalysts above room temperature induces migration of the $Rh(C_2H_2)$ complex to those F centers which have not been populated upon deposition of Rh atoms. On F centers the $Rh(C_2H_2)$ complex is stably trapped with a binding of 2.8 eV. The computed diffusion barrier for $Rh(C_2H_2)$ from a step (≈1 eV) is considerably lower than the binding energy of acetylene to Rh (1.76 eV) and consequently the whole $Rh(C_2H_2)$ complex is detaching from the step edges and migrates on the surface. Interestingly, when Rh is bound at an F center, the binding (1.06 eV) and activation [d(C–C) = 1.274 Å; (H–C–C): 149] of the acetylene molecule is considerably weakened with respect to the step case. This is one of the reasons why Rh can now adsorb a second acetylene molecule from residual reactants in the vacuum chamber by an energy gain of 0.96 eV. At this stage of the process, the sample temperature is high enough (>300 K) that the complex is immediately transformed to the metallopentacycle, $Rh(C_4H_4)$, with a gain of 0.91 eV and overcoming a barrier of 1.18 eV. This complex binds a third acetylene molecule with a binding energy of 1.07 eV and the product is subsequently formed in an exothermic process (3.2 eV). The involved activation energies for the formation of the metallopentacyle ($\Delta E^{\#} = 1.18$ eV) and the formation of benzene ($\Delta E^{\#} = 1.07$ eV) are of the same order but are about 20% higher than for Pd. This explains the higher reaction temperature for Rh. Notice that these are the rate-determining steps as, due to strong Pauli repulsion, benzene desorption requires very little energy (<0.2 eV) for both Pd/MgO(F) and Rh/MgO(F). Assuming the Redhead equation with a pre-exponential factor of 10^{13}, the computed barriers correspond to a reaction temperature of about 300 K, which is lower than the temperature of maximal benzene formation at 420 K. In addition, the activation energy for benzene formation is similar to the barrier for detaching the $Rh(C_2H_2)$ complex from step edges. Thus, only Rh atoms initially trapped at F centers contribute to the initial formation of benzene at 350 K and the main contribution at higher temperature originates from Rh atoms first trapped at step edges and subsequently turned active after diffusion to the F centers. The existence of a distribution of Rh atoms at F centers and at step edges prior to reaction, therefore, explains the rather broad desorption peak of benzene in contrast to the narrow peak observed for Pd, which populates only one kind of defect centers already at the deposition stage. The situation is even more complex as it could be shown that Rh can detach at around 450 K from the F centers and form larger clusters [491]. In this context, it is important to note that larger Rh_n clusters with $n > 10$ are inert for the cyclotrimerization reaction.

In summary, the specific electronic configuration of Ag renders these atoms inert for the polymerization of acetylene both as free or supported atoms. The reactivity of Pd and Rh is strongly influenced by their adsorption and by their diffusion dynamics on the MgO surface. Pd atoms are turned into active catalysts for the cyclotrimerization reaction only when adsorbed on F centers as charge donation from the defect site to the atom occurs upon bonding. Finally, the low-activation barriers of the process on free Rh atoms are substantially increased for Rh adsorbed on an MgO surface, and the cyclotrimerization is only catalyzed when trapped on F centers as otherwise steric effects, which are especially marked for supported atoms and small clusters, prevent the adsorption of a second or third acetylene molecules on step edges.

The Selectivity of Polymerization of Acetylene on Pd_n Clusters. Larger Pd_n clusters also reveal an interesting behavior and the corresponding TPR spectra of the different polymerization products of acetylene on small supported, monodispersed palladium clusters are shown in Fig. 1.103. Striking atom-by-atom size-dependent reactivities and selectivities are observed. Only the three reaction products C_6H_6, C_4H_8, and C_4H_6 are detected. Remarkably, no C_3H_n, C_5H_n, and C_8H_n are formed, indicating the absence of C–C bond scission as already observed on Pd single crystals [492] and Pd particles [484]. Up to Pd_3, only benzene is catalyzed reflecting a high selectivity for the cyclotrimerization of acetylene. Pd_n ($4 \leq n \leq 6$) clusters reveal a second reaction channel by catalyzing in addition the formation of C_4H_6, which desorbs at around 300 K. The third reaction product, C_4H_8, desorbing at a rather low temperature of 200 K, is clearly observed for Pd_8. For this cluster size, the abundance of the three reaction products is similar. For even larger clusters ($13 < n < 30$) the formation of C_6H_6 is increasing with cluster size, whereas the conversion of acetylene into C_4H_8 reaches a maximum for Pd_{20}. Note that Pd_{30} selectively suppresses the formation of C_4H_6 (the peak in the TPR spectrum of C_4H_6 at 200 K is part of the fragmentation pattern of C_4H_8). For Pd_{20}, the experiments were repeated in the presence of D_2 [483]. D_2 was exposed prior and after C_2H_2. The results clearly indicate that no product containing deuterium is formed. Consequently, D_2 is not involved in the polymerization reaction. However, the presence of D_2 opens a new reaction channel, the hydrogenation of acetylene. In addition, D_2 blocks the active sites on the palladium clusters

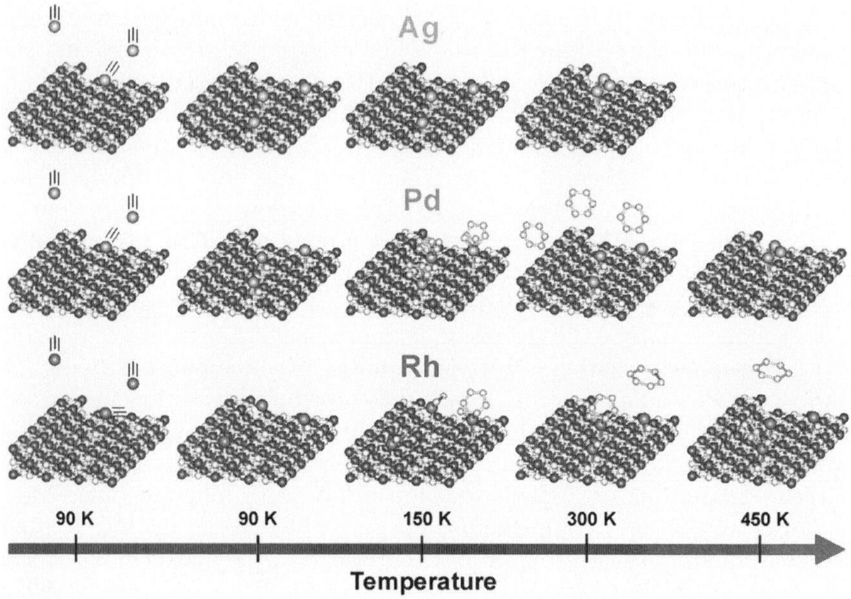

Fig. 1.102. The proposed mechanisms are shown schematically for the three atoms. Ag and Pd atoms are decorating exclusively F centers after deposition whereas Rh is trapped at step edges and F centers at 90 K. Ag atoms do not adsorb acetylene and are therefore inert for the reaction. Pd and Rh are forming benzene only when trapped at F centers. Note that the relatively broad temperature range for the formation of C_6H_6 on Rh originates from the fact that Rh is trapped at two defect sites at 90 K and that the reaction occurs only after diffusion of the $Rh(C_2H_2)$ complexes from steps to F centers

for the polymerization, as the formation of the products is slightly reduced when exposing D_2 prior to C_2H_2. On Pd(111), predosing with H_2 completely suppresses the cyclotrimerization but enhances the hydrogenation of acetylene to form ethylene [489].

If one assumes stoichiometric reactions, as indicated in Fig. 1.104c, and estimates the relative number of C_2H_2 from Fig. 1.104a, one observes a proportional increase of acetylene with the number of palladium atoms per cluster up to Pd_{13}. Surprisingly, at this cluster size, the surface-to-bulk ratio as well as the coordination number of Pd in the cluster is changing as at this size one Pd atom sits completely in the cluster. In addition, according to the free stochiometric chemical reactions, each reaction requires a minimum number of Pd atoms, which are 3, 4, and 6. The experimental results are surprisingly consistent, that is C_4H_6 is formed for Pd_n with $n \geq 4$ and C_4H_8 for cluster sizes with $n \geq 6$.

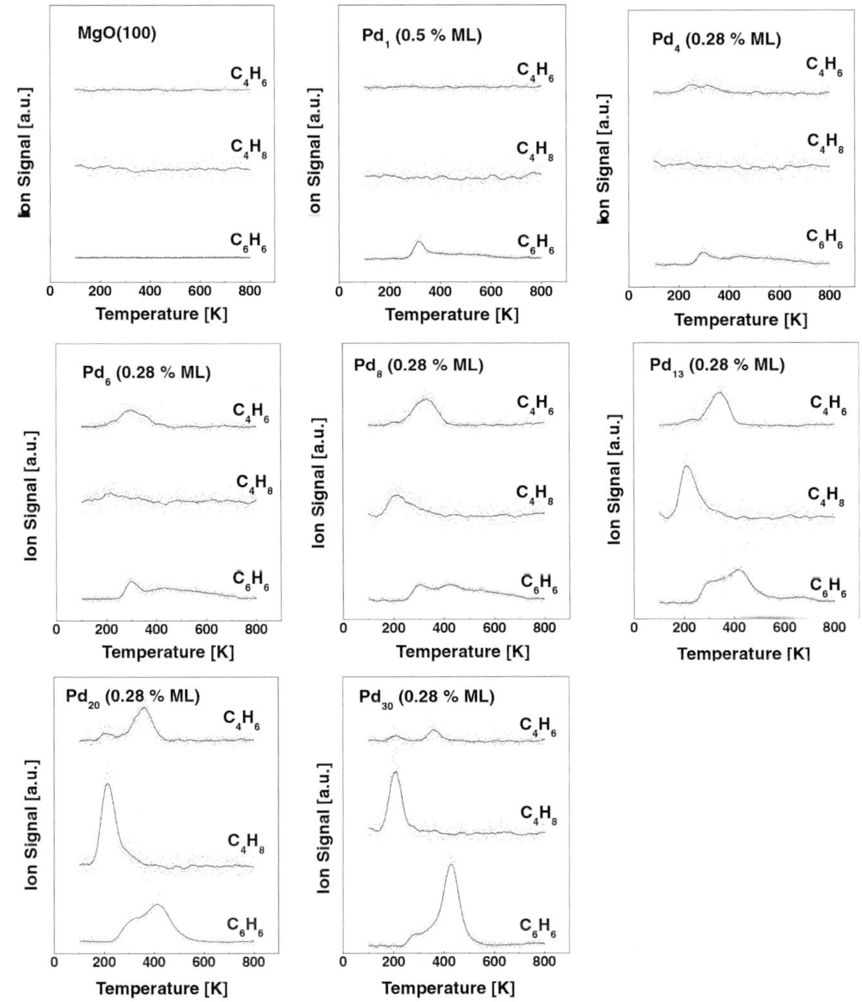

Fig. 1.103. TPR spectra of the catalytic formation of C_6H_6, C_4H_6, and C_4H_8 for a defect-rich MgO thin film, Pd_1, Pd_4, Pd_6, Pd_8, Pd_{13}, Pd_{20}, and Pd_{30}. The relative ion intensities are corrected with the relative detection efficiencies of the experiment, and scale with the number of formed product molecules per cluster

Analyzing the products formed on small size-selected Pd_n ($1 \leq n \leq 30$) clusters deposited on MgO(100) thin films indicates that the surface intermediate C_4H_4 is being produced efficiently on all cluster sizes. Thus at least two acetylene molecules are adsorbed in a π-bonded configuration at the initial stage of the reaction [493]. The observed size-dependent selectivity may then be understood by regarding the influence of the cluster size to steer the reaction either toward the cyclotrimerization to form C_6H_6 or toward

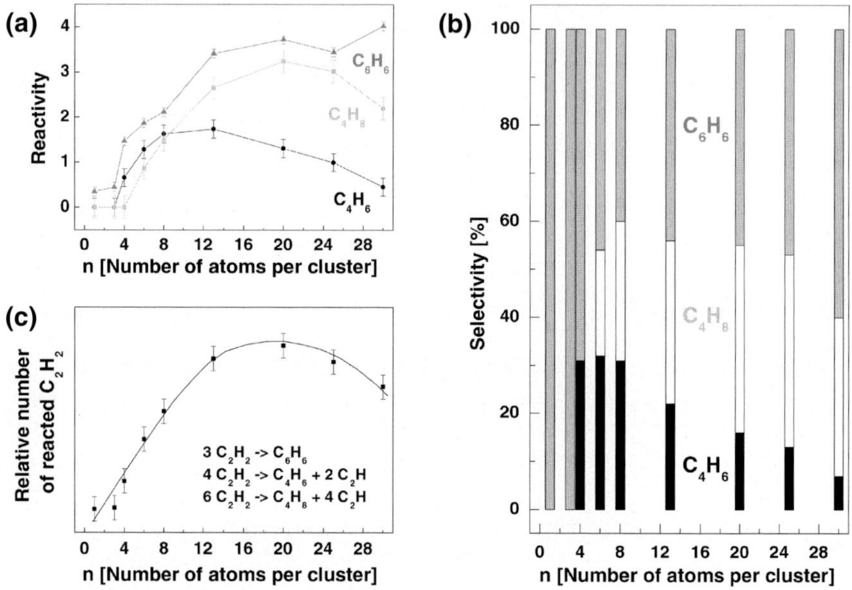

Fig. 1.104. Reactivity (**a**) (expressed as the number of product molecules per cluster) and selectivity (**b**) (expressed as the relative amount in percent) of the polymerization of C_2H_2 on size-selected Pd_n ($n = 1$–30) deposited on defect-rich MgO thin films. Also shown is the relative number of reacted C_2H_2 as function of cluster size

a direct hydrogen transfer from adsorbed C_2H_2 to the C_4H_4 intermediate to catalyze the formation of C_4H_6 or C_4H_8, respectively. Cyclotrimerization is generally observed when a third acetylene molecule is adsorbed in a π-bonded configuration, which results in a change from sp-hybridization toward sp^2-hybridization [494]. This bonding configuration leads to a weak activation of the C–H bond in analogy to ethylene [495]. The hydrogenation of the $Pd_n(C_4H_4)$ metallocycle, on the other hand, is favored by the adsorption of di-σ/π bonded acetylene to three Pd atoms, effecting a more efficient activation of the C–H bond, in analogy to ethylene [495].

As shown above, for Pd atoms adsorbed on defect sites, the $Pd(C_4H_4)$ intermediate is readily formed as shown in Fig. 1.102. A third adsorbed C_2H_2 molecule is purely π-bonded and the activated acetylene molecule reacts with the intermediate to form benzene with a total exothermicity of about 7 eV (Fig. 1.101). The weakly bound C_6H_6 (0.3 eV) then desorbs at low temperature from the nanocatalyst [493]. A second reaction channel, the formation of butadiene, C_4H_6, opens for Pd_4. This channel reveals highest selectivity for Pd_6, in this case a third C_2H_2 molecule can bind in a di-σ/π bond configuration to three Pd atoms. The charge transfer from the substrate to the cluster further enhances the activation of the C–H bonds. For even larger cluster sizes, the adsorption of two di-σ/π-bonded C_2H_2 molecules becomes possible and opens up the third reaction path, the formation of C_4H_8. In our experiments this is clearly observed for Pd_8. Purely geometric arguments (possible adsorption of two di-σ/π-bonded C_2H_2 molecules close to the C_4H_4 intermediate) suggest that this third channel is more pronounced for the larger clusters, and indeed our results show maximal C_4H_8 formation for cluster sizes of 20–25 Pd atoms. For the largest clusters of the measured range, e.g., Pd_{30}, the increased number of metal–metal bonds and the concomitant delocalization of the charge transferred from the substrate to the cluster results in less charge density available for the activation of the C–H bond [496]. Consequently, the cyclotrimerization becomes again more efficient than the hydrogenation of the C_4H_4 intermediate. Going to even larger particles or to Pd(111) single crystals, the cyclotrimerization to benzene is selectively catalyzed.

In conclusion of this chapter, there is no question that a new catalytic chemistry emerges in the nonscalable size regime where every atom counts. This may have important implications for industrial catalysis with respect to low-temperature active catalytic materials and selectivity tuning by size-selection. From a more fundamental point of view, in particular, the combination of gas-phase studies, clusters supported on surface, and ab initio theoretical simulation are fruitful in defining important nanocatalytic factors and concepts. The most important reviewed in this contribution are the tuning of selectivity and activity by electronic quantum size effects and charging, the dynamic structural fluxionality, i.e., the self-selection of the low activation energy reaction path via interconversion of structural isomers, and especially for supported clusters, the tuning of activity through the design of the cluster–support interface and the size-dependent adlineation. Most of these factors are unique for small clusters and not observed for bulk catalytic systems or particles in the scalable size regime as described in Chap. 3.

References

1. Kaldor A, Cox D, Zakin MR (1988) Adv. Chem. Phys. 70:211
2. Whetten RL, Cox DM, Trevor DJ, Kaldor A (1985) Phys. Rev. Lett. 54:1494
3. Holmgren L, Andersson M, Rosen A (1995) Surf. Sci. 331–333:231
4. Berces A, Hackett PA, Lian L, Mitchell SA, Rayner DM (1998) J. Chem. Phys. 108:5476
5. Parks EK, Nieman GC, Kerns KP, Riley SJ (1998) J. Chem. Phys. 108:3731
6. Schulze Icking-Konert G, Handschuh H, Ganteför G, Eberhardt W (1996) Phys. Rev. Lett. 76:1047
7. Shi Y, Ervin KM (1998) J. Chem. Phys. 108:1757
8. Fayet P, Granzer F, Hegenbart G, Moisar E, Pischel B, Wöste L (1985) Phys. Rev. Lett. 55:3002
9. Leisner T, Rosche C, Wolf S, Granzer F, Wöste L (1996) Surf. Rev. Lett 3:1105
10. Xu Z, Xiao F S, Purnell SK, Alexeev O, Kawi S, Deutsch SE, Gates BC (1994) Nature 372:346
11. Heiz U, Vanolli F, Sanchez A, Schneider W-D (1998) J. Am. Chem. Soc. 120:9668

12. Vajda S, Wolf S, Leisner T, Busolt U, Wöste L (1997) J. Chem. Phys. 107:3492
13. Dietz TG, Duncan MA, Powers DE, Smalley RE (1981) J. Chem. Phys 74:6511
14. Siekmann HR, Lüder C, Fährmann J, Lutz HO, Meiwes-Broer KH (1990) Z. Phys. D. 20:417
15. Haberland H (1994) J. Vac. Sci. Technol. A 12:2925
16. Haberland H, Karrais M, Mall M, Thurner Y (1992) J. Vac. Sci. Technol. A 10:3266
17. Keller R, Nöhmeier F, Spädtke P, Schönenberg MH (1984) Vacuum 34:31
18. Cox DM, Reichmann KC, Trevor DJ, Kaldor A (1988) J. Chem. Phys. 88:111
19. Parks EK, Nieman GC, Kerns KP, Riley SJ (1997) J. Chem. Phys. 107:1861
20. Leuchtner RE, Harms AC, Castleman AW Jr (1990) J. Chem. Phys. 92:6527
21. Lian L, Hackett PA, Rayner DM (1993) J. Chem. Phys. 99:2583
22. Ervin KM (2001) Int. Rev. Phys. Chem. 20:127
23. Kaiser B, Bernhardt TM, Kinne M, Rademann K, Heidenreich A (1999) J. Chem. Phys. 110:1437
24. Kappes MM, Staley RH (1981) J. Am. Chem. Soc. 103:1286
25. Schnabel P, Weil KG, Irion MP (1992) Angew. Chem. Int. Ed. Engl. 31:636
26. Berg C, Beyer M, Achatz U, Joos S, Niedner-Schatteburg G, Bondybey VE (1998) Chem. Phys. 239:379
27. Vakhtin AB, Sugawara K (1999) Chem. Phys. Lett. 299:553
28. Oliveira MC, Marcalo J, Vieira MC, Almoster Ferreira MA (1999) Int. J. Mass Spectrom. Ion Proc. 185–187:825
29. Jarrold M, Bower JE (1986) J. Chem. Phys. 85:5373
30. Gerlich D (1992) Inhomogeneous rf-fields: a versatile tool for the study of processes with slow ions. In: Ng C-Y, Baer M (eds.), State-Selected and State-to-State Ion-Molecule Reaction Dynamics, Part 1: Experiment. John Wiley & Sons, Inc., New York, p. 1
31. Fayet P, McGlinchey MJ, Wöste LH (1987) J. Am. Chem. Soc. 109:1733
32. Socaciu LD, Hagen J, Heiz U, Bernhardt TM, Leisner T, Wöste L (2001) Chem. Phys. Lett. 340:282
33. Socaciu LD, Hagen J, Bernhardt TM, Wöste L, Heiz U, Häkkinen H, Landman U (2003) J. Am. Chem. Soc. 125:10437
34. Hagen J, Socaciu LD, Elijazyfer M, Heiz U, Bernhardt TM, Wöste L (2002) Phys. Chem. Chem. Phys. 4:1707
35. Herrmann A, Hofmann M, Leutwyler S, Schumacher E, Wöste L (1979) Chem. Phys. Lett. 62:216
36. Knight WD, Clemenger K, de Heer W, Saunders WA, Chou MY, Cohen ML (1984) Phys. Rev. Lett. 52:2141
37. Heiz U, Vayloyan A, Schumacher E (1997) Rev. Sci. Instrum. 68:3718
38. de Heer WA (1993) Rev. Mod. Phys. 65:611
39. Heiz U, Vanolli F, Trento L, Schneider W-D (1997) Rev. Sci. Instrum. 68:1986
40. Fayet P, Granzer F, Hegenbart G, Moisar E, Pischel B, Wöste L (1986) Z. Phys. D 3:299
41. Eberhardt W, Fayet P, Cox DM, Fu Z, Kaldor A, Sherwood R, Sondericker D (1990) Phys. Rev. Lett. 64:780
42. Fayet P, Patthey F, Roy HV, Detzel T, Schneider W-D (1992) Surf. Sci. 269/270:1101
43. Roy H-V, Boschung J, Patthey F, Fayet P, Schneider W-D (1993) Phys. Rev. Lett. 70:2653

44. Roy H-V, Fayet P, Patthey F, Schneider W-D, Delley B, Massobrio C (1994) Phys. Rev. B 49:5611
45. Schaffner M-H, Jeanneret J-F, Patthey F, Schneider W-D (1998) J. Phys. D: Appl. Phys. 31:3177
46. O'Shea JN, Schnadt J, Andersson S, Patthey L, Rost S, Giertz A, Brena B, Forsell JO, Sandell A, Bjorneholm O, Brühwiler PA, Martensson N (2000) J. Chem. Phys. 113:9233
47. Lau JT, Achleitner A, Wurth W (2000) Chem. Phys. Lett. 317:269
48. Berry RS, Bonačić-Koutecký V, Gaus J, Leisner T, Manz J, Reischl-Lenz B, Ruppe H, Rutz S, Schreiber E, Vajda S, de Vivie-Riedle R, Wolf S, Wöste L (1997) Adv. Chem. Phys. 101:101
49. Bernhardt TM, Hagen J, Socaciu-Siebert LD, Mitrić R, Heidenreich A, Le Roux J, Popolan D, Vaida M, Wöste L, Bonačić-Koutecký V, Jortner J (2005) Chem. Phys. Chem. 6:243
50. Hall SG, Nielsen MB, Robinson AW, Palmer RE (1997) Rev. Sci. Instrum. 69:3335
51. Boyd J, Lapicki A, Aizawa M, Anderson SL (1998) Rev. Sci. Instrum. 69:4106
52. Boyd KJ, Lapicki A, Aizawa M, Anderson SL (1999) Nucl. Instrum. Methods Phys. Res. B 157:144
53. Yamaguchi W, Yoshimura K, Tai Y, Maruyama Y, Igarashi K, Tanemura S, Murakami J (2000) J. Chem. Phys. 112:9961
54. Goldby IM, von Issendorff B, Kuipers L, Palmer RE (1997) Rev. Sci. Instrum. 68:3327
55. Abe H, Schulze W, Tesche B (1980) Chem. Phys. 47:104
56. Francis GM, Goldby IM, Kuipers L, v Issendorf B, Palmer RE (1996) J. Chem. Soc. (Dalton Trans.) 5:665
57. Goldby IM, Kuipers L, von Issendorf B, Palmer RE (1996) Appl. Phys. Lett. 69:2819
58. Kaiser B, Bernhardt TM, Stegemann B, Opitz J, Rademann K (1999) Phys. Rev. Lett. 83:2918
59. Bernhardt TM, Kaiser B, Rademann K (2002) Phys. Chem. Chem. Phys. 4:1192
60. Ganteför G, Siekmann HR, Lutz HO, Meiwes-Broer KH (1990) Chem. Phys. Lett. 165:293
61. Klipp B, Grass M, Muller J, Stolcic D, Lutz U, Ganteför G, Schlenker T, Boneberg J, Leiderer P (2001) Appl. Phys. A 73:547
62. Berry SD (1986) In: Jena P, Rao BK, Khanna SNK (eds.), The Physics and Chemistry of Small Metallic Clusters. Plenum, New York, p. 49
63. Bhaskar NB, Klimcak CM, Früholz R (1990) Rev. Sci. Instrum. 61:366
64. Dawson RH (1995) Quadrupole Mass Spectrometry and its Applications. AIP, Woodbury
65. Farrar JM, Saunders WH Jr (eds.) (1988) Techniques for the Study of Ion-Molecule Reactions. Wiley, New York
66. Schlag EW (ed.) (1994) Time-of-Flight Mass Spectrometry and its Applications. Elsevier, Amsterdam
67. Cotter RJ (1997) Time-of-Flight Mass Spectrometry. American Chemical Society, Washington, DC
68. Fayet P, Woste L (1985) Surf. Sci. 156:135
69. Scoles G (ed.) (1988 and 1992) Atomic and Molecular Beam Methods. Oxford University Press, New York

70. Parent DC, Anderson SL (1992) Chem. Rev. 92:1541
71. Haberland H (ed.) (1994) Clusters of Atoms and Molecules. Springer Series in Chemical Physics, Springer, Berlin
72. Castleman AW Jr, Bowen KH (1996) J. Phys. Chem. 100:12911
73. Heiz U, Schneider W-D (2000) J. Phys. D. 33:R85
74. Röttgen M, Judai K, Antonietti JM, Heiz U, Rauschenbach S, Kern K (2006) Rev. Sci. Inst. 77:013302
75. Begemann W, Meiwes-Broer KH, Lutz HO (1986) Phys. Rev. Lett. 56:2248
76. Wrenger B, Meiwes-Broer KH, Speer O, Garcia ME (1997) Phys. Rev. Lett. 79:2562
77. Klingeler R, Bechthold PS, Neeb M, Eberhardt W (2000) J. Chem. Phys. 113:1420
78. Klingeler R, Bechthold PS, Neeb M, Eberhardt W (2002) Rev. Sci. Instrum. 73:1803
79. Grill V, Shen J, Evans C, Cooks RG (2001) Rev. Sci. Instrum. 72:3149
80. Beck RD, St. John P, Homer ML, Whetten RL (1991) Science 253:879
81. Beck RD, Weis P, Bräuchle G, Rockenberger J (1995) Rev. Sci. Instrum. 66:4188
82. Weinkauf R, Walter K, Weickhardt C, Boesl U, Schlag EW (1989) Z. Naturforsch. 44a:1219
83. Paik DH, Bernhardt TM, Kim NJ, Zewail AH (2001) J. Chem. Phys. 115:612
84. Bernhardt TM, Kaiser B, Rademann K (1997) Z. Phys. D 40:327
85. Christen W, Even U, Raz T, Levine RD (1998) J. Chem. Phys. 108:10262
86. von Issendorff B, Palmer RE (1999) Rev. Sci. Instrum. 70:4497
87. Bernhardt TM (1997) Untersuchungen zur Wechselwirkung größenselektierter Antimon- und Bismutclusterionen mit einer Graphit(0001)-Oberfläche bei Energien bis zu 600 eV. Ph.D. Thesis, Humboldt-University, Berlin
88. Kaiser B, Bernhardt TM, Rademann K (1997) Nucl. Instrum. Meth. Phys. Res. B 125:223
89. Kaiser B, Stegemann B, Bernhardt TM, Opitz J, Rademann K (in preparation)
90. Kaiser B, Bernhardt TM, Rademann K (1998) Appl. Phys. A 66:S711
91. Harbich W (2000) Collision of clusters with surfaces: deposition, surface modification and scattering. In: Meiwes-Broer K-H (ed.), Metal Clusters at Surfaces. Springer Series in Cluster Physics, Berlin, p. 107
92. Petek H, Ogawa S (1998) Prog. Surf. Sci. 56:239
93. Busolt U, Cottancin E, Rohr H, Socaciu L, Leisner T, Wöste L (1999) Appl. Phys. B 68:453
94. Busolt U, Cottancin E, Rohr H, Socaciu L, Leisner T, Wöste L (1999) Eur. Phys. J. D 9:523
95. Busolt U, Cottancin E, Socaciu L, Rohr H, Leisner T, Wöste L (2001) Eur. Phys. J. D 16:297
96. Cleveland CL, Landman U (1992) Science 257:355
97. Cheng H-P, Landman U (1993) Science 260:1304
98. Cheng H-P, Landman U (1994) J. Phys. Chem. 98:3527
99. Broman K, Felix C, Brune H, Harbich W, Monot R, Buttet J, Kern K (1996) Science 274:956
100. Meiwes-Broer K-H (ed.) (2000) Metal Clusters at Surfaces. Springer Series in Cluster Physics, Springer, Berlin Heidelberg New York
101. Ekardt W (ed.) (1999) Metal Clusters. Wiley, Chichester

102. Cheshnovsky O, Yang SH, Pettiette CL, Craycraft MJ, Smalley RE (1987) Rev. Sci. Instrum. 58:2131
103. Taylor KJ, Pettiette-Hall CL, Cheshnovsky O, Smalley RE (1992) J. Chem. Phys. 96:3319
104. Ho J, Ervin KM, Lineberger WC (1990) J. Chem. Phys. 93:6987
105. Ganteför G, Gausa M, Meiwes-Broer KH, Lutz HO (1990) J. Chem. Soc. Faraday Trans. 86:2483
106. Negeshi Y, Nakamura Y, Nakajima A, Kaya K (2001) J. Chem. Phys. 115:3657
107. Pramann A, Rademann K (2001) Chem. Phys. Lett. 343:99
108. Cha C-Y, Ganteför G, Eberhardt W (1992) Rev. Sci. Instrum. 62:5661
109. Boo WD, Ozaki Y, Andersen LH, Lineberger WC (1997) J. Phys. Chem. A 101:6688
110. Handschuh H, Gantefor G, Eberhardt W (1995) Rev. Sci. Instrum. 66:3838
111. Wang LS, Ding CF, Wang XB, Barlow SE (1999) Rev. Sci. Instrum. 70:1957
112. Wiley WC, McLaren IH (1955) Rev. Sci. Instrum. 26:1150
113. Li J, Zahi H-J, Wang L-S (2003) Science 299:864
114. Kroto HW, Heath JR, O'Brian SC, Curl RF, Smalley RE (1985) Nature 318:162
115. Krätschmer W, Lamb LD, Fostiropoulos K, Huffman DR (1990) Nature 347:354
116. Wang XB, Ding CF, Wang LS (1999) J. Chem. Phys. 110:8217
117. Gilb S, Weis P, Furche F, Ahlrichs R, Kappes MM (2002) J. Chem. Phys. 116:4094
118. Furche F, Ahlrichs R, Weis P, Jacob C, Gilb S, Bierweiler T, Kappes MM (2002) J. Chem. Phys. 117:6982
119. Häkkinen H, Moseler M, Landman U (2002) Phys. Rev. Lett. 89:033401
120. Häkkinen H, Yoon B, Landman U, Li XL, Zahi H-J, Wang L-S (2003) J. Phys. Chem. A 107:6168
121. Häkkinen H, Moseler M, Kostko O, Morgner N, Hoffmann MA, von Issendorff B (2004) Phys. Rev. Lett. 93:093401
122. Pyykkö P (1988) Chem. Rev. 88:563
123. Balasubramanian K (1997) Relativistic Effects in Chemistry. Wiley, New York
124. Stolcic D, Fischer M, Ganteför G, Kim YD, Sun Q, Jena P (2003) J. Am. Chem. Soc. 125:2848
125. Sun Q, Jena P, Kim YD, Fischer M, Ganteför G (2004) J. Chem. Phys. 120:6510
126. von Helden G, van Heijnsbergen D, Meijer G (2003) J. Phys. Chem. A 107:1671
127. Fielicke A, Kirilyuk A, Ratsch C, Behler J, Scheffler M, von Helden G, Meijer G (2004) Phys. Rev. Lett. 93:023401
128. van Heijnsbergen D, von Helden G, Duncan MA, van Roij AJA, Meijer G (1999) Phys. Rev. Lett. 83:4983
129. von Helden G, Tielens AGGM, van Heijnsbergen D, Duncan MA, Hony S, Waters LBFM, Meijer G (2000) Science 288:313
130. Asmis KR, Meijer G, Brümmer M, Kaposta C, Santambrogio G, Wöste L, Sauer J (2004) J. Chem. Phys. 120:6461
131. Fielicke A, von Helden G, Meijer G, Pedersen DB, Simard B, Rayner DM (2004) J. Phys. Chem. B 108:14591
132. Kemper PR, Bowers MT (1991) J. Phys. Chem. 95:5134
133. von Helden G, Gotts N, Bowers MT (1993) Nature 363:60
134. Jarrold M (1994) J. Phys. Chem. 98:12819

135. Jarrold MF, Bowers MT (1993) J. Chem. Phys. 98:2399
136. Dugourd P, Hudgins R, Clemmer DE, Jarrold MF (1997) Rev. Sci. Instrum. 68:7722
137. Weis P, Gilb S, Gerhardt P, Kappes MM (2002) Int. J. Mass Spectrom. 216:59
138. De Boer BG, Stein GD (1981) Surf. Sci. 106:84
139. Yokozeki A, Stein GD (1978) J. Appl. Phys. 49:2224
140. Farges J, Feraudy MF, Raoult B, Torchet G (1981) Surf. Sci. 106:95
141. Farges J, Raoult B, Torchet G (1973) J. Chem. Phys. 59:3454
142. Hovick JW, Bartell LS (1998) J. Phys. Chem. B 102:534
143. Bartell LS (1986) Chem. Rev. 86:491
144. Reinhard D, Hall BD, Ugarte D, Monot R (1998) Phys. Rev. B 58:4917
145. Reinhard D, Hall BD, Berthoud P, Valkealahti S, Monot R (1998) Phys. Rev. B 55:7868
146. Maier-Borst M, Cameron DB, Rokni M, Parks JH (1999) Phys. Rev. A 59:R3162
147. Krückenberg S, Schooss D, Maier-Borst M, Parks JH (2000) Phys. Rev. Lett. 85:4494
148. Parks JH, Kappes MM (private communication)
149. Lüttgens G, Pontius N, Bechthold PS, Neeb M, Eberhardt W (2002) Phys. Rev. Lett. 88:076102
150. Broyer M, Delacretaz G, Labastie P, Wolf JP, Wöste L (1986) Phys. Rev. Lett. 57:1851
151. Collings BA, Athanassenas K, Rayner DM, Hackett PA (1994) Chem. Phys. Lett. 227:490
152. de Heer WA, Selby K, Kresin V, Masui J, Vollmer M, Chatelain A, Knight WD (1987) Phys. Rev. Lett. 59:1805
153. Delacretaz G, Grant ER, Whetten RL, Wöste L, Zwanziger JW (1986) Phys. Rev. Lett. 56:2598
154. Knickelbein MB, Menezes WJC (1992) Phys. Rev. Lett. 69:1046
155. Wöste L (1996) Z. Phys Chem. 196:1
156. Kreibig U, Vollmer M (1995) Optical Properties of Metal Clusters. Springer, Berlin
157. Gilb S, Jacobsen KW, Schooss D, Furche F, Ahlrichs R, Kappes MM (2004) J. Chem. Phys. 121:4619
158. Illenberger E, Momigny J (1992) Gaseous Molecular Ions. Steinkopff Verlag, Darmstadt
159. Kaldor A, Cox DM, Zakin MR (1988) Adv. Chem. Phys. 70:211
160. Salisbury BE, Wallace WT, Whetten RL (2000) Chem. Phys. 262:131
161. Knickelbein MB (1999) Annu. Rev. Phys. Chem. 50:79
162. Riley SJ (1992) Nanostructured Mater. 1: 155
163. Kaiser B, Bernhardt TM, Kinne M, Rademann K (1998) Int. J. Mass Spectrom. Ion Proc. 177:L5
164. Kinne M, Bernhardt TM, Kaiser B, Rademann K (1997) Z. Phys. D 40:105
165. Kinne M, Bernhardt TM, Kaiser B, Rademann K (1997) Int. J. Mass Spectrom. Ion Proc. 167/168:161
166. France MR, Buchanan JW, Robinson JC, Pullins SH, Tucker JL, King RB, Duncan MA (1997) J. Phys. Chem. 101:6214
167. Reddy BV, Jena P (1998) Chem. Phys. Lett. 288:253
168. Fielicke A, Rademann K (2000) J. Phys. Chem. A 104:6979

169. Andersson M, Rosén A (2002) J. Chem. Phys. 117:7051
170. Hanley L, Anderson SL (1985) Chem. Phys. Lett. 122:410
171. Jarrold MF, Bower JE, Kraus JS (1987) J. Chem. Phys. 86:3876
172. Hanley L, Ruatta SA, Anderson SL (1987) J. Chem. Phys. 87:260
173. Guo BC, Kerns KP, Castleman AW Jr (1992) J. Phys. Chem. 96:6931
174. Wade K (1976) Adv. Inorg. Chem. Radiochem. 18:1
175. Mingos DMP, Slee T, Zhenyang L (1990) Chem. Rev. 90:383
176. Lauher JW (1978) J. Am. Chem. Soc. 100:5305
177. Alford JM, Weiss FD, Laaksonen RT, Smalley RE (1986) J. Phys. Chem. 90:4480
178. Elkind JL, Weiss FD, Alford JM, Laaksonen RT, Smalley RE (1988) J. Chem. Phys. 88:5215
179. Irion MP, Selinger A (1989) Chem. Phys. Lett. 158:145
180. Irion MP, Selinger A, Wendel R (1990) Int. J. Mass Spectrom. Ion Proc. 96:27
181. Berg C, Schindler T, Niedner-Schatteburg G, Bondybey VE (1995) J. Chem. Phys. 102:4870
182. Dietrich G, Krückenberg S, Lützenkirchen K, Schweikhard L, Walther C (2000) J. Chem. Phys. 112:752
183. Koszinowski K, Schröder D, Schwarz H (2003) J. Chem. Phys. A 107:4999
184. Hess H, Kwiet S, Socaciu L, Wolf S, Leisner T, Wöste L (2000) Appl. Phys. B 71:337
185. Hagen J, Socaciu LD, Heiz U, Bernhardt TM, Wöste L (2003) Eur. J. Phys. D 24:327
186. Bernhardt TM (2005) Int. J. Mass Spectrom. 243:1
187. Steinfeld JI, Francisco JS, Hase WL (1999) Chemical Kinetics and Dynamics. Prentice-Hall, Upper Saddle River, NJ
188. Laidler KJ (1987) Chemical Kinetics. HarperCollins, New York
189. Castleman AW Jr, Weil KG, Sigsworth SW, Leuchtner RE, Keesee RG (1987) J. Chem. Phys. 86:3829
190. Nilius N, Wallis TM, Ho W (2003) Phys. Rev. Lett. 90:186102
191. Judai K, Sera K, Amatsutsumi S, Yagi K, Yasuike T, Yabushita S, Nakajima A, Kaya K (2001) Chem. Phys. Lett. 334:277
192. Vanolli F, Heiz U, Schneider W-D (1997) Chem. Phys. Lett. 277:527
193. Stipe BC, Rezaei MA, Ho W (1998) Science 280:1732
194. Antonietti J-M, Michalski M, Jones H, Heiz U (in preparation)
195. Piednoir A, Perrot E, Granjeaud S, Humbert A, Chapon C, Henry CR (1997) Surf. Sci. 391:19
196. Hojrup Hansen K, Worren T, Stempel S, Laegsgaard E, Bäumer M, Freund H-J, Besenbacher F, Stensgaard I (1999) Phys. Rev. Lett. 83:4120
197. Schaub R, Jödicke H, Brunet F, Monot R, Buttet J, Harbich H (2001) Phys. Rev. Lett. 86:3590
198. Bernhardt TM, Stegemann B, Kaiser B, Rademann K (2003) Angew. Chem. Int. Ed. 42:199
199. Bjornholm O (1990) Contemp. Phys. 31:309
200. Schaffner M-H, Patthey F, Schneider W-D, Pettersson LGM (1998) Surf. Sci. 450:402
201. He J-W, Møller PL (1986) Chem. Phys. Lett. 129:13
202. He J-W, Møller PJ (1986) Surf. Sci. 178:934
203. He J-W, Møller PJ (1987) Surf. Sci. 180:411

204. Alstrup I, Møller PJ (1988) Appl. Surf. Sci. 33/34:143
205. Li Y, Langreth DC, Pederson MR (1995) Phys. Rev. B 52:6067
206. Yudanov I, Pacchioni G, Neyman K, Rösch N (1997) J. Phys. Chem. B 101:2786
207. Musolino V, Selloni A, Car R (1998) J. Chem. Phys. 108:5044
208. Radzig AA, Smirnov BM (1985) Reference Data of Atoms, Molecules and Ions. Springer, Berlin Heidelberg New York
209. Sanchez A, Abbet S, Heiz U, Schneider W-D, Häkkinen H, Barnett RN, Landman U (1999) J. Phys. Chem. A 103:9573
210. Zhukovskii V, Yu F, Kotomin EA, Jacobs PWM, Stoneham AM (2000) Phys. Rev. Lett. 84:1256
211. Conard T, Vohs JM, Thiry PA, Caudano R (1990) Surf. Interf. Anal. 16:446
212. Zhou JB, Gustafsson T (1997) Surf. Sci. 375:221
213. Stumpf R, Scheffler M (1994) Phys. Rev. Lett. 72:254
214. Schaffner M-H, Patthey F, Schneider W-D (1999) Eur. J. Phys. D 9:609
215. Bromann K, Felix C, Brune H, Harbich H, Monot R, Buttet J, Kern K (1997) Surf. Sci. 377:1041
216. Ferrari AM, Pacchioni G (1996) J. Phys. Chem. 100:9032
217. Stirniman MJ, Huang C, Smith RS, Joyce SA, Kay BD (1996) J. Chem. Phys. 105:1295
218. Schaffner M-H, Patthey F, Schneider W-D (1998) Surf. Sci. 417:159
219. Dyke JM, Fayad NK, Morris A, Trickle IR (1979) J. Phys. B: Atom. Mol. Phys. 12:2985
220. Handschuh H, Cha C-Y, Bechthold PS, Ganteför G, Eberhardt W (1995) J. Chem. Phys. 102:6406
221. Fedrigo S, Harbich W, Buttet J (1993) Phys. Rev. B 47:10706
222. Lau JT, Achleitner A, Wurth W (2000) Surf. Sci. 467:L834
223. Schmidt M, Haberland H (1999) Eur. Phys. J. D 6:109
224. Scherer JJ, Paul JB, O'Keefe A, Saykally RJ (1997) Chem. Rev. 97:25
225. Berden G, Peeters R, Meijer G (2000) Int. Rev. Phys. Chem. 19:565
226. Xu S, Sha G, Xie J (2002) Rev. Sci. Instrum. 73:255
227. Antonietti J-M, Michalski M, Heiz U, Jones H, Lim KH, Rösch N, Del Vitto A, Pacchioni G (2005) Phys. Rev. Lett. 94:213402
228. Del Vitto A, Pacchioni G, Lim KH, Rösch N, Antonietti J-M, Michalski M, Heiz U, Jones H (2005) J. Phys. Chem. B 109:19876
229. Feenstra RM (1989) Phys. Rev. Lett. 63:1412
230. First PN, Stroscio JA, Dragoset RA, Pierce TD, Celotta RJ (1989) Phys. Rev. Lett. 63:1416
231. Kuk Y, Jarrold MF, Silverman PJ, Bower JE, Brown WL (1989) Phys. Rev. B: Condens. Matter 39:11168
232. Berndt R, Gaisch R, Gimzewski JK, Reihl B, Schlittler RR, Schneider W-D, Tschudy M (1993) Science 292:1425
233. Nilius N, Ernst N, Freund H-J (2000) Phys. Rev. Lett. 84:3994
234. Marsen B, Lonfat M, Scheier P, Sattler K (2000) J. Electron Spectrosc. Rel. Phen. 109:157
235. Klingeler R, Kann G, Wirth I, Bechthold PS, Eisebitt S, Neeb M, Eberhardt W (2000) In: Proceedings ISSPIC10. Atlanta, GA, USA, p. 12 ff
236. Meiwes-Broer K-H (2000) Electronic level structure of metal clusters at surfaces. In: Meiwes-Broer K-H (ed.), Metal Clusters at Surfaces. Springer Series in Cluster Physics, Berlin, Heidelberg, New York, p. 151 ff

237. Hövel H, Grimm B, Bödecker M, Fieger K, Reihl B (2000) Surf. Sci. 463:L603
238. Valden M, Lai X, Goodman DW (1998) Science 281:1647
239. Niemantsverdriet JW (1995) Spectroscopy in Catalysis. John Wiley and Sons Ltd., New York
240. Christmann K (1991) Introduction to Surface Physical Chemistry. Springer-Verlag, Berlin
241. Hoffmann FM (1983) Surf. Sci. Rep. 3:107
242. Freund H-J, Wolter K, Seiferth O, Libuda J, Kuhlenbeck H, Bäumer M (1998) Surf. Sci. 402–404: 428
243. Libuda J, Meusel I, Hoffmann J, Hartmann J, Piccolo L, Henry CR, Freund H-J (2001) J. Chem. Phys. 114:4669
244. Wörz A, Abbet S, Judai K, Heiz U (2003) J. Am. Chem. Soc. 125:7964
245. Stuckless JT, Frei NA, Campbell CT (1998) Rev. Sci. Instrum. 69:2427
246. Borroni-Bird CE, King DA (1991) Rev. Sci. Instrum. 62:2177
247. Vattuone L, Yeo YY, Kose R, King DA (2000) Surf. Sci. 447:1
248. Duriez C, Henry CR, Chapon C (1991) Surf. Sci. 253:190
249. Piccolo L, Becker C, Henry CR (1999) Eur. Phys. J. D. 9:415
250. Judai K, Abbet S, Wörz A, Heiz U (2003) Int. J. Mass Spectrom. 229:99
251. Röttgen M, Judai K, Abbet S, Wörz A, Henry CR, Heiz U (in preparation)
252. Heiz U, Sanchez A, Abbet S, Schneider W-D (1999) J. Am. Chem. Soc. 121:3214
253. Schumacher E (1997) DETMECH: Chemical Reaction Kinetics Software. In: University of Bern: Chemistry Department
254. Landman U (2005) Proc. Nat. Acad. Sci. USA 102:6671
255. Capelle K (2005) Cond-matt/0211443
256. Born M, Oppenheimer JR (1927) Ann. Phys. 84:457
257. Dreizler RM, Gross EKU (1990) Density Functional Theory. Springer-Verlag, Berlin
258. Kohn W (1999) Rev. Mod. Phys. 71:1253
259. Pople JA (1999) Rev. Mod. Phys. 71:1267
260. Parr RG, Yang W (1989) Density Functional Theory of Atoms and Molecules. Oxford University Press, Oxford
261. Koch W, Holthausen MC (2001) A Chemist's Guide to Density Functional Theory. John Wiley & Sons, New York
262. Jones RO, Gunnarsson O (1989) Rev. Mod. Phys. 61:689
263. Seminario JM (ed.) (1996) Recent Developments and Applications of Modern DFT. Elsevier, Amsterdam
264. Nalewajski RF (ed.) (1996) Density Functional Theory I–IV. Topics in Current Chemistry. Springer, Berlin Heidelberg New York
265. Anisimov VI (ed.) (1999) Strong Coulomb Correlatins in Electronic Structure Calculations: Beyond the Local Density Approximation. Gordon & Breach, New York
266. March NH (1992) Electron Density Theory of Atoms and Molecules. Academic, London
267. Laird BB, Ross RB, Ziegler T (eds.) (1996) Chemical Applications of Density Functional Theory. American Chemical Society, Washington, DC
268. Chong DP (ed.) (1995) Recent Advances in Density Functional Methods. World Scientific, Singapore
269. Joulbert D (ed.) (1998) Density Functionals: Theory and Applications. Springer Lecture Notes in Physics. Springer, Berlin Heidelberg New York

270. Fiolhais C, Nogueira F, Marques M (eds.) (2003) A Primer in Density Functional Theory. Springer Lecture Note in Physics. Springer, Berlin Heidelberg New York
271. Dobson JF, Vignale G, Das MP (eds.) (1998) Density Functional Theory: Recent Progress and New Directions. Plenum, New York
272. Eschrig H (1996) The Fundamentals of Density Functional Theory. Teubner, Leipzig
273. Hohenberg P, Kohn W (1964) Phys. Rev. 136:B864
274. Kohn W, Sham LJ (1965) Phys. Rev. 140:A1133
275. March NH (1975) Self-Consistent Fields in Atoms. Pergamons, Oxford
276. Wang LW, Teter MP (1992) Phys. Rev. B 45:13196
277. Foley M, Madden PA (1996) Phys. Rev. B 53:10589
278. Zhou BJ, Ligneres VL, Carger EA (2005) J. Chem. Phys. 122:044103
279. Lieb EH, Oxford S (1981) Int. J. Quantum Chem. 19:427
280. Chan GK-L, Handy NC (1999) Phys. Rev. A 59:3075
281. Chayes JT, Chayes L, Ruskai MB (1985) J. Stat. Phys. 38:497
282. Ullrich CA, Kohn W (2002) Phys. Rev. Lett. 89:156401
283. Tao J, Perdew JP (2005) J. Chem. Phys. 122:114102
284. Kohn W, Becke AD, Parr RG (1996) J. Phys. Chem. 100:12974
285. Gross EKU, Runge E, Heinonen O (1992) Many Particle Theory. Adam Hilger, Bristol
286. Szabo A, Ostlund NS (1989) Modern Quantum Chemistry. McGraw-Hill, New York
287. Chen J, Krieger JB, Li Y, Iafrate GJ (1996) Phys. Rev. A 54:3939
288. Vosko SH, Wilk L, Nusair M (1980) Can. J. Phys. 58:1200
289. Perdew JP, Zunger A (1981) Phys. Rev. B 23:5048
290. Perdew JP, Wang Y (1993) Phys. Rev. B 45:13244
291. Ceperley DM, Alder BJ (1980) Phys. Rev. Lett. 45:566
292. Perdew JP, Burke K, Ernzerhof M (1996) Phys. Rev. Lett. 77:3865
293. Perdew JP, Burke K, Ernzerhof M (1997) Phys. Rev. Lett. 78:1396(E)
294. Becke AD (1988) Phys. Rev. A 38:3098
295. Lee C, Yang W, Parr RG (1988) Phys. Rev. B 37:785
296. Filippi C, Umrigar CJ, Taut M (1994) J. Chem. Phys. 100:1290
297. Ziesche P, Kurth S, Perdew JP (1998) Comput. Mater. Sci. 11:122
298. Becke AD (1993) J. Chem. Phys. 98:5648
299. Becke AD (1997) J. Chem. Phys. 107:8554
300. Becke AD (1999) J. Comp. Chem. 20:63
301. Tozer DJ, Handy NC (1998) J. Chem. Phys. 108:2545
302. van Voorhis T, Scuseria GE (1998) J. Chem. Phys. 109:400
303. Perdew JP, Kurth S, Zupan A, Blaha P (1999) Phys. Rev. Lett. 82:2544
304. Tao J, Perdew JP, Staroverov VN, Scuseria GE (2003) Phys. Rev. Lett. 91:146401
305. Tao J, Perdew JP, Staroverov VN, Scuseria GE (2003) J. Chem. Phys. 119:12129
306. Tao J, Perdew JP, Staroverov VN, Scuseria GE (2004) J. Chem. Phys. 120:6898
307. Tao J, Perdew JP, Staroverov VN, Scuseria GE (2004) Phys. Rev. B 69:075102
308. Becke AD (1996) J. Chem. Phys. 104:1040
309. Ki YH, Lee IH, Nagaraja S, Leburton JP, Hood RQ, Martin RM (2000) Phys. Rev. B 61:5202

310. Kurth S, Perdew JP, Blaha P (1999) Int. J. Quantum Chem. 75:889
311. Adamo C, Ernzerhof M, Scuseria GE (2000) J. Chem. Phys. 112:2643
312. Hartree DR (1928) Proc. Cambridge Philos. Soc. 24:89
313. Hartree DR (1928) Proc. Cambridge Philos. Soc. 24:111
314. Hartree DR (1928) Proc. Cambridge Philos. Soc. 24:426
315. Svane A, Gunnarsson O (1990) Phys. Rev. Lett. 65:1148
316. Svane A, Gunnarsson O (1994) Phys. Rev. Lett. 72:1248
317. Szotek Z, Temmermann WM, Winter H (1994) Phys. Rev. Lett. 72:1244
318. Vydrow OA, Scuseria GE (2004) J. Chem. Phys. 121:8187
319. Strange P, Svane A, Temmermann WM, Szotek Z, Winter H (1999) Nature 399:756
320. von Barth U, Hedin L (1972) J. Phys. C 5:1629
321. Gunnarsson O, Lundqvist B (1976) Phys. Rev. B 13:4274
322. Oliver GL, Perdew JP (1979) Phys. Rev. A 20:397
323. Capelle K, Libero VL (2005) Cond-matt/0506167
324. Troullier N, Martins JL (1991) Phys. Rev. B 43:1993
325. Pastore G, Smargiassi E, Buda F (1991) Phys. Rev. A 44:6334
326. Thijssen JM (2000) Computational Physics. Cambridge University Press, Cambridge
327. Martin RM (2004) Electronic Structure: Basic Theory and Practical Methods. Cambridge University Press, Cambridge
328. Hellmann H (1937) Einführung in die Quantumchemie, Franz Duetsche, Leipzig
329. Feynman RP (1939) Phys. Rev. 56:340
330. Car R, Parrinello M (1985) Phys. Rev. Lett. 55:2471
331. Barnett RN, Cleveland CL, Landman U (1985) Phys. Rev. Lett. 54:1674
332. Barnett RN, Cleveland CL, Landman U (1985) Phys. Rev. Lett. 55:2035
333. Barnett RN, Landman U (1993) Phys. Rev. B 48:2081
334. Payne MC, Teter MP, Allan DC, Arias TA, Joannopoulos JD (1992) Rev. Mod. Phys. 64:1045
335. Häkkinen H, Landman U (2000) Phys. Rev. B 62:R2287
336. Marcus R (1966) J. Chem. Phys. 45:4493
337. Vineyard GH (1957) J. Phys. Chem. Solids 3:121
338. Berne BJ, Ciccotti G, Coker DF (eds.) (1998) Classical and Quantum Dynamics in Condensed Phase Simulations. World Scientific, Singapore
339. McKee ML, Page M (1993) Reviews in Computational Chemistry, vol. IV. VCH, New York
340. Leach AR (1996) Molecular Modelling: Principles and Applications. Addison-Wesley, Longman, Essex
341. Jonsson H, Mills G, Jacobsen KW (1998) In: Berne BJ, Ciccotti G, Coker DF (eds.), Classical and Quantum Dynamics in Condensed Phase Simulations. World Scientific, Singapore, p. 385
342. Henkelman G, Jonsson H (2000) J. Chem. Phys. 113:9978
343. Henkelman G, Uberuaga BP, Jonsson H (2000) J. Chem. Phys. 113:9901
344. Somorjai GA (1994) Introduction to Surface Chemistry and Catalysis. Wiley, New York
345. Hollemann AF, Wiberg N (1995) Lehrbuch der Anorganischen Chemie. Walter de Gruyter, Berlin
346. Ertl G (1983) Kinetics of Chemical Processes on Well-defined Surfaces. In: Anderson JR, Boudart M (eds.), Catalysis, Science and Technology. Springer, Berlin Heidelberg New York, p. 209

347. Conrad H, Ertl G, Küppers J (1978) Surf. Sci. 76:323
348. Bell AT (2003) Science 299:1688
349. Thomas JM (1994) Angew. Chem. 106:963
350. Hammer B, Nørskov JK (1995) Nature 376:238
351. Hagen J, Socaciu LD, Le Roux J, Popolan D, Bernhardt TM, Wöste L, Mitrić R, Noack H, Bonačić-Koutecký V (2004) J. Am. Chem. Soc. 126:3442
352. Wallace WT, Whetten RL (2002) J. Am. Chem. Soc. 124:7499
353. Koszinowski K, Schröder D, Schwarz H (2003) J. Am. Chem. Soc. 125:3676
354. Koszinowski K, Schröder D, Schwarz H (2003) ChemPhysChem 4:1233
355. Cox DM, Brickman R, Creegan K, Kaldor A (1991) Z. Phys. D 19:353
356. Mills G, Gordon MS, Metiu H (2002) Chem. Phys. Lett. 359:493
357. Wallace W, Leavitt AJ, Whetten RL (2003) Chem. Phys. Lett. 368:774
358. Varganov SA, Olson RM, Gordon MS, Mills G, Metiu H (2003) Chem. Phys. Lett. 368:778
359. Gioumousis G, Stevenson DP (1958) J. Chem. Phys. 29:294
360. Wallace WT, Wyrwas RB, Whetten RL, Mitrić R, Bonačić-Koutecký V (2003) J. Am. Chem. Soc. 125:8408
361. Socaciu LD, Hagen J, Le Roux J, Popolan D, Bernhardt TM, Wöste L, Vajda S (2004) J. Chem. Phys. 120:2078
362. Pacchioni G (2000) Surf. Rev. Lett. 277:7
363. Di Valentin C, Del Vitto A, Pacchioni G, Abbet S, Wörz A, Judai K, Heiz U (2002) J. Phys. Chem. 106:11961
364. Peterka D, Tegenkamp C, Schröder KM, Ernst W, Pfnür H (1999) Surf. Sci. 431:146
365. Illas F, Pacchioni G (1998) J. Chem. Phys. 108:7835
366. Haruta M (1997) Catal. Today 36:153
367. Haruta M, Daté M (2001) Appl. Catal. A 222:427
368. Häkkinen H, Abbet S, Sanchez A, Heiz U, Landman U (2003) Angew. Chem. Int. Ed. 42:1297
369. Yoon B, Häkkinen H, Landman U, Wörz AS, Antonietti J-M, Abbet S, Judai K, Heiz U (2005) Science 307:403
370. Cox DM, Brickman RO, Creegan K, Kaldor A (1991) Mater. Res. Soc. Symp. Proc. 206:43
371. Socaciu LD (2004) Ph.D. Thesis, Freie Universität, Berlin
372. Lee TH, Ervin KM (1994) J. Phys. Chem. 98:10023
373. Jackschath C, Rabin I, Schulze W (1992) Ber. Bunsenges. Phys. Chem. 96:1200
374. Handschuh H, Ganteför G, Bechthold PS, Eberhardt W (1994) J. Chem. Phys. 100:7093
375. Nygren MA, Siegahn PEM, Jin C, Guo T, Smalley RE (1991) J. Chem. Phys. 95:6181
376. Wallace WT, Whetten RL (2000) J. Phys. Chem. B 104:10964
377. Wallace WT, Whetten RL (2001) Eur. J. Phys. D 16:123
378. Balteanu I, Balaj OP, Fox BS, Beyer MK, Bastl Z, Bondybey VE (2003) Phys. Chem. Chem. Phys. 5:1213
379. Yuan DW, Zeng Z (2004) J. Chem. Phys. 120:6574
380. Wolf S, Sommerer G, Rutz S, Schreiber E, Leisner T, Wöste L, Berry RS (1995) Phys. Rev. Lett. 74:4177
381. Leisner T, Vajda S, Wolf S, Wöste L, Berry RS (1999) J. Chem. Phys. 111:1017
382. Häkkinen H, Landman U (2001) J. Am. Chem. Soc. 123:9704

383. Huber H, McIntosh D, Ozin GA (1977) Inorg. Chem. 16:975
384. Bernhardt TM, Socaciu-Siebert LD, Hagen J, Wöste L (2005) Appl. Catal. A 291:170
385. Scheffler M, Stampfl C (1999) In: Horn K, Scheffler M (eds.), Handbook of Surface Science. Elsevier, Amsterdam
386. Knecht J, Fischer R, Overhof H, Hensel F (1978) J. Chem. Soc. Chem. Commun. :905
387. Allers K-H, Pfnuer H, Feulner P, Menzel D (1994) J. Chem. Phys. 100:3985
388. Eichler A, Hafner J (1999) Phys. Rev. B 59:5960
389. Haruta M, Takase T, Kobayashi T (1991) In: Yoshida S, Takezawa N, Ono T (eds.), Catalytic Science and Technology. Kodansha, Tokyo, p. 331
390. Daté M, Haruta M (2001) J. Catal. 201:221
391. Daté M, Ichihashi Y, Yamashita T, Chiorino A, Bocuzzi F, Haruta M (2002) Catal. Today 72:89
392. Daté M, Okumura M, Tsubota S, Haruta M (2004) Angew. Chem. Int. Ed. 43:2129
393. Kung HH, Kung MC, Costello CK (2003) J. Catal. 216:425
394. Costello CK, Yang JH, Law HY, Wang Y, Lin JN, Marks LD, Kung MC, Kung HH (2003) J. Catal. 243:15
395. Bond GC, Thompson DT (1999) Catal. Rev. Sci. Eng. 41:319
396. Bongiorno A, Landman U (2005) Phys. Rev. Lett. 95:106102
397. Yoon B, Häkkinen H, Landman U (2003) J. Phys. Chem. A 107:4066
398. Giordano L, Goniakowski J, Pacchioni G (2003) Phys. Rev. B 67:045410
399. Pireaux JJ, Chtaib M, Delrue JP, Thiry PA, Liehr M, Caudano R (1984) Surf. Sci. 141:211
400. Xu Y, Mavrikakis M (2003) J. Phys. Chem. B 107:9290
401. Lopez N, Norskov JK (2002) J. Am. Chem. Soc. 124:11262
402. Molina LM, Hammer B (2004) Phys. Rev. B 69:155424
403. Smith T (1980) J. Colloid Interf. Sci. 75:51
404. Hintz PA, Ervin KM (1994) J. Chem. Phys. 100:5715
405. Ganteför G, Schulze Icking-Konert G, Handschuh H, Eberhardt W (1996) Int. J. Mass Spectrom. Ion Proc. 159:81
406. Chatterjee B, Akin FA, Jarrold CC, Raghavachari K (2003) J. Chem. Phys. 119:10591
407. Zhou M, Andrews L, Bauschlicher CW Jr (2001) Chem. Rev. 101:1931
408. Brönstrup M, Schröder D, Kretzschmar I, Schwarz H, Harvey JN (2001) J. Am. Chem. Soc. 123:142
409. Goursot A, Papai I, Salahub DR (1992) J. Am. Chem. Soc. 114:7452
410. Ren XL, Hintz PA, Ervin KM (1993) J. Chem. Phys. 99:3575
411. Vajda S, Leisner T, Wolf S, Wöste L (1999) Philos. Mag. B 79:1353
412. Grushow A, Ervin KM (1995) J. Am. Chem. Soc. 117:11612
413. Grushow A, Ervin KM (1997) J. Chem. Phys. 106:9580
414. Spasov VA, Ervin KM (1998) J. Chem. Phys. 109:5344
415. Klopcic SA, Moravec VD, Jarrold CC (1999) J. Chem. Phys. 110:8986
416. Hintz PA, Ervin KM (1995) J. Chem. Phys. 103:7897
417. Li T, Balbuena PB (2001) J. Phys. Chem. B 105:9943
418. Balteanu I, Balaj OP, Beyer MK, Bondybey VE (2004) Phys. Chem. Chem. Phys. 6:2910
419. Kretzschmar I, Fiedler A, Harvey JN, Schröder D, Schwarz H (1997) J. Phys. Chem. A 101:6252

420. Armentrout PB, Halle LF, Beauchamp JL (1982) J. Chem. Phys. 76:2449
421. Weinberg WH (1973) J. Catal. 28:459
422. Avery NR (1983) Surf. Sci. 131:501
423. Kim MH, Ebner JR, Friedman RM, Vannice MA (2001) J. Catal. 204:348
424. Balaj OP, Balteanu I, Roßteuscher TTJ, Beyer MK, Bondybey VE (2004) Angew. Chem. Int. Ed. 43:6519
425. Sault AG, Madix RJ, Campbell CT (1986) Surf. Sci. 169:347
426. Matsushima T (1983) Surf. Sci. 127:403
427. Ohno Y, Matsushima T (1991) Sur. Sci. 241:47
428. Xu J, Yates JT (1993) J. Chem. Phys. 99:725
429. Szabo A, Henderson MA, Yates JT (1992) J. Chem. Phys. 96:6191
430. Freund HJ (1997) Angew. Chem. Int. Ed. Engl. 36:452
431. Goodman DW (1995) Surf. Rev. Lett. 2:9
432. Goodman DW (1995) Chem. Rev. 95:523
433. Henry CR, Chapon C, Duriez C (1991) Surf. Sci. 253:177
434. Zambelli T, Barth JV, Wintterlin J, Ertl G (1997) Nature 390:495
435. Yang SH, Drabold DA, Adams JB, Ordejon P, Glassford K (1997) J. Phys.: Condens. Matter 9:L39
436. Watwe RM, Spiewak BE, Cortright RD, Dumesic JA (1998) Catal. Lett. 51:139
437. Watari N, Ohnishi S (1997) J. Chem. Phys. 106:7531
438. Sachdev A, Masrl RI, Adams JB (1993) Z. Phys. D. 26:310
439. Becker C, Henry CR (1996) Surf. Sci. 352:457
440. Becker C, Henry CR (1997) Catal. Lett. 43:55
441. Piccolo L, Henry CR (2000) Appl. Surf. Sci. 670:162
442. Piccolo L, Henry CR (2001) J. Mol. Cat. 181:167
443. Abbet S, Riedo E, Brune H, Heiz U, Ferrari AM, Giordano L, Pacchioni G (2001) J. Am. Chem. Soc. 123:6172
444. Abbet S, Heiz U, Häkkinen H, Landman U (2001) Phys. Rev. Lett. 86:5950
445. Heiz U, Sanchez A, Abbet S, Schneider W-D (2000) Chem. Phys. 262:189
446. Taylor KJ (1993) Catal. Rev. Sci. Eng 35:457
447. Shelef M, Graham GW (1994) Catal. Rev. Sci. Eng 36:433
448. Nieuwenhuys BE (1999) Adv. Catal. 44:259
449. Zhdanov VP, Kasemo B (1997) Surf. Sci. Rep. 29:31
450. Daté M, Okuyama H, Takagi N, Nishijima M, Aruga T (1995) Surf. Sci. 341:L1096
451. Daté M, Okuyama H, Takagi N, Nishijima M, Aruga T (1996) Surf. Sci. 350:79
452. Pisanu AM, Gigola CE (1999) Appl. Catal. B 20:179
453. Hirsimäki M, Suhonen S, Pere J, Valden M, Pessa M (1998) Surf. Sci. 404:187
454. Rainer DR, Koranne M, Vesecky SM, Goodman DW (1997) J. Phys. Chem. B 101:10769
455. Vesecky SM, Paul J, Goodman DW (1996) J. Phys. Chem. 100:15242
456. Muraki H, Fujitani Y (1986) Ind. Eng. Chem. Prod. Res. Dev. 15:414
457. Xu XP, Goodman DW (1994) Catal. Lett. 24:31
458. Xu X, Chen P, Goodman DW (1994) J. Phys. Chem. B 98:9242
459. Rainer DR, Vesecky SM, Koranne M, Oh WS, Goodman DW (1997) J. Catal. 167:234
460. Vesecky SM, Chen P, Xu X, Goodman DW (1995) J. Vac. Sci. Technol. A 13:1539
461. Honkala K, Pirilä P, Laasonen K (2001) Surf. Sci. 489:72

462. Blagojevic V, Jarvis MJY, Flaim E, Koyanagi GK, Lavrov V, Böhme DK (2004) Angew. Chem. Int. Ed. 42:4923
463. Fayet P, Kaldor A, Cox DM (1990) J. Chem. Phys. 92:254
464. Jigato MP, Somasundram K, Termath V, Handy NC, King DA (1997) Surf. Sci. 380:83
465. Endou A, Yamauchi R, Kubo M, Stirling A, Miyamoto A (1997) Appl. Surf. Sci. 119:318
466. Duarte HA, Salahub DR (1999) Topics Catal. 9:123
467. Citra N, Andrews L (2000) J. Phys. Chem. A 104:8160
468. Judai K, Wörz AS, Abbet S, Heiz U (2003) The chemistry of free and supported metal clusters. In: Russo N, Salahub D (eds.), Metal–Ligand Interactions in Molecular-, Nano-, Micro-, and Macro-Systems in Complex Environments. Kluwer Academic, Netherlands, p. 153
469. Prévot G, Henry CR (2002) J. Phys. Chem. B 106:12191
470. Rainer DR, Xu C, Holmblad PM, Goodman DW (1997) J. Vac. Sci. Technol. A 15:1653
471. Schröder D, Schwarz H (1995) Angew. Chem. Int. Ed. 34:1973
472. Schwarz H, Schröder D (2000) Pure Appl. Chem. 72:2319
473. Trevor DJ, Whetten RL, Cox DM, Kaldor A (1985) J. Am. Chem. Soc. 107:518
474. Trevor DJ, Cox DM, Kaldor A (1990) J. Am. Chem. Soc. 112:3749
475. Achatz U, Berg C, Joos S, Fox BS, Beyer MK, Niedner-Schattenburg G, Bondybey VE (2000) Chem. Phys. Lett. 320:53
476. Klabunde KJ (1994) Free Atoms, Clusters, and Nanoscale Particles. Academic, San Diego
477. Blomberg MRA, Siegahn PEM, Svensson (1992) J. Phys. Chem. 96:5783
478. Cui Q, Musaev DG, Morokuma K (1998) J. Chem. Phys. 108:8418
479. Cui Q, Musaev DG, Morokuma K (1998) J. Phys. Chem. A 102:6373
480. Wesendrup R, Schwarz H (1997) Organometallics 16:461
481. Berg C, Kaiser S, Schindler T, Kronseder C, Niedner-Schattenburg G, Bondybey VE (1994) Chem. Phys. Lett. 231:139
482. Heinemann C, Cornehl HH, Schwarz H (1995) J. Organomet. Chem. 501:201
483. Abbet S, Sanchez A, Heiz U, Schneider W-D (2001) J. Catal. 198:122
484. Ormerod RM, Lambert RM (1990) J. Chem. Soc. Chem. Commun. :1421
485. Holmblad PM, Rainer DR, Goodman DW (1997) J. Phys. Chem. B 101:8883
486. Ormerod RM, Lambert RM (1992) J. Phys. Chem. 96:8111
487. Wörz A, Abbet S, Judai K, Pacchioni G, Heiz U (2003) J. Mol. Cat. 199:103
488. Ferrari AM, Giordano L, Pacchioni G, Abbet S, Heiz U (2002) J. Phys. Chem. B 106:3173
489. Tysoe WT, Nyberg GL, Lambert RM (1983) J. Chem. Soc. Chem. Commun.:623
490. Abbet S, Sanchez A, Heiz U, Schneider W-D, Ferrari AM, Pacchioni G, Rösch N (2000) Surf. Sci. 454–456:984
491. Judai K, Abbet S, Wörz A, Heiz U, Giordano L, Pacchioni G (2003) J. Phys. Chem. B 107:9377
492. Patterson CH, Lambert RM (1988) J. Phys. Chem. 92:1266
493. Abbet S, Sanchez A, Heiz U, Schneider W-D, Ferrari AM, Pacchioni G, Rösch N (2000) J. Am. Chem. Soc. 122:3453
494. Pacchioni G, Lambert RM (1994) Surf. Sci. 304:208
495. Fahmi A, van Santen RA (1996) J. Phys. Chem. 100:5676
496. Burkart S, Blessing N, Ganteför G (1999) Phys. Rev. B 60:15639

2

Theory of Metal Clusters on the MgO Surface: The Role of Point Defects

G. Pacchioni

2.1 Introduction

Oxides surfaces are finding continuous new applications in advanced technologies, like corrosion protection, thermal coating, in catalysis as inert supports or directly as catalysts, in microelectronics for their dielectric properties; films of magnetic oxides are integral components in magnetic recording devices and many microporous materials are based on oxides. For all these reasons, there is a considerable effort to better characterize the surfaces and the interfaces of oxide materials [1,2]. New experimental methods have been developed to grow well-defined films of oxides on metal supports in ultrahigh vacuum (UHV) conditions [3–10]. This approach allows overcoming some difficulties connected to the use of electron spectroscopies for the study of insulating materials, as most of the oxides are. Not surprisingly, the increasing experimental activity has stimulated a parallel computational activity based on high-quality first principle calculations.

As we mentioned, oxide surfaces are important in the field of nanocatalysis by supported metals. In practical applications, the support has the crucial role of stabilizing small metallic particles, which act as the actual catalysts in a chemical process. Once the oxide surface is sufficiently well characterized, one can deposit small metal clusters and study their reactivity as a function of the support, of the metal, of the size of the cluster, etc. In this way, complex catalytic processes can be divided into a series of substeps, which allow a more detailed microscopic characterization. Despite the fact that only recently well-defined metal clusters have been deposited under controlled conditions on oxide surfaces and thin films, great advances have been obtained in the understanding of the mechanisms of adhesion and growth of the metal particles to the oxide surface. In this process, the role of theory is quite substantial.

Important information coming from calculations is the structure of the oxide surface. Oxide surfaces are often heavily reconstructed or relaxed compared to the truncated bulk, and the experimental determination of the surface structure is not always straightforward. To this end, reliable classical potentials have been developed, in particular, for the study of ionic crystals and covalent solids. Nowadays, first principle band structure calculations making use of large supercells are routinely used. Band structure calculations, usually based on plane waves basis sets and on the density functional theory (DFT) approach, represent the most appropriate computational method to study solids. An alternative approach to band structure, the cluster approach, has been developed in the past two decades with the aim of describing in chemical terms interactions occurring at surfaces and other localized phenomena [11].

One of the problems, where the use of cluster models is more appealing, is in the study of point defects. These centers are often the most interesting ones from the point of view of the physical and chemical properties of a material. Several chemical reactions taking place at an oxide surface are directly or indirectly connected to the presence of point and extended defects. Unfortunately, defect centers are elusive species because of their low concentration even in the bulk material, and their identification by spectroscopic methods can be rather difficult. Furthermore, the distinction of surface from bulk defects may be extremely subtle. For all these reasons, the theoretical modeling of defect centers at the surface of oxides is attracting an increasing interest.

2.1.1 Oxide Surfaces: Single Crystals, Powders, Thin Films

There are several strategies to prepare oxide surfaces for cluster deposition [7, 12]. A lot of work has been done using high surface area polycrystalline oxides prepared by chemical means (e.g. decomposition of the corresponding hydroxides or carbonates). Actually, these are the "natural" candidates for practical applications as they are normally used in catalysis as supports for noble metal catalysts. Several attempts have also been done to deposit well-defined clusters onto the surface of polycrystalline oxides starting, for instance, from organometallic precursors [13]. The system is then treated thermally so that the ligands around the organometallic cluster are removed and the metal cage is left on the surface. A recent example is that of an Ir_4 cluster which has been prepared on two substrates, MgO and Al_2O_3, by deposition of $Ir_4(CO)_{12}$ [14]. The existence of the metal unit after the removal of the CO ligands has been proved by extended X-ray absorption fine structure (EXAFS) measurements. These nanoclusters exhibit a specific catalytic activity, in particular in propene hydrogenation. However, it has been found that replacement of the MgO with the Al_2O_3 support boosts the catalytic activity tenfold, showing the important role of the interface bonding on the electronic and chemical properties of the metal particle [14].

Oxide surfaces in normal conditions are generally covered by OH groups as they have the tendency to interact with water from the atmosphere leading to water dissociation and surface hydroxylation. Polycrystalline oxides present a very complex surface morphology, a high number of defects, pores, impurities, etc., and their use for detailed studies of metal–support interactions present problems related to this surface complexity. Therefore, other techniques have been developed in recent years to deal with this problem.

The most direct way to prepare an oxide surface is by in situ cleavage of a single crystal in UHV [1]. For some oxides, in particular cubic oxides, like MgO, NiO, $SrTiO_3$, this works nicely; for other oxides, like SiO_2, Al_2O_3, TiO_2, this process is much more difficult as they tend to form rough surfaces by cleavage. Most of these oxides are good insulators or wide gap semiconductors. This represents an additional problem for their study. To overcome this problem, single crystalline oxide surfaces are prepared by growing epitaxial thin films on single crystal metal supports [7, 12]. Epitaxy (literally "placed" (epi) "arrangement" (taxy) from Greek) is essentially a recrystallization process in which gaseous atoms condense on a template created by a substrate to form a single crystal film. At least three major techniques have been developed in this context [15]. One possibility is to oxidize the substrate by direct exposure of the metal surface to O_2 pressure (e.g., SiO_2 is grown in this way on Si single crystals for microelectronic applications). Often, the direct oxidation of the metal single crystal presents problems connected to the different lattice constants of the metal and its oxide. When the lattice mismatch is larger than a few percents, epitaxial growth is not possible and the film grows in microcrystalline form, with a lot of additional interfaces and grain boundaries where the control of the morphology at microscopic level is low. In some cases, the direct oxidation of the metal substrate leads to good quality films. This is the case for instance of NiAl alloys which are widely used to grow ultrathin, three atomic layers thick films of Al_2O_3 [8–10]. In this case, the segregation of Al toward the surface of the alloy leads to a well-defined interface and a regular Al_2O_3 film.

Probably, the most widely used method to grow oxide thin films in UHV conditions is by metal atoms deposition in an oxygen atmosphere [7]. Here, a metal is vaporized (e.g., in a Knudsen cell) and deposited onto a substrate in a given background O_2 pressure. One grows in this way films of various thicknesses on the substrate whose quality is strongly dependent on a number of parameters (e.g., temperature of the substrate, flux of metal vapor, O_2 pressure, etc.). Films of MgO and NiO on Mo(100), Ag(100), or other metal substrates are usually produced in this way. Recently, a similar technique has been adopted to grow epitaxially crystalline SiO_2 films on Mo(112) [16]. Theoretical studies have shown that the electronic structure of the corresponding single crystal oxide surface is usually recovered after a few oxide monolayers [17–19] (Fig 2.1). It is important to mention that the surface of these films contain more defects than the surface of cleaved single crystals. On the other hand, it is in principle possible to grow films with different concentrations and kinds of defects, allowing in this way to tune another important parameter of metal–oxide interfaces, i.e., the defects concentration.

All the oxide surfaces used to deposit metal clusters in controlled conditions need to be well characterized both structurally and electronically. As we mentioned above, thin oxide films allow one to use a variety of spectroscopies.

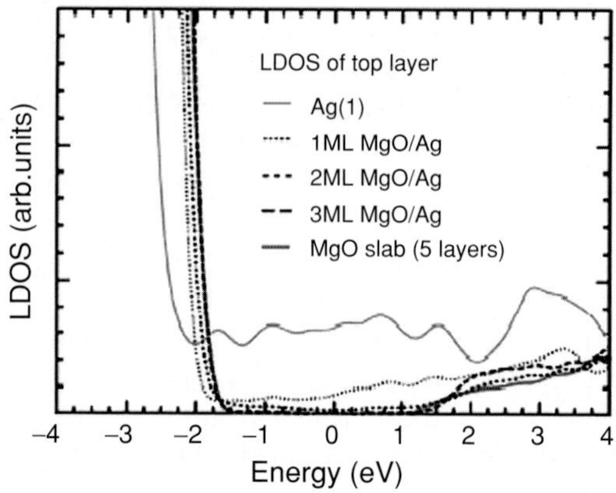

Fig. 2.1. Calculated surface layer LDOS (Local Density of States) for 0- to 3-MgO layers on Ag(100). Reproduced from [19]. Copyright 2001 American Physical Society

Low-energy electron diffraction (LEED) provides information on the long-range order of the film; this can be complemented by short-range order analysis like photoelectron diffraction (PD). Very important for the structural characterization are scanning tunneling microscopy (STM) and atomic force microscopy (AFM). STM is based on the measurement of the tunneling current from a tip to the sample; it provides an image of the empty or of the filled states of the atoms of the surface. When one is dealing with insulating materials, like MgO, SiO_2, or Al_2O_3, the use of STM is not possible unless one makes use of ultrathin films of a few layers thickness.

Once the oxide substrate has been structurally characterized, one has to identify the characteristic features of its electronic properties, and in particular the differences from the corresponding bulk oxides. In fact, the truncation of the bulk Madelung potential (MP) at the surface leads to considerable changes in the chemical bonding at the surface. One direct consequence of this is the reduction of the band gap at the surface; this is even more pronounced at low-coordinated sites, like steps, edges, kinks, and corners, and can lead to a completely different reactivity from that expected for the regular surface. In this respect, several characterization techniques have been applied. Ultraviolet photoemission (UPS) provides a way to investigate the valence band structure of the films. The problem here is that one wants to analyze only the very top layers of the film, which will directly affect the properties of the supported metal particle. UPS, on the other hand, is sensitive to several internal layers, and the features of the surface can be difficult to identify in this way. For this reason, UPS measurements have recently been complemented by another technique, metastable impact electron spectroscopy (MIES) [20] that makes use of energy transfer from He atoms in an excited state to the surface atoms of the films. The technique provides information about the valence band structure of the film but has the advantage of being sensible to the top layer only. In this way, it has even been possible to identify states in the gap associated to surface defects that are not observable with UPS because of the low sensitivity to the surface. Other means to characterize the surface are based on the detection of surface species by vibrational spectroscopies (e.g., to identify the presence of OH groups) or optical transitions (this can be useful in particular in the presence of localized defects).

In general, the good quality of the substrate is essential to analyze in detail the properties of the supported nanoclusters. In fact, as we will discuss below, the cluster–oxide interaction is such to influence the shape and the electronic structure of the deposited cluster. A deep understanding of the cluster properties needs therefore a similarly profound knowledge of the interface bonding. Furthermore, a precise knowledge of the surface structure is essential to design realistic theoretical models.

2.1.2 Metal Particles on Oxides

There are several reasons why supported metal clusters can play a role in nanoscience and nanotechnology [4, 21–23]. The most immediate application is related to catalysis [24]. In typical heterogeneous catalysts as those installed in automobiles, the catalyst consists of a structure covered with porous ceramic materials like alumina. On the surface of the ceramic are deposited small particles of noble metals like palladium, platinum, and rhodium. In order to be active, the metal particle must have a size of a few nanometers. The better the control on the size distribution of the particles, the more efficient the catalyst. However, despite several efforts to better characterize the relationships between surface morphology, size, and shape of the metal particles, a deep understanding of the catalytic mechanisms is often lacking. Also, the nature of the bonding at the interface between the particle and the oxide support is known only very approximately. Ideally, one would like to be able to prepare monodispersed catalysts particles, all having the same number of atoms and possibly of very small size. In fact, the active part of the catalyst is its surface, where the low-coordinated atoms are responsible for most of the chemistry. However, the surface atoms are only a small fraction of the total number of atoms in the particle. Since heterogeneous catalysis often relies on precious metals, reducing the size of the particle immediately results in a better surface-to-volume ratio. In this way, it is possible with a given quantity of material to produce a much larger number of active sites and to increase the efficiency of the catalyst.

The second appealing aspect of a good control on the size and shape of the particles is to achieve a better selectivity of the catalyst. A higher reactivity and a better selectivity are some of the final goals of nanocatalysis, an area aimed at preparing new catalysts at the nanometer scale with good perspectives in practical applications and bright potential for fundamental understanding. One should mention that the preparation of small metal particles supported on an oxide substrate is at the heart of the research on supported metal catalysts. However, what makes the relatively mature field of supported metal catalysts different from nanocatalysis? Basically, it is the level of control on the dimensions, structure, and chemical mechanisms. Nanocatalysis is related to the possibility to produce new systems at the nanoscale under controlled conditions. Some new techniques have been developed in this direction in the last few years (see other chapters of this book).

There are other fields where supported metal clusters will play a role like in the microelectronic technologies and in particular in the production of sensors or magnetic recording. Magnetic clusters provide a link between the magnetism at the atomic level and in the condensed state [25]. Finally, cluster-assembled materials offer new opportunities in material science.

No matter which is the application, in order to be of practical use, the clusters must be deposited and stabilized on a substrate. This leads us to the second aspect of the problem namely that of the atomistic characterization of the supporting material and the defects present on its surface.

2.1.3 The Role of Defects in Nucleation and Growth

A key aspect of the deposition and diffusion of metal atoms and clusters on oxide surfaces is the role played by point and extended defects. There is little doubt that the nucleation and growth of clusters and small metal particles occur at defect sites [26–33]. A recent example is that of atomically resolved STM images of Au atoms on TiO_2 [34] (Fig. 2.2).

Still, relatively little is known about the nature of the defects involved in this process. So far most (if not all) of the attempts to define the site where two metal atoms combine to form a dimer and start the nucleation of the cluster are limited to theoretical analyses [35, 36]. Also, in this case data are extremely scarce and the results are not unambiguous. To make an example, there is general consensus that oxygen vacancies (F centers, see Sect. 2.3.4) at the surface of MgO act as strong traps for the metal atoms diffusing on the surface [26–33]. However, the general assumption that F centers are nucleation

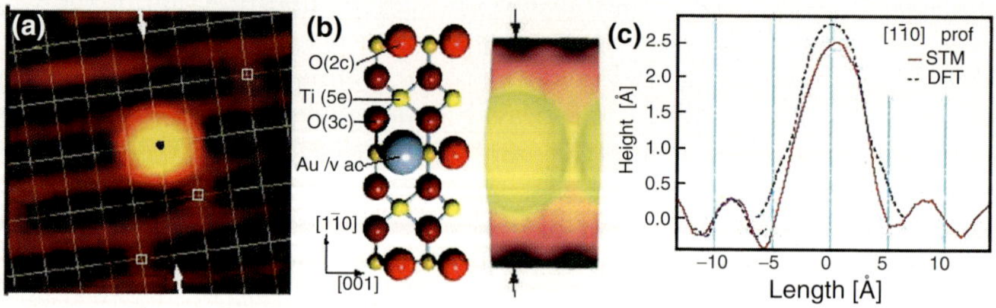

Fig. 2.2. (**a**) STM image of small Au clusters on TiO_2. Vacancies are marked with squares. (**b**) Simulated STM image of a single Au atom trapped in an oxygen vacancy. (**c**) Line profiles comparing DFT theoretical and experimental results. Reproduced from [34]. Copyright 2003 American Physical Society

sites has been recently questioned for the case of Pt and in general of the transition metals (TMs) at the right of the periodic table [35, 36]. It has been suggested that other sites, e.g., a divacancy and an OH group can act as efficient nucleation centers on the MgO surface. On the other hand, very recent results for Pd on MgO show that several defect sites act as dimerization centers with a net energy gain with respect of having two separated Pd atoms, one adsorbed at the defect and one adsorbed on a regular terrace site [37]. Thus, while there is little doubt that the diffusion of metal atoms is prevented once the metal binds to a defect site, it is still unclear which defect sites are most likely involved in the nucleation process.

An accurate answer to this question is probably not possible without the use of a combined theoretical–experimental approach. Based on experimental signatures, one often ends up with a shortlist of possible candidates rather than an unambiguous identification of the defect. The reverse is done by theory. A specific structure is assumed for a defect, and its properties and geometry are calculated at first principle level. If the calculated properties match the experimental ones, there is a high probability that the defect is identified and understood. Some successful examples of combination of theory and experiment to identify point defects and sites where metal atoms are deposited have been reported in the last few years. In this chapter, we report some of these results related to a specific surface, MgO(100), which turns out to be one of the better characterized in the large family of oxide surfaces.

2.2 Theoretical Models

2.2.1 Periodic Models

The electronic structure of solids and surfaces is usually described in terms of band structure. To this end, a unit cell containing a given number of atoms is periodically repeated in three dimensions to account for the "infinite" nature of the crystalline solid and the Schrödinger equation is solved for the atoms in the unit cell subject to periodic boundary conditions [38]. This approach can also be extended to the study of adsorbates on surfaces or of bulk defects by means of the supercell approach in which an artificial periodic structure is created where the adsorbate is translationally reproduced in correspondence to a given super lattice of the host. This procedure allows the use of efficient computer programs designed for the treatment of periodic systems and has indeed been followed by several authors to study defects using either DFT and plane waves approaches [39–41] or Hartree–Fock (HF)-based methods with localized atomic orbitals [42, 43].

The presence of the adsorbate in the surface unit cell, however, results in a periodic repetition of the adsorbate or defect in the two directions of space, hence modeling high coverage. The only way to reduce the adsorbate concentration is to increase the size of the unit cell, a solution that implies a large computational cost. Nowadays, periodic calculations for supercells containing several tens of atoms are routinely done. Even for large supercells containing ≈100 atoms, however, the coverage may still be too large. The supercell approach is therefore based on the assumption that the adsorbates (or defects) do not interact appreciably, except when they are very close to each other, so that rapid convergence is achieved by increasing the size of the supercell. With charged adsorbates (e.g., ions adsorbed at electrochemical cells) or charged defects, quite common in insulators, the supercell approach is feasible but less reliable because of the long-range Coulomb interaction between the charges. Methods to include correction terms to account for these spurious interactions have been proposed [44], but their use is not always straightforward.

2.2.2 Local Cluster Models

An alternative to the periodic band structure methods to study solids is the cluster approach [11,45–47]. Here, one considers explicitly only a finite number of atoms to describe a part of the surface while the rest is treated in a more or less simplified way (embedding). The main conceptual difference is that in the cluster approach one uses molecular orbitals (MO) instead of delocalized functions (e.g., plane waves). The description of the electronic properties is thus done in terms of local orbitals, allowing one to treat problems in solids with the typical language of chemistry, the language of orbitals. This is particularly useful when dealing with surface problems and the reactivity of a solid surface. In fact, the interaction of gas-phase molecules with a solid surface can be described in exactly the same way as the interaction of two molecules. Of course, also the cluster model is not free from limitations. The most serious one is that the effect of the surrounding is often taken into account in a more or less approximate way, thus leading to some uncertainties in the absolute values of the computed quantities. It is also possible that some properties are described differently depending on the size of the cluster used. It is therefore necessary to check the results versus cluster size and shape. The advantages, besides a smaller computational cost, are: (a) a very low coverage of adsorbates or concentration of defects is considered so that no mutual adsorbate–adsorbate or defect–defect interaction is present in the model and (b) that accurate theoretical methods derived from quantum chemistry can be applied (like configuration interaction approaches, see Sect. 2.2.5). This is an important advantage, which should not be underestimated. In fact, in this way it is possible to explicitly include correlation effects in the calculations and to treat exactly the nonlocal exchange as in the HF formalism; in DFT in fact the exchange is taken into account in an approximate way through the exchange-correlation functional (see "Density Functional Theory-Based Methods in Quantum Chemistry"). This second aspect can be particularly important for the description of magnetic systems, radical species, or defects with localized holes or electrons [48].

Therefore, cluster calculations represent an alternative way of describing localized bonds at surfaces as well as defects in ionic crystals. The problem is to introduce in a reasonable way the effect of the rest of the crystal. Completely different strategies can be adopted to "embed" clusters of largely covalent oxides, like SiO_2, or of very ionic oxides, like MgO. In SiO_2 and related materials, the cluster dangling bonds are usually saturated by H atoms [45, 47]. The saturation of the dangling bonds with H atoms is an important aspect of the embedding but not the only one. In fact, in this way one neglects the crystalline Madelung field. While this term is less important in semicovalent materials, like silica, it is crucial in the description of solid surfaces with more pronounced ionic character like MgO.

2.2.3 Embedding Schemes

The very ionic nature of MgO and other ionic oxides implies that the MP is explicitly included. Indeed, several properties of ionic crystals are incorrectly described if the long-range Coulomb interactions are not taken into account [49]. A simple approach is to surround the cluster of ions by a large array of point charges (PC) to reproduce the Madelung field of the host at the central region of the cluster [50]. However, the PCs polarize the oxide anions at the cluster border and cause an incorrect behavior of the electrostatic potential [51]. The problem can be eliminated by placing at the position of the positive PCs around the cluster an effective core potential (ECP) representing the finite size of the cation core [52]. No basis functions are associated to the ECP [53] that accounts for the Pauli or exchange repulsion of the O^{2-} valence electrons with the surrounding. This is a simplified approach to the more rigorous ab initio model potential (AIMP) method [54, 55] but is computationally simple and reliable. In the AIMP approach, the grid of bare charges is replaced by a grid of AIMP, which account not only for the long-range Coulomb interaction but also for the quantum mechanical short-range requirements of exchange and orthogonality without introducing explicitly extra electrons in the model.

The addition of the ECPs to the cluster gives a better representation of the electrostatic potential hence of the electrostatic contribution to the surface bonding. What is still missing from this simplified approach is the polarization of the host crystal induced by an adsorbed species or by the presence of a defect. This effect can be particularly important for charged adsorbates or defects. The polarization, E_{pol}, induced by a charge on the surrounding lattice can be estimated by means of the classical Born formula [56]:

$$E_{\text{pol}} = -\frac{(1 - 1/\varepsilon)q^2}{2R}, \tag{2.1}$$

where ε is the dielectric constant of the material, q the absolute value of the charge, and R the radius of the spherical cavity where the charge is distributed. Since a certain degree of ambiguity remains in the definition of R, this correction is only qualitative. A more refined approach that has been used for the study of the ground state of oxygen vacancies in MgO [57] makes use of the ICECAP program [58]. In this approach, instead of PCs the cluster is surrounded by polarizable ions described according to the shell model approach [59, 60]; in this way, the polarization response of the host is taken into account self-consistently up to infinite distance. In the shell model, an ion is represented by a point core and a shell connected by a spring to simulate its dipole polarizability.

A model based on a mixed quantum–classical approach has been proposed a few years ago [61]. Briefly, the ionic crystal is represented by a large finite nanocluster, which is divided into two regions: region I, which is centered on the defect site and the rest—region II. Region I at the center of the nanocluster includes a quantum-mechanically treated cluster (QM cluster) surrounded by interface ions and a region of classical shell model ions [59]. The remaining part of the nanocluster is represented by PC. The classical ions, both shell model and PCs, interact between themselves via classical interatomic potentials. All quantum–mechanical, interface, and classical ions (both cores and shells) in region I are allowed to relax simultaneously in the course of geometry optimization. Ions outside region I remain fixed and provide accurate electrostatic potential within region I. This approach allows one to take into account the defect-induced lattice polarization of a very large crystal region. In bulk calculations, it can be extended to infinity outside region I using a polarizable continuum model and the Mott–Littleton approach [62]. The interface between QM cluster and classical ions is needed to prevent an artificial spreading of electronic states outside QM cluster. The interface includes cations only; these are represented using a semilocal effective core pseudopotential [63]. The interface atoms interact quantum mechanically with atoms in the QM cluster and classically with other interface atoms and with classical atoms in regions I and II. The interaction between the QM and classical atoms in regions I and II is also included and represented using short-range classical potentials. In this way, it has been possible to include both geometric and electronic relaxations in a region of several angstroms in diameter at a reasonable computational cost [61]. A similar approach has been recently implemented and applied to the study of Pd atoms on MgO [64].

An alternative, more rigorous approach has been developed in recent years by Pisani and coworkers [65–67]. It is named perturbed cluster method and is based on the EMBED computer program [68]. With this approach, the properties of adsorbates at the surface of MgO have been studied at the HF and MP2 levels. The method relies on the knowledge of the one-electron Green function G^{f} for the unperturbed host crystal, which is obtained by means of the periodic program CRYSTAL [69, 70]. A cluster (C) containing the adsorbate or the defect is defined with respect to the rest of the host (H). The molecular solution for the cluster C in the field of H is corrected self-consistently by exploiting the information contained in G^{f} in order to allow a proper coupling of the local wave function to that of the outer region.

2.2.4 Electronic Structure Methods

In this section, we provide a brief account of the different theoretical methods used in the study of electronic structure. This includes two families of methods that arise from the principles of quantum mechanics, the ab initio methods of computation of electronic wave functions and the methods based on DFT. The choice of a particular computational method must contemplate the problem to be solved and in any case is a compromise between accuracy and feasibility. Details of the methods outlined in this section can be found in specialized references, monographs [71], and textbooks [72].

Wave Function–Based Methods

The best attainable approximation to the wave function and energy of a system of N electrons is given by the full configuration interaction (FCI) approach. In the FCI method, the wave function is written as an expansion of Slater determinants with the electrons distributed in the orbitals in all possible ways and the expansion coefficients are given by the corresponding secular equation. This is the exact solution (in a given finite orbital subspace) and is independent on the N-electron basis used. This is a very important property because it means that the total energy and the final wave function are independent of whether atomic or molecular orbitals are used as one-electron basis to construct the Slater determinants. The use of atomic orbitals leads to valence bond (VB) theory [73], whereas use of molecular orbitals leads to the molecular orbital configuration interaction (MO-CI) theory [72]. In the FCI approach, both theories are exactly equivalent and provide the same exact wave function. However, the dimension of the FCI problem grows so fast that practical computations can be carried out only for systems with small number of electrons. Therefore, the FCI method is mostly used to calibrate more approximate methods [74].

The simplest N-electron wave function that can be imagined is a single Slater determinant and the energy is computed as an expectation value. Of course, constraining the wave function to just one Slater determinant largely reduces the variational degrees of freedom of the wave function and the energy is uniquely defined by the one-electron basis used to construct this particular Slater determinant. The only variational degree of freedom concerns the orbital set which is chosen to minimize the energy expectation value. This leads to the HF equations. The optimum orbitals are then named self consistent and HF is synonymous of self-consistent method. Using the HF orbitals has technical advantages because, at least for the ground state wave function, the HF determinant contribution to the FCI wave function is by far the dominant term. The fact that the electronic Hamiltonian includes up to two-electron interactions suggests that double excitations would carry the most important weight in the FCI wave function; this is indeed the case. Therefore, one may design an approximate wave function in which only the reference HF determinant plus the double excited determinants are included. The result is called the doubly excited CI (DCI) method and is routinely used in ab initio calculations. Adding single excitations is important to describe some properties such as the dipole moment (SDCI method) [72].

The truncated CI methods described above are variational and finding the energy expectation values requires the diagonalization of very large matrices. An alternative approach is based on the estimate of the contribution of the excited determinants by using the Rayleigh–Schrödinger perturbation theory up to a given order. This is the basis of the widely used MP2, MP3, MP4, etc., methods which use a particular partition, the Möller–Plesset one, of the electronic Hamiltonian and a HF wave function as zero-order starting point [75, 76]. A disadvantage of perturbation theory is that the perturbation series may converge very slowly or even diverge. Furthermore, its application to metal clusters is not always satisfactory.

The logical extension of the truncated CI expansion is the so-called multireference CI (MRCI) approach where excitations, usually single and double, for a set of reference determinants are explicitly considered [77, 78]; the method is referred to as MR(SD)CI method. Energies and MRCI wave functions are obtained by solving the corresponding secular equation. Again, while the concept is quite simple, solving the corresponding eigenvalue problem is not simple at all. Several ways to solve the problem have been proposed leading to methods, like CIPSI [79–81] and CASPT2 [82–84], where CAS stands for complete active space, the active space being defined once a subset of orbitals is chosen. The use of these methods is restricted to the cluster model approach and is practicable only for relatively small systems (but examples of calculations of solid state problems were up to 100 electrons have been explicitly correlated exists [85]). Still, there is a considerable value in using this kind of approaches in order to estimate the reliability of the results obtained with other, more approximate methods. For instance, these methods can be used to calibrate and validate other approaches, like for instance in the calculation of excited states of defects in solids with time-dependent density functional theory (TD-DFT) (see "Density Functional Theory-Based Methods in Quantum Chemistry") [86]. In this respect, correlated calculations with cluster models represent an important tool also in the study of surface phenomena.

Density Functional Theory–Based Methods in Quantum Chemistry

The Schrödinger equation provides a way to obtain the N-electron wave function of the system and the approximate methods described in the previous section permit reasonable approaches to this wave function. The total energy can be obtained from the approximate wave function as an expectation value and the different density matrices, in particular, the one-particle density matrix, can be obtained in a straightforward way as

$$\rho_0(x) = \int \Psi_0^*(x)\Psi_0(x)\mathrm{d}x, \tag{2.2}$$

where the integration is carried out for the spin and space coordinates of all electrons but one.

In 1964, Hohenberg and Kohn proved a theorem that states that the inverse of this proposal does also hold [87]. They proved that for a nondegenerate ground state the one-electron density determines the external potential and hence the electronic Hamiltonian, the ground state energy, and the electronic wave function. They have proved that a universal functional exists so that the total energy is a functional of the density; i.e., given a density there is a mathematical rule that permits to obtain the exact ground state energy. The resulting theoretical framework is nowadays referred to as density functional theory or simply DFT. Since the functional is unknown, one may think that DFT is useless. However, Hohenberg and Kohn also proved a variational theorem stating that the ground state energy is an extreme (a minimum) for the exact density and, later, Kohn and Sham proposed a general framework that permits the practical use of DFT [88].

In the Kohn–Sham formalism, one assumes that there is a fictitious system of N noninteracting electrons experiencing the real external potential and this has exactly the same density as the real system. This reference system permits to treat the N-electron system as the superposition of N one-electron systems and the corresponding N-electron wave function of the reference system will be a Slater determinant. This is important because in this way DFT permits to handle both discrete and periodic systems. To obtain a trial density one needs to compute the energy of the real system and here it is when a model for the unknown functional is needed. To this purpose, the total energy is written as a combination of terms, all of them depending on the one-electron density only:

$$E[\rho] = T_\mathrm{s}[\rho] + V_\mathrm{ext}[\rho] + V_\mathrm{coulomb}[\rho] + V_\mathrm{XC}[\rho], \tag{2.3}$$

$T_\mathrm{s}[\rho]$ is the kinetic energy of the noninteracting electrons, $V_\mathrm{ext}[\rho]$ accounts for the contribution of the external potential, $V_\mathrm{coulomb}[\rho]$ corresponds to the classical Coulomb interaction of noninteracting electrons and $V_\mathrm{XC}[\rho]$ accounts for the exchange part due to the Fermi character of electrons and the correlation contribution due to the fact that electron densities are correlated. Obviously, the success of DFT is strongly related to the ability to approximate E_XC in a sufficiently accurate way. Now (2.3) plus the Hohenberg–Kohn variational theorem permit to vary the density by varying the orbitals, which are now referred to as Kohn–Sham orbitals. In addition, the Kohn–Sham orbitals can be expressed in a given basis set as in the usual LCAO (Linear Combination of Atomic) approach. The overall procedure is then very similar to the HF one and the orbitals minimizing (2.3) are those satisfying a one-electron eigenvalue problem. Here, however, the one-electron operator contains the exchange and correlation effective potentials. Mathematically, the exchange and correlation potentials are the functional derivatives of the corresponding energy contributions in (2.3). Once $E_{XC}[\rho]$ (or more precisely $E_X[\rho]$ and $E_C[\rho]$) is known, the effective potentials are also known and solving the Kohn–Sham equations is similar to solving the HF equation with the important difference that here one may find the exact solution, if $E_{XC}[\rho]$ is the exact one.

Several approaches to $E_{XC}[\rho]$ have been proposed in the last years with increasing accuracy and predictive power [89]. However, in the primitive version of DFT, the correlation functional was ignored and the exchange part approximated following Slater's $\rho^{1/3}$ proposal (the method was known as Xα). In 1980, Vosko et al. [90] succeeded in solving the electron correlation for a homogeneous electron gas and in establishing the corresponding correlation potential. The resulting method including also the exchange part is nowadays known as the local density approximation (LDA) and has been successful in the description of metals, although has experienced more difficulties in the description of molecules and ionic systems (for instance LDA incorrectly predicts NiO to be a metal [91, 92]). The Kohn–Sham equations were initially proposed for systems with a closed shell electronic structure. The study of open shell systems can be carried out using a spin-unrestricted formalism. In this case, different spatial orbitals are used for alpha- and beta-spin orbitals. In the case of LDA, the resulting formalism is known as the local spin density approximation (LSDA). This is similar to the well-known unrestricted Hartree–Fock (UHF) formalism and suffers from the same drawbacks when dealing with open shell systems [72]. This is especially important when attempting to study magnetic systems with many open shells. The unrestricted Kohn–Sham determinant corresponds in fact to a mixture of different possible multiplets.

In spite of the inherent simplifications, LDA (and LSDA) predictions on molecular geometries and vibrational frequencies are surprisingly good. However, binding energies are much less accurate and require going beyond this level of theory. The density functional (DF) methods that go beyond LDA can be grossly classified in gradient corrected (GC) and hybrid methods. In the first set, the explicit calculation of the $E_{XC}[\rho]$ contributions involves not only the density, ρ, but also its gradient $\nabla\rho$. The number of GC functionals is steadily increasing but among the ones widely used we quote the Becke exchange functional (B) [93], and the Perdew–Wang (PW) [94, 95], exchange-correlation functional [PW is usually referred to as the generalized gradient approximation (GGA)]. Another popular GC correlation functional is the one proposed by Lee–Yang–Parr, LYP) [96]. In some cases, one uses B for the exchange and PW for the correlation part—This is usually referred to as BP—or B for the exchange and LYP for the correlation part giving rise to the BLYP method. The term hybrid functionals is used to denote a family of methods based on an idea of Becke [97]. This approach mixes DF and Fock exchange and local and GC correlation functionals in a proportion that is obtained from a fit to experimental heats of formation for a wide set of molecules. The most popular hybrid method is B3LYP, where the number indicates that three parameters are fit to experiment [97].

Before closing this paragraph, it is useful to mention that successful examples of applications of DFT to excited state problems have been reported in recent years. This is based on the TD-DFT method. In this method, excitation energies are calculated from the poles of the frequency-dependent polarizability and the oscillator strengths from the residues. Recent studies show that TD-DFT is more accurate than the configuration interaction single (CIS) method [98] and that it can provide transition energies with accuracy similar to that of MRCI or coupled-cluster methods [86]. However, Rydberg states [99] and in general transitions to states that are close to the conduction band edge [86] are poorly reproduced. For a more detailed description of GC and hybrid methods, the reader is referred to general textbooks [100] or to a more specialized literature [89].

2.2.5 Density Functional Theory Versus Wave Function Methods: Cu on MgO

Aside the great advances in the development of new exchange-correlation functionals, the question of which functional provides the best chemical accuracy is still under discussion. In the wave function–based methods, it is possible to check the accuracy of a given level of theory by systematic improvement on the basis set and on the level of treatment of electron correlation. Unfortunately, this is not completely feasible in the framework of DFT. One can improve the basis set but there is no way to systematically improve the $E_{XC}[\rho]$ term. Therefore, one needs to establish the accuracy of a given approach by comparing several choices for $E_{XC}[\rho]$. For many systems, this choice is not critical and the use of several functionals permits to add error bars to the computed quantities. However, in the systems of interest in the present chapter, namely the description of the metal–oxide interface, the choice of the functional is crucial [101]. An extreme case is that of Cu adsorption on MgO: The reported adsorption energies range from a practically unbound Cu atom at the HF level to moderate adsorption, 0.35–0.90 eV, at gradient corrected DF level, to strong adsorption, about 1.5 eV, using LDA. Ranney et al. [102] have extrapolated the adhesion energy of a single copper atom on MgO from their microcalorimetric measures of the heat of adsorption and have found a value of 0.7 eV. The comparison of this value with the computed adsorption energies of [101] indicates that, among the currently used approximations of the exchange-correlation functional, the BP and BLYP ones seem to provide the best answer, while hybrid functionals slightly underestimate the adsorption energy.

2.3 Defects on MgO

Point defects in oxides determine the optical, electronic, and transport properties of the material and usually dominate the chemistry of its surface. A detailed understanding and a control at atomistic level of the nature (and concentration) of point defects in oxides are therefore of fundamental importance to synthesize new materials with well-defined properties. Of course, before to be created in controlled conditions point defects have to be known in all aspects of their physicochemical properties. The accurate theoretical description of the electronic structure of point defects in oxides is essential for the understanding of their structure–properties relationship.

MgO is a particularly well-studied oxide; the structure of the (100) single crystal surface is extremely flat, clean, and stoichiometric. Both relaxation, $-0.56 \pm 0.4\%$, and rumpling, $1.07 \pm 0.5\%$, are extremely small [103]. However, no real crystal surface consists of only idealized terraces. A great effort has been undertaken to better characterize the MgO surface, in particular for polycrystalline or thin film forms, which in some cases exhibit a heterogeneous surface due to the presence of various sites. All these sites can be considered as defects.

In the following, we provide a summary of the most important kinds of defects at the MgO surface (Table 2.1). In a broad classification, one can recognize at least 12 major kinds of point defects:

Low-coordinated cations (Sect. 2.3.1). These are Mg^{2+} ions with a number of neighbors lower than on the flat (100) terraces. To this category belong four-coordinated ions located at step and edge sites, $Mg^{2+}{}_{4c}$, and the three-coordinated ions located at corners, kinks, etc., $Mg^{2+}{}_{3c}$.

Low-coordinated anions (Sect. 2.3.2). The O^{2-} sites exhibit a completely different chemistry when located in the bulk, at the five-coordinated terraces, $O^{2-}{}_{5c}$, or at irregular sites with lower coordination, $O^{2-}{}_{4c}$ and $O^{2-}{}_{3c}$, where the MP (and the basicity) of the site change dramatically.

Hydroxyl groups (Sect. 2.3.3). H_2O is almost ubiquitous and easily reacts with low-coordinated sites to form OH groups at the surface of MgO. These centers induce asymmetries in the surface electric field and exhibit a classical Brønsted acid behavior.

Oxygen vacancies (Sect. 2.3.4). These are usually called color or F centers from the German word for color, *Farbe*. The vacancies can have different formal charges. The removal of a neutral O atom results in a neutral F center; the removal of an O^- ion in an F^+ center (paramagnetic); the removal of an O^{2-} ion in an F^{2+} center. Since the surface O ions can be located at terraces, steps, edges, corners, etc., also the corresponding vacancies can in principle form at five-, four-, and three-coordinated sites. In the following, the subscript s distinguishes a surface F center, $F_s{}^{n+}$, from the bulk counterpart, F^{n+}.

Cation vacancies (Sect. 2.3.5). These are often classified as V_s centers. Cation vacancies in MgO correspond to the removal of Mg, Mg^+, or Mg^{2+}

Table 2.1. Summary of most important surface point defects in MgO

Defect	Symbol	Schematic description
Low-coordinated cation	${Mg^{2+}}_{nc}$ $(n = 3, 4)$	Coordinatively unsaturated cation
Low-coordinated anion	${O^{2-}}_{nc}$ $(n = 3, 4)$	Coordinatively unsaturated anion
Hydroxyl group	(OH)	Proton attached to O^{2-}
Anion vacancy	${F^{m+}}_{nc}$ $(m = 0, 1, 2; n = 3, 4, 5)$	Missing oxygen with trapped electrons
Cation vacancy	${V^{m-}}_{nc}$ $(m = 0, 1, 2; n = 3, 4, 5)$	Missing cation with holes at O neighbors
Mg and O divacancy	V_{MgO}	Cation and anion vacancy
Impurity atoms	$M^{n+}/O^{n-}; Mg^{2+}/X^{2-}$	Substitutional cation (M) or anion (X)
Oxygen radical	${O^{-}}_{nc}$ $(n = 3, 4, 5)$	Hole trapped at O anion
Anion vacancy aggregate	M^{m+} $(m = 0, 1)$	Two (or more) adjacent oxygen vacancies
Shallow electron traps	$MgO_{corner}(e^{-})$	Trapped electrons at exposed cations
Deep electron traps	$(M^{+})(e^{-})$ pair	Cation (H^{+}, Na^{+}, etc.) and electron pairs
(111) microfacets	None	Small ensemble of Mg_{3c} or O_{3c} ions

species and result in V, V^{-} and V^{2-} defects. The V and V^{-} centers are paramagnetic.

Divacancies (Sect. 2.3.6). Closely related to the O and Mg vacancies are the divacancies, defect centers created by removing a neutral MgO unit. The formation of a divacancy on a (100) terrace results in a mini step, a site where some of the surface atoms are four coordinated.

Impurity atoms (Sect. 2.3.7). The presence of substitutional atoms can lead to modified chemical centers on the surface. The replacement of Mg ions by Ni ions, as in MgO–NiO solid solutions, introduces TM atoms in a MgO matrix and can alter the local properties of the material. Even more effective is the replacement of a divalent Mg^{2+} cation by a monovalent dopant, like Li^{+}. In order to compensate the charge, an O^{2-} anion at the surface becomes O^{-}, a highly reactive paramagnetic species.

Oxygen radical anions (Sect. 2.3.8). The O^{-} center can be formed by various means on the surface and not only by doping the material with alkali metals ions. O^{-} exhibits a high chemical reactivity and can be located at low-coordinated sites as well as at regular terrace sites.

Anion vacancy aggregates (Sect. 2.3.9). Under some conditions, aggregates of O vacancies can form in the bulk of MgO. The smallest aggregate is the M center, consisting of two adjacent O vacancies.

Shallow electron traps (Sect. 2.3.10). Some specific sites at the MgO surface exhibit electron affinities of about 1 eV and can bind electrons. Among these sites, there are low-coordinated cations, like an $Mg^{2+}{}_{3c}$ corner site, divacancies, and other morphological sites like the reverse corner.

$(M^+)(e^-)$pairs (Sect. 2.3.11). The number of paramagnetic centers on polycrystalline MgO can be increased by chemical means, like for instance alkali metal deposition or exposure to molecular hydrogen followed by UV irradiation. In both cases, irregular sites at the surface, like steps, corners, reverse corners, etc., can promote the stabilization of $(M^+)(e^-)$ pairs (M = H or an alkali metal). These neutral defects are color centers with characteristic EPR (Electron Paramagnetic Resonance) signals.

(111) microfacets (Sect. 2.3.12). Small ensembles of three-coordinated atoms with a pyramidal structure resembling a reconstructed (111) polar surface can form and have been indeed been proposed as the centers where oxygen-exchange reactions occur.

2.3.1 Low-Coordinated Cations

Adsorption of CO on MgO is probably the most studied case of molecular adsorption on an oxide [104]. CO adsorbs on Mg cations with the molecular axis normal to the (100) MgO surface. Infrared (IR) measurements at 77 K of the C–O vibrational frequency on polycrystalline MgO have shown a small shift of about $+10\,\mathrm{cm}^{-1}$ with respect to free gas-phase CO [105]. The adsorption energy of CO on MgO powders [106] is quite low, 0.15–0.17 eV, indicating weak adsorption. This is confirmed by experiments on single crystal MgO [107,108]. CO thermal desorption spectra (TDS) show a peak at 57 K corresponding to an adsorption energy, 0.14 eV, very close to the earlier data of CO on MgO powder [106]. When adsorbed on low-coordinated cations, the properties of adsorbed CO change significantly. Part of the interaction of CO with the cation sites of MgO is of electrostatic nature [109], and there is a direct relationship between the surface electric field and the CO stretching frequency [109]. Less-coordinated cations generate a larger electric field, which interacts with the CO dipole and multipole moments giving rise to an increase of ω_e. The adsorption of CO on low-coordinated Mg^{2+} cations at edge and corner sites leads to larger binding energies and larger CO vibrational shifts [110]. For instance, CO frequency shifts of +31, +55, and $+97\,\mathrm{cm}^{-1}$ were predicted at the HF level for CO adsorbed on $Mg^{2+}{}_{5c}$ (terrace), $Mg^{2+}{}_{4c}$ (edge), and $Mg^{2+}{}_{3c}$ (corner) sites [110], with a similar trend to that experimentally observed in IR for CO on various sites of MgO smokes, +13, +21, and $+60\,\mathrm{cm}^{-1}$ [111]. Accurate cluster model calculations on CO/MgO including correlation effects [112] have shown that the binding energies of CO adsorbed on Mg^{2+} cations at terraces, edge, and corners of MgO are 0.08, 0.18, and 0.48 eV; the frequency shifts are of +9, +27, and $+56\,\mathrm{cm}^{-1}$, respectively [112]. However, it has been found recently that the correlation between the electric field generated by the surface cations, the strength of the MgO–CO bonding, and the frequency shift

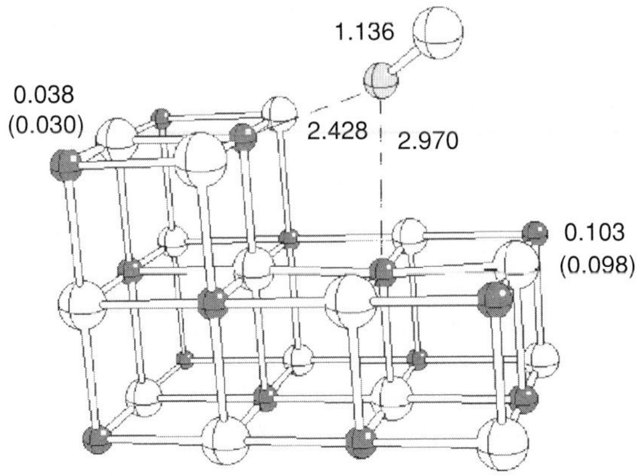

Fig. 2.3. Optimal structure of CO adsorbed on a step on the MgO surface. Bond lengths and inward displacements (Å) of the adsorbing cation(s) with and without (in parentheses) adsorbates are given. In this configuration, the CO molecule is relatively strongly bound but exhibits almost not shift of the vibrational frequency. Reproduced from [113]. Copyright 2000 Elsevier

of the adsorbed CO molecule is not always so simple: On some specific sites (e.g., a step, Fig. 2.3), the nonuniform field results in a relatively large binding energy (0.2 eV) despite a negligible CO vibrational shift [113].

2.3.2 Low-Coordinated Anions

The activation of an adsorbed molecule is often connected to the transfer of electronic charge to its antibonding orbitals, with consequent rupture or weakening of the bond. In this respect, the O anions can play a very important role in the activation of adsorbed species. A simple example is that of the interaction of CO_2 and SO_2 with MgO. These two molecules can form surface carbonates, $MgO + CO_2 \rightarrow MgCO_3$, and sulfites, $MgO + SO_2 \rightarrow MgSO_3$. Ab initio cluster model studies [114, 115] have shown that the five-coordinated O^{2-} ions at the (100) MgO terraces are unreactive toward CO_2 and SO_2, while the $O^{2-}{}_{4c}$ ions at the step sites are very reactive and spontaneously form carbonates and sulfites. The $O^{2-}{}_{5c}$ ions at the (100) terraces of CaO exhibit a similar reactivity to that of the low-coordinated anions of MgO, and stable chemisorbed species form at the terraces of CaO. This result is consistent with the trend of surface basicity, MgO < CaO < SrO < BaO. The surface basicity is the capability of the oxide to donate charge to an adsorbed molecule (Lewis acid). A general explanation of this behavior has been proposed [114]. The different reactivity of the terrace, step, and corner O^{2-} ions is not due to a different charge of the anions that is similar in high- and low-coordinated sites [116]. The O^{2-} anion is unstable in gas phase where it dissociates into $O^- + e^-$. O^{2-} anions are stable in ionic solids because of the Madelung potential (MP). A larger MP results in a more stable

O^{2-} ion (the corresponding filled 2p levels are shifted to lower energy). The MP is smaller at low-coordinated sites: On MgO, the MP is 23.9 eV for a bulk $O^{2-}{}_{6c}$, 23.0 eV for a terrace $O^{2-}{}_{5c}$, 21.8 eV for a step $O^{2-}{}_{4c}$, and 18.4 eV for a corner $O^{2-}{}_{3c}$ [45]. Thus, an O^{2-} at a corner site is less stable than the same ion at a terrace site. CaO has the same cubic structure of MgO but a larger lattice constant (2.399 Å in CaO versus 2.106 Å in MgO); in a full ionic model, the MP at the (100) terraces of CaO is 20.2 eV, about 3 eV smaller than in MgO. This means that the donor levels of an $O^{2-}{}_{5c}$ ion (terrace) of CaO are at a similar energy of the corresponding levels of an O^{2-} ion at the step and corner sites of MgO. The oxide basicity is therefore a direct function of the electrostatic potential at a given site: A lower potential results in a less stable anion and larger tendency to donate the electronic charge to an adsorbate. The different reactivity of MgO versus CaO surfaces predicted theoretically [114] has been confirmed by metastable impact electron spectroscopy (MIES) experiments on CO_2 adsorption on MgO [117] and CaO [118] as well as by synchrotron based photoemission spectroscopy on the same systems [119, 120].

The enhanced activity of the low-coordinated ions of the MgO surface has stimulated specific studies aimed at characterizing the nature of these centers [121–123]. We mentioned already that one important difference is the value of the MP at these sites and the consequent shift in position of the filled O 2p levels and stabilization of the empty Mg 3s levels. This results in a narrowing of the band gap in correspondence of these sites as found in CI cluster studies [123] of the optical absorption energies of O^{2-} ions at regular and low-coordinated sites of the MgO. A decrease of the transition energy from 7.1 eV for a terrace site to 5.7 and 4.5 eV for edge and corner sites, respectively, has been found [123].

2.3.3 Hydroxyl Groups

OH groups can easily form at the surface of MgO, even in UHV conditions or after thermal treatment of polycrystalline samples due to the presence of humidity or traces of H_2 in the environment. This may lead to the formation of a small but not negligible number of OH groups on the surface with possible consequences on the study of chemisorption processes. On MgO various types of OH groups can form, depending on the sites involved in the chemisorption process. Quite intensive theoretical efforts have been done to elucidate the mechanism of dissociation of H_2O molecules on MgO (for a complete account see a recent review [124]). This process is of fundamental importance for the understanding of the hydroxylation–dehydroxylation processes occurring on the surface and to identify the different OH groups present. The methods used to study H_2O adsorption and dissociation range from ab initio calculations on molecular $(MgO)_n$ clusters [125], to periodic band structure [126], and Car-Parrinello molecular dynamics [127–129] calculations. All methods show that isolated H_2O molecules are physisorbed on the MgO(100) terraces without dissociating while dissociation occurs rapidly at stepped surfaces or low-coordinated sites, in particular corner sites, a conclusion which is in complete agreement with the experimental observations. However, at high coverages water dissociation can occur also at the MgO terraces, thanks to the hydrogen bonding interactions between adjacent molecules [129].

2.3.4 Anion Vacancies

In MgO thin films, O vacancies can be generated by electron bombardment or sputtering with Ar^+. In polycrystalline MgO samples, O vacancies can be created by thermal treatment of hydroxylated surfaces. The dehydroxylation occurs at the expense of a lattice O^{2-} anion with consequent formation of a surface vacancy or F center. Anion vacancies on the MgO surface, the ${F_s}^{2+}$ centers corresponding to the removal of O^{2-}, are difficult to observe, as these centers do not have specific spectral properties. However, they can act as electron traps and their existence can be deduced by doping the material with excess electrons to form the corresponding ${F_s}^+$ and F_s centers which can be detected either by EPR (${F_s}^+$) [130–133] or by optical (${F_s}^+$ and F_s) [134–137] spectroscopies. In the F^+ or F centers one or two electrons are associated with the defect; they are localized in the vacancy [39–43, 50] by the MP.

In 1970, Tench suggested that a surface F center has a similar structure as in the bulk [130, 131] and proposed a model consisting of a missing O ion from a (100) terrace of MgO [130, 131]. The resulting vacancy can be classified as F_{5c} to indicate the fivefold coordination of the missing O ion. Recently, new accurate EPR measurements [133, 138] have shown that the large majority of the paramagnetic F^+ centers at the MgO surface obtained by irradiation in hydrogen are associated with the presence of a nearby proton. These centers are named $F_s(H)^+$ centers, i.e., an oxygen vacancy in the vicinity of an adsorbed hydrogen. The EPR signal of a surface paramagnetic center consistent with the Tench model, i.e., of the ${F_s}^+$ site, in absence of the coupling with the proton has not been observed yet. Furthermore, based on IR and EPR spectroscopies, as well as of ab initio calculations it has been suggested that the paramagnetic centers form preferentially at low-coordinated sites (steps, edges, corners) [138]. Theoretical estimates of the formation energies support the view of a lower cost to remove oxygen atoms from the low-coordinated sites [41, 139]. Several theoretical studies have been dealing with the hyperfine coupling constants of F^+ centers in MgO [50, 133, 140]. In general, excellent agreement is obtained for the bulk where a firm assignment is possible. For the surface, despite a satisfactory numerical agreement between theory and experiment a firm assignment is not yet possible.

An important question when one addresses the stability of charged defects in insulators is that of the relative position of the corresponding energy levels with respect to the vacuum level. This problem has been considered for the MgO(100) surface [61]. Using cluster models embedded in polarizable ions and PCs, the ionization energies and electron affinities of oxygen vacancies located at the terraces and corners of the MgO surface have been determined.

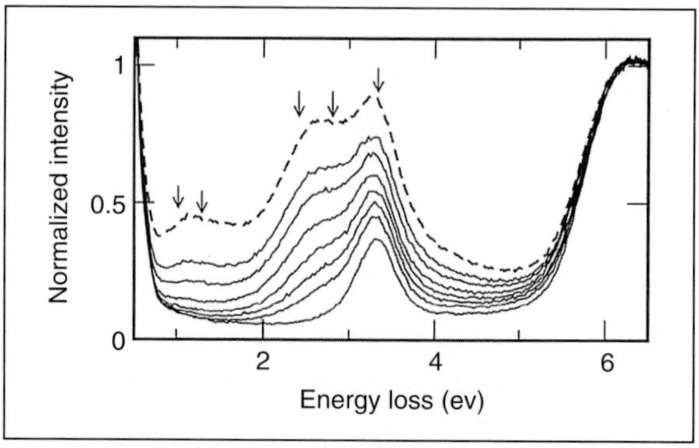

Fig. 2.4. EEL spectra of an MgO thin film after electron bombardment to create oxygen vacancies on the surface. Reproduced from [137]. Copyright 2002 Elsevier

A neutral F_{5c} center gives rise to an impurity level in the gap which is about 3 eV above the top of the valence band. F^+ centers give rise to states closer to the valence band (for a terrace $F^+{}_{5c}$ the level is about 1 eV above the O 2p states). The vertical ionization energy of a neutral F_{5c} defect is 3.4 eV; that of an $F^+{}_{5c}$ center 5.6 eV. Similar IPs (Ionization Potential) are found for F centers located at a corner site: The IPs of F_{3c} and $F^+{}_{3c}$ centers are 3.4 and 6.6 eV, respectively [61], showing that not only electrons trapped at low-coordinated sites may exist but they are also more strongly bound than at the corresponding terrace sites.

Another indication of the location of F centers in MgO comes from the analysis of optical spectra. Bulk F and F^+ centers give rise to an intense absorption band around 5 eV with two components, one due to F^+ centers at 4.96 eV and one to neutral F centers at 5.03 eV [141]. Much less is known on the optical properties of surface F centers. Optical spectra of MgO surface F centers have been measured on polycrystalline materials [138, 142], single crystals [134], and ultrathin MgO films [135–137, 143]. Optical measurements on fine powder samples of MgO using diffuse reflectance technique showed a feature at 2.05 eV attributed to $F^+{}_{5c}$ centers [142]. A peak at 2.3 eV has been observed by Henrich et al. [134] on MgO single crystals using electron energy loss spectroscopy (EELS). The feature was tentatively assigned to surface F centers. By reactive deposition of Mg in oxygen atmosphere or by electron bombardment of MgO thin films, Pfnür and coworkers [136, 137] have been able to study the optical spectra of MgO films with large oxygen deficiency (Fig. 2.4). They observed characteristic loss peaks at 2.4, 2.8, and 3.4 eV, which can be attributed to color centers due to surface oxygen vacancies located at various sites [137], in close agreement with theoretical predictions (see below). It is interesting to note that optical absorption bands in the same regions, 380 and 520 nm, have been observed in the UV–Vis spectra of polycrystalline MgO [138]. Only recently accurate CI calculations have been reported on this problem [85, 144]. The estimated transitions energies and the corresponding assignments are: Bulk F_{6c} and $F^+{}_{6c} \approx 5\,eV$, terrace F_{5c} and $F^+{}_{5c} \approx 3\,eV$, step F_{4c} and $F^+{}_{4c} \approx 2.2$–$2.5\,eV$, corner F_{3c} and $F^+{}_{3c} \approx 2.1$–$2.2\,eV$ [85, 144]. No evidence of transitions below 2 eV due to F centers has been found. The computed optical transitions imply that in the EELS experiment [136,137] not only five-coordinated F centers but also low-coordinated oxygen vacancies are present, in agreement with the results of other spectroscopic investigations [138]. Recent EELS measurements on stepped MgO surfaces fully confirm the assignment of low-energy transitions (around 2 eV) to low-coordinated F centers [145]. It is important to mention in this context that the optical absorptions measured by Pfnür and coworkers on MgO thin films [136,137,145] combined with the theoretical estimates of the transition energies associated to F centers [85, 144] not only allows a firm assignment for these transitions but also provides the only direct proof of the existence of F centers on the surface of MgO thin films grown in UHV on metal substrates.

2.3.5 Cation Vacancies

Among intrinsic defects, the cation vacancy or V center is of particular interest because it favors the formation of trapped holes at neighboring O ions. At the surface, this results in particularly active O^- species [146]. There are several studies dedicated to the V center in the bulk both experimental [147,148] and theoretical [40, 41, 149, 150], as reviewed by Chen and Abraham [151]. This center can exist in three charged states, V^0, V^-, V^{2-}. The ground state of V^0 corresponds to two holes located on oxygen ions at opposite sides of the vacancy, and the ground state is a triplet. The singly charged V^- defect is very stable and experimentally one of the best characterized, while theoretical information mostly exists on the diamagnetic V^{2-} center. Compared to the F centers, much less work has been devoted to the V_s center. One reason is that these centers are supposed to be present in lower concentrations than the O vacancies [152,153] due to their high formation energy (of the order of 20–30 eV for a V^{2-} center [40]). The electrons associated to the V^- and V^{2-} surface centers are occupying levels near the top O 2p valence band and the associated holes are completely localized on the oxygen atoms around the vacancy [50], in analogy with the bulk electronic structure. A combined periodic supercell and embedded cluster study on the electronic structure of the cation vacancy at the MgO surface [154] has shown that the formation energy of the neutral V^0 defect in the bulk, 15 eV, is higher than at the surface, 13.5 eV, suggesting that migration to the surface should be thermodynamically favored.

2.3.6 Divacancies

Early experiments on the EPR spectrum of paramagnetic centers at the surface of MgO by Lunsford and Jayne [155] have been interpreted in terms of an electron trapped at a surface divacancy because of the analogy with the spectrum of an $F^+{}_{5c}$ center, attributed to a divacancy with a trapped electron in bulk MgO [156]. There are several arguments in favor of the existence of divacancies [157]: (a) they are easier to form than a pair of isolated cationic and anionic vacancies; (b) due to the mobility of cationic vacancies, recombination of isolated vacancies should occur at high temperatures, $> 1,000\,K$, the typical temperatures reached in the thermal treatment of polycrystalline MgO; (c) the divacancy is intrinsically asymmetric, a fact that would easily explain the anisotropy of the EPR signal; (d) the divacancy is a neutral center and can be present in high concentrations without the need for compensating charges.

In the last few years, there has been a growing interest around this center [157, 158], leading to the conclusion that divacancies are relatively stable defects [157] with an appreciable electron affinity of almost 1 eV so that they can act as shallow electron traps (see Sect. 2.3.10). Recent experiments on the formation of surface O^- ions by bleaching color centers with N_2O point to the important role of these defects [146]. The study of the hyperfine coupling constants, however, resulted in a too small anisotropy of the signal compared to experiment. In this respect, the assignment of the observed EPR spectra [133] to surface divacancies with a trapped electron is not conclusive [157]. A detailed theoretical study of the reaction of a surface divacancy with H_2 aimed at the identification of the $F_s{}^+(H)$ color center observed in EPR also come to the conclusion that the calculated magnetic parameters show discrepancies with the experiment [158]. Thus, while the divacancy is an attractive center from a thermodynamic point of view, there is no clear spectroscopic evidence of its existence.

2.3.7 Impurity Atoms

Impurity atoms in MgO can be either cations replacing Mg^{2+} or anions substituting O^{2-}. In either case, the surface reactivity is significantly modified by their presence. At least two kinds of cationic impurities can be distinguished. If Mg^{2+} is replaced by another cation with the same formal charge, e.g., $Ca^{2+}, Ni^{2+}, Co^{2+}$, etc., the main changes in the electronic structure are connected to the different size of the ion, to small modifications in the chemical bonding with local increase or decrease of the covalent character and to the presence of partially filled d orbitals. On the other hand, when a monovalent cation, like Na^+ or Li^+ substitutes Mg^{2+}, an adjacent anion changes its formal charge from -2 to -1 [159]; the O^- species is a radical with a chemistry fundamentally different from that of the corresponding fully reduced species (Sect. 2.3.8). M^+/O^- pairs have been extensively studied in the bulk of alkaline-earth oxides by optical studies, EPR and ENDOR (Electron Nuclear Double Resonance) measurements [160, 161] as well as by embedded cluster calculations [162].

MgO can dilute ions of similar size, as for instance Ni^{2+} or Co^{2+} forming NiO–MgO and CoO–MgO solid solutions with an infinite range of composition.

The effect of progressively replacing Mg^{2+} by Ni^{2+} or similar cations (Co^{2+}, Cu^{2+}) on the surface properties has been investigated both experimentally [111, 163, 164] and theoretically [165, 166]. The presence of Ni^{2+} cations diluted in the MgO matrix results in an efficient catalyst for nitrous oxide, N_2O, decomposition probably because of the different bond strength of the Ni–O and Mg–O bonds at the surface [163]. Also NO, which on pure MgO is weakly bound [167], on Ni-doped MgO forms relatively strong bonds with the Ni^{2+} cations [168].

Less is known about anion substitution. A recent study of the O^{2-}–S^{2-} exchange reaction on MgO has been reported [169]. The reaction involves adsorption of CS_2 on MgO powders and the subsequent exchange reaction with formation of COS and S^{2-} ions located at the low-coordinated sites. The basicity of the MgO surface doped with sulfur ions is drastically modified with respect to that of pure MgO [169]. The substitution of low-coordinated O^{2-} anions of the MgO surface with sulfur anions completely suppresses the chemical activity of the sample.

2.3.8 O^- Radical Anions

O^- anions located at the surface of ionic crystals are capable of cleavage processes, such as hydrogen abstraction from methane and hydrocarbon molecules adsorbed on the surface. They can also facilitate UV-induced homolytic splitting of hydrogen. The reactivity of the O^- anions has been studied extensively by means of EPR and TPD (Temperature Programmed Desorption) [170]. Some of the most efficient catalysts to promote the conversion of methane into higher hydrocarbon derivatives are based on alkali-doped metal oxides, including Na/CaO and Li/MgO and their activity is attributed to the presence of M^+/O^- pairs. The O^- ions can also be produced on the MgO surface by other methods, e.g., by excitation of low-coordinated surface anions, O_{lc}, by UV light [146, 171–173] or filling anionic vacancies with paramagnetic O^- species [146, 174]. In the first case, the process involves ionization or charge transfer processes:

$$O_{lc}^{2-} + h\nu \rightarrow O_{lc}^{-} + e^{-}, \tag{2.4}$$

where the electron is released to an electron acceptor. In UHV conditions, short lived O^-–Mg^+ pairs form in the excitation step. The energy required to ionize O^{2-} anions from the MgO surface from DFT cluster calculations [61] is 6.7 eV, in excellent agreement with an experimental estimate of 6.7 ± 0.4 eV [20]. On an O_{3c} anion at a corner site, the IP decreases to 5.6 eV [61]. Thus, it is conceivable that UV irradiation will preferentially produce O^- radicals at low-coordinated sites.

Another mechanism to generate O^- radicals on the MgO surface implies the creation of color centers and then their bleaching with N_2O:

$$(e^-)_{trapped} + N_2O \rightarrow O^-_{surface} + N_2 \tag{2.5}$$

Clearly, the presence of O^- species at the surface of MgO is not disconnected from the existence of other defects, low-coordinated anions or O vacancies. In fact, the complex interconversion of one center into another one is one of the reasons for the difficult identification of defect centers on oxide surfaces.

2.3.9 M Centers (Anion Vacancy Aggregates)

In a recent paper by Pfnür and coworkers [137], O vacancies have been created on MgO thin films by electron bombardment and the corresponding optical properties have been measured by EELS (see Sect. 2.3.4). A band at 1.2 eV has been tentatively assigned to surface M centers, an aggregate of two adjacent F centers. Neutral and charged M centers in the bulk and on the surface of MgO have been studied by means of explicitly correlated cluster model calculations (CASPT2 approach) [175]. The ground state of the M center is characterized by the presence of two doubly occupied impurity levels in the gap of the material (Fig. 2.5); in M^+ centers, the highest level is singly occupied. The allowed electronic transitions have been computed and the results show that bulk M and M^+ centers give rise to intense absorptions at about 4.4 and 4.0 eV. Another less intense transition at 1.3 eV has also been found for the M^+ center. On the surface, the transitions occur at 1.6 eV (M^+) and 2 eV (M). While the computed transition energies for the surface ${M_S}^+$ center are close to the observed band in EELS experiments, this is not sufficient to conclusively demonstrate the presence of M centers. Still, high-temperature annealing of MgO samples can result in the migration of vacancies and formation of vacancy pairs or larger aggregates, a process which has not yet been carefully investigated.

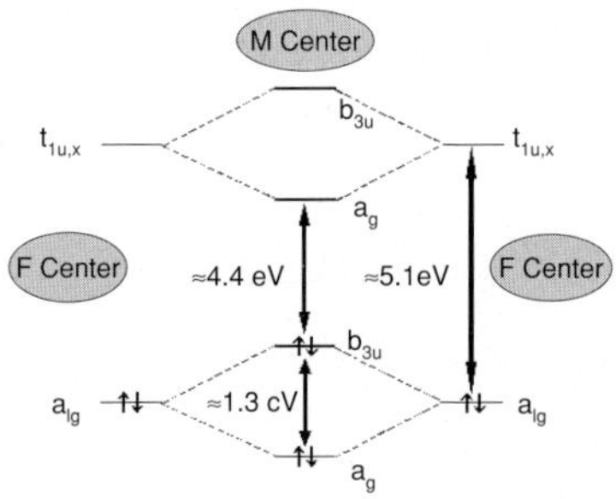

Fig. 2.5. Schematic representation of the energy levels involved in the electronic transitions of bulk F and M centers. The energies reported refer to the computed excitations energies for the bulk centers at the CASPT2 level [175]. Copyright 2003 American Physical Society

2.3.10 Shallow Electron Traps

We have mentioned in Sect. 2.3.6 that a possible precursor site for electron trapping is a Mg and O divacancy on the MgO terraces, V_{MgO}; recent theoretical calculations have confirmed this hypothesis [157, 176]. However, these are not the only shallow traps existing on the MgO surface. In fact, low-coordinated Mg sites, such as kinks and corners, can serve as surface electron traps [177] and even divacancies located at steps or corners are able to trap electrons [176]. All these centers are classified as shallow electron traps since they bind quite loosely one electron and are thermally unstable, at least at temperatures higher than 300 K. Still, the fact that on the MgO surface there is a large number of morphological sites which can act as electron traps has been recognized only recently. This fact could be of some importance to explain charging effects observed in AFM measurements on vacuum cleaved single crystal MgO samples.

2.3.11 $(M^+)(e^-)$ Centers

Although several experimental and theoretical studies point to the existence of surface oxygen vacancies (F_s centers, Sect. 2.3.4) [40–42, 178–181], their exact nature is still under discussion. The natural abundance of F centers (or at least of trapped electrons) in as-prepared high surface area MgO samples is too low to be monitored by UV–Vis spectroscopy and even by the highly sensitive EPR technique. The number of populated centers can be significantly increased using chemical or photochemical treatments. One way of achieving that is to deposit low dosage of alkali metals on the surface [133, 182, 183]. Another approach is based on the adsorption of molecular hydrogen, which dissociates at specific sites into H^+ and H^- species. Under UV irradiation, the H^- species transforms into $H + e^-$, where the electron is trapped at the surface producing an EPR-active center:

$$MgO + H_2 \rightarrow MgO(H^+)(H^-) \tag{2.6}$$

$$MgO(H^+)(H^-) + h\nu \rightarrow MgO(H^+)(e^-)_{trapped} + H \tag{2.7}$$

Recently it has been proposed that some morphological sites, like a corner Mg^{2+} cation, MgO_{corner}, Fig. 2.6, a reverse corner, MgO_{RC}, or a step, MgO_{step}, are the precursors of $MgO(H^+)(e^-)_{trapped}$ centers and that these are the actual structures of the $F_s(H)^+$ paramagnetic defect centers [184, 185] (see Sect. 2.3.4). The new sites can be considered as the prototypes of an entire family of surface traps. They represent abundant "natural" morphological features at real surfaces. Their existence does not require the high formation energy of other traps like the oxygen vacancies.

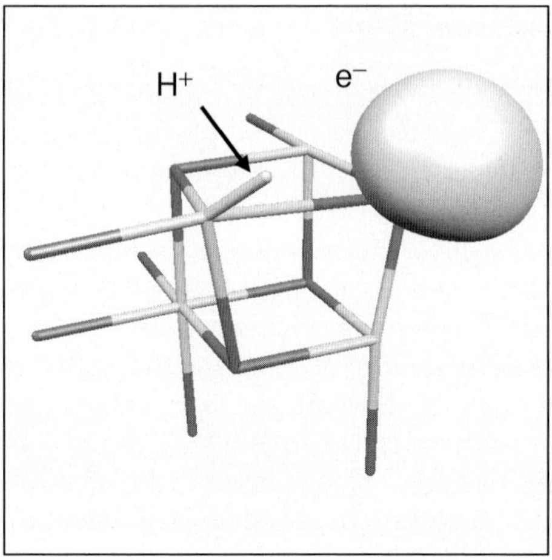

Fig. 2.6. Schematic structure of a $(H^+)(e^-)$ pair (paramagnetic color center) at an Mg^{2+} corner ion; the H^+ ion is bound to a four-coordinated O^{2-} anion

These centers in particular fit a number of observations [184, 185]:

1. They dissociate the H_2 molecule with a nonactivated reaction and an energy release, 0.5 eV, close to the measured one.
2. The structure of the resulting $MgO(H^+)(H^-)$ center is consistent with an heterolytic dissociation of H_2 and the computed vibrations are in full agreement with the observation.
3. The removal of a neutral H atom from the $MgO(H^+)(H^-)$ center has a cost consistent with the use of UV light for the process and results in a trapped electron at the MgO surface, $MgO(H^+)(e^-)_{trapped}$.
4. The analysis of the isotropic hyperfine coupling constants (in particular the existence of a small coupling with H) provides strong support to the proposed model [185]. New, highly resolved EPR spectra clearly show in fact three different ^{25}Mg hyperfine coupling constants at about 10–12, 30, and 60 G (Fig. 2.7). It has been suggested that the three signals arise from $MgO(H^+)(e^-)_{trapped}$ centers formed at steps, reverse corners, and corner sites, respectively, thus providing a direct proof of the large number of surface traps at the MgO surface and of the variety of coordination modes [185]. The results also provide a direct tool to titrate low-coordinated Mg sites at the surface of polycrystalline MgO samples.
5. The computed transition energies for the trapped electron, about 2 eV, are close to values reported for optical excitations at the surface of polycrystalline MgO where $F_s(H)^+$ defect centers have been created according to reactions (2.6) and (2.7).

Similar centers can be formed by alkali and other adsorbed atoms M, which can donate electrons to the surface and stabilize them by the Coulomb field created by the corresponding M^+ cation [186]. The centers illustrated

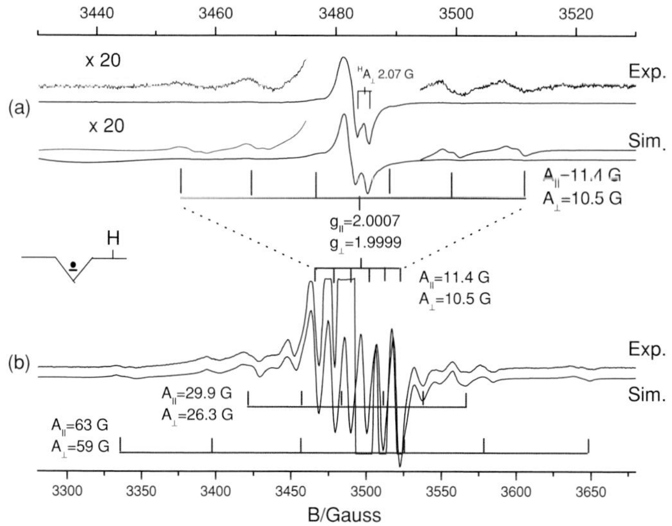

Fig. 2.7. Experimental and simulated EPR spectra of $MgO(H^+)(e^-)_{trapped}$ centers. The different values of the A matrix indicate the presence of a variety of centers. In particular, the features at 10–11 G are attributed to electrons trapped at steps, the features at 26–30 G at electrons trapped at reverse corners, and the weak features at 59–63 G at electrons trapped at exposed corner sites [185]. Copyright 2003 Wiley-VCH

in this section explain and rationalize a number of observations into a single coherent picture. It should be stressed at this point that these centers have been created and observed on the surface of polycrystalline MgO samples, characterized by a rich surface morphology and that their existence or formation has not been reported for MgO(100) single crystal surfaces nor for MgO thin films.

2.3.12 (111) Microfacets

In contrast to nonpolar (100) surfaces of rock-salt structured ionic materials, which are thermodynamically stable, (111) polar surfaces are unstable if they remain in the bulk terminated structure due to the diverging surface potential [2, 187] and undergo substantial surface reconstruction. It has been predicted that the (111) surfaces of NaCl, NiO, MgO, etc. reconstruct from the p(1 × 1) bulk terminated into a p(2 × 2) structure [187]. This kind of reconstruction has been indeed observed on NiO(111) grown on Au(111) [188]. Polar surfaces can also be stabilized by the presence of oxygen vacancies, impurities, hydroxyl groups [189] or in general molecular adsorbates [190, 191]; on NiO, by simple heat treatment, the OH groups are desorbed and the surface reconstructs exhibiting a diffuse p(2 × 2) structure [189]. The surface morphology and faceting of MgO(111) surfaces has been studied with atomic force, scanning, and electron microscopies [192, 193]. It has been suggested that oxygen trimers reminiscent of cyclic ozone appear on the reconstructed surface [193]. However, no further reports of similar structures have been reported and the formation of cyclic ozone has been questioned by some authors.

A related important defect in MgO consists in a missing cation from a cube corner, leaving a (111) microface with three oxygen ions and one hole. The electronic structure of this center was then investigated theoretically in a pioneering study of Kunz and Guse [194]. The involvement of (111) microfacets was also invoked in a thermal desorption and isotope exchange reactions study of CO and CO_2 on MgO where the condition needed for double exchange was identified by means of HF cluster calculations in small ensembles of three-coordinated O ions [195, 196].

2.4 Metal Deposition on MgO

2.4.1 Transition Metal Atoms on MgO(001)

Understanding the bonding mechanism of individual metal atoms with the surface of a simple oxide, like MgO, is essential for the theoretical study of more complex systems, like clusters or organometallic fragments, supported on oxide substrates. The interaction of various transition metal (TM) atoms, Cr, Mo, W, Ni, Pd, Pt, Cu, Ag, and Au, with the oxygen anions of the MgO(001) surface has been investigated theoretically [197] and provide a first basis for the identification of some trends. We restrict the analysis to the O anion sites as it is known from other theoretical investigations that the Mg cations of the surface are rather unreactive. Using cluster models and the DFT-BLYP functional [93, 96], it has been possible to divide the metal atoms in two groups according to their more or less pronounced tendency to bind to the MgO surface.

The presence of a filled d shell and singly occupied s orbital prevents the Cu, Ag, and Au atoms ($d^{10}s^1$ atomic ground state) from easily changing their configuration during bond formation. On the other hand, Cu, Ag, and Au form relatively strong bonds with oxygen atoms of not completely reduced oxide surfaces, like TiO_2, by partial transfer of their outer electron to the substrate [198]. On MgO, the coinage metal atoms retain their $n\mathrm{d}^{10}(n+1)\mathrm{s}^1$ atomic configuration and only a very small mixing of the metal σ orbitals with the O $2p_\sigma$ orbital occurs. As a consequence, these atoms are weakly bound to the surface with adsorption energies of 0.2–0.3 eV. This means that these atoms are likely to diffuse freely on the surface even at very low temperatures. Their interaction with the substrate becomes appreciable only in correspondence to surface defects.

There is little or no charge transfer from the Cu, Ag, or Au atoms to the substrate, and the bonding is mainly due to the polarization of the metal

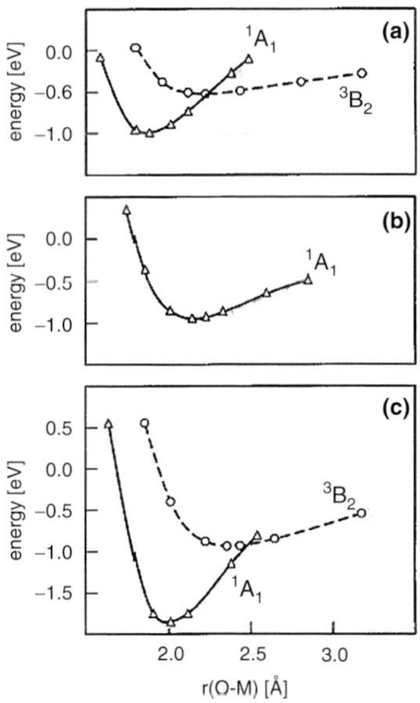

Fig. 2.8. Potential energy curves for the $^3B_2$ and $^1A_1$ states of (**a**) Ni, (**b**) Pd, and (**c**) Pt atoms interacting with an O atom of the MgO surface. Reproduced from [197]. Copyright 1997 American Chemical Society

electrons by the surface electric field plus a small covalent interaction. This result suggests that the oxide anions of the MgO surface are highly charged with little tendency to ionize the adsorbed coinage atoms. The relatively weak bonding found for Cu, Ag, and Au is therefore the consequence of the large adatom size (due to the singly occupied valence s orbital), which determines the long surface–adsorbate distance.

The interaction of Ni, Pd, and Pt atoms with MgO is considerably more complex than that of the coinage metal atoms because of the interplay between the d^{10}, the d^9s^1 and, at least for Ni, the d^8s^2 atomic configurations. The interaction of free atoms with the MgO gives rise to several states; the lowest triplet and singlet states, $^3B_2$ and $^1A_1$, correlate at infinite adatom-surface separation with $MgO + M(d^9s^1)$ and $MgO + M(d^{10})$, respectively. The magnetic state, $^3B_2$, features the lowest energy for long surface–metal distances. At shorter distances, however, a crossing of the $^3B_2$ with the $^1A_1$ state occurs (Fig. 2.8). At short distances, the $^1A_1$ state becomes lower in energy, although the detailed nature of the ground state is not completely established [199]; the potential energy curve exhibits a deeper minimum and a rather high adsorption energy of about 1 eV or more depending on the metal. The curve crossing implies that a magnetic quenching accompanies the formation of the bond. The key mechanism for the bonding is the formation of s–d hybrid orbitals which can conveniently mix with the O $2p_\sigma$ orbitals; in fact,

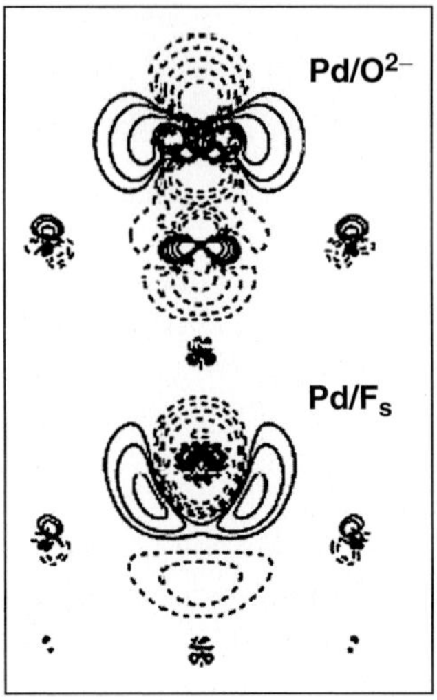

Fig. 2.9. Electron density difference plots for a Pd atom bound at a regular O^{2-} site (top) and at an F center (bottom). $\Delta\rho = \rho(\mathrm{Pd/MgO}) - \rho(\mathrm{Pd}) - \rho(\mathrm{MgO})$. Reproduced from [64]. Copyright 2001 American Institute of Physics

in the $^1A_1$ configuration the metal s, d_σ and the O $2p_\sigma$ orbitals are strongly hybridized.

The bonding with MgO has a somewhat different character for Ni than for Pd and Pt. Ni exhibits a non-negligible polarity of the bond while Pd and Pt form a more covalent bond (Fig. 2.9). This is consistent with the fact that Ni is easier to ionize. However, by no means the bonding can be viewed as a charge transfer from the metal atom to the oxide surface. The orbital analysis clearly shows a substantial mixing (hybridization) of metal nd and O 2p orbitals. The fact that all three atoms in the group form relatively strong bonds with the surface, compared for instance to the coinage metals, can be explained by the more pronounced tendency of the group 10 atoms to form s–d hybrid orbitals and a more direct involvement of their valence d orbitals in the bond with the oxygen. This tendency can be generally related to the s → d (or d → s) transition energies which are considerably smaller for Ni, Pd, and Pt than for Cu, Ag, and Au.

The interplay between high- and low-spin states is of great importance in the interaction of Cr, Mo, and W atoms with MgO. These atoms have high-spin atomic ground state configurations, $n\mathrm{d}^5(n+1)\mathrm{s}^1$ ($^7A_1$ in C_{4v} symmetry); their valence d shell is therefore half filled, and a comparison of their interaction with that of the late TM atoms is particularly instructive. For Cr, the ground state of the M/MgO cluster, $^7A_1$, correctly dissociates into Cr and MgO ground states and exhibits a very shallow potential energy curve ($r_e = 2.8$Å, $D_e = 0.34$ eV). Also for Mo, the $^7A_1$ state is lowest in energy ($r_e = 3.0$Å, $D_e = 0.33$ eV). At adsorption heights below 2.2–2.3 Å, the low-spin $^5A_1$ state becomes energetically preferred. The two atoms, Cr and Mo, also have similar bonding properties. The bonding is largely due to polarization and dispersion, with little, if any, chemical mixing of the metal orbitals with the surface electronic states. A completely different bonding arises from the interaction of W with MgO. The interaction of W in the high-spin d^5s^1 configuration results in a flat curve very similar to those computed for Cr and Mo. This state, however, is not the lowest one, even for long surface-adsorbate distances. The low-spin $^5A_1$ state exhibits a deep minimum near 2.15 Å with $D_e = 0.72\,eV$, a chemical bond of similar strength than that calculated for the Ni triad. In this state, the d^4s^2 atomic configuration is mixed not only with the d^5s^1 state but also with the O 2p orbitals. This is consistent with the fact that the s–d excitation energy is smallest in W where the s^1d^5 and s^2d^4 states are almost isoenergetic in contrast to the situation for Cr and Mo. The easy s–d hybridization is actually the reason for the strong bonding of W. As for adsorbed Ni, the bond with W exhibits a considerable polar character.

In summary, the Ni triad is the only one for which strong interface bonds are formed among the atoms considered here. The adhesion energy of these atoms on top of surface oxygens is about 1 eV per atom or more. The only other metal atom which exhibits a tendency to form strong bonds with the surface is W. For W, in fact, the bond strength is comparable to that of the Ni triad. The tendency to form strong bonds is connected to the fact that metal s and d orbitals hybridize and that these hybrid orbitals mix with the p orbitals of the surface oxygen. Pd is somewhat special in the Ni triad as it has the smallest binding energy. This reflects the general tendency toward a nonmonotonous behavior in many chemical properties as one moves down a TM group; in particular, second-row TM atoms often exhibit weaker interactions than the isovalent first- and third-row atoms. To some extent, this trend is observed also for the other two triads considered. The bonding of Cr, Mo, Cu, Ag, and Au to a MgO substrate, however, is weak, arising mainly from polarization and dispersion effects with only minor orbital mixing with the surface oxygen orbitals. This explains the very long-bond distances found in some cases, and the flat potential energy curves which result in very small force constants.

Thus, the TM atoms considered can be classified into two groups, atoms which tend to form relatively strong chemical bonds with the surface oxygen anions of MgO (Ni, Pd, Pt, and W) and atoms which interact very weakly with the surface with adsorption energies of the order of one-third of an electron volt or less (Cr, Mo, Cu, Ag, and Au). The interaction does not imply a significant charge transfer from the metal to the surface. This is an important conclusion, connected to the highly ionic nature of the MgO surface where the surface oxygen atoms have their valence almost saturated. To a first-order approximation, MgO can be described in terms of classical ionic model, $Mg^{2+}-O^{2-}$. This means that the oxygen centers at the regular surface sites of MgO(001) are almost completely reduced and are not able to oxidize adsorbed metal atoms. This conclusion is valid for the regular surface sites but of course adsorption phenomena at oxide surfaces can be substantially different when occurring at defect sites. In Sect. 2.4.2, we consider some of these sites and their interaction with deposited metal atoms.

A final comment is required about the problem of the description of isolated TM atoms on oxide surfaces in general. TM atoms have complex spin states which are not properly described within the DFT approach. Spin-polarized calculations provide a way to take into account the spin properties of the system but some details of the interaction may be described incorrectly. In this respect, the use of wave function-based methods is particularly important for benchmarks and comparisons. This problem is less severe when one considers small clusters where several electronic states exist separated by small energies. But one should be well aware of the fact that the treatment of isolated atoms, dimers, and very small aggregates with DFT methods requires spin-polarized approaches and special care in evaluating the results [199, 200].

2.4.2 Small Metal Clusters on MgO(001)

The interaction of TM metal clusters with the surface of MgO has been studied with ab initio methods with both cluster [201–204] and slab [205–207] models but mostly small supported clusters, containing less than 10 atoms, have been considered. The theoretical study of these systems is rather complex because of the very many isomers that one can have. The number of these isomers increases exponentially with the cluster nuclearity making an accurate evaluation with ab initio methods of the relative stabilities very difficult and extremely time consuming. The metals studied so far are Co, Ni, Cu, Pd, Ag, and Au. Some of isomers studied for Co, Ni, Cu, Pd, Ag tetramers are shown in Fig. 2.10. They can be classified into three main groups: Planar or nearly planar structures with the cluster plane "parallel" to the MgO surface, Fig. 2.10a–c, planar structures with the cluster plane "normal" to the surface, Fig. 2.10d–f and tetrahedral or distorted tetrahedral structures (Fig. 2.10g–i).

The results show that nanoclusters adsorbed on the regular MgO(001) surface do not necessarily tend to adhere to the surface with the largest possible number of metal atoms (surface wetting) but rather that they keep some bond "directionality." This results from the balance of various terms, the energy gain due to the bond formation with the O anions, the Pauli repulsion with the surface, and the loss of metal–metal bonding within the cluster due to distortions of the metal frame.

When the metal–oxygen bond is sufficiently strong, e.g., as for Ni and Pd, the formation of new interface bonds may compensates the loss of metal–metal bonds and the cluster distorts from its gas-phase geometry (Fig. 2.10i). For weaker metal–oxide bonds, as found for Co, Cu, and Ag, the metal–metal interactions prevail over the metal–MgO ones so that the cluster maintains, to a large extent, its electronic and geometric structure (Fig. 2.10f). Ni_4 and Pd_4, which form relatively strong bonds with MgO, tend to adhere with more metal atoms to the surface leading to a distorted tetrahedron while in the

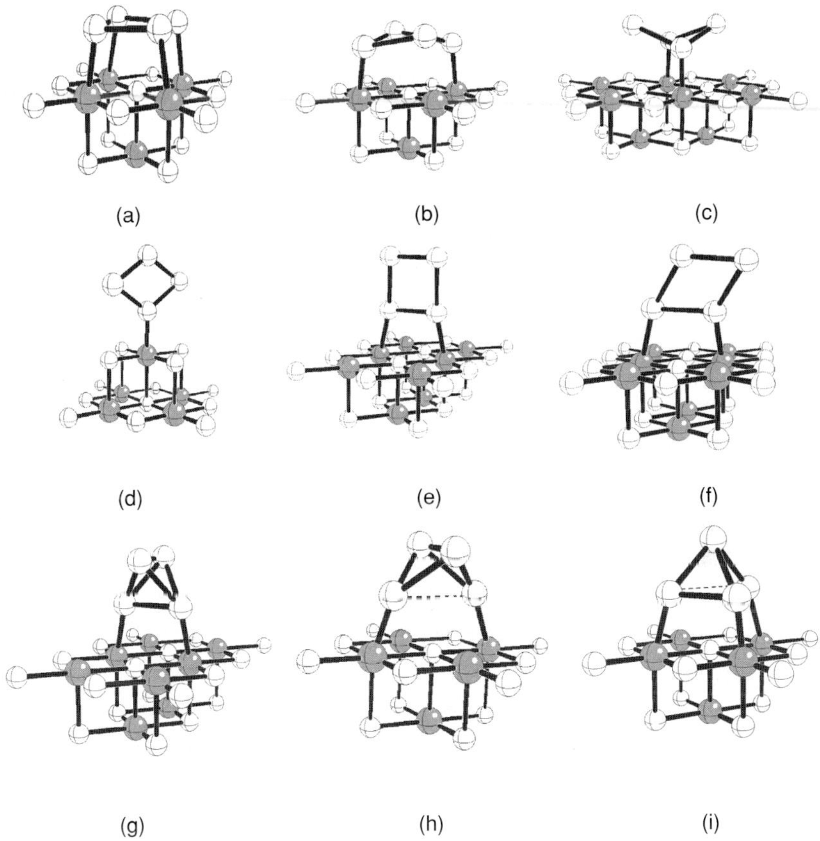

Fig. 2.10. Optimal structures of Co, Ni, Cu, Pd, Ag tetramers adsorbed on the regular MgO(100) surface

gas phase they both assume a tetrahedral shape (Fig. 2.10i). Ag_4 and Cu_4 clusters which interact weakly with MgO, assume a rhombic form with the cluster plane normal to the surface (Fig. 2.10f); their structure on the surface is not too different from what they have in gas phase. Co_4, which forms bonds of intermediate strength between Ag and Cu on one side, and Ni and Pd on the other, exhibits several surface isomers with different structures but similar stabilities.

The magnetization on the Ni and Co clusters is largely unchanged also in the supported species. In some cases, however, there is a partial quenching of the magnetic moment which is generally restricted to the metal atoms in direct contact with the oxide anions [203]. Thus, despite the relatively strong MgO/M_4 bonds (Co_4 is bound on MgO by 2.0 eV, Ni_4 by 2.4 eV), the electronic structure of supported transition metal moieties is only moderately perturbed. These conclusions are valid only for an ideal defect-free surface;

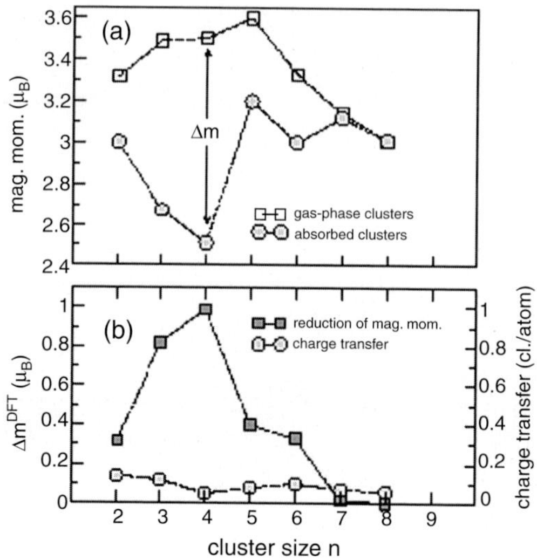

Fig. 2.11. (**a**) Magnetic moment per atom versus cluster size for gas phase (*squares*) and adsorbed (*circles*). (**b**) Reduction of magnetic moment and surface to cluster charge transfer as determined from DFT calculations. Reproduced from [208]. Copyright 2003 American Physical Society

investigations of the interaction of metal clusters with surface defects may lead to quite different conclusions [200]. Similar results have been recently obtained on Fe clusters containing up to 8 atoms deposited on MgO flat terraces [208]. Also in this case, it was found that upon deposition the clusters tend to preserve their gas-phase structure. Compared to gas phase, significant reductions of the magnetic moments were found for Fe_n clusters with $n < 6$ (Fig. 2.11). The reduction of the magnetization was attributed, as for Ni and Co clusters [203], to the interaction with the 2p orbitals of the surface O anions rather than to structural deformations or charge transfer from the surface [208]. The conservation of the magnetization seems to be general as ab initio molecular dynamics simulations have found that also clusters of a nonmagnetic metal, like Pd, have magnetic ground states in the gas phase and retain the magnetic moment when supported on the MgO surface [207].

Only very few ab initio molecular dynamic studies on the structure and growth of supported metal clusters have been reported, and the only case studied is that of Cu clusters on the regular MgO(100) surface [205–207]. It has been found that Cu–Cu intra-cluster interactions are stronger than the cluster-surface interactions, so that three-dimensional (3D) structures are largely preferred with respect to two-dimensional (2D) ones. This indicates a Volmer–Weber growth mode for Cu on MgO in agreement with experimental observations. It has also been found that the adsorbed clusters diffuse by "rolling" and "twisting" motions with barriers that may be smaller than for the diffusion of the isolated atoms (Fig. 2.12).

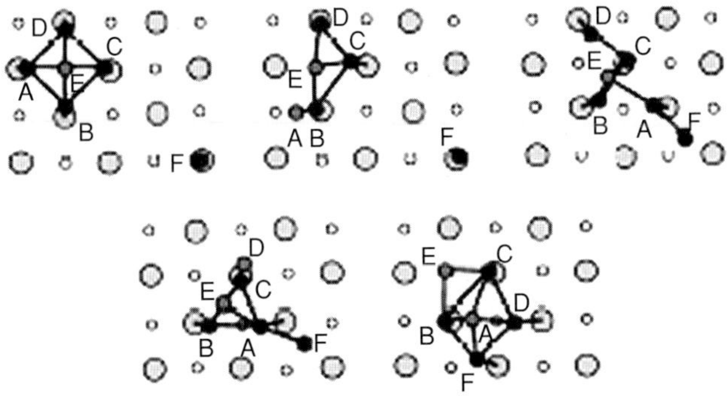

Fig. 2.12. Snapshots (top view) from a molecular dynamics simulation of the coalescence of an adsorbed pentamer (ABCDE) with a single adatom (F). Reproduced from [207]. Copyright 1999 American Physical Society

2.4.3 Metal Atoms on MgO: Where Are They?

A key question for the analysis of the reactivity, nucleation, and growth of small metal aggregates is where the atoms adsorb preferentially and which are the nucleation centers? A clear answer to this question is very difficult. Atoms arrive from the gas phase with a small but nonzero kinetic energy and can be reflected from the surface, can stick on a given adsorption site, or can diffuse on the surface until they become trapped at some strongly binding site (usually a defect). STM images show that small metal particles grow homogeneously on the surface at low temperature but is very difficult to image single metal atoms and on some substrates, like MgO, this is simply impossible unless the film thickness is kept below three monolayers.

We have seen above (Sect. 2.3), how many different defect sites exist even on the surface a simple oxide like MgO. Each of these sites can react in a different way with an adsorbed species and is a potential candidate for the nucleation and growth of metal particles. The complexity of the problem is increased by the fact that the concentration of the defects is usually low making their detection by integral surface sensitive spectroscopies very difficult. A microscopic view of the metal–oxide interface and a detailed analysis of the sites where the deposited metal atoms or clusters are bound become essential in order to rationalize the observed phenomena and to design new materials with known concentrations of a given type of defects.

For the specific case of isolated Pd atoms, a number of these sites have been investigated theoretically and a classification of the defects in terms of their adhesion properties is possible [32]. We first consider the case of Pd interacting with anion sites, O_{5c}, O_{4c}, or O_{3c}. The binding energy of a Pd atom with these sites increases monotonically from ≈ 1 eV (O_{5c}) to ≈ 1.5 eV (O_{3c}) and consequently the distance of the Pd atoms from the surface decreases. This is connected to the tendency of low-coordinated anions on the MgO surface to behave as stronger basic sites as the coordination number decreases (see Sect. 2.3.2). However, the difference in binding energy between an O^{2-} at a terrace site or a corner site, 0.5 eV, is not too large. Much larger binding energies have been found for other defect sites like the oxygen vacancies or F centers. On these sites, the bonding of Pd is about three times stronger than on the O anions [209, 210]. There is no large difference in Pd atom adsorption energy when the F center is located at a terrace, at a step, or at a corner site ($E_b \approx 3.5 \pm 0.1$ eV). Notice that the binding energy of Pd to an F center turns out to be the same with different computational methods. In fact, cluster model studies [32] and periodic plane wave calculations [210] give almost exactly the same energy and geometry for Pd on an F_{5c} center. Thus the Pd atoms are likely to diffuse on the surface until they become trapped at defect sites like the F centers where the bonding is so strong that only annealing at high temperatures will induce further mobility. Another group of defect sites is that of the paramagnetic F^+ centers (see Sect. 2.3.4). Formally, these centers are positively charged as a single electron replaces an O^{2-} ion in the lattice. On these sites, the bond strength of Pd is between that of the O_{nc} sites and F centers. For instance, on $F^+{}_{5c}$, Pd is bound by 2.10 eV, while E_b for Pd on O_{5c} is 0.96 eV and that for F_{5c} is 3.42 eV [32]. Recently, the adsorption of a Pd atom on a model of a surface OH group on MgO has shown that the binding energy at this site, 2.6 eV [36], is in between that found for neutral and charged F centers suggesting that hydroxyl groups at the oxide surface are good candidates for metal nucleation and growth.

Although this list is far from being complete, it provides strong evidence that in the initial phases of metal deposition the defect centers play a crucial role in stabilizing the metal atoms and in favoring nucleation and growth. It should be mentioned at this point that the use of CO as a probe molecule to check the adsorption site of the metal atom has proved to be very efficient. Its use will be discussed here with a concrete example.

An indirect yet powerful tool to monitor the metal adsorption sites is to study CO adsorption. CO is widely used as a probe molecule because of its low reactivity and quite sensitive vibrational frequency: Small changes in the substrate electronic structure reflect into measurable changes in CO stretching frequency. The study of CO desorption temperature from a metal covered MgO film and of the IR bands associated to the formation of metal–carbonyl complexes provides an useful tool to identify the surface sites involved in the stabilization of the M–CO species [31]. However, for an atomistic view of the sites involved, it is essential to combine the experimental evidences with the results of ab initio calculations. Recently, this has been done for a series of transition metal atoms, Rh, Pd, and Ag, all belonging to the second transition series but characterized by quite different valence structures (Rh, d^9; Pd, d^{10}; Ag, $d^{10}s^1$) [211, 212].

Thermal desorption spectra (TDS) of ^{13}CO for the three metal atoms (Ag, Pd, and Rh) deposited on thin MgO(100) films have been compared with the ^{13}CO-desorption spectra for the clean thin films. CO desorption at low temperature (120 K) originates from CO adsorbed on low-coordinated cation sites (most likely steps) of the MgO film (Sect. 2.3.1). For the case of Ag atoms, a modest contribution to the CO desorption after metal deposition is observed in the temperature range of 130–200 K, and the corresponding binding energy estimated using the Redhead approximation is 0.3–0.5 eV. The CO desorption from Pd atoms reveals two desorption peaks at around 230 K (0.6 eV) and 370 K (1.0 eV), respectively. The CO desorption from deposited Rh atoms is very similar to that of Pd and exhibits two desorption peaks at 180 K (0.5 eV) and 390 K (1.1 eV). These results show distinct differences in the interaction of CO with the three metal atoms. CO hardly interacts with the Ag atoms ($E_b = 0.3$–0.5 eV) while for Pd and Rh the CO interaction is stronger (E_b(Pd–CO) = 0.6 and 1.0 eV, E_b(Rh–CO) = 0.5 and 1.1 eV). The presence of two desorption peaks, however, could have at least three different origins: (a) subsequence desorption of two or more CO molecules bound to the same metal atom, (b) desorption of CO molecules bound to metal atoms adsorbed on different surface sites, and (c) desorption of CO molecules from metastable atoms/clusters (larger clusters can be formed on the surface upon heating during the TDS experiment).

To disentangle the origin of the double peak in the CO TPD spectra, IR spectroscopy provides valuable information. For Ag, however, almost no change in the CO frequency has been observed going from MgO(100) to Ag_1/MgO(100). On the contrary, the results for Pd and Rh clearly indicate that more than one CO molecule is bound to the metal atom and that complex diffusion processes occur as the temperature is increased. For instance, for Pd a band at 2,045 cm^{-1} with a shoulder at 2,010 cm^{-1} has been observed at 95 K, but at 180 K only a single peak at 2,010 cm^{-1} remains, and at 300 K the CO frequency shifts to 1,830 cm^{-1}, typical of bridge bonded CO, indicating nucleation of small Pd clusters. On the contrary, no indication is found of formation of small Rh aggregates. To interpret these results, extensive ab initio calculations have been performed within the cluster model approach [211].

Deposition of Ag atoms represents the simplest case due to the low-adsorption energy of Ag to MgO and to the absence of reactivity of the deposited Ag atoms. The calculations exclude the formation of stable Ag–CO complexes at any of the surface sites considered (regular sites or point defects like F centers or steps). The small changes in the TDS and FTIR (Fourier Transform Infra Red) spectra before and after Ag deposition must thus be attributed to indirect effects, like local changes in the surface potential induced by the presence of the Ag atoms or the formation of van der Waals complexes.

For Pd, at low temperature ($T < 300$ K), there is evidence from the IR experiment of the formation of $Pd(CO)_2$ species [32] while at higher temperatures Pd aggregation occurs. The calculations show that, most likely, the Pd atoms at 95 K possess a sufficient mobility on the surface to diffuse and become stabilized at strong trapping sites before CO exposure. This is connected to a relatively low barrier for diffusion (< 0.5 eV) on the terrace sites

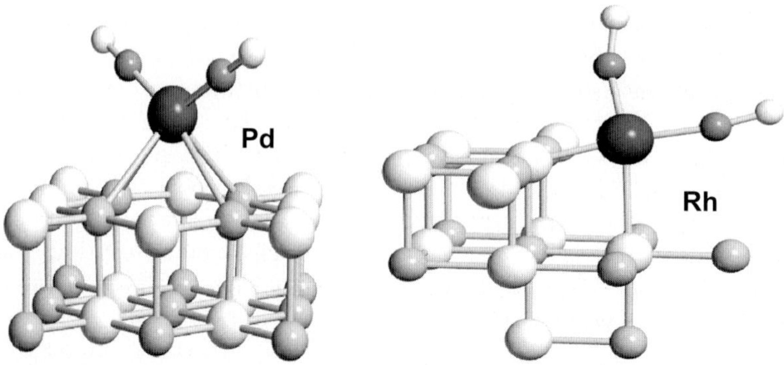

Fig. 2.13. Optimal geometry of $M(CO)_2$ complexes adsorbed on the MgO surface. (left) $Pd(CO)_2$ complex formed at an oxygen vacancy (F center); (right) $Rh(CO)_2$ complex formed at a step

but also from steps to terraces. In fact, the steps of the MgO surface are not strong binding sites for the Pd atoms. The best candidates for trapping the Pd atoms are the oxygen vacancies (F centers). Thus, by exposure of the sample to CO, Pd(CO), and $Pd(CO)_2$ complexes form on F centers where the Pd atoms have been trapped during the diffusion process. The CO molecules are desorbed from these sites at relatively low temperature according to the calculations: Both CO molecules desorb from a Pd atom bound at an F center with $E_b < 0.6$ eV, consistent with a desorption temperature of 230 K). Above 370 K, few residual Pd atoms or Pd(CO) units, bound at other defect sites (steps) diffuse on the surface until they aggregate to form larger particles. The calculations exclude the possibility that the Pd atoms are bound to oxide anions of the surface either terraces or steps. In fact, on these sites the Pd–CO binding energy is very high, more than 2 eV, a value which is totally incompatible with the measured desorption temperatures. The F centers have been found to be the only sites where the computed binding energies for CO fit with the experimental picture [32] (Fig. 2.13).

According to the calculations, Rh atoms can diffuse on the flat terraces very easily, even more than Pd, but they bind quite strongly to the step sites (much more than Pd). The binding energy of a Rh atom at a step is about twice that on a terrace. This means that the step sites are decorated by Rh atoms at low temperature (95 K) and only a small fraction of Rh atoms populates the F centers. The exposure to CO leads to the formation of relatively stable $Rh(CO)_2$ complexes at steps, Fig. 2.13, at variance with Pd. The first CO desorption from the $Rh(CO)_2$ units formed at steps occurs around 180 K (computed E_b(RhCO–CO) = 0.7 eV) and leaves on the surface a stable Rh(CO) unit. The calculations show that the Rh–CO bonding is strong (E_b(Rh–CO) = 2.6 eV) and that this species is rather mobile. It is likely that this is the species which diffuses on the surface until it becomes trapped at F centers. From these sites, a further increase in the temperature leads to a second desorption of CO from Rh atoms (390 K, computed E_b(Rh–CO) = 1.2 eV) [211].

The experimental results and the tentative explanations clearly indicate that the deposition of metal atoms on the MgO films is followed by complex diffusion mechanisms, which depend quite strongly on the nature of the metal–MgO bond. The use of CO as a probe molecule provides invaluable information to identify the metal adsorption sites but also contributes to complicate the diffusion mechanisms as new M–CO species, quite strongly bound in the case of Rh and Pd, can become mobile on the surface and compete with the mobility of the individual metal atoms.

2.5 Reactivity of Supported Metal Atoms: The Role of Defects

We have seen above that a great variety of defect centers can form at the surface of MgO (Table 2.1). Each surface defect has a direct and characteristic effect on the properties of adsorbed species. This becomes particularly important in the analysis of the chemical reactivity of supported metal atoms and possibly of small clusters as well [23, 24]. In fact, the defects not only act as nucleation centers in the growth of metal islands or clusters, but they can also modify the catalytic activity of the deposited metal by affecting the bonding at the interface [213–215]. The role of point defects at the surface of MgO in promoting or modifying the catalytic activity of isolated metal atoms or clusters deposited on this substrate has been recently investigated by considering the cyclization reaction of acetylene to form benzene, $3C_2H_2 \rightarrow C_6H_6$ [214, 215]. This process has been widely studied on single crystal surfaces [216–218] and it has been shown unambiguously that the reaction proceeds through the formation of a stable C_4H_4 intermediate, resulting from addition of two acetylene molecules [219]. This intermediate has been characterized experimentally [220] and theoretically [221] and its structure is now well established. Once the C_4H_4 intermediate is formed, it can combine with a third acetylene molecule to form benzene that then desorbs from the surface at a temperature of about 500 K [216, 217].

The reaction has been studied on size-selected, soft-landed Pd atoms and clusters deposited on MgO thin films [214, 215]. The clusters assembled materials obtained in this way exhibit peculiar activity and selectivity in the polymerization of acetylene to form benzene and aliphatic hydrocarbons [222]. Here, we restrict the analysis to the reactivity of isolated Pd atoms. On a clean MgO(100) surface, no benzene is produced at the same experimental conditions. However, even 0.2 monolayers of deposited Pd atoms produce benzene for temperatures of about 215 K. This is an important result which has allowed to investigate the activity of the Pd atoms as function of the support where it is deposited or of the sites where it is bound.

The possibility to catalyze the acetylene trimerization depends critically on the ability of the metal center to coordinate and activate two C_2H_2 molecules and then to bind a C_4H_4 intermediate according to reaction:

$$Pd + 2C_2H_2 \rightarrow Pd(C_2H_2)_2 \rightarrow Pd(C_4H_4). \qquad (2.8)$$

The activation of the acetylene molecules is easily monitored, for instance, by the deviation from linearity of the HCC angle due to a change of hybridization of the C atom from sp to sp^2, or by the elongation of the C–C distance, d(C–C), as a consequence of the charge transfer from the metal 4d orbitals to the empty π^* orbital of acetylene. The level of activation of acetylene is clearly larger when a single C_2H_2 is adsorbed, consistent with the idea that the electron density on the metal is essential for promoting the molecular activation. A key point is therefore the ability of the Pd atom to bind more than one acetylene molecule. Once two acetylene molecules are bound to the Pd atom, in fact, the formation of the C_4H_4 intermediate occurs with a significant energy gain. The following step, $C_4H_4 + C_2H_2 \rightarrow C_6H_6$, requires the capability of the metal center to coordinate and activate a third acetylene molecule according to reaction:

$$Pd(C_4H_4) + C_2H_2 \rightarrow Pd(C_4H_4)(C_2H_2) \rightarrow Pd(C_6H_6). \qquad (2.9)$$

DFT calculations show unambiguously that the third C_2H_2 molecule interacts very weakly with $Pd(C_4H_4)$ in gas phase [214, 215]. Therefore, an isolated Pd atom is not a catalyst for the cyclization process, at variance with the experimental observation for Pd_1/MgO.

The role played by the support becomes therefore a critical aspect of the interaction. A simple model study where the electronic charge on a Pd atom has been artificially augmented has shown that the excess of charge on Pd reinforces the bonding and the activation of acetylene [214]. This result shows that the increase of the electron density on Pd is a key mechanism to augment the catalytic properties of the metal atom. The role of the substrate is just that of increasing the "basic" character of the Pd atoms (or clusters). As we have seen above, the Pd–MgO bonding is not characterized by a pronounced charge transfer. However, things change when one considers defect sites.

When Pd is deposited on an MgO site like an O^{2-} ion at a terrace, a step, or a corner site the activation of C_2H_2 is much more efficient than for an isolated Pd atom. The structural distortion of adsorbed C_2H_2 follows the trend Pd–$O^{2-}{}_{3c}$ > Pd–$O^{2-}{}_{4c}$ > Pd–$O^{2-}{}_{5c}$ > free Pd atom [215]. The donor capability of Pd increases as a function of the adsorption site on MgO in the order terrace < edge < corner. However, even on the low-coordinated anions the Pd atom is not an active catalyst for the cyclotrimerization. The increase of electron density on Pd due to the bonding with the MgO substrate at these irregular sites does not account for the observed reactivity.

Things are different when the Pd atom sits on the F and F^+ centers. In this case, one can form stable $Pd(C_4H_4)$ complexes, and bind and activate

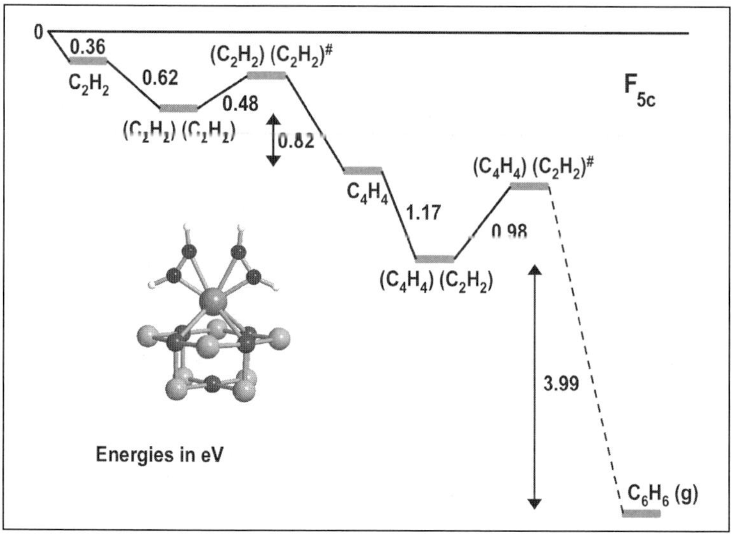

Fig. 2.14. Energy profile for the cyclization reaction of acetylene to benzene occurring on a Pd atom supported on an F_{5c} center, Pd_1/F_{5c} (DFT-BP results). Reproduced from [223]. Copyright 2000 American Chemical Society

the third acetylene molecule. In addition, the calculation of the energy barriers involved in the process shows that these are broadly consistent with the observed desorption temperature (the barriers are of the order of 1 eV, Fig. 2.14 [223]).

Much larger barriers are found for sites different from the F centers. The results show that only in the presence of F centers, an isolated Pd atom becomes an active catalyst for the reaction and that the substrate has a fundamental role in the reactivity of the supported metal. The main reason is that the F centers delocalize in a very effective way the two electrons trapped in the cavity over the adsorbed Pd atom, thus increasing substantially its ability to bind and activate adsorbed hydrocarbon molecules. Therefore, the F centers (oxygen vacancies) not only act as trapping sites for the Pd atoms but they also contribute to modify their chemical activity.

2.6 Summary

Our understanding of the behavior of isolated metal atoms and small metal clusters adsorbed on oxide surfaces has increased enormously in the last few years, thanks to spectacular advances on the experimental side but also to the high level of reliability reached by first principle calculations. Not surprisingly, this is a complex problem which cannot be solved from scratch by using one or another technique. Rather, from the combination of several results and different approaches one can get a deeper insight at an atomistic level of the interaction of a metal atom or cluster with the support.

After some initial skepticism, it is now universally accepted that a key role in determining the adsorption, diffusion, stabilization, and reactivity of a metal deposit is played by the surface defects, in particular the point defects. These are present on every surface, independent on the form of the material, single crystal, powder, thin film. However, it is likely that the relative abundance of the various defects differs substantially on the form of the material as well as on the preparation and on the thermal treatment of the surface. This makes a direct comparison of results obtained on the same material in different forms difficult and sometimes confusing.

In this chapter, we have concentrated on MgO, one of the most studied and better understood oxide materials. We have shown that even on such a simple nontransition metal oxide about a dozen of different surface defect centers have been identified and described in the literature. Each of these centers has a somewhat different behavior toward adsorbed metal atoms. It becomes immediately clear that the precise assignment of the defect sites involved in the interaction, nucleation, and growth of the cluster is a formidable task. Nevertheless, thanks to the combined use of theory and experiment, the progress in this direction has been particularly significant and promising. For instance, a lot of evidence has been accumulated that points toward the role of the oxygen vacancies, the F centers. At the moment, these sites seem the most likely sites for nucleation and growth of small metal clusters.

The theoretical results have also indicated that when metal atoms are bound to specific defects their chemical activity may change, in particular can increase. This is likely to be true also for small metal clusters. This has not been fully appreciated so far. In fact, even "inert" supports, like silica, alumina, or magnesia, can interact strongly with the supported metal if this is bound at a defect site and can have a direct role in the chemistry of the supported species. Some preliminary calculations on supported clusters, however, suggest that the effect of the defect on the cluster electronic structure is restricted to very small, really nanometric clusters of about ten atoms size [224]. Should the size of active catalysts in real applications go down to this size, the specific interaction with the substrate could no longer be ignored in the interpretation of the catalytic activity.

Acknowledgments

I am indebted to my coworkers Sivia Brazzelli, Fabrizio Cinquini, Annalisa Del Vitto, Cristiana Di Valentin, Emanuele Finazzi, Anna Maria Ferrari, Livia Giordano, and Davide Ricci and who have contributed over the last few years to the study of metals deposited on oxide surfaces. I also thank for the fruitful discussions and ongoing collaboration Thomas Bredow, Hajo Freund, Elio Giamello, Jacek Goniakowski, Ueli Heiz, Francesc Illas, Konstantin Neyman, Notker Rösch, Alex Shluger, and Peter Sushko. This work has been supported by the Italian Istituto Nazionale per la Fisica della Materia through the PRA-ISADORA project.

References

1. Henrich VE, Cox PA (1994) The Surface Science of Metal Oxides. Cambridge University Press, Cambridge
2. Noguera C (1996) Physics and Chemistry at Oxide Surfaces. Cambridge University Press, Cambridge
3. Colbourn EA (1992) Surf. Sci. Rep. 15:281
4. Henry CR (1998) Surf. Sci. Rep. 31:231
5. Renaud G (1998) Surf. Sci. Rep. 32:1
6. Bonnel DA (1998) Prog. Surf. Sci. 57:187
7. Chambers SA (2000) Surf. Sci. Rep. 39:105
8. Freund HJ (1999) Faraday Discuss. 114:1
9. Freund HJ, Kuhlenbeck H, Staemmler V (1996) Rep. Prog. Phys. 59:283
10. Bäumer M, Freund HJ (1999) Prog. Surf. Sci. 61:127
11. Pacchioni G, Bagus PS, Parmigiani F (eds.) (1992) Cluster Models for Surface and Bulk Phenomena NATO ASI Series B., vol. 283, Plenum, New York
12. Woodruff DP (ed.) (2001) The Chemical Physics of Solid Surfaces—Oxide Surfaces, vol 9. Elsevier, Amsterdam
13. Lamb HH, Gates BC, Knözinger H (1988) Angew. Chem. Int. Ed. Engl. 27:1127
14. Argo AM, Odzak JF, Lai FS, Gates BC (2002) Nature 415:623
15. The surface science of metal oxides (1999) Faraday Discuss. 114:1172
16. Schroeder T, Giorgi JB, Bäumer M, Freund HJ (2002) Phys. Rev. B 66:165422
17. Giordano L, Goniakowski J, Pacchioni G (2003) Phys. Rev. B 67:04541018.
18. Casassa S, Ferrari AM, Busso M, Pisani C (2002) J. Phys. Chem. B 106:12978
19. Schintke S, Messerli S, Pivetta M, Patthey F, Libioulle L, Stengel M, De Vita A, Schneider WD (2001) Phys. Rev. Lett. 87:276801
20. Kantorovich LN, Shluger AL, Sushko PV, Günster J, Stracke P, Goodman DW, Kempter V (1999) Faraday Discuss. 114:173
21. Schmid G, Bäumle M, Geerkens M, Heim I, Osemann C, Sawitowski T (1999) Chem. Soc. Rev. 28:179
22. Freund HJ (2002) Surf. Sci. 500:271
23. Campbell CT (1997) Surf. Sci. Rep. 27:1
24. Lambert RM, Pacchioni G (eds.) (1997) Chemisorption and Reactivity of Supported Clusters and Thin Films, NATO ASI Series E. vol. 331, Kluwer, Dordrecht
25. Shi J, Gider S, Babcock K, Awschalom DD (1996) Science 271:937
26. He J, Moller PJ (1986) Surf. Sci. 129:13
27. Stracke P, Krishok S, Kempter V (2001) Surf. Sci. 473:86
28. Kim YD, Stulz J, Wei T, Goodman DW (2002) J. Phys. Chem. 106:6827
29. Campbell CT, Starr D (2002) J. Am. Chem. Soc. 124:9212
30. Zhukovskii YF, Kotomin EA, Jacobs PWM, Stoneham AM (2000) Phys. Rev. Lett. 84:1256
31. Frank M, Bäumer M, Kühnemuth R, Freund HJ (2001) J. Phys. Chem. B 105:8569
32. Abbet S, Riedo E, Brune H, Heiz U, Ferrari AM, Giordano L, Pacchioni G (2001) J. Am. Chem. Soc. 123:6172
33. Haas G, Menck A, Brune H, Barth JV, Venables JA, Kern K (2001) Phys. Rev. B 61:11105
34. Wahlström E, Lopez N, Schaub R, Thostrup P, Rønnau A, Afich C, Laegsgaard E, Nørskov JK, Besenbacher F (2003) Phys. Rev. Lett. 90:026101

35. Bogicevic A, Jennison DR (1999) Surf. Sci. 437:L741
36. Bogicevic A, Jennison DR (2002) Surf. Sci. 515:L481
37. Giordano L, Di Valentin C, Pacchioni G, Goniakowski J (2004) Phys. Rev. Lett. 92:0961
38. Cox PA (1987) The Electronic Structure and Chemistry of Solids. Oxford Science Publications, Oxford
39. Gibson A, Haydock R, LaFemina JP (1993) Appl. Surf. Sci. 72:285
40. Gibson A, Haydock R, LaFemina JP (1994) Phys. Rev. 50:2582
41. Kantorovich LN, Holender JM, Gillan MJ (1995) Surf. Sci. 343:221
42. Castanier E, Noguera C (1996) Surf. Sci. 364:1
43. Orlando R, Millini R, Perego G, Dovesi R (1997) J. Molec. Catal. A119:253
44. Leslie M, Gillan MJ (1985) J. Phys. C 18:973
45. Sauer J, Ugliengo P, Garrone E, Saunders VR (1994) Chem. Rev. 94:2095
46. Pacchioni G (1995) Heter. Chem. Rev. 2:213
47. Bagus PS, Illas F (1998) The surface chemical bond. In: Schleyer PV, Allinger NL, Clark T, Gasteiger J, Kollman PA, Schaefer HF, Schreiner PR (eds.), Encyclopedia of Computational Chemistry, vol. 4. Wiley, Chichester, UK, p. 2870
48. Pacchioni G, Frigoli F, Ricci D, Weil JA (2001) Phys. Rev. B 63:054102
49. Pacchioni G, Ferrari AM, Marquez AM, Illas F (1997) J. Comp. Chem. 18:617
50. Ferrari AM, Pacchioni G (1995) J. Phys. Chem. 99:17010
51. Ferrari AM, Pacchioni G (1996) Int. J. Quant. Chem. 58:241
52. Nygren MA, Pettersson LGM, Barandiaran Z, Seijo L (1994) J. Chem. Phys. 100:2010
53. Stevens WJ, Basch H, Krauss M (1984) J. Chem. Phys. 81:6026
54. Barandiaran Z, Seijo L (1988) J. Chem. Phys. 89:5739
55. Luaña V, Pueyo L (1989) Phys. Rev. B 39:11093
56. Born M (1920) Z. Physik 1:45
57. Pandey R, Vail JM (1989) J. Phys.: Condens. Matter 1:2801
58. Harding JH, Harker AH, Keegstra PB, Pandey R, Vail JM, Woodward C (1985) Physica B & C 131:151
59. Dick BG, Overhauser AW (1958) Phys. Rev. 112:90
60. Catlow CRA, Dixon M, Mackrodt WC (1982) In: Catlow CRA (ed.), Computer Simulation of Solids. Springer, Berlin Heidelberg New York, p. 130
61. Susko PV, Shluger AL, Catlow CRA (2000) Surf. Sci. 450:153
62. These techniques have been reviewed in a special issue: Catlow CRA, Stoneham AM (eds.) (1989) J. Chem. Soc. Faraday Trans. II 85
63. Stevens W, Bach H, Krauss J (1984) J. Chem. Phys. 81:6026
64. Nasluzov VA, Rivanenkov VV, Gordienko AB, Neyman KM, Birkenheuer U, Rösch N (2001) J. Chem. Phys. 115:8157
65. Pisani C (1993) J. Mol. Catal. 82:229
66. Pisani C, Corà F, Nada R, Orlando R (1994) Comput. Phys. Commun. 82:139
67. Pisani C, Birkenheuer U (1996) Comput. Phys. Commun. 96:152
68. Pisani C, Birkenheuer U, Corà F, Nada R, Casassa S (1996) EMBED96 User's Manual. Università di Torino, Torino
69. Pisani C, Dovesi R, Roetti C (1988) In: Hartree–Fock ab-initio treatment of crystalline systems. Lecture Notes in Chemistry, vol. 48. Springer, Berlin Heidelberg New York
70. Saunders VR, Dovesi R, Roetti C, Causà M, Harrison NM, Orlando R, Zicovich-Wilson CM (1998) CRYSTAL98 User's Manual. Università di Torino, Torino

71. Schleyer PV, Allinger NL, Clark T, Gasteiger J, Kollman PA, Schaefer HF, Schreiner PR (eds.) (1998) Encyclopedia of Computational Chemistry. Wiley, Chichester, UK
72. Szabo A, Ostlund NS (1982) Modern Quantum Chemistry: Introduction to Advanced Electronic Structure Theory. Macmillan, New York
73. Weeny RMc (1992) Methods of Molecular Quantum Mechanics. Academic, London
74. Taylor PR (1992) Accurate calculations and calibration. In: Roos BO (ed.), Lecture Notes in Quantum Chemistry, vol. 58. Springer, Berlin, Heidelberg New York, p. 325
75. Pople JA, Binkley JS, Seeger R (1976) Int. J. Quant. Chem. Symp. 10:1
76. Raghavachari K, Pople JA, Replogle ES, Head-Gordon M (1990) J. Phys. Chem. 94:5579
77. Buenker RJ, Peyerimhoff SD, Butscher W (1978) Mol. Phys. 35:771
78. Peyerimhoff SD, Buenker RJ (1972) Chem. Phys. Lett. 16:235
79. Huron B, Rancurel P, Malrieu JP (1973) J. Chem. Phys. 75:5745
80. Evangelisti S, Daudey JP, Malrieu JP (1983) Chem. Phys. 75:91
81. Illas F, Rubio J, Ricart JM, Bagus PS (1991) J. Chem. Phys. 95:1877
82. Andersson K, Malmqvist PÅ, Roos BO, Sadlej AJ, Wolinski K (1990) J. Phys. Chem. 94:5483
83. Andersson K, Malmqvist PÅ, Roos BO (1992) J. Chem. Phys. 96:1218
84. Serrano-Andrés L, Merchán M, Nebot-Gil I, Roos BO, Fülscher M (1993) J. Chem. Phys. 98:3151
85. Illas F, Pacchioni G (1998) J. Chem. Phys. 108:7835
86. Raghavachari K, Ricci D, Pacchioni G (2002) J. Chem. Phys. 116:825
87. Hohenberg P, Kohn W (1964) Phys. Rev. B 136:864
88. Kohn W, Sham LJ (1965) Phys. Rev. A 140:1133
89. Koch W, Holthausen MC (2002) A Chemist Guide to Density Functional Theory. Wiley, Weinheim
90. Vosko SH, Wilk L, Nusair M (1980) Can. J. Phys. 58:1200
91. Terakura K, Oguchi T, Williams AR, Klüber J (1984) Phys. Rev. B 30:4734
92. Shen ZX, List RS, Dessau DS, Wells BO, Jepsen O, Arko AJ, Barttlet R, Shih CK, Parmigiani F, Huang JC, Lindberg PAP (1991) Phys. Rev. B 44:3604
93. Becke AD (1988) Phys. Rev. A 38:3098
94. Perdew JP, Wang Y (1992) Phys. Rev. B 45:13244
95. Perdew JP, Chevary JA, Vosko SH, Jackson KA, Pederson MR, Singh DJ, Fiolhais C (1992) Phys. Rev. B 46:6671
96. Lee C, Yang W, Parr RG (1988) Phys. Rev. B 37:785
97. Becke AD (1993) J. Chem. Phys. 98:5648
98. Matxian JM, Irigoras A, Fowler JE, Ugalde JM (2000) Phys. Rev. A 63:013202
99. Shieh JC, Chang JL, Wu JC, Li R, Mebel AM, Handy NC, Chen YT (2000) J. Chem. Phys. 112:7384
100. Levine IN (2000) Quantum Chemistry, 5th ed. Prentice-Hall, New Jersey
101. Lopez N, Illas F, Rösch N, Pacchioni G (1999) J. Chem. Phys. 110:4873
102. Ranney JT, Starr DE, Musgrove JE, Bald DJ, Campbell CT (1999) Faraday Discuss. 114:195
103. Robach O, Renaud G, Barbeir A (1998) Surf. Sci. 401:227
104. Pacchioni G (2000) Surf. Rev. Lett. 7:277
105. Escalona-Platero E, Scarano D, Spoto G, Zecchina A (1985) Faraday Discuss. Chem. Soc. 80:183

106. Furuyama S, Fuijii H, Kawamura M, Morimoto T (1978) J. Phys. Chem. 82:1028
107. Wichtendahl R, Rodriguez-Rodrigo M, Härtel U, Kuhlenbeck H, Freund HJ (1999) Surf. Sci. 423:90
108. Wichtendahl R, Rodriguez-Rodrigo M, Härtel U, Kuhlenbeck H, Freund HJ (1999) Phys. Stat. Sol. (a) 173:93
109. Pacchioni G, Cogliandro G, Bagus PS (1992) Int. J. Quant. Chem. 42:1115
110. Pacchioni G, Minerva T, Bagus PS (1992) Surf. Sci. 275:450
111. Scarano D, Spoto G, Bordiga S, Coluccia S, Zecchina A (1992) J. Chem. Soc. Faraday Trans. 88:291
112. Pettersson LGM, Nyberg M, Pascual JL, Nygren MA (1997) In: Lambert RM, Pacchioni G (eds.), Chemisorption and Reactivity on Supported Clusters and Thin Films NATO ASI Series E. vol. 331, Kluwer, Dordrecht, p. 425
113. Soave R, Pacchioni G (2000) Chem. Phys. Lett. 320:345
114. Pacchioni G, Ricart JM, Illas F (1994) J. Am. Chem. Soc. 116:10152
115. Pacchioni G, Clotet A, Ricart JM (1994) Surf. Sci. 315:337
116. Sousa C, Mejias JA, Pacchioni G, Illas F (1996) Chem. Phys. Lett. 249:123
117. Ochs D, Brause M, Braun B, Maus-Friedrichs W, Kempter V (1998) Surf. Sci. 397:101
118. Ochs D, Braun B, Maus-Friedrichs W, Kempter V (1998) Surf. Sci. 417:406
119. Carrier X, Doyle CS, Kendelewicz T, Brown GE (1999) Surf. Rev. Lett. 6:1237
120. Doyle CS, Kendelewicz T, Carrier X, Brown GE (1999) Surf. Rev. Lett. 6:1247
121. Nada R, Hess AC, Pisani C (1995) Surf. Sci. 336:353
122. Goniakovski J, Noguera C (1995) Surf. Sci. 340:191
123. Shluger AL, Sushko PV, Kantorovich LN (1999) Phys. Rev. B 59:2417
124. Brown GE, Henrich VE, Casey WH, Clark DL, Eggleston C, Felmy A, Goodman DW, Grätzel M, Maciel G, McCarthy MI, Nealson KH, Sverjensky DA, Toney MF, Zachara JM (1999) Chem. Rev. 99:77
125. Anchell JL, Hess AC (1996) J. Phys. Chem. 100:18317
126. Scamehorn CA, Harrison NM, McCarthy MI (1994) J. Chem. Phys. 101:1547
127. Langel W, Parrinello M (1994) Phys. Rev. Lett. 73:504
128. Langel W, Parrinello M (1995) J. Chem. Phys. 103:3240
129. Giordano L, Goniakowski J, Suzanne J (1998) Phys. Rev. Lett. 81:1271
130. Tench AJ, Nelson RL (1968) J. Colloid Interface Sci. 26:364
131. Tench AJ (1971) Surf. Sci. 25:625
132. Giamello E, Murphy D, Ravera L, Coluccia S, Zecchina A (1994) J. Chem. Soc. Faraday Trans. 90:3167
133. Giamello E, Paganini MC, Murphy D, Ferrari AM, Pacchioni G (1997) J. Phys. Chem. 101:971
134. Henrich VE, Dresselhaus G, Zeiger HJ (1980) Phys. Rev. B 22:4764
135. Wu MC, Truong CM, Goodman DW (1992) Phys. Rev. B 46:12688
136. Peterka D, Tegenkamp C, Schroder KM, Ernst W, Pfnür H (1999) Surf. Sci. 431:146
137. Kramer J, Ernst W, Tegenkamp C, Pfnür H (2002) Surf. Sci. 517:87
138. Paganini MC, Chiesa M, Giamello E, Coluccia S, Martra G, Murphy DM, Pacchioni G (1999) Surf. Sci. 421:246
139. Pacchioni G, Pescarmona P (1998) Surf. Sci. 412–413:657
140. Pacchioni G, Ferrari AM, Ieranò G (1997) Faraday Discuss. 106:155
141. Chen Y, Williams RT, Sibley WA (1969) Phys. Rev. 182:960

142. Nelson RL, Hale JW (1971) Trans. Faraday Soc. 52:77
143. Tegenkamp C, Pfnur H, Ernst W, Malaske U, Wollschlager J, Peterka D, Schroder KM, Zielasek V, Henzler M (1999) J. Phys. Condens. Matter 11:9943
144. Sousa C, Pacchioni G, Illas F (1999) Surf. Sci. 429:217
145. Kramer J, Tegenkamp C, Pfnür H (2003) Phys. Rev. B 67:235401
146. Sterrer M, Diwald O, Knözinger E (2000) J. Phys. Chem. B 104:3601
147. Chen Y, Abraham MM, Templeton LC, Unruh WP (1975) Phys. Rev. B 11:881
148. Halliburton LE, Kappers LA, Cowan DL, Dravnieks F, Wertz JE (1973) Phys. Rev. B 8:1610
149. Foot JD, Colbourn EA, Catlow CRA (1988) J. Phys. Chem. Solids 49:1225
150. Eglitis RI, Kotomin EA, Borstel G (1998) Phys. Stat. Sol. (b) 208:15
151. Chen Y, Abraham MM (1990) J. Phys. Chem. Solids 51:747
152. Tench A, Duck MJ (1973) J. Phys. C: Solid State Phys. 6:1134
153. Abraham MM, Chen Y, Unruh WP (1974) Phys. Rev. B 9:1842
154. Baranek P, Pinarello G, Pisani C, Dovesi R (2000) Phys. Chem. Chem. Phys. 2 3893
155. Lunsford JH, Jayne JP (1966) J. Phys. Chem. 70:3464
156. Wertz JE, Orton JW, Auzins P (1961) Discuss. Faraday Soc. 31:140
157. Ojamäe L, Pisani CJ (1998) Chem. Phys. 109:10984
158. D'Ercole A, Pisani C (1999) J. Chem. Phys. 111:9743
159. Yang Z, Liu C, Wu R (2002) Phys. Rev. B 65:235432
160. Abraham MM, Unruh WP, Chen Y (1974) Phys. Rev. B 10:3540
161. Rius G, Herve A (1974) Solid State Commun. 15:399
162. Zuo J, Pandey R, Kunz AB (1991) Phys. Rev. B 44:7187
163. Izumi Y, Shimiuzu T, Kobahashi T, Aika K (2000) Chem. Commun. :1053
164. Zecchina A, Scarano D, Bordiga S, Ricchiardi G, Spoto G, Geobaldo F (1996) Catal. Today 27:403
165. Colbourn EA, Mackrodt WC (1982) Surf. Sci. 117:571
166. Neyman KM, Rösch N (1993) Chem. Phys. 177:561
167. Di Valentin C, Pacchioni G, Chiesa M, Giamello E, Abbet S, Heiz U (2002) J. Phys. Chem. B 106:1637
168. Di Valentin C, Pacchioni G, Bredow T, Dominguez-Ariza D, Illas F (2002) J. Chem. Phys. 117:2299
169. Scarano D, Bertarione S, Zecchina A, Soave R, Pacchioni G (2002) Phys. Chem. Chem. Phys. 4:366
170. Ito T, Kawanami A, Toi K, Shirakawa T, Tokuda T (1988) J. Phys. Chem. 92:3910
171. Garrone E, Zecchina A, Stone F (1980) Philos. Mag. B 42:683
172. Iwamoto M, Lunsford JH (1979) Chem. Phys. Lett. 66:48
173. Che M, Tench AJ (1983) Adv. Catal. 23:1
174. Williamson NB, Lunsford JH, Naccache C (1971) Chem. Phys. Lett. 9:33
175. Dominguez-Ariza D, Sousa C, Illas F, Ricci D, Pacchioni G (2003) Phys. Rev. B 68:054101
176. Ricci D, Pacchioni G, Sushko PV, Shluger AL (2002) J. Chem. Phys. 117:2844
177. Sushko PV, Gavartin JL, Shluger AL (2002) J. Phys. Chem. B 106:2269
178. Scorza E, Birkenheuer U, Pisani C (1997) J. Chem. Phys. 107:9645
179. Sharma RR, Stoneham AM (1976) J. Chem. Soc. Faraday Trans. 2:913
180. Finocchi F, Goniakowski J, Noguera C (1999) Phys. Rev. B 59:5178
181. Pisani C, Corà F, Dovesi R, Orlando R (1994) J. Electron Spectrosc. 96:1

182. Giamello E, Ferrero A, Coluccia S, Zecchina A (1991) J. Phys. Chem. 95:9385
183. Murphy D, Giamello E (1995) J. Phys. Chem. 99:15172
184. Ricci D, Di Valentin C, Pacchioni G, Sushko PV, Shluger AL, Giamello E (2003) J. Am. Chem. Soc. 125:738
185. Chiesa M, Paganini MC, Giamello E, Di Valentin C, Pacchioni G (2003) Angew. Chemie Int. Ed. 42:1759
186. Brazzelli S, Di Valentin C, Pacchioni G, Giamello E, Chiesa M (2003) J. Phys. Chem. B 107:8498
187. Wolf D (1992) Phys. Rev. Lett. 68:3315
188. Ventrice CA, Bertrams T, Hannemann H, Brodde A, Neddermayer H. (1994) Phys. Rev. B 49:5773
189. Rohr F, Wirth K, Libuda J, Cappus D, Bäumer M, Freund HJ (1994) Surf. Sci. Lett. 315:L977
190. Schönnenbeck M, Cappus D, Klinkmann J, Freund HJ, Petterson LGM, Bagus PS (1996) Surf. Sci. 347:337
191. Matsumoto T, Kubota J, Kondo JN, Hirose C, Domen K (1999) Langmuir 15:2158
192. Plass R, Feller J, Gajdardziska-Josifovska M (1998) Surf. Sci. 414:26
193. Plass R, Egan K, Collazo-Davila C, Grozea D, Landree E, Marks LD, Gajdardziska-Josifovska M (1998) Phys. Rev. Lett. 81:4891
194. Kunz AB, Guse MP (1977) Chem. Phys. Lett. 45:18
195. Huzimura R, Yanagisawa Y, Matsumura K, Yamabe S (1990) Phys. Rev. B 41:3786
196. Yanagisawa Y, Takaoka K, Yamabe S (1994) J. Chem. Soc. Faraday Trans. 90:2561
197. Yudanov I, Pacchioni G, Neyman KM, Rösch N (1997) J. Phys. Chem. B 101:2786
198. Giordano L, Pacchioni G, Bredow T, Fernandez-Sanz (2001) J. Surf. Sci. 471:21
199. Markovits A, Skalli MK, Minot C, Pacchioni G, López N, Illas F (2001) J. Chem. Phys. 115:8172
200. Lopez N, Paniagua JC, Illas F (2002) J. Chem. Phys. 117:9445
201. Pacchioni G, Rösch N (1996) J. Chem. Phys. 104:7329
202. Ferrari AM, Xiao C, Neyman KM, Pacchioni G, Rösch N (1999) Phys. Chem. Chem. Phys. 1:4655
203. Giordano L, Pacchioni G, Ferrari AM, Illas F, Rösch N (2001) Surf. Sci. 473:213
204. Moseler M., Häkkinen H., Landman U. Phys. (2002) Rev. Lett. 89 176103
205. Musolino V, Selloni A, Car R (1998) J. Chem. Phys. 108:5044
206. Musolino V, Dal Corso A, Selloni A (1999) Phys. Rev. Lett. 83:2761
207. Musolino V, Selloni A, Car R (1999) Phys. Rev. Lett. 83:3242
208. Sljivancanin Z, Pasquarello A (2003) Phys. Rev. Lett. 90:247202
209. Yudanov I, Vent S, Neyman K, Pacchioni G, Rösch N (1997) Chem. Phys. Lett. 275:245
210. Goniakovski J (1998) Phys. Rev. B 58:1189
211. Judai K, Abbet S, Wörz AS, Heiz U, Giordano L, Pacchioni G (2003) J. Phys. Chem. B 107:9377
212. Giordano L, Del Vitto A, Pacchioni G, Ferrari AM (2003) Surf. Sci. 540:63
213. Sanchez A, Abbet S, Heiz U, Schneider WD, Häkkinen H, Barnett RN, Landmann U (1999) J. Phys. Chem. A 103:9573
214. Sanchez A, Abbet S, Heiz U, Schneider WD, Ferrari AM, Pacchioni G, Rösch N (2000) J. Am. Chem. Soc. 122:3453

215. Sanchez A, Abbet S, Heiz U, Schneider WD, Ferrari AM, Pacchioni G, Rösch N (2000) Surf. Sci. 454–456:984
216. Tysoe WT, Nyberg GL, Lambert RM (1983) J. Chem. Soc., Chem. Commun. .623
217. Holmblad PM, Rainer DR, Goodman DW (1997) J. Phys. Chem. B 101:8883
218. Abdelrehim IM, Pelhos K, Madey TE, Eng J, Chen JG (1998) J. Mol. Catal. A 131:107
219. Hoffmann H, Zaera F, Ormerod RM, Lambert RM, Yao JM, Saldin DK, Wang LP, Bennett DW, Tysoe WT (1992) Surf. Sci. 268:1
220. Ormerod RM, Lambert RM, Hoffmann H, Zaera F, Yao JM, Saldin DK, Wang LP, Bennet DW, Tysoe WT (1993) Surf. Sci. 295:277
221. Pacchioni G, Lambert RM (1994) Surf. Sci. 304:208
222. Abbet S, Sanchez A, Heiz U, Schneider WD (2001) J. Catal. 198:122
223. Ferrari AM, Giordano L, Rösch N, Heiz U, Abbet S, Sanchez A, Pacchioni G (2000) J. Phys. Chem. B 104:10612
224. Giordano L, Pacchioni G, Illas F, Rösch N (2002) Surf. Sci. 499:73

3

Catalysis by Nanoparticles

C.R. Henry

3.1 Introduction

What is the meaning of nanocatalysis? Is it not a pleonasm? Indeed, industrial catalysis generally takes place at the nanoscale (or sub-nanoscale): most of the catalysts are made of metal particles of a few nanometers in size and *in fine* all the elementary reaction steps occur at the atomic (or molecular) scale. Thus, catalysis seems to be intrinsically a nanoscale phenomenon. The word nanocatalysis, in fact, does not apply to the catalytic phenomenon itself but to the intrinsic properties of the catalysts, which may change in the nanoscale. As we will see, some properties of nanometer sized, supported metal particles directly affect their catalytic activities and these properties are, in size range of up to a couple of hundred atoms, not scalable from bulk properties.

First, we can look at the properties of atoms and small clusters. The most important property which influences catalysis is the electronic structure. Figure 3.1 shows the evolution of binding energies of the 4s and 3d electrons of free Cu clusters as a function of size. These results, from the group of Smalley [1], clearly show the electronic structure (influenced by the onset of the s and d bands) to evolve in a non-monotonic way as function of size. In fact, it depends on the exact number of atoms in the cluster. When the clusters, however, reach a size of about 30–50 atoms, the electronic structure evolves smoothly towards the bulk limit. Experiments by several groups [2–5] have shown that the chemical properties of small free metal clusters indeed evolve in a discrete manner with the exact number of atoms. Moreover, beautiful experiments on size-selected clusters softlanded on MgO thin films, performed in the Heiz group [6–9], have shown that the same trend is observed for supported small cluster, paraphrasing the title of a paper of Heiz, I would state: each atom counts [7].

Figure 3.2 exemplifies remarkably the fact that the reactivity for the CO oxidation of Pt clusters shows a strong dependence with the number of atoms in the cluster. After reaching a size of about 25–30 atoms, the reactivity varies more smoothly. This behaviour is a direct consequence of the discrete

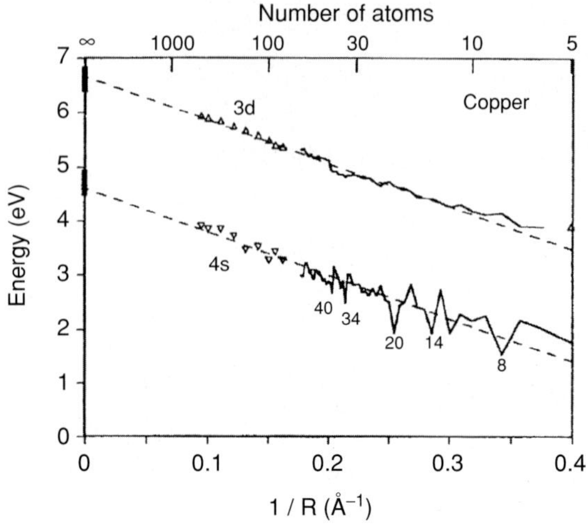

Fig. 3.1. Evolution of the electronic properties of small free Cu clusters as a function of their size (from [1])

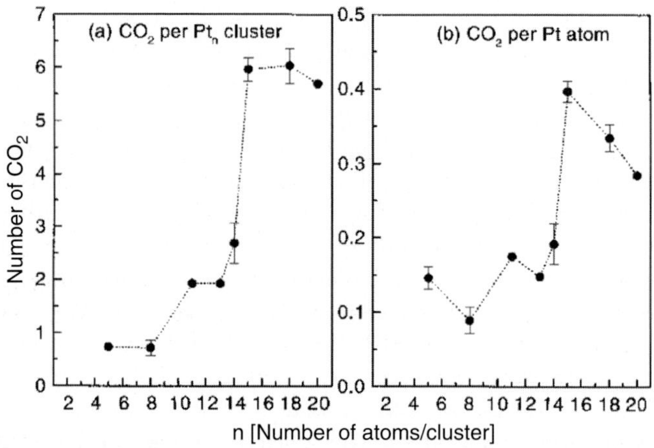

Fig. 3.2. Evolution of the activity of small Pt clusters towards CO oxidation as a function of the number of atoms (from [7])

evolution of the electronic properties. Now we could expect that the catalytic properties of the metal cluster will evolve smoothly after say 50 atoms, and they should be scalable up to bulk limit. In fact, this is not true, very often the catalytic properties of supported nanometer metal particles are not directly predictable from experiments on bulk metal surfaces [10, 11]. In fact, nanometer-sized supported particles, which constitute the majority of heterogeneous catalysts, have specific physical properties that differ from their bulk counterpart. The most striking example is given by gold, which is the

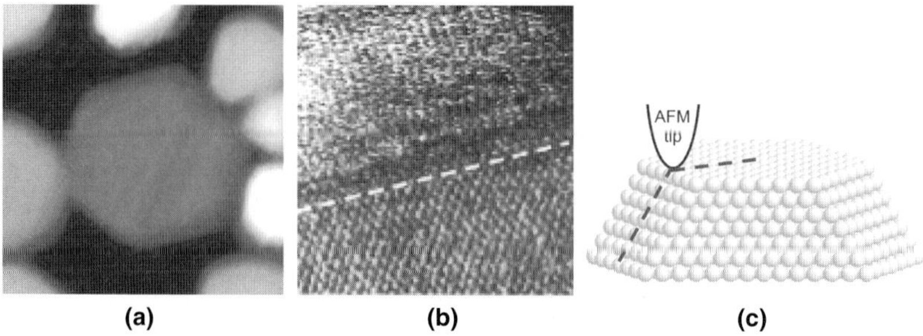

Fig. 3.3. AFM image of a 10-nm gold nanoparticle (middle of the (a)-image) supported on mica (001) showing the (111) and (100) crystal planes constituting the crystal habit. (**a**) Top view AFM image, (**b**) high-resolution AFM image close to an edge (dotted line) showing the lattice of the (111) top facet and of a (100) lateral facet, (**c**) ball model of the Au particle showing a line scan of the AFM tip (from [14])

noblest metal and thus non-reactive in the bulk state. This observation, by the way, is very well understood theoretically [12]. In the nanometer size range, however, gold turns into a very active catalyst for several reactions [13]. This specificity of the nanometer size range in catalysis is due to several reasons that we will emphasize in the following. First, small metal particles (1–20 nm) reveal generally crystalline structures exhibiting facets constituted by dense planes. As shown in Fig. 3.3, metal particles (here Au nanoparticles supported on mica [14]) present well-defined facets (in this specific case (111) and (100) planes) that minimize the surface energy. Their lattice can directly be imaged by atomic force microscopy (AFM). It is known that the reactivity of crystal planes is often strongly dependent on their structure [15]. Metal particle having different sizes will have, as we will see in Sect. 3.2, different proportions of facets that may react completely differently. This effect will not only depend on the percentage of the different facets but also on the diffusion barriers of adsorbed species to move from one facet to another [16]. As imaged by scanning tunnelling microscopy (STM) [17], edges separating two crystal facets and corresponding to low-coordinated atoms have different local densities of states than highly coordinated atoms from the facets. Thus, coordination also plays an important role for the catalytic action. This edge effect will be only important for particles smaller than about 5 nm, for which the ratio of edge atoms to surface atoms is important.

Going further to smaller sizes, other dramatic changes appear. A typical example is depicted in Fig. 3.4, which displays an atomically resolved image of a Pd nanoparticle supported on a $MoS_2(0001)$ surface [17]. This particle is only two-layer high, all the surface atoms are imaged. From this, we can deduce easily that it contains 27 atoms exactly, 20 in the first layer and 7 in the second one. It is interesting to see that compared to the larger particle (roughly 400 atoms) represented in Fig. 3.5, this small particle presents very open sites. One of those marked by an **X** is a B_5 site, which was predicted long time

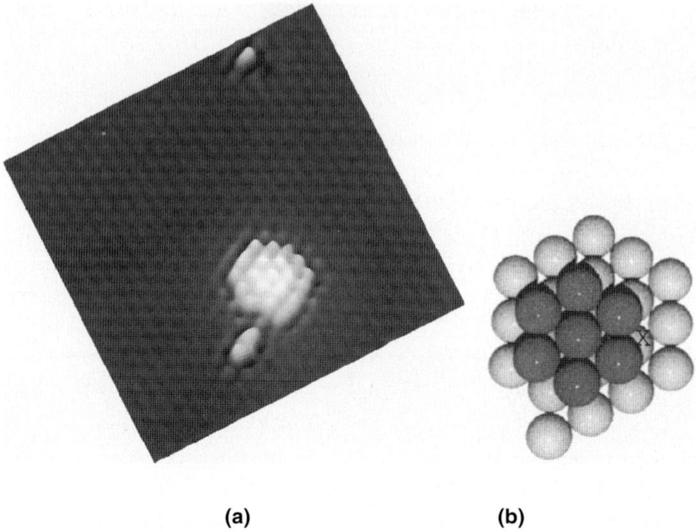

Fig. 3.4. STM image of a Pd cluster supported on MoS_2 containing 27 atoms (**a**) On the ball model (**b**) the sign **X** indicates a B_5 site (from [17])

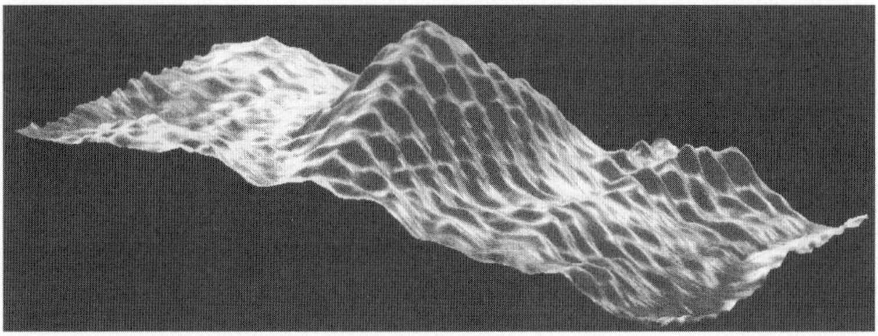

Fig. 3.5. STM image at the atomic scale of a 4-nm Pd particle supported on a graphite (0001) substrate. The STM image reveals unambiguously that this particle reveals (100) facets forming a regular tetrahedron [17]

ago in a famous paper by Van Hardeveld and Hartog [18]. It corresponds to a very active site and is responsible for strong size effect in catalysis. By closer inspection of Fig. 3.4, one can see that around the Pd cluster, the sulphur atoms of the substrate (identified as protrusions in the STM images [19, 20]) seem to be raised above the substrate plane. In fact, this is an apparent effect due to electronic modification of the substrate atoms by the metal atoms from the interface. Conversely, the electronic structures of metal atoms at the interface are also modified by the substrate, which by itself may have a strong influence on their catalytic properties. For very small sizes, eventually, all the atoms of the metal cluster can be modified by the substrate especially on defects resulting in a charged cluster, which modifies drastically the catalytic properties [21].

The presence of the substrate, even if it does not directly take part in the reaction, can have a profound influence on the reaction kinetics. Molecules physisorbed on the substrate are, besides the gas phase molecules, a source of reactants for the metal particles. This second source of reactants depends on particle size and density, as we will see in Sect. 3.1. This phenomenon [22–24] is called reverse-spillover in reference to the well-known spillover phenomenon, which represents the diffusion of adsorbed species from the metal particles to the support [25].

Another effect of the nanometer size in reaction kinetics of catalytic reaction is the disappearance of the bi-stability in the CO oxidation as recently evidenced by molecular beam experiments on supported model catalysts [26].

In this chapter, we will examine the change of the physical properties of metal particles in the nanometer size range and the consequence it has for the evolution of the catalytic properties with size. We will also see that by controlling the fabrication of the metal particles, it will become possible to design new catalysts with tuned reactivities and possibly enhanced selectivities.

3.2 Specific Physical Properties of Free and Supported Nanoparticles

3.2.1 Surface Energy and Surface Stress

Surface energy results from the different environment of surface atoms relatively to the bulk ones. In a simple description, one can consider that the surface energy corresponds to the number of broken bonds of surface atoms relatively to the bulk situation. For a compact structure there are 12 neighbours. The work necessary to increase the surface by adding atoms is proportional to the increment of the area dA, the proportionality factor is the surface free energy, γ:

$$\mathrm{d}W = \gamma \mathrm{d}A. \tag{3.1}$$

One can also increase a surface by stretching it and keeping the number of atoms constant, then we define the surface stress g_{ij}:

$$\mathrm{d}W = g_{ij}\mathrm{d}A. \tag{3.2}$$

Here g_{ij} is a tensor, its relationship with the surface free energy is given by the following equation:

$$g_{ij} = \delta ij + (\delta\gamma/\delta uij) \tag{3.3}$$

where u_{ij} is the strain tensor of the deformation and δ_{ij} the Kronecker symbol. Few accurate experimental values of the surface energies [27] are available and even less for the surface stress [28]. However, realistic values of surface energies can now be obtained by numerical calculations [29]. The question arises now, how far these bulk values can be extended to small nanoparticles? Recently, calculations have been performed by using EAM many-body potentials [30]. These calculations have shown both the surface energy and the surface stress to increase for particles below 2–3 nm.

3.2.2 Lattice Parameter

The surface stress has a direct influence on the lattice parameter of the nanoparticles. Indeed, surface stress induces a pressure (ΔP) towards the core of a spherical nanoparticle that is given by

$$\Delta P = \frac{2g}{R}, \tag{3.4}$$

where R is the particle radius and g the surface stress when considering the case of a simple cubic structure. This increase of pressure induces a contraction of the lattice (assuming a positive surface stress, which is generally the case). The variation of the lattice parameter ($\Delta a/a$) is given by the following equation [31]:

$$\frac{\Delta a}{a} = \frac{(-2/3)\chi g}{R}, \tag{3.5}$$

where χ is the compressibility. We see the decrease of the lattice parameter to be inversely proportional to the radius of the particles. Contraction has been observed for many particles and is generally of the order of a few percents. In Fig. 3.6, the variation of the lattice parameter of Cu particles, measured by EXAFS and SEELFS is represented.

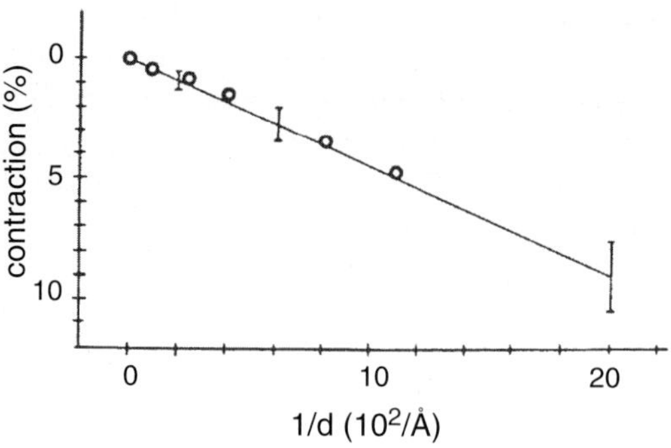

Fig. 3.6. Variation of the lattice parameter of free Cu clusters as a function of their reciprocal diameter. *Circles* represent SEELFS measurements (from [32]), while the *straight line* corresponds to EXAFS measurements (from [33])

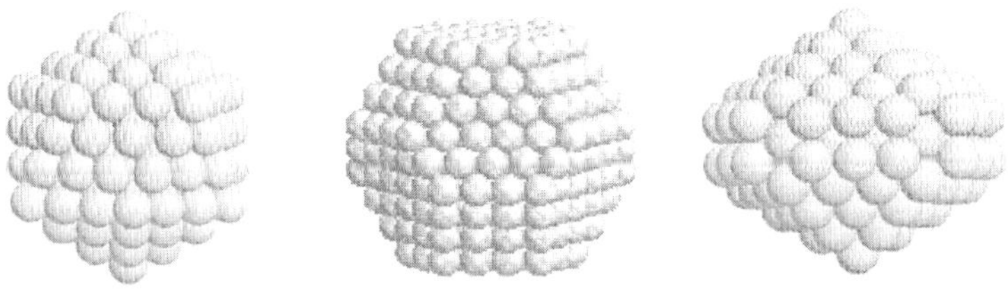

Fig. 3.7. Most common stable shapes of nanoparticles: (**a**) icosahedron, (**b**) truncated octahedon (Wulff shape), (**c**) Marks decahedron

When the particles are in epitaxy with a crystalline support, the lattice of the small particles can be accommodated to the substrate lattice. If the lattice parameter of the support is larger than those of the particle, its lattice may expand. This has been observed by HRTEM for 2-nm Pd particles supported on MgO(100), where the expansion is 8% [34]. However, when the size of the particles increases, they progressively recover their bulk lattice parameter by the introduction of dislocations as shown experimentally [35] and simulated by molecular dynamics [36].

3.2.3 Equilibrium Shape

The equilibrium shape of a macroscopic crystal is an old problem first addressed by Wulff [37], who showed the equilibrium shape at 0 K to be a polyhedron. At the equilibrium, the surface energy is given by the famous Wulff's theorem:

$$\frac{\gamma_i}{h_i} = \text{const.}, \tag{3.6}$$

which expresses the surface energy of a given facet to be inversely proportional to the central distance to this facet. For an FCC crystal, it is a truncated octahedron that exhibits hexagonal (111) facets and square (100) facets (see Fig. 3.7b). These two planes have the smallest surface energy. The anisotropy of surface energy $\gamma_{(100)}/\gamma_{(111)}$ is $2/\sqrt{3}$. However, the validity of these findings to very small particles is questionable for several reasons: We have seen that below 2–3 nm, the surface energy increases, an effect which is not negligible. Furthermore, below about 5 nm, the fraction of edge sites increases, another finding which is neither taken into account in the Wulff theorem. Marks was the first to take into account for the effect of edges in the equilibrium shape of nanoparticles [38]. Using a simple first neighbour bonding model, he showed the (100) facets to vanish at small sizes, and the equilibrium shape of an FCC particle to become an octahedron. Another point, which has to be taken into consideration is: at very small sizes the bulk structure is rarely the most stable one. It is known from many molecular dynamic calculations that often the most stable structure of a 13 atoms metal cluster is the icosahedron [39,40]. The icosahedron structure (see Fig. 3.7a) can be stable up to relatively large sizes, it generally transforms not directly to the bulk structure but transits towards another peculiar structure called Marks decahedron, which is in fact a truncated decahedron [41] (see Fig. 3.7c).

The transition to the bulk structure appears for several metals at relatively large sizes, 17,000 atoms for Ni [39] and about 50,000 atoms for Cu [42]. It is important to notice that the energy difference between these structures is often very small. This explains the very rapid fluctuations between different structures observed by HRTEM [43] that we call quasi-melting [44].

If the crystal is lying on a support, the equilibrium shape is modified by the interaction with the substrate. This problem has been solved independently by Kaichew [45] and by Winterbottom [46]. The equilibrium shape is expressed by the Wulff–Kaichew theorem represented by the following equation:

$$\frac{\Delta h}{h_s} = \frac{\beta}{\gamma_s}. \tag{3.7}$$

In this case, the Wulff shape is truncated at the interface by an amount Δh, which is proportional to the adhesion energy β. The latter represents the work to separate the supported crystal from the substrate at an infinite distance. h_s and γ_s are the central distance and the surface energy of the facet parallel to the interface, respectively. In particular, this theorem shows that the stronger the particle–substrate interaction (given by β) is the flatter is the supported particle. Equation (3.7) offers a simple way for determining the adhesion energy of a supported crystal from TEM pictures of supported particles observed in a profile view [47].

Figure 3.8 shows an example of a 10-nm Pd particle supported on MgO(100). Its shape is a truncated octahedron seen in the [110] direction. Assuming a surface energy of $1.64\,\mathrm{J\,m^{-2}}$ [48] for the top (100) facet, one derives an adhesion energy of $0.90\,\mathrm{J\,m^{-2}}$ [49], which is in good agreement with molecular dynamic simulations giving an adhesion energy of $0.83\,\mathrm{J\,m^{-2}}$ [36].

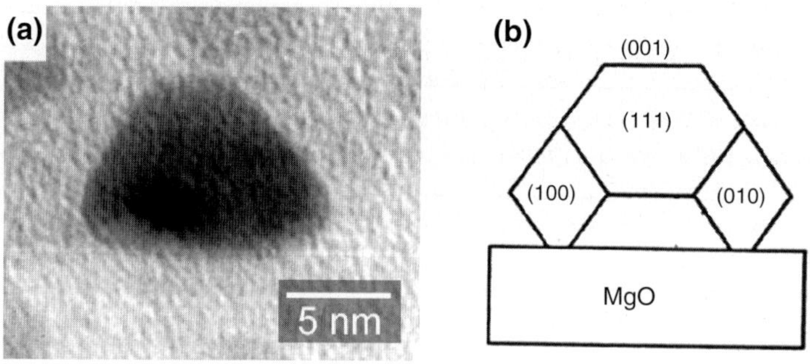

Fig. 3.8. Picture of a 10-nm Pd particle supported on MgO(100) seen in profile view. (**a**) TEM picture in the [110] direction, (**b**) schematic representation of the 3D shape

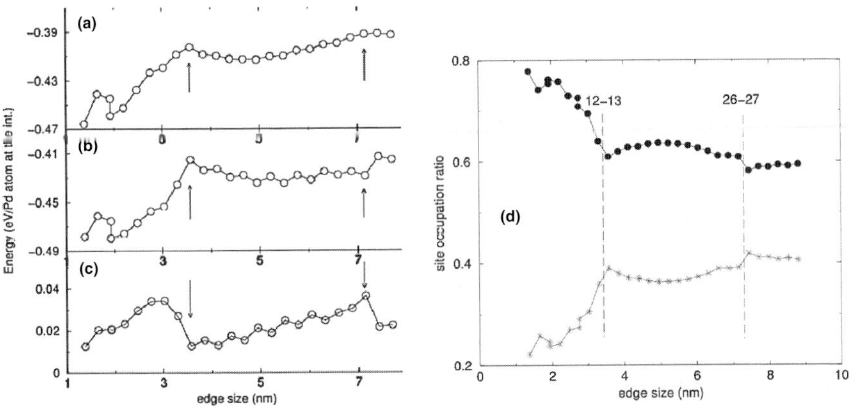

Fig. 3.9. Adhesion energy and fraction of occupation of oxygen sites as a function of the size of Pd particles supported on MgO(100) calculated from molecular dynamics simulation (from [36]). The adhesion energy (**a**) is decomposed into two parts: the Pd–MgO bonds (**b**) and the Pd–Pd bonds (**c**) at the interface. (**d**) Fraction of substrate oxygen (*filled circle*) and magnesium (*asterisk*) occupied by Pd atoms at the interface

In fact, the application of the Wulff–Kaichew theorem implicitly assumes that the supported particle is fully relaxed. This is true for 10 nm or larger particles, for particles smaller than about 7 nm, however, the first layers from the Pd/MgO interface are dilated in order to accommodate, at least partially, the lattice misfit of 8% [35]. This strain energy has to be accounted in the minimization of the total energy. This situation leads to a decrease of the aspect ratio of the Pd particle (or an increase of the truncation at the interface) [35], which is equivalent to an increase of the adhesion energy. This result can be understood from molecular dynamics simulations [36] showing that, at very small sizes, the first plane of Pd atoms in contact with the MgO substrate is dilated. In addition, the number of Pd atoms sitting on top of the oxygen ions is maximized as the binding energy is higher at these sites. This situation maximizes the adhesion energy. When particle size increases, the Pd lattice starts to relax towards its bulk value and less direct Pd–O bonds are formed, and finally, interfacial dislocations appear that correspond to the full relaxation of the Pd lattice. Figure 3.9 depicts the evolution of the adhesion energy with the size of the Pd particles. The adhesion energy decreases (in absolute value) when particle size increases. It can be decomposed into two contributions: the Pd–Pd bonds (it corresponds to a decrease of the bonding relative to the free particle) and the Pd–O bonds at the interface, the latter is by far the major contribution. The minimum in the absolute value of the adhesion energy corresponds in fact to the minimum of the Pd–O bonds corresponding to a partial relaxation of the Pd lattice by the introduction of dislocations.

Another factor, which has a strong effect on the equilibrium shape of crystals is the adsorption. Indeed, from Gibbs we know that adsorption reduces the surface energy (the number of broken bonds decreases):

$$\mathrm{d}\gamma = -\sum \Gamma_i \mathrm{d}\mu_i, \tag{3.8}$$

where Γ_i and μ_i are the concentration and the chemical potential of the ith species, respectively. As generally the equilibrium concentrations of the adsorbed species are different from one facet to another, the decrease in the surface energy is facet dependent. If the surface energy anisotropy changes upon adsorption, the equilibrium shape of the particle will change. The change of the equilibrium shape has been studied theoretically by Shi [50]. Experimentally, it has been shown that Pd particles supported on MgO(100) become flatter upon heating under oxygen due to an increase of the (100) facets at the expense of the (111) facets [49]. Calculations showed that already an oxygen coverage of 0.2 changed significantly the equilibrium shape [49]. This phenomenon occurs already in bulk crystal, but the kinetics of the process is very slow and it is expected at a temperature close to the melting point. It has been shown that the relaxation time depends inversely to the fourth power of the radius of the crystal [51]:

$$\tau = \frac{2kTR^4}{v^{4/3}D\gamma}, \tag{3.9}$$

where k is the Boltzmann constant, T the temperature, v the atomic volume, D the surface self-diffusion coefficient and γ the (average) surface energy. Thus, for particle in the nanometer size range, the change of the equilibrium shape occurs in a time of the order of a few minutes, as observed by environmental HRTEM [52].

3.2.4 Melting Temperature

Another physical property, which strongly depends on particle size, is the melting point. It has been known since a long time that the melting point $T(R)$ of a crystal decreases with the inverse of its radius (R). This relation is expressed in the Pawlow law [53]:

$$\frac{T}{T_\mathrm{m}} = 1 - \frac{2(\gamma_\mathrm{s}\rho_\mathrm{s}^{-2/3} - \gamma_\mathrm{l}\rho_\mathrm{l}^{-2/3})}{LR\rho_\mathrm{s}^{1/3}}. \tag{3.10}$$

T_m is the bulk melting temperature, γ and ρ are the surface energy and the density of the solid phase and the liquid phase noted by the subscripts s and l and L is the latent heat of fusion. This phenomenon is due to the increase in the fraction of surface atoms. The Pawlow law is verified experimentally for particles larger than 5 nm and more sophisticated thermodynamic models have been developed which can be applied down to 2-nm particles [54]. During catalytic reactions, particles are generally solid, but in the case of carbon nanotubes synthesized by CVD, the catalyst particles could melt at temperatures lower than the bulk melting point.

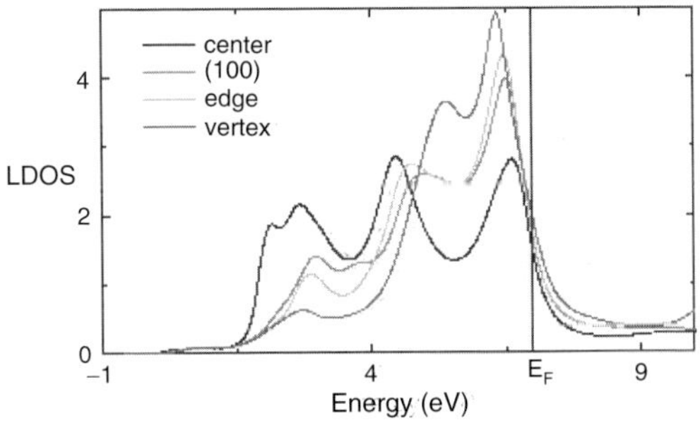

Fig. 3.10. Evolution of the local density of valence states on various sites on Pd particles containing 3,871 atoms, calculated in the tight-binding approximation (from [40])

3.2.5 Electronic Band Structure

Metal particles larger than about 100 atoms present an electronic band structure like in the bulk state, when the proportion of surface atoms, however, becomes non negligible, several differences appear in the band structure. First, the width of the valence band is reduced and, second, its centre of gravity is shifted towards the Fermi level [55, 56]. This evolution is a consequence of the reduction of the coordination that is equivalent to an increase in the localization of the valence electrons. This becomes more dramatic if we consider the local density of states on low-coordinated sites like edge and corner atoms. Figure 3.10 shows the calculated density of states on various atoms from a cubo-octahedron Pd cluster containing 3,871 atoms (equivalent to a radius of 5.7 nm) [40].

The shift of the d-band centre on the various Pd atoms is of particular importance for the reactivity because it is now well established, essentially from the work of the group of Nørskov, that a shift of the centre of the d band towards the Fermi level leads to an increase of the adsorption energy and a decrease of the dissociation barriers for adsorbed molecules [57].

3.3 Reactivity of Supported Metal Nanoparticles

3.3.1 Support Effect: Reverse-spillover

The spillover of reactant is a well-known phenomenon in heterogeneous catalysis. It is due to the diffusion of atomic or molecular intermediates, formed, e.g. by dissociation, from the catalyst to the support material [25]. The reverse-spillover is the opposite process and it corresponds to the diffusion of molecular species adsorbed on the support towards the catalyst's particles (see a complete discussion of this phenomenon in [58]). This phenomenon was first

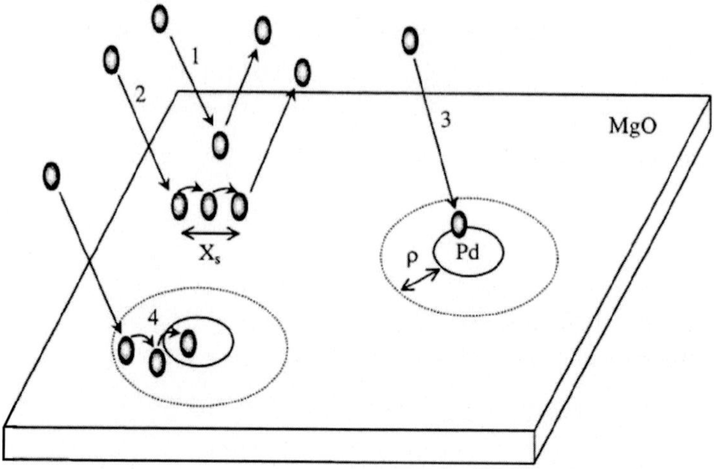

Fig. 3.11. Schematic representation of the reverse-spillover for gas molecules impinging on a Pd/MgO model catalyst. (**1**) Reflection from the MgO, (**2**) adsorption–desorption from the MgO, (**3**) direct chemisorption on the Pd cluster, (**4**) capture by a Pd cluster of a molecule physisorbed on the MgO

suggested by Tsu and Boudart in 1961 [59]. It was later used to explain size effect in CO oxidation on model Pd catalysts [22, 23]. A quantitative model for the reverse-spillover was given by Henry and applied to the CO oxidation, assuming that the catalyst particles were distributed on a regular lattice [24]. Later, Zhdanov and Kasemo have extended the model for the case where the catalyst particles are randomly distributed [60].

Within this model, the capture of adsorbed molecules physisorbed on the support is a second channel for chemisorption of reactants on the catalyst's particles, with the direct impact from the gas phase being the first channel (see Fig. 3.11).

In a simple approach, one defines a capture zone around each cluster of width ρ, where all physisorbed molecules will become chemisorbed on the metal cluster by diffusion. In the first degree of approximation, ρ is equal to the mean diffusion of a molecule physisorbed on the substrate:

$$X_{\mathrm{s}} = (D\tau)^{1/2}, \tag{3.11}$$

where D is the diffusion coefficient and τ the mean lifetime of a molecule on the substrate. This is true for an isolated cluster. In the case of an assembly of clusters of radius R, which are distributed on a square lattice of parameter $2L$, ρ is given by the following equation representing the capture area A [61]:

$$A = \pi(\rho^2 + 2\rho R) = 2\pi R X_{\mathrm{s}} P\left(\frac{R}{X_{\mathrm{s}}}, \frac{L}{X_{\mathrm{s}}}\right), \tag{3.12}$$

where $P(x, y)$ is defined by:

$$P(x, y) = \frac{I_1(y)K_1(x) - K_1(y)I_1(x)}{I_0(x)K_1(y) + K_0(x)I_1(y)}. \tag{3.13}$$

I_i and K_i are the modified Bessel functions of ith order. The number of molecules joining a cluster per second is:

$$F_{\text{diff}} = \alpha JA, \tag{3.14}$$

where α is the adsorption probability of a molecule on the substrate and J the impinging flux of molecules. The total rate of molecules joining a cluster (by diffusion and by direct impingement) will be

$$F_{\text{tot}} = F_{\text{diff}} + \pi R^2 J, \tag{3.15}$$

assuming a unity sticking probability on the metal particles.

We can define a parameter α_g that represents the fraction of the flux of molecules impinging on $1\,\text{cm}^2$ of sample that becomes chemisorbed on the metal clusters:

$$\alpha_g = \frac{nF_{\text{tot}}}{J}, \tag{3.16}$$

where n is the density of metal clusters. α_g is called global adsorption probability and it can be directly measured in a molecular beam experiment as shown in Fig. 3.12 [58,61].

The global adsorption probability has been directly measured for CO and NO on Pd/MgO(100) model catalysts [61,62]. Figure 3.13 presents such measurements for NO [62].

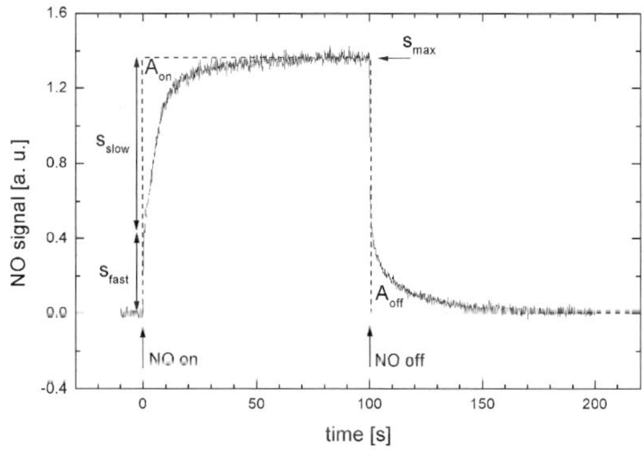

Fig. 3.12. Principle of the measurement of the global adsorption probability in a molecular beam experiment. The desorption pulse is composed by a fast component (S_{fast}) corresponding to molecules reflected or desorbing from the substrate and a slow component (S_{slow}) corresponding to (chemisorbed) molecules desorbing from the metal clusters. The relative amplitude of the slow component gives the global adsorption probability α_g

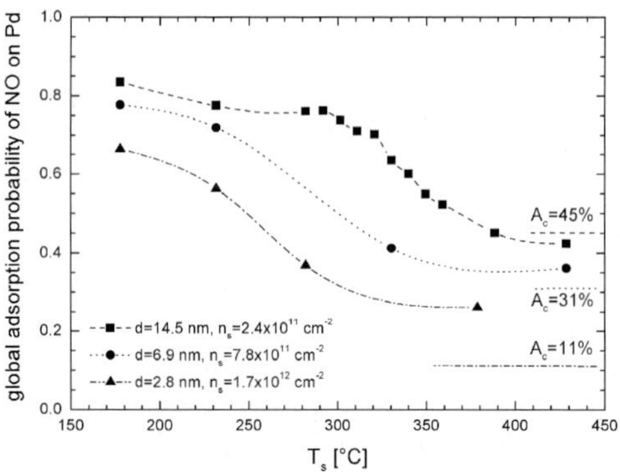

Fig. 3.13. Variation with temperature of the global adsorption probability of NO on three different Pd/MgO(100) model catalysts (from [62])

We see the global adsorption to increase when the temperature decreases. At high temperature it is equal to A_c, the fraction of the substrate covered by the metal clusters. In that case, the width of the capture zone, ρ, is zero and only molecules directly impinging on the clusters are chemisorbed and α_g is equal to $\pi R^2 n$. When the temperature decreases, the mean diffusion length, X_s, grows and ρ increases. Finally, the global adsorption probability reaches saturation when the capture zones overlap. The maximum value of the adsorption probability is given by:

$$\alpha_g^{max} = (1 - A_c) + A_c . \quad (3.17)$$

The global adsorption is a function of three parameters: the substrate temperature, the cluster density and the cluster size. For example, in the case of the smallest clusters in Fig. 3.13, the maximum value of the global adsorption probability is six times larger than A_c, meaning that the total flux joining the clusters is six times larger than the direct flux. This phenomenon can affect strongly the reaction rate of a catalytic reaction, especially in cases where the reaction is limited by the adsorption of the reactants. This is true for the CO oxidation in the oxygen-rich regime (i.e. high temperature) [22–24,63]. In this regime, the TON is proportional to the CO coverage [58], which itself is proportional to the effective flux of CO joining the Pd clusters. Then, the TON can be written by the following expression:

$$\mathrm{TON} = \frac{P_{CO} S(T)[1 + (\alpha X_s/R) P(x,y)]}{N_o \sqrt{(2\pi m k T)}} . \quad (3.18)$$

$S(T)$ is the reaction probability that is assumed to be the same as on a Pd extended surface, N_o the density of surface atoms, R the cluster radius, $P(x,y)$ is given by (3.13) and X_s by (3.11). When the capture zones overlap (at low temperature) the TON becomes maximum and has a simple expression:

$$\mathrm{TON} = \frac{P_{\mathrm{CO}} S(T)[1 - \alpha/2 + \alpha/(2\pi R^2 n)]}{N_{\mathrm{o}}\sqrt{(2\pi mkT)}}. \tag{3.19}$$

In the reduction of NO by CO at low temperature, on Pd/MgO(100) model catalysts, the reaction rate is independent of CO pressure but increases with NO pressure [64,65], then the TON of the reaction is modified by the reverse-spillover effect and, in particular, depends on particle size. In that case, it is no longer possible to compare the TON value measured on different particle sizes, it is more appropriated to calculate the reaction probability [64], which takes into account for the real flux of molecules that reach the clusters which can be measured by molecular beam experiments. Reverse-spillover effects have also been recently observed for the CO oxidation on size-selected Pd clusters soft-landed on MgO epitaxial films [66].

3.3.2 Morphology Effect

Morphology effects appear for reactions that are structure sensitive. A good example is the reduction of NO by CO on Pd, which shows the highest activity on (111) planes and lower activity on open surfaces [67]. However, in the case of supported particles it is difficult to control the particle shape independently of particle size [68]. However, by using a careful TEM characterization it is possible to disentangle size and morphology effects. An example is given by the NO reduction of CO on Pd/MgO(100) model catalysts [64].

Figure 3.14 shows the reaction probability as a function of temperature for three different samples having particle sizes of 2.8, 6.9 and 15.6 nm. In the low-temperature regime, the reaction is limited by the dissociation of NO, we see that the largest particles are the least reactive ones, while the smallest

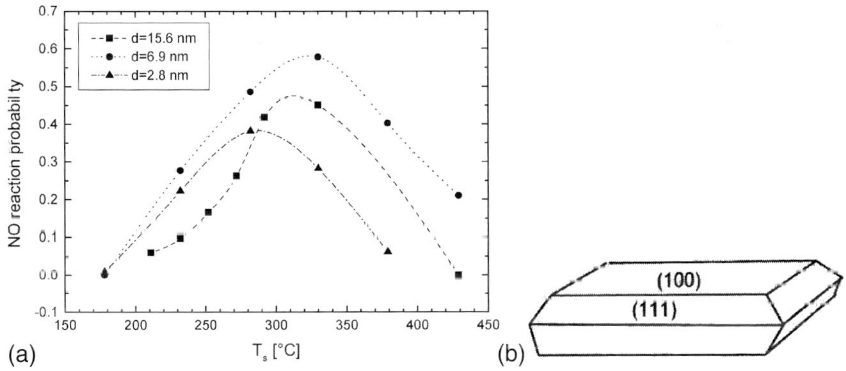

Fig. 3.14. NO reduction by CO. (**a**) Steady-state NO-reaction probability for three different samples with different particle sizes. (**b**) The shape of the large particles is schematically represented, other particles have a truncated octahedron shape as in Fig. 3.8

ones are less active than the middle sized ones. The same evolution for the three different samples is also observed for the NO dissociation rate measured on the same samples [62]. One can ask the question: Why does this evolution not follow the size evolution? The answer comes from the shape of the particles. The largest particles have coalesced and present mainly (100) facets, which are the least active for the reaction (see Fig. 3.14b). The medium sized particles, which have a truncated octahedron shape that mainly presents (111) facets, are the most active ones (see Fig. 3.8). Finally, the smallest particles, which reveal also a majority of (111) facets, are a little less active because for that size (2.8 nm) the proportion of edge sites is no longer negligible. These sites are, however, low-coordinated atoms, which reveal low activity. The origin of the low activity on open surfaces (and low-coordinated sites) has been explained by Goodman and co-workers [69]. They have shown that for open surfaces there is a second type of adsorbed nitrogen atom, resulting from the dissociation of NO, and which are strongly bound and thus do not desorb at the reaction temperature. In other words, they poison the catalyst.

3.3.3 Effect of the Edges

As we have seen in Sect. 3.1, the proportion of edge sites on the catalyst particles no longer to be negligible below a size of about 4–5 nm. These low-coordinated sites have different electronic structures (see Sect. 3.2.5) that generally induce higher binding energies and lower dissociation barriers for the adsorbed molecules. An example of this effect has been observed on Pd particles for the binding energy of CO at low coverages. The lifetime of CO molecules has been measured as a function of temperature by molecular beams methods [58].

Figure 3.15 displays the evolution of the binding energy of CO at a coverage of a few percent of a monolayer as a function of size of the Pd particles supported on MgO(100) [70]. The Pd particles were obtained by epitaxial growth at high temperature. From TEM observation, the shape of the particles was shown to be of truncated octahedral shape [68]. By decreasing particle size, the adsorption energy is constant (about 30 kcal mol^{-1}), but it increases for particles below 5 nm to values of up to 38 kcal mol^{-1}. This follows the evolution of the proportion of edge sites. A more quantitative account of this effect can be given from theoretical calculations. We have seen in Sect. 3.2.5, from the calculation of Mottet et al. [40], the evolution of the local density of d states for various sites on the Pd particle. Going from a site on a (111) facet to a site on a (100) facet, the d band shifts towards the Fermi level by 0.19 eV, when going to an edge site, however, the shift is 0.22 eV and finally going to a vertex site the shift is 0.45 eV. From Hammer and Nørskov [57], we know that the upward shift of the d band corresponds to a proportional increase of the binding energy of CO. The proportionality factor has been determined by ab initio calculation [57]. We can then calculate the increase

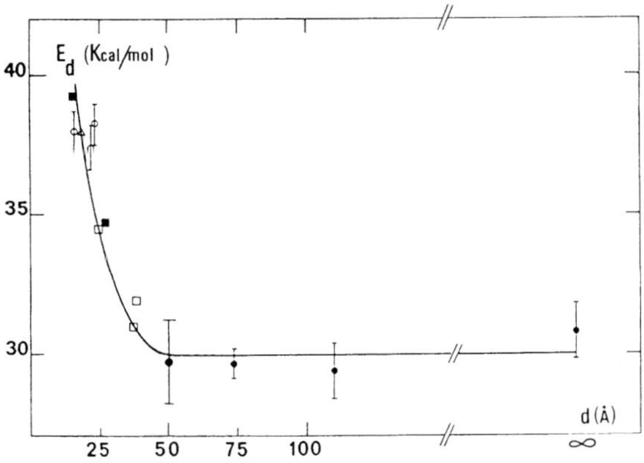

Fig. 3.15. Size effect in the adsorption of CO at low coverage on Pd particles supported on MgO(100). On CO adsorption energy measured at low coverage by a molecular beam relaxation spectroscopy (MBRS) (from [70])

of the adsorption energy, which is 3.4, 3.9 and 8.1 kcal mol^{-1} for (100) facets, edges and vertices sites, respectively. These values are in fairly good agreement with the measurements. This proves that when particle size decreases, the low-coordinated sites become more and more important for catalysis.

3.3.4 The Peculiar Case of Gold Nanoparticles

Gold presents, probably, the most remarkable genuine "nanosize effect". In fact, whereas bulk gold is the most inert metal in the bulk state [12], it becomes chemically very active when it is dispersed as nanometer-sized particles [13]. The catalytic properties of nanogold were already discovered in the 1970s [71], but nobody paid attention since the work of Haruta in the late 1980s [72]. A complete chapter in this book is devoted to catalytic properties of gold [13]. Modestly, in this part, I want to present some recent results obtained by surface science techniques that answer at least partially some open questions concerning the catalysis by gold.

From the numerous results published on catalytic gold, it is clear that for particles larger than about 5 nm the activity is very low. Haruta and co-workers [73] showed that for the CO oxidation at RT on gold catalysts prepared by chemical methods, the activity is maximum for sizes of around 3 nm (see Fig. 3.16). This result was later confirmed by Goodman and co-workers [74] on Au particles grown by vacuum evaporation on $TiO_2(110)$.

Many hypotheses have been proposed to explain the peculiar behaviour of gold nanoparticles. One of the first explanations attributes these observations to a quantum size effect. Indeed, Goodman and co-workers [74] observed on gold supported on TiO_2 by STS the opening of a band gap for particles of 4 nm or smaller (see Fig. 3.16). For size-selected clusters soft-landed on MgO

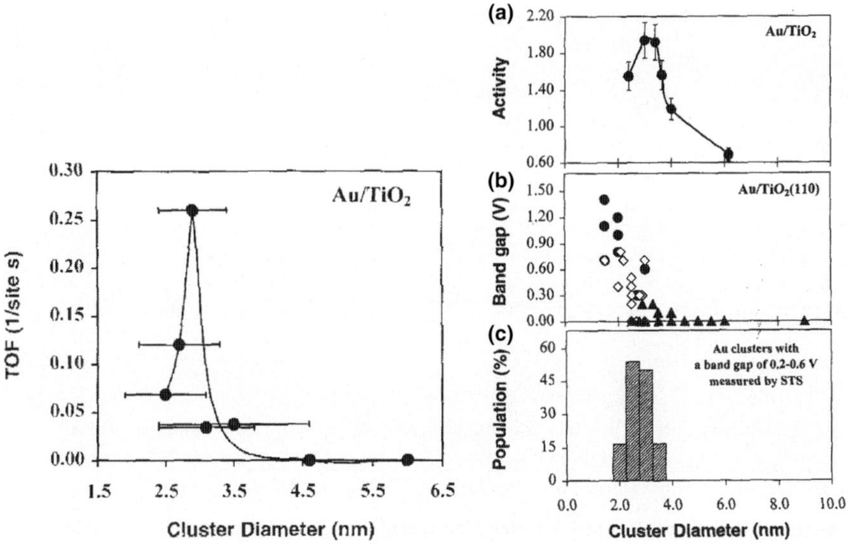

Fig. 3.16. CO oxidation on Au/TiO_2 powder catalysts prepared by deposition precipitation (from [73]) on the left and on a model catalyst $Au/TiO_2(110)$ curve (**a**), panels (**b**) and (**c**) represents band measured by STS and the proportion of particles presenting a bandgap, as a function of particle size (from [74])

thin films, the onset of catalytic activity for eight-atom gold clusters was explained from ab initio calculations by a charge transfer between an F-centre on the substrate and the gold cluster [21]. The presence of oxidized gold at the Au/oxide interface was further postulated by Bond and Thompson [75]. The role of edges as reaction sites was put forward by the group of Nørskov to explain the size effect [76]. The maximum of activity should be due to the deactivation of the interface gold atoms, which have electronic properties modified by the substrate [76]. The existence of special sites on tiny gold clusters was inferred from ab initio calculations by Lopez and Nørskov [77]. Finally, a specific role of gold atoms at the interface with the oxide was postulated by several authors [78, 79]. In this line, Molina and Hammer [79] have shown by ab initio calculations for Au/MgO(100) that CO is only adsorbed on particle edges and that an oxygen molecule is stabilized at the interface between the gold particle and the MgO support in the re-entrant corner (see Fig. 3.17). The lowest reaction barrier is obtained when the edge on the gold particles is in the second gold layer.

The mechanism of the reaction for CO oxidation (the most studied reaction) is still an open question. Some papers conclude that the reaction occurs between adsorbed oxygen atoms resulting from the dissociation of an oxygen molecule [76, 80, 81] while others claim that adsorbed oxygen molecules react directly with adsorbed CO [21, 75, 78, 79] or eventually that both mechanisms occur [82].

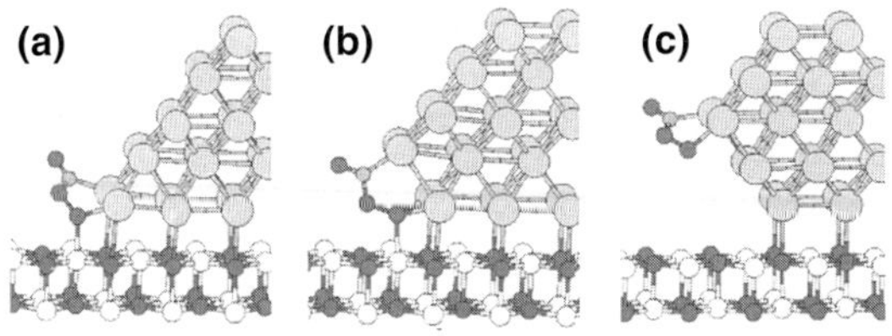

Fig. 3.17. Ab initio calculation of the CO oxidation on Au nanoparticles supported on MgO(100) with a truncated octahedron shape with different truncation at the interface. The most favourable situation corresponds to case (**b**), where the CO, bound on the edge, is close to the interface where the oxygen molecule is stabilized in the re-entrant corner of the Au particle (from [79])

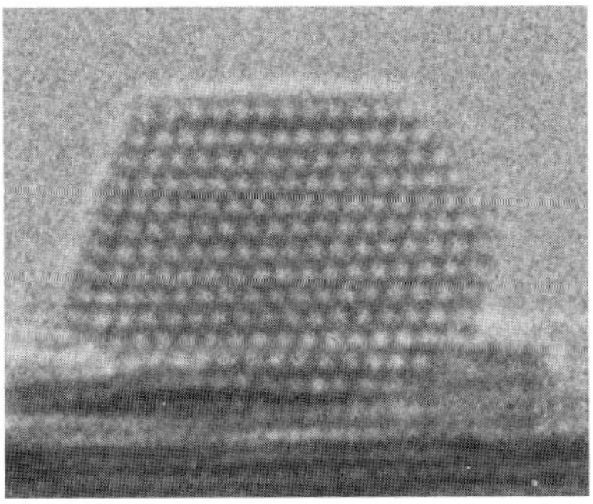

Fig. 3.18. HRTEM profile view image of a gold nanoparticle in the equilibrium shape on MgO(100)

In a recent study, we have investigated the reactivity of gold nanoparticles prepared by UHV-deposition on in situ cleaved MgO surfaces [83,84]. Particles grown at 200°C or higher were in epitaxy on the MgO surface and presented truncated octahedron shapes, which is the equilibrium shape [85]. The profile of such a particle observed by HRTEM is presented in Fig. 3.18. The aspect ratio is 0.6.

At RT, the gold particles are no longer in epitaxy and besides truncated octahedron FCC particles multiple twinned particles with decahedron and icosahedron shapes were observed [84]. CO adsorption was studied by MBRS and by IR spectroscopy. The molecular beam experiments showed that on 1.5-nm gold clusters, some CO were adsorbed revealing a rather strong adsorption energy (up to 17 kcal mol^{-1}). Two stretching frequencies, 2,113 and 2,123 cm^{-1}, were measured by IR spectroscopy, which were attributed to strongly and loosely adsorbed CO, respectively [84]. The CO oxidation was studied at low pressure with an isotropic pressure of O_2 and a molecular beam of CO. Figure 3.19 shows steady-state desorbed pulses of unreacted CO and CO_2 from 1.5-nm gold clusters on MgO(100) at RT [83,84].

For the first time, a steady-state production of CO_2 is observed at low pressure on a gold catalyst. The CO conversion is very high: 50%. On 4-nm

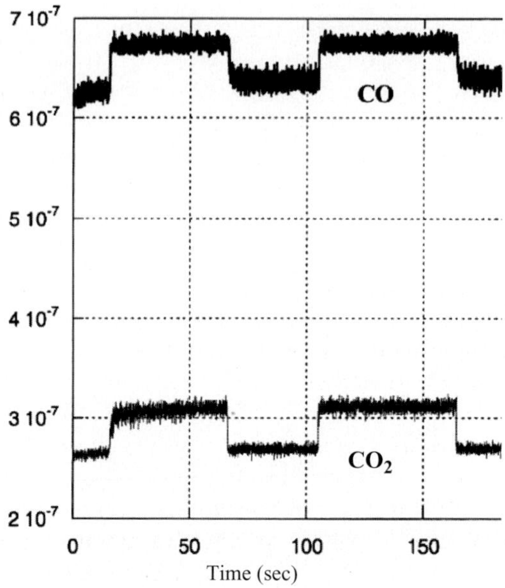

Fig. 3.19. Steady-state CO oxidation on 1.5-nm Au clusters supported on MgO(100) at RT studied by molecular beam reactive scattering. The equivalent in the CO beam is: 5×10^{-8} mb, the isotropic O_2 pressure is: 1×10^{-8} mb (from [84])

particles, the conversion is only 5% and not measurable on 6-nm particles. To know if oxygen was dissociated or not on the gold particles, the sample with 1.5-nm gold particles was exposed to 600 L of molecular oxygen and after pumping off the oxygen, CO pulses were sent to the sample to titrate adsorbed oxygen that would immediately react with CO to form CO_2. No CO_2 was observed, proving oxygen not to be dissociated on the gold/MgO model catalyst.

These results are in good agreement with the calculation of Molina and Hammer [79]. First, molecular oxygen reacts with CO, second, the fact that 1.5-nm particles are very active is compatible with the proposed model (see Fig. 3.17). Indeed, taking the truncated-octahedron equilibrium shape (see Fig. 3.18) observed by HRTEM with an aspect ratio of 0.6, the size giving the position of the edge in the second layer from the interface would be 1.4–1.6 nm.

However, we have to assume that only FCC particles are active. In fact, from this experiment we cannot exclude that the particles with a fivefold symmetry axis are active.

3.4 Conclusions and Future Prospects

We have seen in this brief review that metal nanoparticles have catalytic properties that cannot be directly extrapolated from either the clusters of a few atoms in size or the bulk of single crystal surfaces. The specificity of nanoparticles origins from several reasons: the presence of low-coordinated sites, lattice distortions due to surface stress or accommodation with the substrate, the presence of different types of facets and the presence of the substrate. These findings open the way towards an understanding of size effects in catalysis. Furthermore, it is now possible to use this knowledge to design better catalysts in term of activity and more important in term of selectivity. Several possibilities can be envisaged. First, the properties can be modified by changing the elemental composition. The simplest approach is the design of alloyed particles. A more sophisticated way is to make core–shell structures, the properties of the shell being modified by the presence of the core either by lattice distortion [86] or by electronic effect [87]. Another possibility is to use the preferential segregation of one type of (inactive) atoms to specific sites (like edges) to avoid an unwanted reaction [88]. Several groups are already working along these directions [89–91].

For the basic studies on model catalysts it is important to decrease the size dispersion. One way to reach this goal is to grow regular arrays of nanoparticles, for example by using (naturally or artificially) nanostructured substrates as templates [92, 93]. The use of regular arrays of nanoparticles also allows for taking into account the reverse-spillover effect, exactly. Nanolithography [94] with its rapidly increasing spatial resolution will also play a role, not only on model catalysts but possibly also for the preparation of sophisticated catalysts that could be used with micro-reactors working in parallel. It is also very important to combine experimental works with theoretical calculations that help not only to understand the experimental results but also to find, by accurate simulations, better ways to reach a desired catalytic property [95].

References

1. Taylor KJ, Pettiette-Hall CL, Cheshnovsky O, Smalley RE (1992) J. Chem. Phys. 96:3319
2. Whetten R, Cox DM, Kaldor A (1985) Phys. Rev. Lett. 14:1494
3. Hintz PA, Ervin KM (1995) J. Chem. Phys. 103:7897
4. Vajda S, Wolf S, Leisner T, Busolt U, Wöste LH (1997) J. Chem. Phys. 107:3492
5. Holmgren L, Andersson M, Rosén A (1998) J. Chem. Phys. 109:3232
6. Sanchez A, Abbet S, Heiz U, Schneider WD, Häkkinen H, Barnett RN, Landman U (1999) J. Phys. Chem. A103:9573
7. Heiz U, Sanchez A, Abbet S, Schneider WD (1999) J. Am. Chem. Soc. 121:3214
8. Abbet S, Sanchez A, Heiz U, Schneider WD, Ferrari AM, Pacchioni G, Rösch N (2000) J. Am. Chem. Soc. 122:3453
9. Judai K, Abbet S, Wörz AS, Heiz U, Henry CR (2004) J. Am. Chem. Soc. 101:21
10. Henry CR (2003) In: Savinova ER, Wieckowski A, Vayenas C, Dekker M (eds.), Catalysis and Electrocatalysis at Nanoparticle Surfaces. New York, p. 239
11. Freund HJ, Bäumer M, Libuda J, Risse T, Rupprechter G, Shaikhutdinov S (2003) J. Catal. 216:223

12. Hammer B, Nørskov JK (1995) Nature 376:238
13. Carabineiro S, Thompson D, Chapter 6, this volume
14. Ferrero S, Piednoir A, Henry CR (2001) Nanoletters 1:227
15. Somorjai GA (1994) Introduction to Surface Chemistry and Catalysis. Wiley, New York
16. Zhdanov VP, Kasemo B (1998) Phys. Rev. Lett. 81:2482
17. Piednoir A, Perrot E, Granjeaud S, Humbert A, Chapon C, Henry CR (1997) Surf. Sci. 39:119
18. Van Hardeveld R, Hartog F (1969) Surf. Sci. 15:189
19. Perrot E, Humbert A, Piednoir A, Chapon C, Henry CR (2000) Surf. Sci. 44:5407
20. Altibelli A, Joachim C, Sautet P (1996) Surf. Sci. 367:209
21. Sanchez A, Abbet S, Heiz U, Schneider WD, Häkkinen H, Barnett RN, Landman U (1999) J. Phys. Chem. A 103:9573
22. Matolin V, Gillet E (1986) Surf. Sci. 166:L115
23. Rumpf F, Poppa H, Boudart M (1988) Langmuir 47:22
24. Henry CR (1989) Surf. Sci. 223:519
25. Eriksson M, Petersson LG (1994) Surf. Sci. 311:139
26. Johánek V, Laurin M, Grant AW, Kasemo B, Henry CR, Libuda J (2004) Science 304:1639
27. Tyson WR, Miller WA (1977) Surf. Sci. 62:267
28. Solliard C, Flueli M (1985) Surf. Sci. 156:487
29. Vitos L, Ruban AV, Skriver HL, Kollar J (1998) Surf. Sci. 411:186
30. Swaminarayan S, Najafabadi R, Srolovitz DR (1994) Surf. Sci. 306:367
31. Henry CR (1998) Cryst. Res. Technol. 331:119
32. De Crescenzi M, Diociaiuti M, Lozzi L, Picozzi P, Santucci S (1987) Phys. Rev. B 35:5997
33. Apai G, Hamilton JF, Stöhr J, Thomson A (1979) Phys. Rev. Lett. 43:165
34. Giorgio S, Chapon C, Henry CR, Penisson JM (1990) J. Cryst. Growth 100:254
35. Graoui H, Giorgio S, Henry CR (2001) Philos. Mag. B 81:1649
36. Vervisch W, Mottet C, Goniakowski J (2002) Phys. Rev. B 65:245411
37. Wulff G (1901) Z. Krist. 34:449
38. Marks LD (1985) Surf. Sci. 150:358
39. Cleveland CL, Landman U (1991) J. Chem. Phys. 94:7376
40. Mottet C, Tréglia G, Legrand B (1997) Surf. Sci. 383:L719
41. Marks LD (1984) Philos. Mag. A 49:81
42. Mottet C, Goniakowski J, Baletto F, Ferrando R, Tréglia G (2004) Phase Transitions 77:101
43. Iijima S, Ichihashi T (1986) Phys. Rev. Lett. 56:616
44. Ajayan PM, Marks LD (1988) Phys. Rev. Lett. 60:585
45. Kaichew R (1952) Arbeitstagung Festkörper Physik, Dresden, p. 81
46. Winterbottom WL (1967) Acta Met. 15:303
47. Giorgio S, Graoui H, Chapon C, Henry CR (1999) In: Braunstein P (ed.), Metal Clusters in Chemistry. Wiley-VCH, Weinheim, p. 1994
48. Liu C, Cohen JM, Adams JB, Voter AF (1991) Surf. Sci. 253:334
49. Graoui H, Giorgio S, Henry CR (1988) Surf. Sci. 417:350
50. Shi A (1987) Phys. Rev. B 36:9068
51. Kern R (1987) In: Sunagawa I (ed.), Morphology of Crystals. Terra, Tokyo, p. 77
52. Giorgio S, Sao Joao S, Nitsche S, Chaudeson D, Sitja G, Henry CR (2006) Ultramicroscopy 106:503

53. Pawlow P (1909) Z. Phys. Chem. 65:545
54. Kofman R, Cheyssac P, Garrigos R (1990) Phase Transitions 24–26:283
55. Cini M, De Crescenzi M, Patella F, Motta N, Sastry N, Rochet M, Pasquali R, Balzarotti A, Verdozzi C (1990) Phys. Rev. B 41:5685
56. Mottet C (1997) Ph.D. thesis, University of Aix-Marseille II
57. Hammer B, Nørskov JK (2000) Adv. Catal. 45:71
58. Henry CR (2003) In: Woodruff DP (ed.), The Chemical Physics of Solid Surfaces, Surface Dynamics, vol. 11. Elsevier, Amsterdam, p. 247
59. Tsu K, Boudart M (1961) Actes du 2ème Congrès International de Catalyse, vol. 1. Editions Technip, Paris, p. 593
60. Zhdanov VP, Kasemo B (1997) J. Catal. 170:377
61. Henry CR, Chapon C, Duriez C (1991) J. Chem. Phys. 95:700
62. Piccolo L, Henry CR (2000) Surf. Sci. 452:198
63. Becker C, Henry CR (1996) Surf. Sci. 352–354:45
64. Piccolo L, Henry CR (2001) J. Mol. Catal. A 167:181
65. Prévot G, Henry CR (2002) J. Phys. Chem. B 106:12191
66. Abbet S, Röttgen M, Judai K, Wörz A, Heiz U, Henry CR (submitted for publication)
67. Vesecky SM, Chen P, Xu X, Goodman DW (1995) J. Vac. Sci. Technol. A 13:1539
68. Henry CR (1998) Surf. Sci. Rep. 31:231
69. Rainer DR, Vesecky SM, Koranne M, Oh WS, Goodman DW (1997) J. Catal. 167:234
70. Henry CR, Chapon C, Goyhenex C, Monot R (1992) Surf. Sci. 272:283
71. Cha DY, Parravano G (1970) J. Catal. 18:200
72. Haruta M, Yamada N, Kobayashi T, Iijima S (1989) J. Catal. 115:301
73. Bawenda GR, Tsubota S, Nakamura T, Haruta M (1997) Catal. Lett. 44:83
74. Valden M, Lai X, Goodman DW (1998) Science 281:1647
75. Bond G, Thompson DT (2000) Gold Bull. 33:41
76. Mavrikakis M, Stoltze P, Nørskov JK (2000) Catal. Lett. 64:101
77. Lopez N, Nørskov JK (2002) J. Am. Chem. Soc. 142:11262
78. Haruta M (2002) CatTech 6:102
79. Molina LM, Hammer B (2003) Phys. Rev. Lett. 90:206102
80. Bondzie VA, Parker SC, Campbell CT (1999) Catal. Lett. 63:143
81. Valden M, Pak S, Lai X, Goodman DW (1998) Catal. Lett. 56:7
82. Stiehl JD, Kim TS, Mc Clure M, Mullins CB (2004) J. Am. Chem. Soc. 126:13574
83. Meerson O (2004) Ph.D. thesis, University of Aix-Marseille II
84. Meerson O, Sitja G, Henry CR (2005) Eur. Phys. J. D 34:119
85. Giorgio S, Henry CR, Chapon C, Nihoul G, Penisson JM (1991) Ultramicroscopy 38:1
86. Hermann P, Guigner JM, Tardy B, Jugnet Y, Simon D, Bertolini JC (1996) J. Catal. 163:69
87. Rodriguez JA, Goodman DW (1992) Science 257:897
88. Rousset JL, Renouprez AJ, Cadrot AM (1998) Phys. Rev. B 58:2150
89. Carlsson AF, Heemeier M, Naschitzski M, Bäumer M, Freund HJ (2003) Angew. Chem. Int. Ed. 41:4073
90. Sao-Joao S, Giorgio S, Chapon C, Bourgeois S, Penisson JM, Henry CR (2005) J. Chem. Phys. B 109:342

91. Santra AK, Yang F, Goodman DW (2003) Surf. Sci. 56:122
92. Degen S, Becker C, Wandelt K (2003) Fraday Discuss. 125:1
93. Hamm G, Becker C, Henry CR (2006) Nanotechnology 17:1943
94. Baldelli S, Eppler AS, Anderson E, Shen YR, Somorjai GA (2000) J. Chem. Phys. 113:5432
95. Besenbacher F, Chorkendorff I, Clausen BS, Hammer B, Molenbroek AM, Nørskov JK, Stensgaard I (1998) Science 279:1913

4

Lithographic Techniques in Nanocatalysis

L. Österlund, A.W. Grant, and B. Kasemo

4.1 Introduction

A challenge in heterogeneous catalysis research and development is to establish a constructive feedback loop between the knowledge obtained under idealized conditions in the laboratory, and the practical conditions met in industrial or environmental applications. One of the difficulties with the *reductive* approach is how far one can go in simplifying the catalyst, and the reaction conditions, for a particular catalytic system, without loosing the essentials of the addressed problem. This difficulty has led to a long-standing discussion about the so-called *pressure gap* and *structure (or materials) gap* in heterogeneous catalysis, and how to best *bridge* these gaps [1–6].

The pressure gap concerns the different gas (partial) pressure conditions used in ultrahigh vacuum (UHV) studies (of the order 10^{-10} to 10^{-7} mbar) and technical applications ($\geq$1,000 mbar), leading to differences in, for example, surface coverages, absolute turnover rates, heat dissipation per site and unit time, and gas-phase heat and mass transport properties. Recent advancements in surface-sensitive, high-pressure techniques suggest that the surface science approach to achieve high surface adsorbate coverage, that is, to use low substrate temperatures, indeed is a valid approach [3, 7, 8]. However, it is also clear that certain conditions prevailing at real reaction conditions at high pressures can never be reached at UHV pressures. The pressure gap is not addressed explicitly here, although it is implicitly connected with the main focus of this chapter—namely, the structure gap, and how the so-called *nanocatalysis* approach can help bridging this gap.

The structure gap concept derives from the difficulty of knowing to what extent idealized catalysts are representative of the results obtained with *real-life* catalysts. The most idealized *catalysts* expose only one well-defined single crystal plane with surface areas of the order of 1 cm^2 and are most often studied under UHV conditions. In contrast to such *simple* single crystal model systems, real-life catalysts normally consist of small-supported nanoparticles buried in a porous support material. For example, in emission cleaning

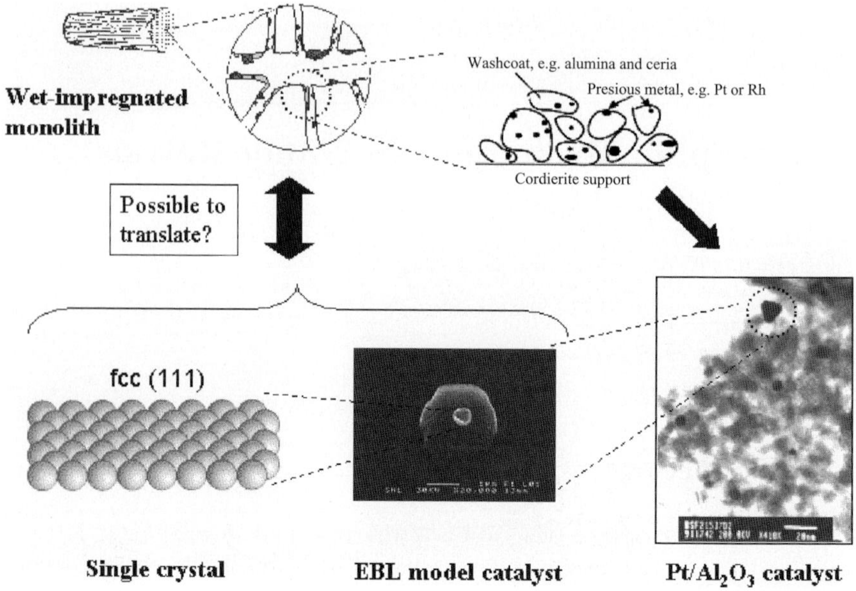

Fig. 4.1. Illustration of the structure gap for an automotive three-way catalysts (TWC) depicted by a schematic drawing of a monolithic, supported catalyst. The lower right image shows a TEM picture of aged automotive catalyst: 12-nm Pt on Al_2O_3. Note, the presence of small Pt single crystals. Well-defined model catalysts such as those made by electron-beam lithography (EBL; lower middle image) are used to mimic real catalyst and bridge the gap between single crystal studies (bottom left) and *real-life* catalysts

catalysis, nanoparticles of noble metals (Pt, Rh, or Pd) are dispersed on a 3D porous oxide support, such as alumina, ceria, and zirconia, in order to achieve as high catalyst surface area as possible [1]. The latter structure constitutes a complicated 3D architecture where the oxide support has several additional functions. This includes stabilization to obtain resistance to thermal sintering/deactivation and chemical compositions that resist poisoning (chemical deactivation) of the nanoparticles, as well as added chemical functions, such as a source/sink of oxygen in the case of ceria [9] or nitrogen oxides in the case of barium oxide [10]. In the following, we restrict ourselves to the subclass of real catalysts called *supported catalysts*, consisting of nanoparticles distributed on the usually very porous, high surface area of a carrier material—the support.

Some key aspects of the structure gap are illustrated in Fig. 4.1, which schematically illustrates the structure of a real automotive emission cleaning catalyst and also includes the simplified system consisting of a single crystal. The key question is to what extent there is translatability of results between the two catalyst structures. These difficulties are articulated with the help of Fig. 4.2. It shows, on a logarithmic length scale, the different sizes of catalysts

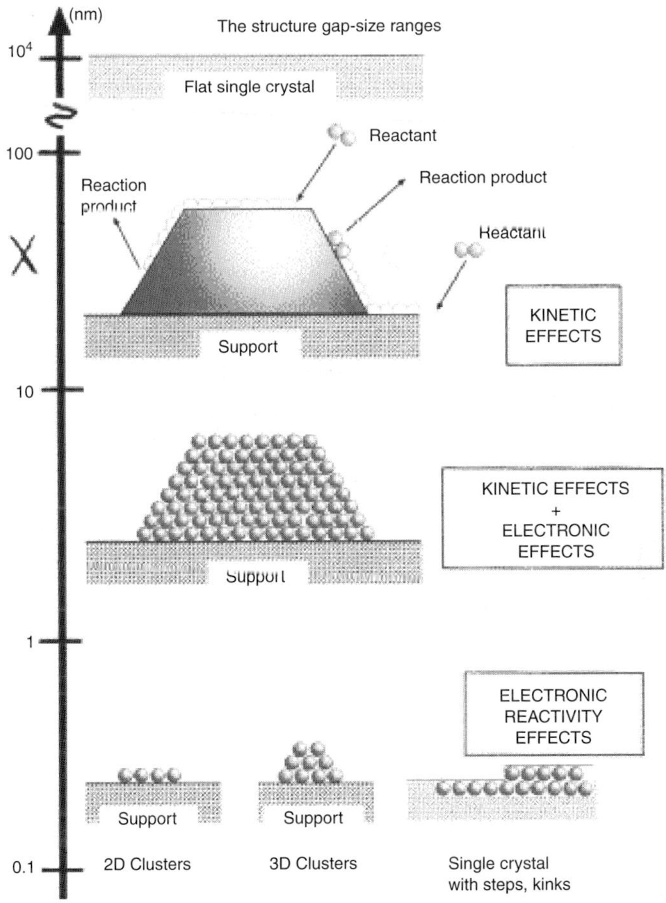

Fig. 4.2. Electronic and structural effects as an ingredient of the structure gap in heterogeneous catalysis going from small to large particles

that one can use in research and practice, and which are used to study the various aspects and details of catalytic reactions. At the top, we have single crystals, which from a practical standpoint have infinite extension in two directions and are confined only in the direction of the solid–gas (or liquid) interface. At the other end of the length scale, i.e., at the bottom of the graph, we have very small clusters of a few atoms. These clusters may be 2D or 3D. From detailed studies of both free clusters and supported clusters, it is well known that their reactive properties, including catalytic properties, are often very different from those of single crystals and depend on the particle size, as discussed in more detail below and elsewhere in this book [*please Ref to Claude's chapter in this book here*]. These differences are related to the difference in electron structure and local atomic structure, which for these few-atom systems are still very different from those for extended systems.

Moving to somewhat larger structures, roughly $d > 5\,\mathrm{nm}$ but somewhat depending on system, bulk electronic properties start to be well established, and at first thought one might anticipate that single crystals are good models of such particles. This is both right and wrong; the properties of the terraces on these relatively large nanoparticles are likely to be quite similar to those of extended single crystals; however, there are also edges between the different terraces, and in addition, there are edge and perimeter atoms in contact with the support. Both types of sites are likely to exhibit new unique reactivity compared to single crystals. (Single crystals with deliberately prepared atomic steps and kinks, etc., can be used to explore some, but not all, of these properties).

Another new ingredient is the possibility of new kinetic effects by *communication* between facets; a reactant molecule can adsorb on one facet and then diffuse to an adjacent facet where it reacts. This can give rise to entirely new kinetics, not present on an extended single crystal. Figure 4.3 shows an example of such facet communication [11]. A Monte Carlo simulation was performed for an $A + B_2$ reaction on a particle with the shape of a *truncated pyramid.* The simulations essentially mimic the $CO + O_2 \rightarrow CO_2$ reaction on a Pt particle with a (100) top facet and (111) side facets. The difference in *game rules* for the two types of facets was that the sticking coefficient for the O_2 particle was twice as high on the (100) facet. The second important rule was that CO diffusion was very rapid, so that CO particles made many diffusion jumps before reacting. Most importantly, these diffusion jumps were also allowed to occur over facet boundaries. Thus, a CO particle visited many sites after adsorption, on both types of facets, before it reacted and was removed as CO_2. It is evident from Fig. 4.3 that facet communication has profound influence of the turn over number (TON) and coverage distribution on the Pt particle.

Based on the above, it is not surprising that the structure gap has received so much attention; it is a reality. One major reason that the structure gap has proved so hard to bridge lies in the difficulty in preparing structurally and chemically well-defined model catalysts that can be used to elucidate the above and related questions. A second reason is the increased complexity of possible surface reactions when more elaborate surface geometries are considered.

The most ideal catalysts from a modeling viewpoint are thus single crystals, and a majority of the most fundamental knowledge that we have today of catalytic surface reactions, originates from UHV studies on single-crystal surfaces [1,2]. The tremendous developments of surface-sensitive methods during the past 30 years, combined with parallel progress in conceptual and quantitative theoretical descriptions, have paved the way for this state of the art, and provided unprecedented detailed information about catalytic reactions. This is particularly true for single component metallic catalysts, which facilitate the use of electron spectroscopy and are theoretically much simpler than compound catalysts.

Real-life catalysts are thus obviously less accessible for detailed characterization and kinetic model experiments than single-crystal model catalysts.

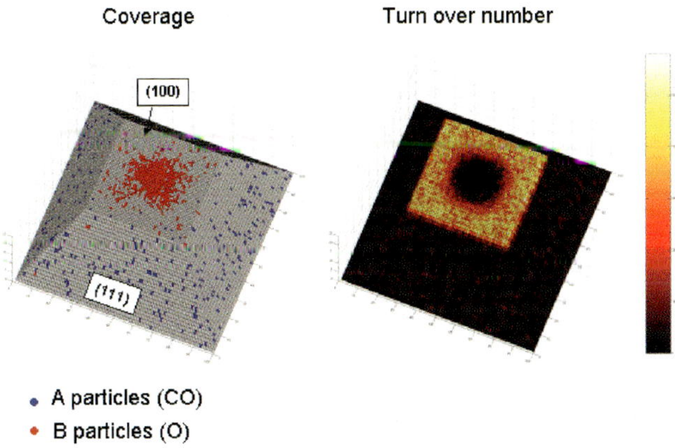

Fig. 4.3. Results from kinetic Monte Carlo simulations depicting snapshots of the coverage and TON at steady-state conditions for $p = 0.36$ on a particle mimicking a Pt particle exposing (100) and (111) facets, where $p = P_{\mathrm{A}}/(P_{\mathrm{A}} + P_{\mathrm{B}_2})$ is the ratio of partial pressure of reactant, A. The picture to the left is a coverage map; red is B (oxygen), blue is A (CO), and gray are empty sites. The picture to the right shows the TON per site. Dark means low, red intermediate, and yellow high TON (from [11])

Simultaneously, as articulated above, it is realized that intrinsically new phenomena may occur on the nanoparticles, which cannot be deduced from the knowledge obtained from single crystals, due to changes in the electronic structure [3], morphology [3, 6], and kinetics of nanometer scale particles [4] (Fig. 4.3). This is in essence what the structure gap in catalysis is about. Clearly, it is desirable to devise better methods to prepare model catalysts, which not only maintain the critical aspects of their original function but also provide a well-defined structure that allows for detailed scrutiny with a large range of analytical techniques (experimental and theoretical).

This leads us to the concept called *nanocatalysis, and specifically to nanofabricated model catalysts*, as an approach to bridge the structure gap. In Fig. 4.4, some examples of planar model structures of increasing complexity are depicted, which fulfill these criteria. At the top, there is a simple array of catalyst particles on an inactive support. The inactive support can be replaced by an active support (second picture from the top), meaning a support that significantly affects the properties of the nanoparticles via particle–support interactions (a clear distinction between *inactive* and *active* is not easy or not even possible—there is always some influence of the support on the supported particle). In some cases, the size of the support particle has an influence on the overall catalytic activity. This is, for example, the case when there is a *spillover* or *capture zone* for reactants or intermediates, which move by diffusion from the catalyst nanoparticle to the support or vice versa. In order to study such effects, one may want to systematically vary the radius of the

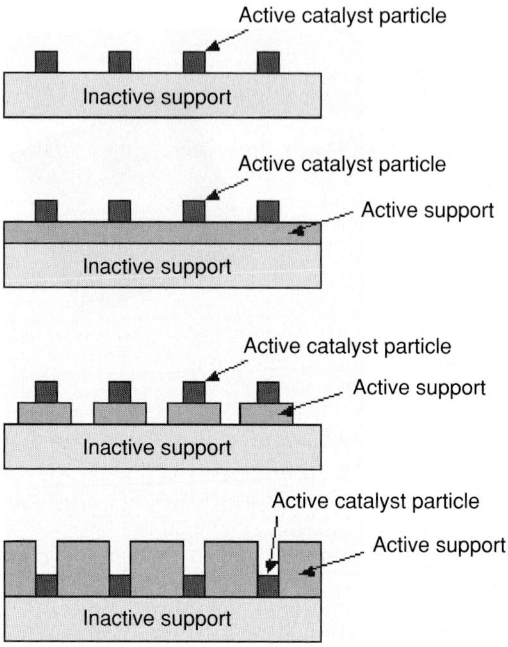

Fig. 4.4. Schematic representations of different possible nanofabricated structures made by lithographic techniques

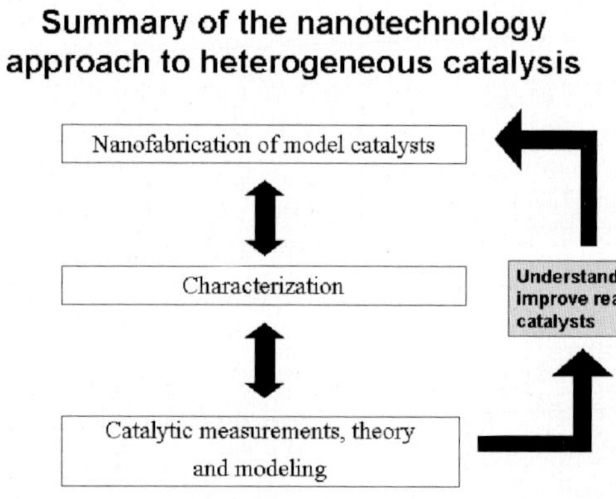

Fig. 4.5. Summary of the nanotechnology approach to heterogeneous catalysis

support material on which the catalyst particle is sitting (next lowest picture in Fig. 4.4 and a real example in Fig. 4.1). Finally, in advanced model catalysts, one may also want to fabricate well-defined pores, in order to investigate the effect of pore diffusion (bottom picture in Fig. 4.4). This generalized approach is shown schematically by the loop in Fig. 4.5.

In this chapter of the book, we will illustrate some components of this approach. It will be done by examples primarily from our own work but with reference to other related work in other laboratories. The examples are chosen to give examples of the three *corner stones* of the approach; fabrication of model nanocatalyst samples, characterization of such samples, catalytic reaction studies with such samples. The treatment is thus exemplifying rather than a comprehensive review of all activities in the field. For other review articles in this area that complement our treatment, see, e.g., the reviews by Freund et al. [3], Henry [6], and Zhdanov and Kasemo [4], and references therein.

4.2 Methods to Make Model Nanocatalysts: A Brief Overview

In Sect. 4.1, we discussed the structure gap in catalysis. From this discussion, it is obvious that a range of model systems are needed to reliably bridge the gap. For supported catalysts, this involves fabrication–characterization–evaluation of supported model catalyst systems according to the loop in Fig. 4.5. There is no single type of model catalyst that can satisfy all these needs; a range of preparation and characterization methods come into play, each with its own merits. Before plunging into the details of lithographic techniques, which is the main theme of this chapter, it is appropriate to also recollect other relevant approaches to make model catalysts. In this section, we briefly review some of the fabrication methods in a more general context, and describe their merits and shortcomings. We give some introductory comments on lithographic techniques as a prelude to Sect. 4.2.1. Next, we discuss the so-called surface science approach to heterogeneous catalysis. The motif for this diversion is simply that this has been the natural platform to analyze and extract detailed information from well-defined model catalysts – something that we repeatedly will come back to in this chapter. Finally, we briefly comment on some of the other techniques, namely, spin coating and self-assembly. This is motivated, since the lithography-based methods are complementary to these latter techniques (and others) and in many cases combinations of the techniques described above are possible (as one example, we will later discuss in detail colloidal lithography). The chapter is by no means comprehensive; it is rather picking a few examples to illustrate the diversity of preparation methods that come into play.

4.2.1 Lithographic Techniques: An Introduction

One of the attractive ways to make such model structures is based on lithographic techniques. We devote the next section to two important lithographic fabrication model catalyst methods, namely, electron-beam lithography (EBL) and colloidal lithography (CL). Lithography can be defined as a patterning process whereby an initial pattern is designed as some type of dataset which is subsequently written on the surface of a substrate as an array or ordered assembly of features [12]. The common feature in lithography is the printing of a predefined pattern onto a substrate (*lithography*, from Greek: *lithos* stone; *grafein* write)[1]. Considering the versatility of lithographic techniques for catalysis research, their impact is expected to be dramatic when the loop indicated in Fig. 4.5 is improved and the main bottlenecks eliminated (see later). A few examples in this chapter illustrate the potential benefits. In view of the broad range of approaches in this area, it is difficult to foresee which developments will have the largest impact. For example, the fully in situ growth and characterization and evaluation of evaporated model catalysts have the advantage of best control on cleanness and access to in situ control. However, there is a limitation with this approach, e.g. in the independent control of particle size and separation. EBL has the latter control but currently lacks in precision to make very small crystalline particle arrays, and large areas are still too costly and demanding to make, to allow general, broad use. However, the EBL technique is continuously developing and may in a few years be even more competitive. The self-assembly approach has many advantageous features but still suffer from, e.g., lack of variety of materials, ways to obtain really clean materials, and restrictions concerning the upper size range.

4.2.2 The Surface Science Approach: In Situ Vapor Deposition Methods

At present, the most detailed knowledge we have of supported gas–solid heterogeneous catalysts originates from surface science–based methods and the progress in studies of metal particle deposition on well-defined oxide surfaces in vacuo – ultimately single crystals with one single exposed facet. This is part of what is sometimes called the surface science approach to catalysis and utilizes such idealized surfaces in order to employ the powerful arsenal of surface spectroscopy and microscopy techniques in vacuo. These studies have proved that relevant model systems can be tailor-made, which simulate critical aspects of real-life catalysts. A key to the success of these studies has been the fruitful interplay between careful experiments and ab initio calculations (primarily density functional theory calculations) well adapted to simulate these idealized catalyst surfaces [13]. Most of these studies deal with binary oxide surfaces, such as silica (SiO_2), alumina (Al_2O_3), titania (TiO_2), iron oxides (Fe_2O_3, FeO_2, or Fe_3O_4), magnesium oxides (MgO), and a wide range of different transition metal and noble metals deposits have been studied on these oxide surfaces [6,14,15]. More recently also oxide [16,17] and sulfide catalysts [18] have been studied in detail as the active catalyst, which constitute a very important class of catalysts. In a few cases, the structure gap problem has been addressed; particle size effects have been confirmed for some systems (for example, CO oxidation on Pt/Al_2O_3 [19,20], and Au/TiO_2 [21–25]). Indeed, for particles <10 nm, nearly all reactions show some particle size effect.

[1] It was originally developed as a method of color printing in 1798 by using wax as a mask on a limestone stone, which was then dipped in water-based inks.

In many cases, conflicting results have been reported (such as about hydrogenation of CO, which has been reported to be structure sensitive but more recently was concluded to be structure insensitive [26]). The origin of size effects are, however, in most cases still not resolved, and may involve both electronic [3, 24] and kinetic [4, 27] properties of nanometer particles. The situation becomes increasingly more complicated if the support material is active and has to be included in the analysis. Even though it has long been recognized that the presence of the support material can alter the overall catalytic response significantly, relatively few studies have provided insight into the fundamental mechanisms behind these effects. This includes geometric, electronic, and kinetic effects. Kinetic effects are manifested in spillover phenomena, where surface diffusion of surface species occurs over the catalyst particle–support boundary [28,29], or over particle facets [4]. Electronic modification of nanometer scale particles, which are evident due to quantum effects and/or a relative increase of atoms with lower coordination, can be further changed at the particle–support boundary due to electronic modifications, or chemical interactions leading to the formation of new species at the boundary sites.

A wide range of methods has been employed to prepare well-defined supported model catalysts for surface science studies in UHV conditions [5,6,30]. The most common have hitherto used well-defined oxide-support surfaces produced in vacuo either by in situ cleavage [6] or by in situ growth [26, 31]. Single crystal bulk samples impose, however, several experimental problems such as in situ cleaning possibilities to restore the initial surface state, and electrostatic charging when applying electron spectroscopy, or scanning tunneling spectroscopy. In some cases (such as TiO_2 and ZnO), it is possible to sputter and/or anneal in vacuo the single-crystal oxide to create sufficient bulk charge carrier density to facilitate electron spectroscopy [6, 14, 15]. A question is, however, whether these modifications alter the oxide surface in a significant manner compared to real-life (porous) oxide supports. Another strategy, which can overcome many of these problems are homoepitaxial and heteroepitaxial growths of thin oxide films onto conducting substrates. A wide range of well-ordered binary transition metal oxide thin films has been prepared in this way [3, 5, 6, 32]. A benefit of thin oxide films is that they can be made sufficiently thin (<50 nm) to allow for transmission electron microscopy (TEM) (cf. Sect. 4.4). Nanoparticles are subsequently deposited onto the oxide surface in vacuo, usually in a line-of-sight thermal evaporation process [physical vapor deposition (PVD)]. Chemical vapor deposition (CVD) methods have also been used [30]. In favorable cases, using appropriate substrate temperature and particle flux, vapor deposition allows for some control of particle size, shape, and interparticle distance, especially in the small particle regime (<1 nm) [6, 15, 31]. In general, the size distribution is, however, rather broad and depends on the specific system under investigation, but there are also examples that narrow size distributions can be achieved. Recent advances in size-selected soft-landing of clusters may be a way around the former problem. This approach is reviewed in a separate chapter of this book by Heiz et al. [*please Ref to chapter by Heiz et al in this book here*]. The main advantage with in vacuo preparation techniques is the high degree of control of surface cleanliness, the accessibility of a wide range of surface science techniques for planar model catalysts systems, and that the whole preparation – characterization – evaluation loop (Fig. 4.5) is performed in situ.

4.2.3 Spin Coating

The *spin coating* technique has attracted interest, since it maintains many aspects of technical catalysts prepared by pore volume or incipient wetness impregnation, and simultaneously allows the interpretation and analysis in a similar way as the more well-defined model systems discussed above [30]. Here, a solution of the desired catalyst precursor is dropped onto a wafer covered with an oxide film, which is spun on a rotor to create a liquid layer of uniform thickness in order to mimic traditional wet impregnation preparation of catalysts. Control of the catalyst loading and particle size is to some degree achieved by varying the rotation speed, concentration, and vapor pressure of the solute. Still the method suffers, however, from many of the drawbacks associated with wet-impregnated model catalysts, which imparts detailed mechanistic studies.

4.2.4 Self-Assembly

Self-assembly of molecules and nanoparticles to build well-defined structures, constitutes another approach to make model catalysts [33, 34]. Here, nanostructured surfaces are made from nanoscale building blocks that are synthesized from atoms and molecules by chemical means. There has been a tremendous development in this field during the past decade, which includes a number of different strategies, including microemulsions [33], (micellar) block copolymers [35,36], and template CVD growth [37]. Relatively little work has, however, so far been directed toward heterogeneous catalysis in the sense described in this chapter, i.e., to make supported catalysts [38]. There are many reports on preparations but relatively much fewer on evaluations of catalytic activity, trends, or reactivity versus particle size, etc. A main issue for model catalysts prepared by self-assembly is whether they maintain the well-defined character after, e.g., template removal and calcinations and other pretreatment steps, before they can be used as model catalysts.

4.3 Fabrication of Supported Model Catalysts by Lithography

The lithographic approach to make model nanocatalysts has emerged in the wake of the microelectronic boom almost half a century ago. The continued endeavor for miniaturization (and associated problems) predicted by Moore in his classic paper *Cramming more components onto integrated circuits* [39] has now reached the nanometer world, and, with dimensions (line widths) reaching below 10 nm, lithographic techniques have become genuinely relevant for catalysis applications. Application of these techniques in catalysis is, however, still somewhat limited, and much of the published studies so far deal with the principles of the *fabrication process* and associated physical and chemical and physical *charactcrization*. Several approaches to apply lithographic techniques in catalysis have been developed (Table 4.1). Many of these techniques have been translated to catalysis applications from their original use in semiconductor microelectronics industry. Many ideas and concepts have been transferred directly, while others have been adjusted to comply with the special requirements in catalysis applications. Some new lithography techniques use quite different methodologies for the pattern transfer but share the evaporation and liftoff steps depicted in Fig. 4.6.

A common feature that has put some severe limitations on lithographic approaches, and has stalled the progress in nanocatalysis research is the special requirement of very small nanoscale features (in the range 1–10 nm). These small dimensions are essential to model many real catalysts. Conventional photoresist techniques have been used to prepare Pd micron size particles on silica [40]. The diffraction limit in conventional photolithography (employing UV light), which restricts feature sizes to the >0.1-μm range has been considered to restrict the applicability in catalysis. Inventive variations of the photolithography have, however, been used to prepare *pillar-shaped* microstructures, each consisting of alternating Ni and SiO_2 layers of 4–10-nm thickness. The latter microfabricated catalyst was found active for hydrogenolysis of ethane at rates comparable to supported Ni/SiO_2 catalysts [41]. Recent work demonstrate fabrication of 2- to 4-nm Au islands on TiO_2(110) inside micron-sized patches fabricated with photolithography [42]. Progress in photolithography has included alternative light sources with shorter wavelengths (for example X-ray lithography [43]), laser interference lithography [44, 45], and phase-shifting lithography [46], which eventually may also spillover into catalysis research. With phase-shifting photolithography, whereby you expose the substrate twice (or more), it is possible to make 50- to 200-nm structures depending on spacing of the lines, etc. It has been used to make ZnO fibers grown on Au islands, not catalysis but clearly a faster technique than, e.g., EBL, albeit less versatile [46].

Scanning probe techniques can, in principle, provide spatial control down to single molecule level using scanning tunneling microscopy (STM) [47, 48] but the processing speed for lithography with scanning probes is slow. Several variations of scanning probe lithography have been described in the literature, and involve electrochemical [49, 50] and physical methods (displacements [51] or indentation [52]) using STM or atomic force microscopy (AFM). Variations of these methods include (spatially confined) chemically induced reactions with catalyst-coated AFM tips [53], and the so-called dip-pen

Table 4.1. Description of lithographic fabrication methods reported in the literature

Method	Principle	Advantage	Disadvantage	Recent progress
Electron-beam lithography	Electron-sensitive mask	Good control of structure, multi-component structures	Slow, residual chemicals	Smaller line width, resists
Photolithography	Photon-sensitive mask	Parallel, fast, multi-component structures	Large structures, residual chemicals	New light sources
Scanning probe techniques	Tip manipulation	Atomic-scale precision, no resist, solvents, etc.	Very slow for single molecule control (serial)	Parallel methods, micromachining
Nanoimprint lithography	Molding of pattern	Parallel, fast, no resist, solvents, etc.	Large structures, residual chemicals	Smaller structures, improved pattern transfer
Soft lithography	Molding of pattern	Parallel, fast, no resist, solvents, etc.	Large structures	Smaller line width, improved pattern transfer
Colloidal lithography	Self-assembled mask	Parallel, fast, multi-component structures, no resist, solvents, etc.	Large particles, no long-range order	Smaller colloidal particles, new mask removal methods
Block co-polymers	Self-assembled mask	Parallel, fast, small size, no resist, solvents, etc.	Material constraints, no long-range order, cleanliness	New amphiphilic, diblock copolymers, multifunctional structures

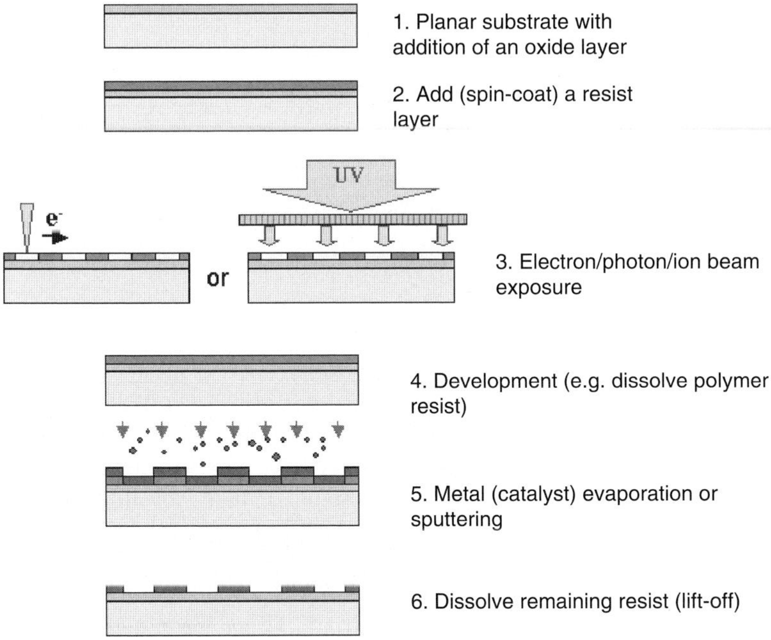

Fig. 4.6. General lithographic fabrication procedure, applicable to both electron-beam lithography and colloidal lithography

nanolithography [54], where, for example, alkanethiols (or even DNA [55]), the *ink*, are transferred from an AFM tip to a gold substrate. Typical line width using these latter scanning probe lithography techniques is of the order of 100 nm or larger. In few cases, a line width down to 10 nm has been reported [50]. Progress in scanning probe lithography to achieve high throughputs involves development of parallel [56] and micromachining SPM technology [57]. Some of these approaches are based on the idea of using large arrays of STM or AFM tips for the pattern writing.

Recent progress in nanoimprint lithography [58], and/or some kind of compression molding to create a thickness contrast pattern in a thin-resist film (soft lithography) [59], which is carried onto a substrate, constitute further options to achieve nanoscale patterning of surfaces, which may offer high throughput processing of materials. Structures down to 10 nm have been reported by these methods [60–62].

All ex situ methods discussed above have one important aspect in common. As in microelectronic applications, contamination issues are of extreme importance in catalysis. All ex situ preparation steps, which involve additions of resist layers, liftoff steps, etc., constitute potential contamination risks. This problem is shared with self-assembly methods, which usually involve some organic carrier of the inorganic nanoparticles to be deposited.

Any new method to make model catalysts must carefully address this issue, and prove that no influence of contamination is at hand or, if present, can be removed in some after-treatment or *activation* process. Since active catalytic sites very well can have orders of magnitude higher turnover than other surface sites, only a few sites can often dominate the observed overall reaction response. It has, for example, long been recognized that low-coordinated sites, such as kink or steps are often very reactive [2]. Blocking or modifying these sites even by very small traces of impurities can dramatically alter the catalytic process. Illustrative examples of this effect include O_2 dissociation over stepped Pt surfaces [63], N_2 dissociation over stepped Ru surfaces [64]. In both cases, blocking of step sites (which are present at the 1% level) by Ag and Au deposition, respectively, decreases the dissociative sticking probability by several orders of magnitude. This can also be applied to new types of products: Application of this phenomenon is utilized in the NO_x sensor developed by NGK Insulators, Ltd. [65]. These considerations emphasize the need for careful evaluation of the process steps for model catalyst fabrication, and may partly explain why progress in the development of new fabrication methods has been comparably slow.

We consider below recent progress in EBL and colloidal lithography (CL) to make well-defined planar model catalysts. The former method has been used for almost a decade in various model studies in catalysis mainly by us and Somorjai and coworkers at Berkeley [66–71], and must be considered as an *old* lithographic method, at least in comparison with the many other methods discussed above. Colloidal lithography, however, represents a new method that brings together ideas from surface chemistry of self-assembly and lithographic methods in terms of process versatility and cleanliness. These two methods represent slow serial (EBL) and fast parallel (CL) fabrication of model nanocatalysts.

4.3.1 Electron-Beam Lithography

Electron-beam lithography (EBL) has been used to make model catalysts because of its flexibility in terms of pattern, material, and process options [29, 66–70, 72]. Simultaneously, EBL is a mature technique in microelectronic industry (it is, for example, commonly used to make the masks for photolithography), which has continuously improved over the years, and is today capable of producing features with line widths of ~10 nm (and even better in some cases) [73, 74]. EBL is probably the most controlled and versatile technique to make model catalysts in the 10- to 100-nm range, with high potential for further reduction in size, due to instrumentation and method developments. The latter is primarily motivated by the needs in micro- and nanoelectronics, and by no means by the needs in (nano-)catalysis, i.e., one could say it is pushed by the so-called *Moore's law.* Thus, the EBL-nanocatalysis approach just has to adjust to the offerings given by the instrumentation development driven by other technology and science areas. We refer to textbooks

Fig. 4.7. The JEOL 9300FS EBL system situated in the MC2 Laboratory at Chalmers University of Technology, Sweden (taken from www.mc2.chalmers.se)

for a more comprehensive treatment of the technical aspects of EBL [75, 76]. There are also a few review articles on the application of EBL in catalysis research [30, 67, 77]. In the following, we will focus on certain technical aspects relevant for making model catalysts along with a detailed discussion on characterization of model catalysts prepared by EBL.

The early EBL technology was based on modified scanning electron microscopes (SEM), and consists of *writing* a computer generated pattern with a highly focused electron beam. The pattern is written, by the electron beam, into an electron sensitive polymeric film (the resist). Today state-of-the-art EBL machines (Fig. 4.7) use field emission electron guns with the ability to operate at 100 kV, producing current densities which are 10^3 to 10^4 times more intense than in the initial systems, and with greater stability due to improvements in the electron optics [74]. These intense current densities, with the smallest spot sizes of ∼3–4 nm, increase the speed at which the computer-generated patterns can be written into the polymeric films, by a factor of 10^3 or more over earlier models, allowing larger area patterns to be produced at the same high resolution. The major features of EBL are: (1) It is capable of extremely high resolution (down to the nanometer level). (2) It is inherently flexible which is compatible with a variety of materials and an almost infinite variety of patterns (if they can be reduced to basic geometric shapes). This flexibility includes independent control over particle size, lateral shape, and lateral separation. (3) It is a serial technique and thus slower than parallel lithographic techniques, such as photolithography or colloidal lithography (see Sect. 4.3.2). (4) It is quite expensive and complex, requiring a specialized environment, well-trained personnel and a huge initial capital investment [75], which explains why there are only a limited number of sites worldwide that have complete EBL setups.

In EBL, a pattern is written into a resist with the electron beam. The electron beam changes the solubility of the resist by changing the structure or the bonding of the polymer chains (Fig. 4.6). Resists are categorized as either positive or negative, depending on whether those areas exposed to the electron beam are rendered more soluble (positive resists) or less soluble (negative resists). Examples of positive resists typically employed in EBL are polymethyl methacrylate (PMMA) and Zep-520 (Nippon Zeon Company). Zep is a copolymer of α-chloromethacrylate and α-methylstyrene. Most EBL-fabricated model catalysts are made with positive resists. With negative resists, the unexposed sections of the resist are removed, often allowing for patterns to be created out of resists. Examples of negative resists are SAL 601 (Shipley Inc.), which consists of three components, a base polymer, an acid generator, and a cross-linking agent. Other examples are inorganic resists, such as SiO_2, where the exposed area acts as a catalyst for HF etching [73].

Due to the extremely short wavelengths of the accelerated electrons, the resolution of the instrument is not due to the radiation diffraction limits but to other effects such as multiple scattering of the electrons in the resist layer and substrate surface, and secondary electron emission [73, 75, 78]. These, mainly resist dependent, effects are hard to identify in detail but can be controlled to some extent during the exposure. However, it is not just the resolution of the resist which is the controlling factor in the ultimate particle size reached, rather it depends on exposure conditions (dose, beam intensity, etc.), the type of developer and processing technique used, and the way in which the final pattern is transferred to the substrate [79].

Fabricating model catalysts with EBL provides precise control over—particle size (both diameter and height), interparticle spacing, and component composition (bimetallic, metal on oxide, or metal oxide). Figure 4.8 shows an electron microscopy image of Cu/SiO_2 pattern of $\sim$50 nm Cu dots with interparticle spacing of 100, 200, and 500 nm, demonstrating the versatility of EBL. The smallest particle sizes for nanocatalysis applications achieved thus far (with an electron-beam diameter of $\sim$3 nm) is 28-nm Pt dots on an alumina support with an interparticle spacing of 100 nm [68, 80]. Interparticle spacing of less than $\sim$50 nm leads to the collapse of the resist and low yields [81]. The majority of the questions which can be asked with such fairly large particles concern issues related to *aged* (working) catalysts, i.e., particle sizes in the range of 10–100 nm in diameter. Apart from structural concerns (see "Reaction-Induced Restructuring") and sintering effects (studies for which the EBL technique is very suitable), from a reactivity point of view such questions deal with support spillover phenomena and support–particle or interparticle transport effects (the latter by varying interparticle separation)

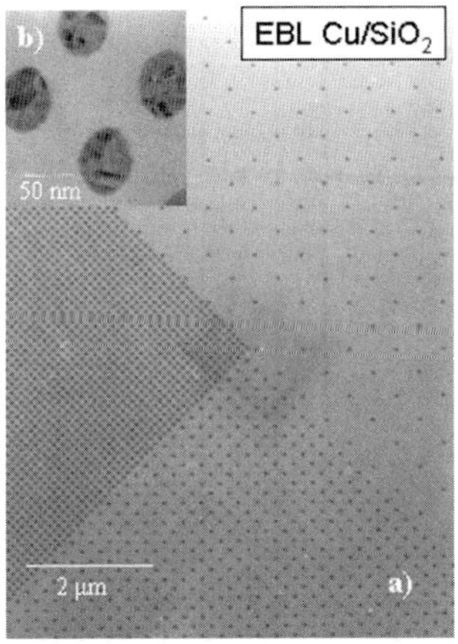

Fig. 4.8. EBL-fabricated Cu/SiO_2 with ~50-nm particles with three different interparticle spacing (center-to-center) of 100, 200, and 500 nm on a TEM membrane. The inset is a higher resolution image of the 100-nm interparticle spacing

(see "CO Oxidation on Pt/silica and Pt/ceria"), and with facet communication on the particles themselves (see "CO Oxidation on Pd/SiO_2: Molecular Beam Studies").

Preparation Procedures

Successful EBL patterning of model catalysts consists typically of the following process steps, which are shown schematically in (step numbering as in Fig. 4.6):

1. (Optional) Addition of an oxide layer to the substrate, e.g., CeO_2 or Al_2O_3 (step 1).
2. Preparation of the surface for resist coating, which usually consists of wet chemical cleaning and plasma cleaning, sometimes priming for resist adhesion.
3. Spin-coating of one or two layers of an electron sensitive polymer film or resist (step 2).
4. *Soft* bake of the resist to remove residual solvents in the resist, using either a hot plate or an oven.
5. Exposing a predesigned pattern in the resist with a scanning electron beam (step 3).
6. Development of the resist (dissolve the exposed parts of the positive resist) in an appropriate solvent (step 4).
7. Vapor deposition of a catalytically active metal(s) and/or oxide (step 5).
8. *Liftoff* of the remaining undeveloped resist with a different solvent (step 6).

The most frequently used substrates are Si(100) wafers, where the top surface has either been wet-oxidized or thermally oxidized creating an SiO_2 film with a thickness of 100–400 nm. However, in no way are the substrates limited to SiO_2 films. For example, other thin films have been deposited on the Si wafers, such as CeO_2 [29, 82] and Al_2O_3 [68, 80]. The CeO_2 films were vapor deposited with electron-beam heating resulting in a nonstoichiometric film, resembling the mixed oxide found in catalytic converters (i.e., CeO_x, where $x < 2$). Insulating substrates, e.g., sapphire wafers and glass substrates have also been used (unpublished results). When depositing onto an insulating substrate, an additional step is required where a thin metal film (~10 nm), such as Au, is deposited onto the resist layer before exposure, to minimize sample charging and beam deflection. This metallic layer must be removed before development.

Once the substrate has been prepared, the next step consists of cleaning the wafer and priming it for the resist adhesion. Usually, this consists of rinsing in methanol, or other solvents, in an ultrasonic bath, or alternatively, in heated methanol. In addition, plasma cleaning with O_2 can help to both activate the surface and also *clean* the surface, by very effectively removing hydrocarbon residues/contamination on the surface. With some resist/substrate combinations, it is necessary to use a primer to increase the adhesion of the resist [i.e., polydimethylglutarimide (PMGI)]. The thickness of the resist depends on the speed of the spin coating (between 1,000 and 6,000 rpm) and the viscosity of the resist; the slower the spin rate, the thicker the resist. It is also common to dilute the resist, reducing the viscosity, in order to obtain thinner layers, obtaining higher resolution. For instance, Zep-520 (Nippon Zeon Co.) diluted 1:2 at 6,000 rpm makes a layer 60-nm thick, which has been used to make sub-50 nm features. After the resist has been deposited and baked, the sample is mounted into the e-beam exposure chamber. In order to obtain the sharpest exposures, it is necessary to be able to focus the electron beam at the same height as the resist to ensure it is within the same depth of focus. Once the correct current for the exposure is chosen and the electron column is aligned and focused, the electron beam is stepped through the predesigned pattern, exposing specific parts of the polymer, changing its solubility.

Next, the pattern is developed by dissolving the exposed parts of the pattern in the correct solvent. The quality of the pattern (contrast) depends on both the exposure and the development time, both of which must be experimentally determined for different resist systems and substrates. Once the pattern has been exposed and developed, the actual catalytically active metal(s) or oxide(s) is vapor deposited both within the *holes* in the polymer and on top of the polymer (step 5). This metal (or oxide) film should be discontinuous at the pattern boundaries in order to make possible the liftoff the remaining resist. This requires that the resist layer is thicker than the deposited film, and that the developed resist features have an undercut or *negative* profile

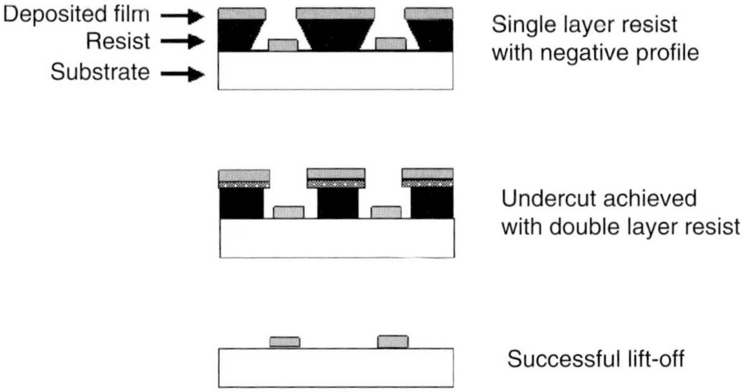

Fig. 4.9. Schematics for successful liftoff procedures showing a negative profile in a single-layer resist and the undercut achievable with a two-layer resist

(Fig. 4.9). In a single-layer resist system this is achieved by either a slight overexposure or a slightly longer development time.

One of the benefits of EBL is that it is possible to vary the aspect ratios of the nanoparticles (i.e., height-to-width ratios). However, in order to obtain good liftoff when particles are thick (typically >50 nm) or if a nonplanar substrate is being used, it is necessary to use a two-layer resist system. A higher sensitivity resist is used as a bottom layer and a less sensitive resist is used as a top layer [75]. Two frequently used examples are low and high molecular weight PMMA or PMGI/Zep-520. By using a higher sensitivity resist on the bottom layer (especially if it is insoluble in the developer of the top layer), it is possible to develop an undercut (see Fig. 4.9) to facilitate liftoff. Such techniques have been used in our group to create aspect ratios from 0.1 to 1.0. In addition, it is possible to use trilayer systems, where a barrier, such as Ti, SiO_2, or Ge is used in-between two polymer layers [75]. This allows the top layer to be developed and subsequently used as an etch mask to etch the barrier layer before the bottom layer is developed. This can allow the size of the original hole to be reduced depending on the etching characteristics. This might be an excellent way to decrease the lower limit of particle sizes obtainable with EBL, perhaps down to 1–10 nm. However, such techniques have not been employed in nanocatalysis as of yet.

To create more complex patterns, such as shown in the lower middle image in Fig. 4.1, where two or more layers are present at the same location, or two or more different substances are located adjacent to each other, it is necessary to *mark* locations close to the patterns. These marks are either created by etching depressions into the substrate or by adding a high-Z metal, such as Au or Pt. This enables the EBL machine to locate the marks during the exposure and place the patterns in the desired locations with respect to both the marks and to each other with ~50 nm tolerance (in practice) depending on the distance from the marks to the pattern.

Characterization of Model Catalysts

Chemical Properties

One of the major concerns with EBL-fabricated model catalysts is contamination from the processing steps described above, with remaining resist being the largest concern, leading primarily to C- and N-containing compounds. The primary way employed to remove these contaminants has been to treat the catalysts with relatively high-temperature oxidation and reduction cycles, typically oxidation in O_2 at 500°C for 1 h, followed by reduction at 550°C for 1 h (one may compare this treatment with the calcination–reduction treatments of real, e.g., wet-impregnated catalysts). This has been shown to reduce the C 1s XPS) signal by 60–70% and completely remove other contaminants below detection-limit in XPS. Further, cleaning steps are necessary before the catalyst is active. For Pd/SiO_2 nanoparticles, an additional exposure to 4×10^{-4} mbar O_2 for 2 h at 650 K followed by CO reduction was required to obtain an active catalyst for CO oxidation, as measured by the expected CO_2 production based on the Pd surface area [83]. In addition, guided by experience with Pt wire catalysts and supported Pt/alumina catalysts in flow reactors [84], mixtures of H_2 and O_2 diluted in Ar have been used to pretreat/clean samples before H_2 and CO oxidation experiments [29, 85] (see Sect. 4.5.1). Other techniques have involved sputtering the samples with 1-kV Ne^+ (ion current of 20 μA), annealing in O_2 at 500°C, and annealing at 650°C in UHV to reduce the particles [86]. This process removed 99% of the contaminants as determined by AES. Furthermore, it has been reported that treating Pt/Al_2O_3 catalysts with a strong oxidant (NO_2) at 300°C oxidizes the Pt, which can be reduced with a 0.1-L dose of CO at room temperature followed by a flash anneal to 300°C [68]. This has been shown to render these catalysts active to ethylene hydrogenation [68], whereas oxidizing and reducing treatments in O_2 and H_2, respectively, rendered the catalyst inactive.

Structure

The EBL structures, as fabricated, resemble towers, cylinders, or disks (see, e.g., Fig. 4.1). SEM gives a qualitative view of the surface topography, especially when viewed at an off-normal angle. It is the best way to get a quick and good picture of the geometry of the surface structure. The resolution limit of most SEM instruments is ~5 nm (the diameter of the electron beam), although higher resolution instruments become more and more common. The nanoparticle diameter can be obtained with an SEM micrograph taken from the surface normal [68]. Quantitative height information can be gathered from AFM; however, particle radius cannot easily be determined for small particles because the lateral resolution is then influenced by tip deconvolution effects, which are nontrivial to account for. In addition, AFM can give some information about the adhesion strength of the particle to the substrate, via the force that is required to remove the particles [86]. It has been reported that as-prepared Pt particles supported on SiO_2/Si can be removed with an ~10-nN lateral force, but Pt particles subjected to heat treatment could not be removed by the applied force of the AFM tips [86].

Detailed scrutiny of particle structures (especially crystallinity) typically requires transmission electron microscopy (TEM) analysis. Samples with particles of sufficiently small diameter (<30 nm from a practical viewpoint) depend on TEM or AFM even for basic analysis. In TEM applications, it is necessary that the particles are either situated on a very thin substrate (<100 nm) of suitable materials, such that the electrons are able to transmit through the film with high enough phase contrast, or cross-sectional analysis can be done for particles situated at edges of grids or on large catalyst particles. In the latter case, the planar geometry and its advantages are lost. In the former case, the supports normally need to be mechanically thinned before TEM analysis can be performed [87], which is a rather elaborate procedure requiring months of practice to optimize, even for an experienced user. In Sect. 4.4, a novel technique based on silicon micromachining to overcome this labor-intensive step is described, which eliminates the need of any TEM preparation step after the nanofabrication steps.

4.3.2 Colloidal Lithography

Colloidal lithography, as we define it here, is a hybrid method [88–92]. Other names for this technique or closely related techniques are *Colloidal Monolayer Lithography* [93], *Natural Lithography* [94], *and Nanosphere Lithography* [95]. It combines the advantages of the lithographic methods borrowed from the microelectronic fabrication technology, in terms of processing purity and versatility, and the colloidal chemical methods in terms of self-assembling properties and process speed. Colloidal particles have a number of attractive features, which can be used for nanofabrication of planar model catalysts. The most important property is their ability to self-assemble and form ordered structures over larger surface areas. Furthermore, colloidal particles can be obtained with a broad class of chemical and physical properties and can readily be functionalized for the intended purpose. Colloidal particles are normally cheap. Combined with the parallel nature of the surface assembly process, a versatile and cost effective nanofabrication process can be realized with the possibility to prepare larger surface areas (typically cm^2) in short processing time. Considerable interest has previously been directed toward self-assembly of hexagonally close-packed structures of colloidal particles, employing spin-coating [95], electric field layering [96], or capillary-based deposition methods [97], producing uniform but fairly small domains (order of 100 μm^2) of ordered structures.

In this section, we will describe a different approach, which utilizes adsorption of charged colloidal particles onto an oppositely charged surface. This leads to an electrostatic attraction between the particles and the surface and electrostatic repulsion between the particles, resulting in a quite uniform particle–particle distance on the surface. This approach has been demonstrated in a number of studies to make nano- and microstructured surfaces for biomaterials [88–90], catalysis [91, 98], and electrocatalysis applications [99]. The quasi-uniform adsorption of charged nanoparticles circumvents many of the difficulties otherwise encountered in nucleation and growth that may influence (or even completely determine) the adsorption process, i.e., defects, stacking faults, and grain boundaries, are difficult to control. For the purpose of producing monodispersed model catalysts, there are some obvious choices of primary colloidal particle model systems. Different kinds of monodisperse polystyrene (PS) particles are commercially available from ∼20 nm and upwards. However, as the particles become smaller, they become more polydispersed. Small silica colloids can, however, be made quite monodispersed by additional autoclaving with the expense of an increased particle size (∼10–15 nm). The primary particles can be obtained with different chemical properties and charges or may be appropriately functionalized. In the present review, we show the results for self-assembly of negatively charged sulfate polystyrene (latex) particles on positively charged metal oxide surfaces (alumina, ceria, and titania) and on glassy carbon.

The adsorption by electrostatic interaction is irreversible and localized, resulting in a successive exclusion of surface area due to electrostatic repulsion exerted by already adsorbed particles on particles arriving form the liquid phase. These features are well accounted for in the random sequential adsorption models (RSA) [100, 101], which is the prototype case for sequential adsorption processes [102]. A detailed theoretical account of the RSA models can be found in [103]. Several experimental studies of protein (100) and colloidal particle adsorption [90, 101, 104, 105] at liquid–solid interfaces support the 2D RSA model. The main features of the RSA model are: (a) the existence of a jamming saturation coverage, θ_∞. The maximum coverage in 2D at infinite adsorption times is $\theta_\infty \equiv \rho_\infty \pi a^2 = 0.547$, where ρ_∞ is the surface particle density at infinite adsorption times and a the particle radius. This is significantly lower than the close-packed density $\theta_\infty = 1$, if the particles are allowed to restructure (diffuse) between adsorption steps; (b) slow kinetics when approaching the jamming limit ($\theta_\infty - \theta(t) = 1/\sqrt{t}$ in 2D); (c) short range ordering; (d) different adsorption configurations than equilibrium systems; and finally (e) an absence of thermodynamic phase transitions. The latter is a consequence of the irreversible adsorption, or in other words, the infinite memory of the adsorption process. The RSA model can be extended to include particle–particle or particle–surface interactions, bulk diffusive transport, polydispersity, and multilayer formation [89]. These extensions do not change the qualitative features of the RSA model.

The electrostatic screening realized in an electrolyte can be used to control the particle–particle interactions. This interaction potential, $V(r)$, is characterized by the Debye length κ^{-1}, which in the simplest mean field models describes the electrostatic screening of ionic particles interacting via Coulomb forces in a dielectric continuum, viz $V(r) \propto \exp(-\kappa r)$. It is convenient to

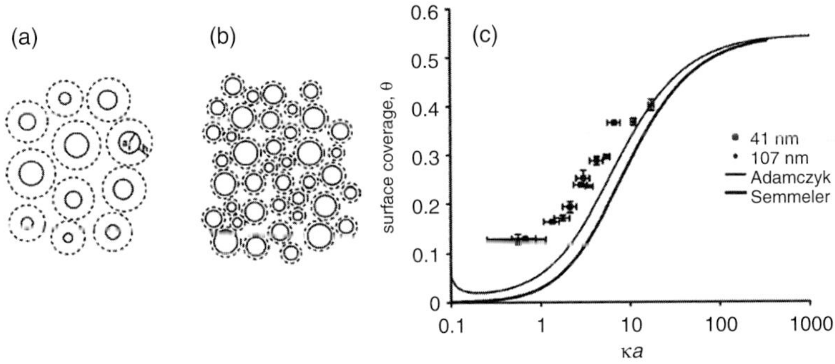

Fig. 4.10. Sketch of adsorption of polydisperse particles at (**a**) low and (**b**) high salt concentrations. *Dotted lines* show the effective particle radius (or interaction distance). (**c**) Surface coverage of the polystyrene particles versus κa (κa is a dimensionless screening parameter, where κ is the inverse Debye length and a the particle diameter). The more polydisperse particles (41 versus 107 nm) have a slightly increased coverage at high κa. *Solid curves* are approximations derived from the effective hard sphere model (see [89] for further details)

characterize the electrostatics by the dimensionless parameters κa. Many different adsorption systems have experimentally been observed to follow the same dependence of θ_∞ on κa [90, 101, 106]. With decreasing salt concentration, κ^{-1} increases, which results in larger interparticle separations and hence a smaller saturation coverage (Fig. 4.10). Varying the salt concentration in the electrolyte is thus a simple way to control the nearest-neighbor (interparticle) separation. Below, we describe in more detail the colloidal lithography process adapted to fabrication of planar model catalysts.

Preparation Procedures

Process Description: Overview

The basic fabrication steps in colloidal lithography are described in Fig. 4.11 [91]. At the heart of the process lies the idea of creating a colloidal mask with the desired properties (chemical composition, size, shape, charge, etc.), which can be used to shield the surface beneath the colloidal particles during the etching (step 1), which subsequently can be removed to expose the underlying (masked) surface (step 2). The mask is finally removed, for example, by an O_2 plasma step (step 3). Alternatively, the colloidal template may be used to create controlled corrugated films or to create a template for film deposition in-between the colloids, thus allowing holes to be fabricated (after removal of the colloidal particles) [107, 108]. In the limit of hexagonally close-packed colloids, deposition into the voids between the colloids creates nanoparticles in the 20–400-nm size range with a triangular footprint, which is typically referred to as nanosphere lithography [93, 95, 109, 110]. It should be noted

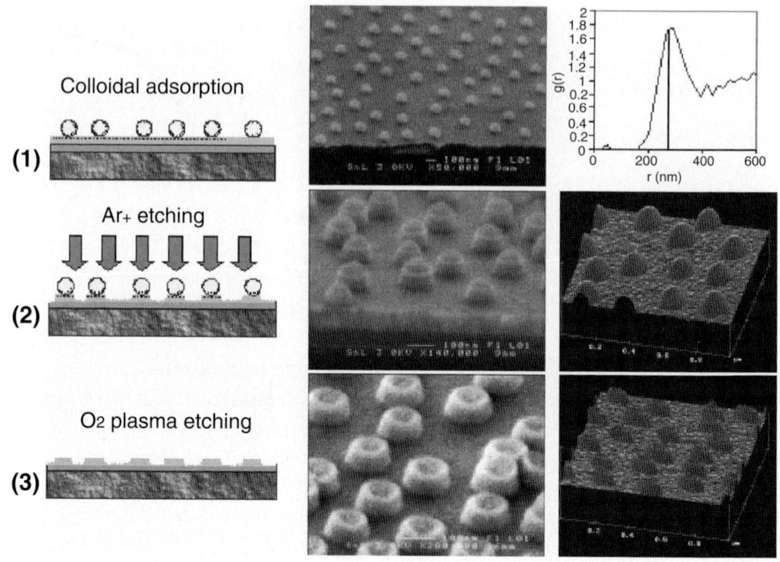

Fig. 4.11. The basic steps in the fabrication of a model catalyst (depicted by Pt/CeO_x as an example) by colloidal lithography, with corresponding SEM (middle row) and AFM micrographs (end row). (**1**) Deposition of Pt and CeO_x and the adsorption of colloidal particles (polystyrene, $d = 107$ nm), which self-assemble into pattern with short-range order as shown by the corresponding radial distribution function, $g(r)$, showing the nearest neighbor distance between the polystyrene particles of 270 nm (top right corner). (**2**) Removal of the unprotected Pt by directed Ar^+ etching. (**3**) Removal of the polystyrene particles by O_2 plasma etching (UV ozone is also used) (from [92])

that colloidal lithography does not require flat surfaces; the method can in principle be applied to rough, uneven, or curved substrates. The potential requirement of flat surfaces is then transferred to the film deposition step. Below, we describe each step in detail for fabrication of planar model catalysts with Pt/ceria as a pedagogical example.

Step 1: A 50-nm thick ceria film is evaporated on an Si(100) wafer with a thermally grown SiO_2 layer. The wafers are cut into pieces of the desired size and shape with a diamond saw, either prior to or after the colloidal lithography process. A 30-nm thick Pt film is subsequently evaporated on top of the ceria film. A solution of 10% wt/wt (% wt/wt, percentage by weight) aluminum chloride hydroxide (ACH) is added on the Pt surface to positively charge it by aluminum ions, adsorbed on the surface (this is the simplest procedure to charge the surface, see "The Drying Problem" for more advanced methods). Afterwards, the surface is rinsed in water and blown dry with nitrogen gas. A water suspension of 0.1% wt/wt 107-nm PS particles is then rapidly poured over the Pt surface to immediately cover the whole surface with the colloidal suspension. After a few minutes, the sample is again rinsed in water and blown dry with nitrogen gas. The negatively charged PS particles adsorb irreversibly on the positively charged Pt surface. (During drying there are strong capillary forces on the PS particles, which may cause agglomeration if the adhesion to the substrate is not strong enough; there is, however, a way to improve the adhesion—see below).

The electrostatic repulsion prevents the particles from adsorbing closer to a neighboring particle than a certain minimum distance, the latter influenced by the (screening) salt concentration. The adsorbed colloids have short-range order with a well-defined nearest-neighbor distance, with little or no long-range ordering, as can be seen in the radial distribution function, $g(r)$, as a function of interparticle distance in the upper right hand corner in Fig. 4.11.

Step 2: The surface is etched by directed Ar^+ sputtering, which removes the Pt, which is not shadowed by adsorbed PS particles. The etching is continued 10 nm into the ceria layer (with about half the etch-rate) to ensure that all exposed Pt is removed.

Step 3: The PS particles are removed by an O_2 plasma treatment (2 min at 250 W, and 500 mTorr). In "Chemical Properties" we present XPS results, which describe the effect of O_2 plasma treatment regarding removal of the PS particles by O_2 plasma and oxidation of Pt. It should be noted that it is not possible to dissolve the PS particles in acetone after the Ar^+ etching process, which is believed to be due to ion-induced crosslinking of the polymer chains during ion etching (111), making them resistant to normal solvents for PS. The radial distribution function, $g(r)$, from the initial colloidal adsorption step is preserved throughout the nanofabrication procedure.

Controlling Interparticle Separation

The interparticle distance may readily be changed by adjusting the salt concentration in the electrolyte. Figure 4.12 shows Pt dots on a ceria support using 107-nm PS particles as templates and NaCl concentrations between 0 and 1 mM in the suspensions. The insets in Fig. 4.12 show $g(r)$ at each salt concentration. The projected geometrical coverage of the Pt dots increases from 13% to 40% (corresponding to a particle density of $3.6 \times 10^{10}\,cm^{-2}$ and $1.1 \times 10^{11}\,cm^{-2}$, respectively), with increasing NaCl concentration in the colloidal suspension. The corresponding nearest neighbor distance decreases from 270 to 180 nm. It is difficult to obtain more densely packed colloids than 180 nm in this case due to agglomeration. The larger separation (270 nm) corresponds to the largest interparticle distance that is possible to obtain (no salt added). With no added salt, the disruption of the ordered pattern of the adsorbed colloids during drying may be a problem for larger particles (>200 nm), which is due to the lower ratio between the attractive forces between the substrate and the particles and the strong lateral (capillary) forces during drying. There are, however, ways to avoid this. The simplest and most effective way is to dip the sample into boiling water for 30 s after the adsorption of the colloidal PS particles [90]. (It is important not allowing the sample surface to

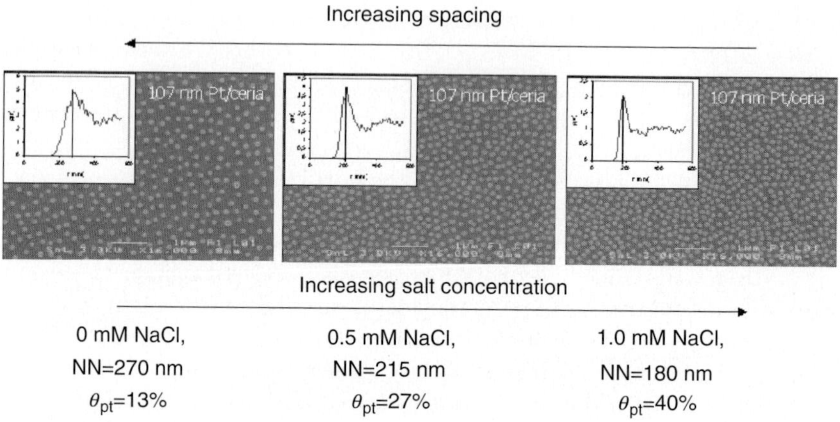

Fig. 4.12. Pt/ceria catalysts made by colloidal lithography using suspensions of $d = 107$ nm polystyrene particles with three different NaCl concentrations, resulting in different surface coverages and interparticle separations. *NN* refers to the nearest neighbor distance, as measured by the main peak in the radial distribution function, $g(r)$, shown in the inset of each SEM picture (from [92])

dry before the immersion in the hot water bath). The glass transition temperature for PS is about 100°C, so the boiling water will partly melt the PS particles and *glue* them to the substrate, thus immobilizing the particles and also counteracting agglomeration. After the boiling step, the sample is again rinsed in water and dried as described above.

Controlling Particle Shape

The SEM and AFM images of the Pt particles after the final step in the colloidal lithography process in Fig. 4.11 (step 3) show a crater- or cup-like shape, which is evident also in AFM (Fig. 4.13). It arises when Pt, which is sputtered away from the surface by argon ions, redeposits underneath the adsorbed PS particles, schematically illustrated in Fig. 4.14a. In Fig. 4.14b, it is seen how a Pt cup has started to form around an adsorbed 530-nm PS particle during the Ar^+ etching. If planar Pt dots are desired, the adsorbed PS particles can be melted to half-spheres prior to the Ar^+ etching, thus preventing redeposition of Pt (Fig. 4.14c). Figure 4.14d shows 130-nm flat Pt dots on ceria made by adsorption of 107-nm PS particles (0.2-mM NaCl in the particle suspension), which were melted to hemispheres (2 min at 112°C), prior to the Ar^+ etching. The melting increases the diameter of the adsorbed particles to 130 nm, which also leads to fusion of some adjacent particles [91].

Controlling Particle Size

The general interest in catalysis is typically particles of <10 nm in diameter. This requirement introduces additional difficulties in the colloidal lithography process. The size of nanoparticles resulting from the colloidal lithography

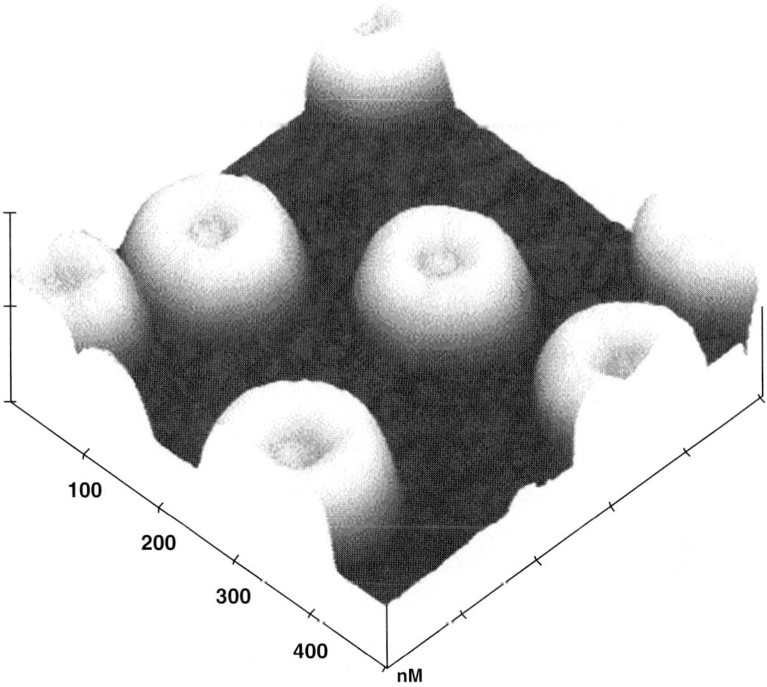

Fig. 4.13. AFM line scan of 110-nm Pt/CeO_2 showing the crater structure of the Pt particles due to the Ar^+ etching processing of spherically shaped polystyrene particles

process can simply be controlled by using colloidal particles with the desired size. Polystyrene colloids are commercially available in sizes from about 20 nm up to several μm. Silica colloids are commercially available with diameters from 5 nm and upward, and can be made quite monodisperse by further autoclave treatments (with a resulting diameter of the order 10–15 nm due to Ostwald ripening). In Fig. 4.15 is shown adsorbed PS colloids on ceria and titania, respectively, with diameters of 530 nm (titania only), 107 nm and 41 nm, as specified by the manufacturer (which are almost identical to the values we have measured) [89]. It is notable that the colloids with 41 nm specified diameter have a significantly broader size distribution than those with 107 and 530 nm. This originates primarily from the manufacturing process of the PS colloidal particles, which gives more polydispersed particles, at the smaller sizes. The polydispersity is, however, accentuated by the adsorption process, as shown by Hanarp et al. [89] making the size distribution of the adsorbed particles broader and shifted toward smaller diameters compared to the size distribution in the suspension. This is because the small particles diffuse faster in the suspension, and also may adsorb between larger particles, where the adsorption of larger particles is prevented. Thus, the requirement of strict size homogeneity of the colloidal particle suspension is accentuated at smaller

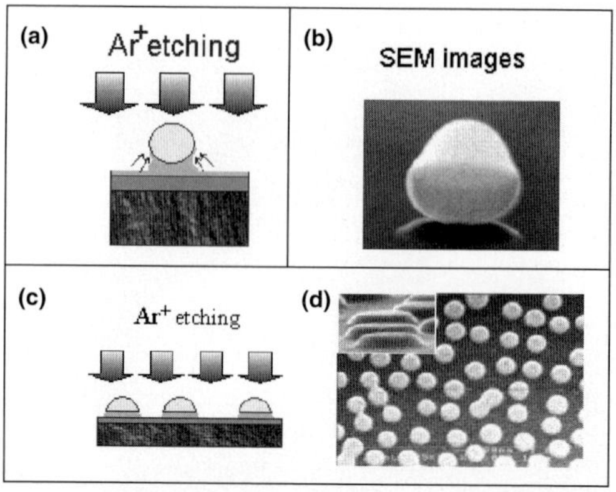

Fig. 4.14. The mechanism of crater formation during colloidal lithography. (**a**) Schematic of Pt redeposition underneath a polystyrene particle during Ar^+ etching, forming a cup-like structure. (**b**) SEM image showing of a 530-nm polystyrene particle with the Pt cup surrounding the bottom. (**c**) By partially melting the polystyrene particle to a hemisphere before Ar^+ etching, redeposition is avoided and flat Pt dots are formed. (**d**) SEM image of flat Pt particles formed by melting the PS particles before etching. This melting leads to an increase in particle size by ~18%, and may cause some fusing of adjacent particles which must be taken into account in the preparation procedure (from [92])

particle sizes. One interesting approach to overcome this problem would be to use proteins, which are completely monodispersed, as masking particles. One candidate is the spherically shaped protein ferritin, which has a diameter of about 13 nm, with an iron oxide core of about 8 nm (100). Si nanorods have been produced using adsorbed ferritin molecules as a Cl-etch mask [112].

If lower surface coverages and/or smaller particles are desired than what the colloidal template permits, one method is to adsorb larger particles and then shrink them to the desired size by a mild O_2 plasma treatment prior to Ar^+ etching [90, 91]. Using this method, ~40-nm Pt/ceria [91] and ~20-nm Pt/C have been fabricated. The largest degree of particle shrinking achieved by this approach is so far about 80%. A more straightforward approach is to use extended Ar^+-etching times, i.e., longer times than necessary to etch through the Pt layer, thus etching the PS particles to the desired size, utilizing the fact that the PS is etched at about half the rate as compared to Pt. This method also opens up an efficient way to make more complex 3D structures, which we discuss in "More Advanced Structures". With this approach, 65% smaller particles than the original PS particles have been made [92].

More Advanced Structures

Colloidal lithography allows for fabrication of advanced structures, such as 3D structures similar to those shown in Fig. 4.4. Here, we describe three different strategies to make advanced Pt/ceria structures.

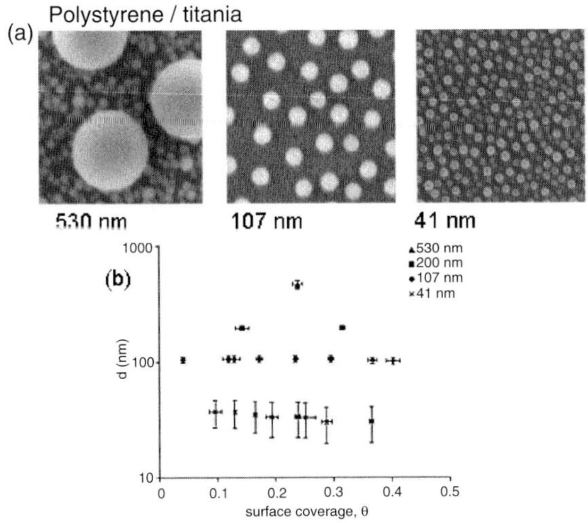

Fig. 4.15. (**a**) Polystyrene particles of diameters 530, 107, and 41 nm imaged with the same magnification, and with the same coverage, 0.24. Small Si particles are seen in-between the 530-nm polystyrene particles. (**b**) Overview of particle sizes and surface coverages. All data points represent well-controlled particle films with separated particles and short-range order (note the logarithmic scale). The large error bars on the smallest particle sizes are a function of the polydispersity (from [89])

Shrinking of the Polystyrene Colloids by O_2 Plasma. The idea here is to etch through the Pt and ceria layer in two steps, with an intermediate shrinking of the masking colloidal particles as depicted in Fig. 4.16. Initially, the process follows the same steps as when flat Pt particles on a ceria film are made. Polystyrene colloids are first adsorbed on a Pt and a ceria film on an oxidized silicon wafer and melted to hemispheres. Using the experimentally determined time–size relation calibration curve for shrinking PS particles by O_2 plasma [90], the PS hemispheres are shrunk to desired size after the first etching step. In the second Ar^+-etching step, the exposed Pt area is removed and at the same time the ceria layer is etched through to the underlying silica layer. Finally, the remaining PS residues are removed by an O_2 plasma or UV-ozone treatment. In the choice of process parameters (film thickness, particle size, O_2 plasma parameters), it must be taken into account that also the PS particles are etched down during the Ar^+ sputtering, which affects the final structure. That is the reason why the Pt particles in Fig. 4.16 are smaller than the original masking PS hemispheres.

Extended Ar^+-Etching Times. Here, a route to make 3D multicomponent units is presented, without the need to divide the Ar^+ etching into two steps, as discussed above. Instead, the fact that the PS particles are reduced in size during the Ar^+ etching is utilized. The Ar^+ etching is continued through both the Pt and oxide support layer. During etching the size of the PS hemispheres

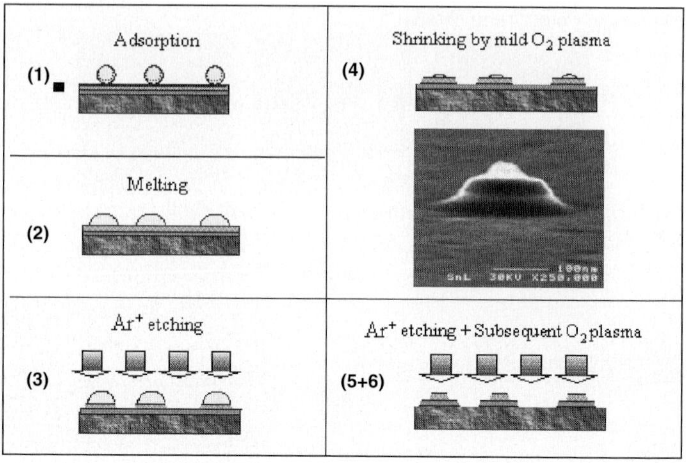

Fig. 4.16. Fabrication process for making Pt dots on ceria disks: (**1**) adsorption of polystyrene particles on top of a Pt film on a ceria film, (**2**) melting of the colloids to hemispheres, (**3**) Ar^+ etching through the Pt film and partially into the ceria film, (**4**) shrinking of the polystyrene hemispheres by a mild O_2 plasma. The SEM image shows a 40-nm polystyrene hemisphere on top of a 160-nm Pt particle on a ceria film, (**5**) continued Ar^+ etching into the ceria can be seen by the raised structure in the SEM image, and finally (**6**) removal of the PS by O_2 plasma (from [91])

is also reduced. By carefully choosing the thickness of the Pt and oxide layer and using appropriate etching times, the structure of the resulting 3D unit may be controlled. Figure 4.17 shows 40-nm Pt dots on alumina and ceria made according to this procedure. In Fig. 4.17, it is seen that the 40-nm Pt particles sit on 170-nm ceria islands, which in turn are situated on the bare silica surface. In contrast, the alumina layer is not etched through despite that the oxide layer thickness and etching times are identical. This is due to a smaller etching rate of alumina compared to ceria. In the latter case, the elevation in the alumina layer around each Pt dot indicates the size of the original PS particle. Hence, if a continuous support film is desired (instead of a 3D structure as in Fig. 4.17), it is important to make it thick enough so that it is not etched through during the extended Ar^+-etching step, used to shrink the PS particles. By varying the size of the adsorbed particles, the degree of melting (thickness of the melted PS particles), the thickness and materials of the evaporated films, and the Ar^+-etching times many different structures can be made.

Intermediate PS Melting and Postdeposition. Structures where the particle is physically separated from the oxide support provide means to, for example, study mobility of suboxide species in a controlled manner. Here, we describe the fabrication of such Pt/ceria structures, where a lateral spacing of alumina is added in-between the Pt particles and the ceria layer (Fig. 4.18). Beginning with melted PS hemispheres adsorbed on a Pt/alumina double layer, deposited

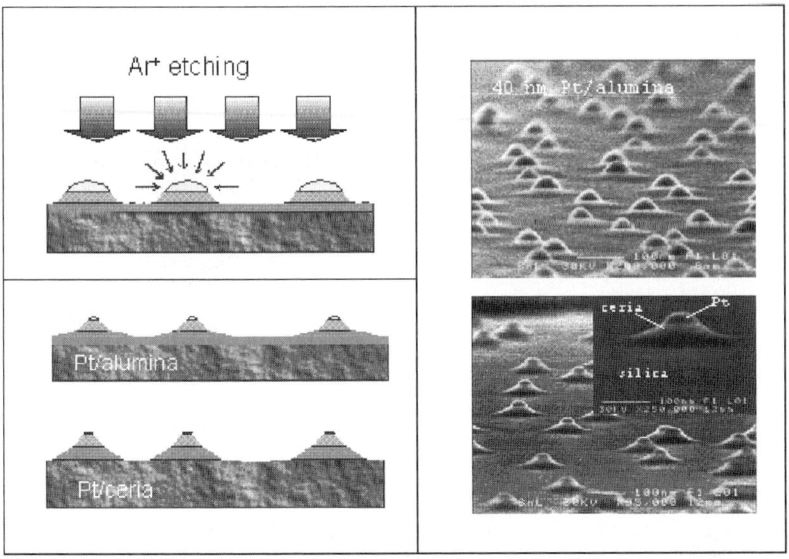

Fig. 4.17. By utilizing extended Ar^+-etching times, the adsorbed, premelted PS particles are shrunk during the etching, resulting in smaller diameter Pt particles than the original PS particle, allowing lower coverage regimes. Pt/alumina and Pt/ceria/silica model catalysts are shown both schematically and in SEM images (from [92])

on a silica terminated Si wafer, the Pt layer is etched through to the alumina film, except where the Pt is protected by the PS particles. As a next step, the PS hemispheres are subsequently further melted, so that they spread out over the exposed alumina surface close to the Pt particle. A ceria layer is then evaporated on the sample. In the final step, the PS particles covered by ceria are removed by immersion of the sample in hot ultrasonic acetone bath followed by an O_2 plasma treatment.

The Drying Problem. A critical step during the colloidal lithography process is to preserve the colloidal particles at their original position obtained in the initial adsorption process [91]. Failure to control the particle (im-)mobility during drying can ruin the whole process. Capillary forces act between the adsorbed colloidal particles during drying. Larger particles exhibit a higher tendency for agglomeration than smaller ones, which depends on the ratio between the lateral capillary forces and the attractive forces between the colloidal particles and the support. The lateral forces originate from the receding waterfront and the water meniscus, which forms between two adjacent particles during the drying process. The extension of the meniscus depends on the contact angle between the water and the surface. A small contact angle, i.e., a hydrophilic sample surface, allows a water meniscus to form between two particles further away from each other, than if the surface is more hydrophobic, which increases the tendency for particle agglomeration on a more hydrophilic

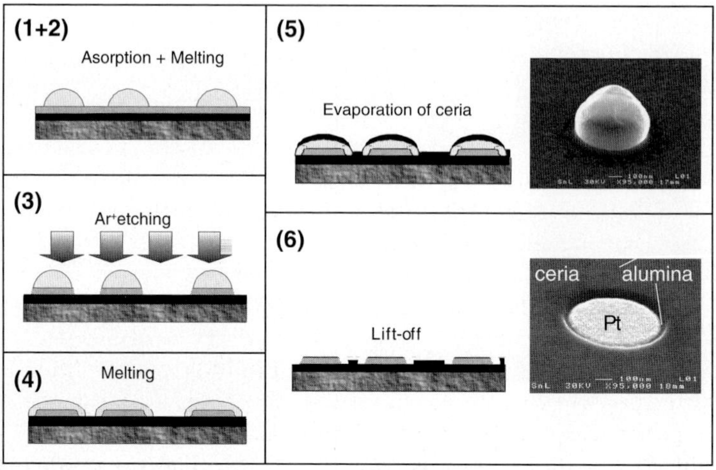

Fig. 4.18. Procedure to make Pt particles separated from Pt dots by an alumina interface: (**1**) Adsorption of 530-nm PS particles on top of a Pt film on an alumina film, (**2**) melting of PS colloids to hemispheres, (**3**) Ar^+ etching through the Pt film to the alumina layer, (**4**) further melting of the PS hemispheres, (**5**) deposition of a ceria layer, and finally, (**6**) liftoff of the PS spheres by a combination of a hot acetone bath and O_2 plasma. The SEM image shows a 540-nm Pt particle surrounded by ceria with a thin separating alumina interface (from [91])

surface. A hydrophobic surface, on the other hand, prevents the adsorption of aluminum ions during the ACH treatment of the sample surface (step 1 in Fig. 4.11). An optimal contact angle seems to be around 50°. Depending on the choice of material and the pretreatment (for example, O_2 plasma, which increases the hydrophilicity for most materials), it may be difficult to achieve an ideal contact angle. An alternative procedure to circumvent this problem is to employ a triple-layer film instead of only a single-layer ACH film when charging the surface. The triple-layer film is made by treating the surface in the following way: (a) 2 wt/wt poly(diallyl-dimethylammonium) (PDDA, MW 200,000–350,000) rinse in water and dry with N_2; (b) 2 wt/wt poly(sodium 4-styrenesulfonate) (PSS, MW 70,000), rinse and dry with N_2; and finally, (c) 5 wt/wt ACH followed by rinse in water and dry in N_2. The thickness of the PDDA/PSS/ACH layer obtained in this manner is reported to be ~1 nm [113]. The triple-layer film gives the surface a contact angle close to the optimal 50°, independent of which substrate is used. All colloidal lithography fabrication on alumina and ceria presented in this chapter was done with this method.

Characterization of Model Catalysts

Structure

The Pt particles formed after etching and O_2 plasma following the preparation route described in Fig. 4.11 have a cup-like shape due to Pt atom redeposition underneath the PS particles during Ar^+ etching. The depth of the craters in the 107-nm Pt nanoparticles shown in Figs. 4.11 and 4.13 is about 10–15 nm. The depth can be varied with etching time by changing Pt film thickness so that more or less Pt is redeposited around the PS template particle (Fig. 4.14). The crucial point here is that unless tricks are used to avoid this (such as melting PS spheres to hemispheres), cup-like shapes are inherent to the procedure in Fig. 4.11. However, an important aspect is that the particles formed are usually modified during reactions or annealing, in the direction of their minimum energy configuration (Wulff shape), and posttreatment steps to accomplish this in a controlled manner may be employed (see Sect. 4.5.2).

Metal films made by directed vapor deposition are in general polycrystalline. The Pt particles maintain the polycrystalline structure throughout the colloidal lithography process, since all steps, except adsorption of the colloidal mask, are performed at room temperature in vacuo; the polycrystallinity has been confirmed by diffraction analysis (Fig. 4.19). The O_2 plasma-cleaning step may, however, promote crystallization by formation of mobile Pt–O species (see Sect. 4.5.2). Samples that have gone through the O_2 plasma step may, therefore, have a larger average grain size (after reducing the particles—chemically or by annealing). TEM images of 120-nm Pt/SiO_2 particles, which are 10-nm thick, show that cracks develop in the large Pt particles (Fig. 4.19). This is most likely due to the particle–support interaction, which builds up strain in the Pt film during deposition and locks the Pt particles in a metastable structure. The strain is relieved and the grain size is increased by either annealing and/or reaction pretreatments (Fig. 4.20). These findings are similar to those for Pt particles fabricated by electron-beam lithography [72] (see also "Reaction-Induced Restructuring"). The oxide-support films formed by electron-beam heating of pressed tablets of metal oxides are known to be amorphous or quasi-amorphous (see [6], and references therein). Naturally, the

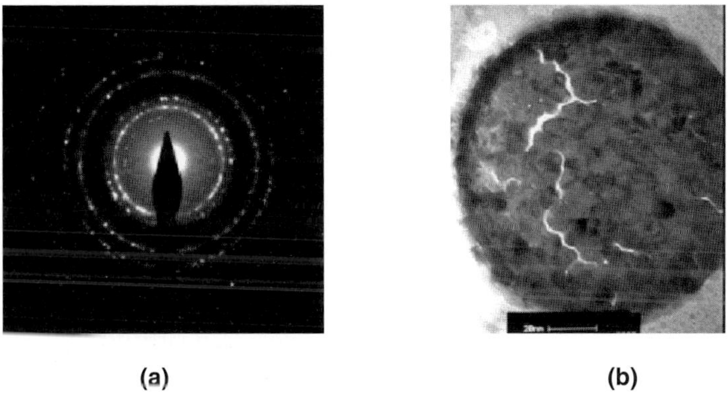

Fig. 4.19. Diffraction pattern and bright field TEM images of as-prepared 120-nm Pt particles (10-nm thick) on SiO_2 fabricated with colloidal lithography. The images were taken by Tomas Liljenfors, Chalmers University of Technology

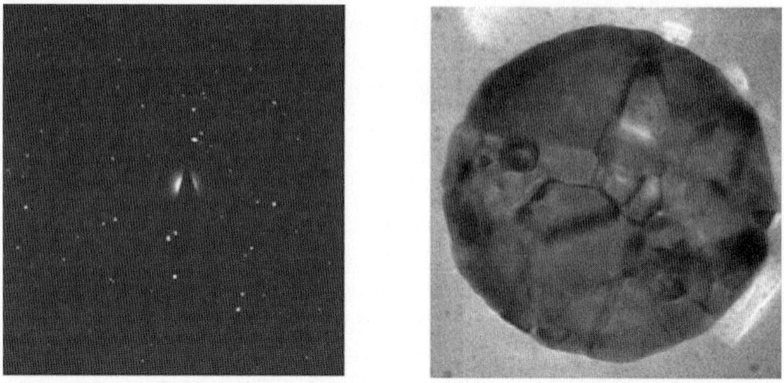

Fig. 4.20. Diffraction pattern and bright field TEM images of 120-nm Pt particles on SiO_2 fabricated with colloidal lithography. The particles were pretreated at 773 K with a stoichiometric gas mixtures of H_2 and O_2 diluted with Ar ($P_{H_2}/(P_{O_2} + P_{H_2}) = 0.67$). The images were taken by Tomas Liljenfors, Chalmers University of Technology

amorphous structure of the oxide film influences the growth of the metal film. Similarly, reaction-induced structural modification of the support material is translated into the metal particle structure. In "Reaction-Induced Restructuring on Pt/Ceria", we will exemplify this.

Chemical Properties

There are a number of potential contamination sources in the colloidal lithography process. The critical steps are steps 2 and 3 in Fig. 4.11. In step 2, the surface with preadsorbed colloidal particles is inserted into a vacuum chamber for Ar^+ etching. It is paramount that all metal is removed in between the PS colloidal particles. Dynamic secondary ion mass spectroscopy (SIMS) depth profiling of a Pt film shows that the Pt signal drops sharply under the detection limit, when the ceria layer is reached and shows that Pt is not buried in the ceria layer (at the detetion limit of SIMS) during the Ar^+ etching procedure [91]. Thus, after step 2, only Pt, which originates from Pt underneath the PS particles, remains on the surface. In step 3, it must be ensured that the remains of the colloidal particles after the Ar^+-etching step are removed. It is important that no chemical traces are left, since even submonolayers of PS traces may alter the catalytic activity substantially. One procedure to remove PS particles is to clean the sample in mild O_2 plasma. XPS shows that a mild O_2 plasma treatment effectively removes all PS. The efficiency of this process is most likely due to the oxygen radicals in the plasma, which attack H and C atoms in the PS, forming water, CO and CO_2 (maybe also SO_2 from the sulfate groups of PS).

The O_2 plasma also causes the sample to be deeply oxidized, however. Figure 4.21 shows XPS spectra of the Pt 4f region from three samples from each category were analyzed: (a) one sample as-prepared, (b) one which was heated in vacuum to 400°C with a temperature ramp of 5 K min^{-1}, held at 400°C for 2 min and cooled down and introduced into the XPS analysis chamber without being exposed to atmosphere, and finally, (iii) one sample reduced by H_2 plasma (250 W, 500 mTorr, 2 min). The Pt film exposed to O_2 plasma exhibits a more oxidized Pt surface than the 130-nm Pt/ceria sample. Ceria is known to stabilize Pt oxides from being reduced to metallic Pt, but conversely it also appears to stabilize PtO from being oxidized to PtO_2. A simple in situ annealing step in connection with a catalytic activity measurement (Fig. 4.21b), or a H_2 plasma treatment (Fig. 4.21c) at the end of the fabrication process efficiently reduces Pt. The former is in good agreement with the thermal stability of PtO_2. The latter is in agreement with Pt film data, which shows that an oxidized Pt film is reduced to metallic by H_2 plasma [114].

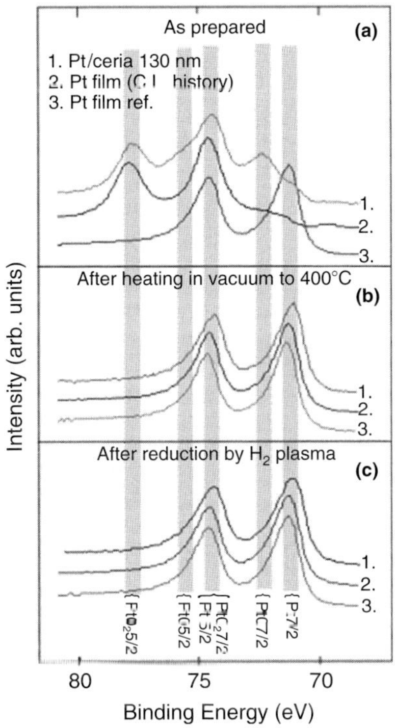

Fig. 4.21. Normalized XPS spectra of the Pt 4f region for: (1) 130-nm Pt dots on ceria, (2) Pt film exposed to all the steps in the colloidal lithography process, (3) untreated Pt film, as reference. The samples were analyzed (**a**) as-prepared; (**b**) after heating in vacuum to 400°C and cooled to room temperature; (**c**) after reduction in H_2 plasma (500 mTorr, 250 W, 2 min). The binding energies were referenced to adventitious carbon at 284.8 eV (from [92])

Finally, XPS shows no traces of other elements. In particular, neither Cl nor S were detected, which is important since the PS particles contain sulfate groups and NaCl was added to the particle suspension during the preparation; all of these elements can influence the catalytic activity dramatically [1]. In circumstances where the O_2 plasma may be causing irreversible damage to the sample, care has to be taken, for example, to avoid volatilization of the catalyst (as with Ru). A milder way of removing the PS mask is to employ an UV/ozone treatment, i.e., irradiation of the sample, kept in pure O_2 or air, with a UV lamp creating ozone [108].

4.4 Microfabrication of TEM Membrane *Windows*

Transmission electron microscopy (TEM) is probably the most powerful technique for obtaining structural information of supported nanoparticles [115–118]. Complementary methods are STM, AFM, and SEM. Both the latter and TEM analysis provide more or less detailed size, shape, and morphology information, i.e., imaging in real space. TEM has the great additional advantage to provide information in Fourier transform space, i.e., diffraction information, which can be transformed to crystal structure information. From a practical point of view, considering the kinds of planar model catalysts discussed above, STM, AFM, and SEM are more easily applied for analysis than TEM, since the former three can be applied without additional sample preparation, once the model catalyst is made. In contrast, TEM usually requires one or more additional preparation steps. In this section, we concentrate on recent developments of microfabrication methods to prepare flat TEM membrane supports, or *windows*, by lithographic methods, which eliminate the requirement of postfabrication preparation of model catalysts for TEM analysis. For a more comprehensive treatment of other, more conventional, procedures to make flat TEM supports, and also similar microfabrication procedures as described here, we refer to previous reviews [118–120].

From a TEM viewpoint, extremely thin single crystal films (~50 nm) of light elements are the preferred TEM substrates because they scatter electrons less and do not produce artificial phase contrast features (the diffraction spots of these films should not coincide with those from the material to be analyzed). Amorphous films, on the other hand, do not produce diffraction spots but diffuse rings, simplifying the interpretation. However, during bright-field imaging, they produce a speckled contrast in the background which must be imaged initially at the same focus [119, 121]. However, it is normally extremely demanding to prepare and handle the ultrathin samples (<50 nm) required for TEM analysis. Typical preparation techniques require that the sample be thinned mechanically, which can potentially alter fragile nanostructures [80, 86, 122]. The use of a supported membrane for both nanofabrication and analysis avoids these difficulties [123, 124]. Information from TEM about the nanostructures can then be used for both information about the fabrication procedure and also for studying the nanostructures themselves.

For high process control in nanofabrication, it is critical to be able to image the nanostructure under preparation, after each individual preparation step and after completion. For example, in nanoelectronics several manufacturing steps are usually employed to place, e.g., source, drain, and gate structures, in the right geometric arrangement. The same holds for complex model nanocatalysts [77, 80, 86]. The iterative use of TEM combined with nanofabricated model catalysts is potentially an extremely enabling tool for such control and research. The TEM analysis is valuable both to follow and optimize the fabrication steps, and to follow morphological and crystallinity changes, sometimes also chemical changes, during catalytic reactions. Both microscopy and reactivity measurements can be done iteratively with exactly the same sample (see [125] and Sect. 4.3). TEM membranes fulfill all the requirements above in that they are robust enough for nanofabrication, characterization, and reactivity measurements. Furthermore, large quantities of identical TEM membranes can be made from the same batch. Finally, we note that such membranes are compatible with the other methods that are usually required to obtain complementary information, such as AFM, STM, SEM, XPS, and SIMS.

4.4.1 Preparation Procedures

The microfabrication procedure for the TEM membrane windows is shown schematically in Fig. 4.22. A more detailed outline of the procedure is described in [124, 125]. The windows are typically fabricated on 3-in. double side polished, n-type Si(100) wafers. In the first step, the wafers are cleaned with the two-step standard cleaning procedure used in the semiconductor industry. This procedure ensures the removal of all the organic particles by oxidation, metals by forming soluble complexes, and the native oxide with HF as shown schematically in step 1 of Fig. 4.22 (these steps are hazardous and

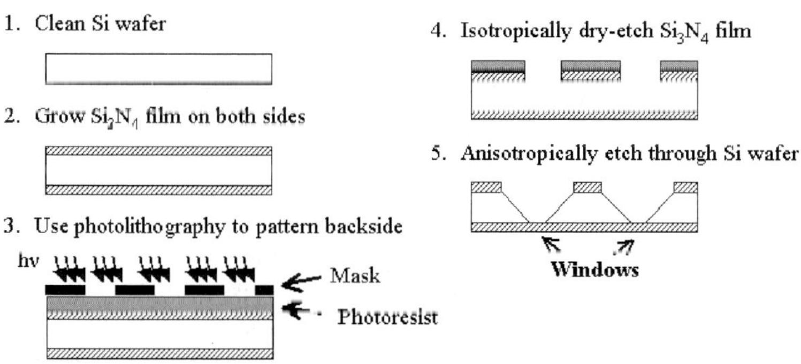

Fig. 4.22. Schematic fabrication procedure for the preparation of TEM windows

require trained personnel). Stoichiometric silicon nitride films are deposited after cleaning on both sides of the cleaned Si wafer using low pressure chemical vapor deposition (LPCVD) as shown in step 2 in Fig. 4.22. Si_3N_4 thin films grown with LPCVD are known to be amorphous [126]. The thickness of the Si_3N_4 window determines the quality in high-resolution transmission electron microscopy (HR-TEM) analyses because background electron scattering from the membrane increases with the thickness of the membrane, degrading the attainable resolution and contrast. The amorphous nature of the film (opposed to a polycrystalline), simplifies the interpretation of the HR-TEM images and diffraction patterns, eliminating Moiré patterns originating from the substrate (artificial background phase contrasts), but as noted above the amorphous films gives more background scattering than a suitable light element single-crystal film. In any case, we judge that the practical advantages of working with amorphous films of light elements make them the preferred choice in most cases.

A photolithography mask is designed to cover the 3-in. wafer with a matrix of more than 700 individual TEM window membranes, each with an area of $2.4 \times 2.4\,\text{mm}^2$. These individual TEM windows of 5.76-mm^2 area fit well into standard 3-mm diameter cups of TEM sample holders. The TEM windows are separated by rectangular patterns, which form trenches (at least as deep as half the wafer thickness) after wet-etching, allowing for efficient cleavage of the individual supports. In the center of each individual support, there is a square pattern, which will become the $100 \times 100 - \mu\text{m}^2$ membrane window after wet-etching on the nonpatterned side of the wafer. With this mask, one side of the Si_3N_4 coated Si wafers is patterned with photolithography (step 3 in Fig. 4.22).

After developing the exposed photoresist layer, the Si_3N_4 layer is isotropically dry-etched in a reactive ion etching system with CF_4, operated with a low flow and low power. This removes the Si_3N_4 only in the patterned areas as shown schematically in step 4 of Fig. 4.22. After removing the remaining resist with N-methyl-2-pyrrolidone, the patterned side of the wafer serves as an etch mask for etching in concentrated (25 wt%) tetramethyl ammonium hydroxide (TMAH) at 90°C. Stirring and reflux ensure a constant concentration during the ~9-h etch time. The etch-rate varies with the age and supplier of the TMAH solution and is $\sim$40–45 $\mu\text{m}\,\text{h}^{-1}$ at this temperature and concentration [127, 128]. TMAH is a relatively new wet etchant, compared to KOH, and has been shown to produce structures with low surface roughness [129]. The etching results in tip-truncated pyramidal shaped holes separated by a square network of trenches, which act as cleavage planes. The walls of the pyramids are (111) planes at an angle of 54.74° from the (100) plane. The tip of the pyramidal shapes is a thin Si_3N_4 membrane, which is thin enough (~40 nm) to be electron transparent. Figure 4.23 shows a light microscope image of ~300 membrane windows. The cleavage planes appear as dark lines, whereas the membrane windows appear as bright squares at both the centers and at the corners of the 5.76-mm^2 supports. The inset in Fig. 4.23 shows

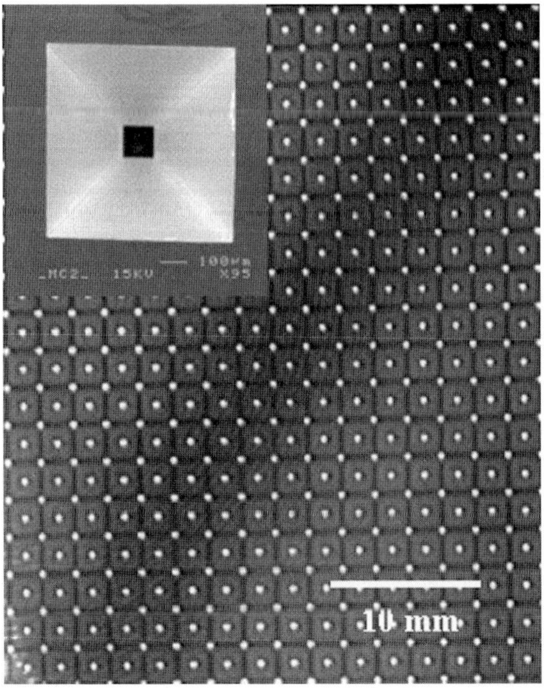

Fig. 4.23. Two images of the backside of the TEM membrane windows. The larger scale image, taken with a light microscope, is part of a 3-in. wafer containing ~300 windows. The light squares are the TEM membranes ($2.4 \times 2.4 = 5.76\,mm^2$). The dark lines are the cleavage planes. The inset shows a SEM pictures showing the pyramidal etch pit with the electron transparent window at the bottom

an SEM image of a pyramidal etch pit showing the thin Si_3N_4 membrane as a dark square at the bottom of the pit. A schematic of the side view of the support and membrane is shown in Fig. 4.24. Nanofabrication of model catalysts on top of the support is similar to that on the more usual Si wafers (see Sect. 4.2). Note that the entire area of the 5.76-mm^2 support can be covered with nanoparticles (with the center $100 \times 100\,\mu m$ suitable for TEM), which can thus be characterized by the variety of available surface science techniques, such as XPS, AES, and AFM. They can also be used in reactivity measurements at atmospheric pressures and subject to further TEM analysis to investigate reaction-induced morphological changes, such as crystallization, shape changes, and sintering (see Sect. 4.3).

Another benefit of creating an entire wafer of TEM windows is that arrays of windows (such as 3×3 matrices of windows or larger) can be used for further micro- or nanofabrication, which allows either the same nanostructure to be fabricated on all the windows in a parallel fashion or several different nanostructures to be fabricated on different windows at the same time in a serial fashion (as with EBL). The individual 5.76-mm^2 windows can be cleaved

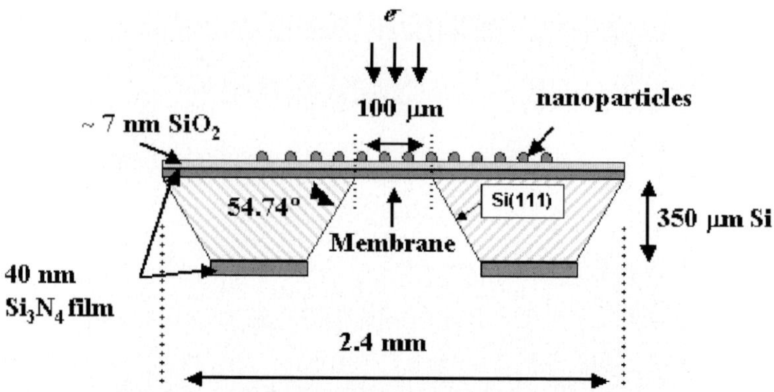

Fig. 4.24. Schematic view of a cross-section of the TEM membrane window showing both the membrane, the oxidized top surface of the membrane, and the Si support with the pyramidal etch along the (111) direction. Nanoparticles can be place over the entire substrate (5.76 mm^2), as shown

for TEM analysis after the nanostructures have been fabricated. On a practical level, larger arrays of windows are easier to manipulate through the multiple processing steps involved with nanofabrication. In addition, reproducibility is ensured if identical nanocatalysts are fabricated on an array of windows at the same time.

Regarding the choice of support material, one is not restricted to Si_3N_4. If another support material is desired it can be deposited as a thin film (preferably only 10–20 nm or less) on top of the nitride. Depending on the crystallinity and electron scattering strength of the added film, it might worsen the contrast and resolution but if the film is thin enough these effects are acceptable. Thus, interesting support materials, such as alumina, ceria, barium oxide, and so on, can be used with the TEM windows.

4.4.2 Chemical and Structural Characterization of TEM Windows

In order to create a more catalytically interesting substrate than Si_3N_4, O_2 plasma oxidation can be used to oxidize the top surface to SiO_2. A uniform film of SiO_2 was formed at room temperature using a high power of 250 W at a relatively high O_2 pressure (500 mTorr). In the following, we present results of the structural and chemical properties of the windows with and without this oxide layer.

Structural Properties

A TEM image of a 100-nm thick cross-section of a TEM membrane prepared by microtome slicing is shown in Fig. 4.25 [125]. The membrane thickness

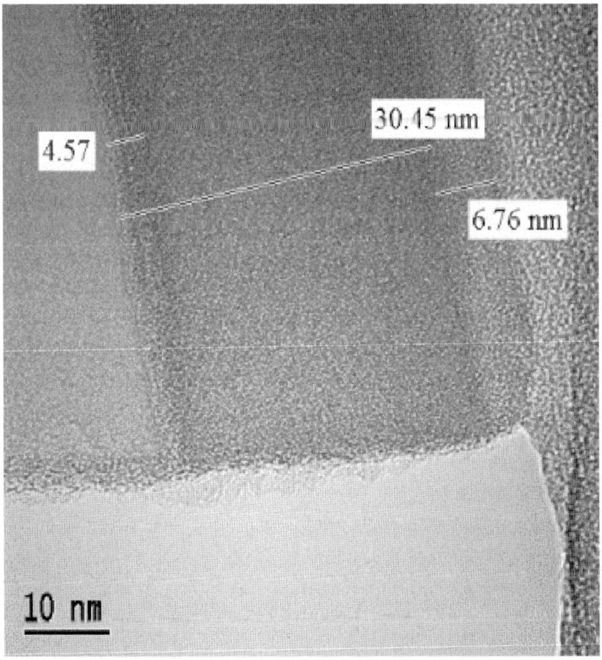

Fig. 4.25. TEM image of a 100-nm thick sliver of a TEM window cross-section prepared by microtome slicing supported on a glassy carbon grid (on the right side of the image). The thicknesses of the Si_3N_4 (30 nm), SiO_2 (7 nm), and the interfacial layer between the Si and Si_3N_4 are shown (from [125])

is ∼42 nm as expected. The topmost ∼7 nm of the membrane is an amorphous SiO_2 film clearly distinguishable from the Si_3N_4, as confirmed with XPS and SIMS, as discussed below. Selected area electron diffraction (SAED), using Au supported on amorphous carbon for calibration, shows a broad and diffuse low-intensity ring of radius 0.71 $Å^{-1}$. This indicates a small degree of short-range order within the film, yet confirms the amorphous nature of LPCVD Si_3N_4 films. The amorphous nature of the SiO_2 film contributes to a diffuse background. If the supported nanoparticles are thin enough (roughly <30-nm thick depending on material) and oriented correctly with respect to the beam, it is possible to obtain HR-TEM images through this substrate.

The surface of the TEM membranes is very smooth. AFM measurements of the surface of these membranes yield a root mean square roughness of <0.5 nm over a 400-μm^2 area (including 3–4 dust particles!). The mean surface roughness and the root mean squared roughness reduces to <0.2 nm over a 4-μm^2 area [125]. Interestingly, it is easy to detect the window edge in AFM because the vibrational characteristics of the membrane versus the support wafer translate to a noticeable edge in the image.

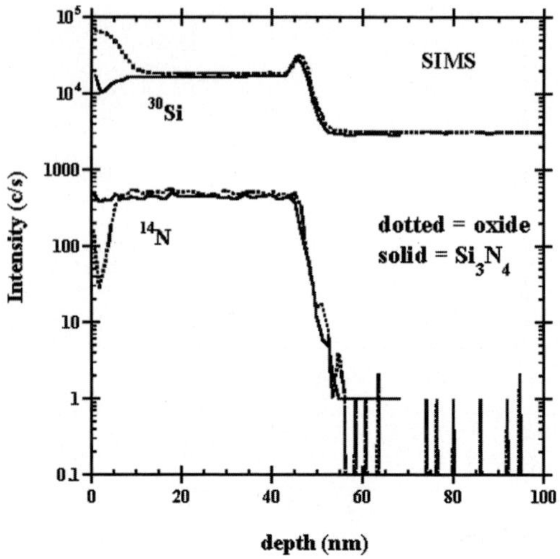

Fig. 4.26. ^{30}Si and ^{14}N SIMS depth profiles comparing a TEM membrane window with native oxide and with a plasma-oxidized layer taken with O^{2+} ions

Chemical Properties

These oxidized TEM windows were characterized with SIMS, XPS, and energy dispersive X-ray spectroscopy (EDS) [125]. Dynamic SIMS was employed to detect the chemical composition of the windows and provide an additional measure of the thickness of silicon oxide layer. A window oxidized with an O_2 plasma and a window with an adventitious oxide layer were compared. The spectra, taken with O_2^+ ions, are shown in Fig. 4.26. The depth scale was calibrated by a comparison to a well-calibrated, 4.2-nm thick oxide layer grown thermally on top of an SiGe structure. It was assumed that the sputtering yields for the SiGe and Si_3N_4 samples were the same. Similar depths of the oxide layer and the Si_3N_4 layer were found as with TEM, which further proves the thickness profile of the SiO_2/Si_3N_4 window.

Figure 4.27 shows the decrease in the normalized N 1s XPS area and the increase in the O 1s XPS area across four different samples; the as-prepared Si_3N_4 windows, windows subjected to O_2 plasma oxidation at both 50 and 250 W, and an Si wafer with a 0.4-μm thick thermal oxide layer. In the heavily plasma oxidized sample, the N content was only 3% of the N content in the as-prepared sample. Through the use of the Auger parameter, it has been unambiguously shown that the chemical state of the surface is SiO_2. The Auger parameter can serve as a *fingerprint* for the chemical state [130]. Both the Si 2p binding energy and the Si(KLL) Auger transition were measured. There was a large shift in the binding energy (BE) of the Si 2p of ~1.7 eV and the kinetic energy (KE) of Si(KLL) of ~4.0 eV between the oxidized and unoxidized membranes. The inset in Fig. 4.27 lists the values of the Auger parameter calculated from the addition of the BE Si 2p and KE Si(KLL)

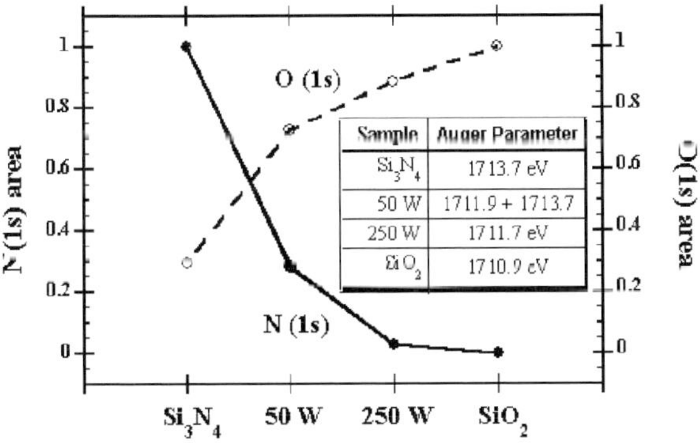

Fig. 4.27. Integrated O 1s and N 1s XPS areas for four different samples, a native oxide on the Si_3N_4 membrane, a low-power (50 W) plasma oxidation, a high-power (250 W) plasma oxidation, and a 0.4-μm thick wet-oxidized Si wafer (SiO_2). The inset table gives the Auger parameter for all four samples. See text for further explanation

for each of the samples. Two peaks were found in the Si(KLL) for the lower power plasma oxidation (50 W) and thus two values are tabulated. Because these samples are insulating samples and no charge compensation was used, all spectra were corrected to a C 1s binding energy of 284.8 eV, the binding energy of adventitious carbon [131]. The tabulated value for the Auger parameter for Si_3N_4 is 1,714.1 eV and for SiO_2 is 1,712.2 eV [130]. This proves that the chemical composition of the outermost film is SiO_2.

A diagonal scan across the surface of the window was recorded with X-ray energy dispersive spectroscopy (XEDS) in the TEM equipment. The atomic ratios of Si:O:N were found to be 43%:13%:44%, which is not the expected stoichiometry of SiO_2 and Si_3N_4 (40%:20%:40%). However, we know from the TEM images and the SIMS data shown below that the thickness of the oxide layer is ~7 nm and the total window thickness is about 42 nm from which we would expect atomic ratios of 42%:8%:50%. This is most likely indicative of a native oxide present on the backside of the windows, enriching the O content of the whole film (as seen with SIMS), since the escape depth of the photons is much greater than the window thickness.

4.4.3 Nanofabrication of Model Catalysts on TEM Windows

It is becoming more and more evident that in situ techniques are indispensable for future development in catalysis research, since catalyst properties during operation conditions can be dramatically different to those inferred from pre- and post-operation analyses [117, 132]. It is well known that catalyst particles change shape, sinter into larger particles, or even may volatilize in the form of, for example, carbonyls, volatile oxides, or organometallic compounds in the course of a catalytic process, which in turn alter the surface kinetics occurring on deposited catalyst particles. Despite this knowledge, most structural studies so far have been performed before and after the catalytic process [121]. The reason for this is primarily the lack of analytical tools that allow for in situ investigations, which currently are being developed [3,7,116,121,132,133]. We present here examples from pre- and postTEM studies, as well as in situ TEM characterization. The latter is enabled due to recent developments in situ TEM techniques [116, 117, 121].

The combination of nanofabrication techniques and in situ TEM provide an ideal platform to study dynamics of adsorbate- and reaction-induced structural changes of model catalysts. There are two types of phenomena that may be envisaged: (a) Particle dynamics induced by exposing the supported model catalyst to reactive gases, at sufficiently high temperature to allow motion of atoms. It is known that particle diffusion may be triggered by gas exposures at high pressure and/or temperatures, but this has only been studied in detail in few cases [134, 135]. (b) Dynamical changes of the particle structure itself can be induced by gas-phase reactions at high pressures and/or temperatures. One of several possible mechanisms behind such adatom-induced diffusion is a surfactant mechanism by which an adatom reduces the binding energy and diffusion barrier of the diffusing specie [136]. The two types of dynamical phenomena may of course be coupled or occur simultaneously.

In the following, we show results from three types of model catalyst fabricated on a thin SiO_2 terminated TEM membrane. The first example is Pt/SiO_2 prepared by ordinary directed metal evaporation of an ultrathin (0.5–2 nm) Pt film, followed by annealing in H_2 to produce a fairly narrow size distribution of nanoparticles [118, 125]. Figure 4.28 shows TEM images of Pt/SiO_2 with an average particle diameter of 6 nm. The size distribution is, however, rather broad, which is inherent to the vapor deposition procedure employed. In the HR-TEM image shown in Fig. 4.28a, it is seen that the edge of the particles appears rather diffuse, which is due to the overlapping elastic

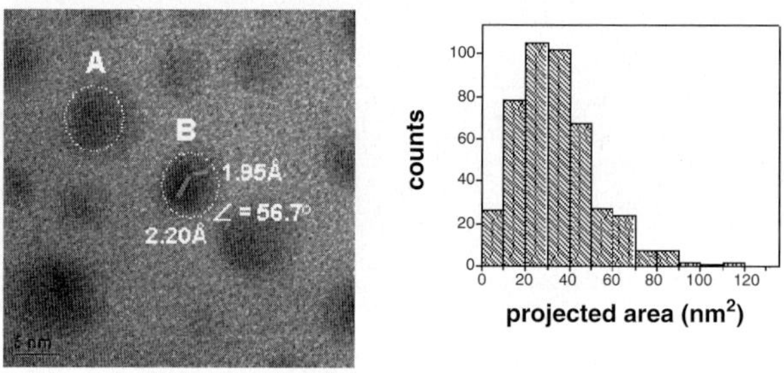

(a) (b)

Fig. 4.28. TEM bright field image and particle size distribution of vapor deposited Pt/SiO_2 after preoxidation at 823 K. The morphology, size, and density of the Pt particles do not change with varying preoxidation temperature up to 923 K. The black *circles* frame Pt particles with a contrast difference due to diffraction (from [125])

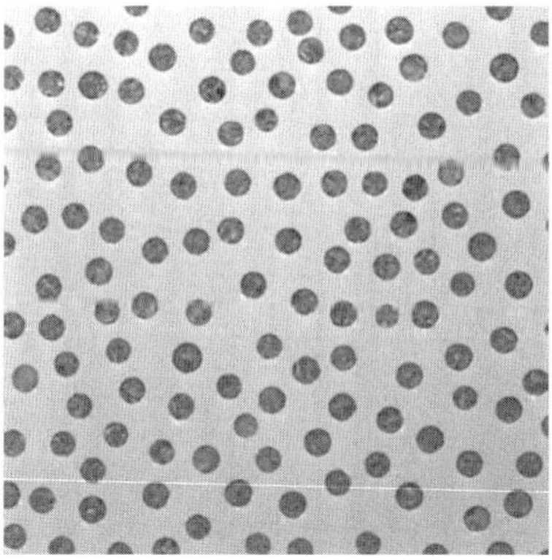

Fig. 4.29. Approximately120-nm Pt particles (10-nm thick) supported on SiO_2 on a 40-nm thick TEM window. The Pt particles were prepared by colloidal lithography. Image taken by Tomas Liljenfors, Chalmers University of Technology

scattering of the electrons in the quite thick 100-nm SiO_2 terminated Si_3N_4 membrane employed here. To resolve the particles on the atomic scale, it is therefore important to reduce the membrane thickness

Figure 4.29 shows 120-nm wide and 10-nm thick Pt particles made by colloidal lithography on a 40-nm thick oxidized TEM membrane, i.e., with SiO_2 as the Pt support. Even though a thinner membrane yields better phase contrasts in TEM, atomic-scale resolution of particle–support boundary sites will normally require cross-sectional analysis. Using colloidal lithography, it is evident that a much narrower size distribution is obtained than by evaporation.

Finally, with EBL it is possible to prepare well-defined arrays of particles on the TEM membranes with both well-defined particle sizes and interparticle separation (both short- *and* long-range ordering). In Fig. 4.8, this is illustrated by an array of ~50-nm wide and 15-nm thick Cu nanoparticles on SiO_2. There are three interparticle distances, 100, 200, and 500 nm, respectively.

To be able to achieve high-resolution images of the deposited metal particles in TEM, it is vital that the particles are not too thick (typically <30 nm; the exact thickness depends on material. Heavier elements scatter the electrons more efficiently). This is a further benefit of lithographic methods, such as EBL and colloidal lithography, where the thickness of the as-prepared particles is precisely determined by the (calibrated) catalyst film thickness from the directed vapor deposition. The thickness of the particles will, of course, change in the course of reactions and heat treatments to adjust to a minimum energy (facetted) configuration (see "Reaction-Induced Restructuring"). Figure 4.30 is an example of particle reshaping induced from gas exposure of an EBL Cu particle array as obtained with in situ TEM. The original arrays consisted of ~70-nm Cu particles, 15-nm thick, with 200-nm interparticle spacing (Fig. 4.30a). After heating at elevated temperatures in an oxidizing atmosphere followed by reduction in H_2 at 400°C, significant particle reshaping is observed (Fig. 4.30b), with small crystallites (~10 nm in

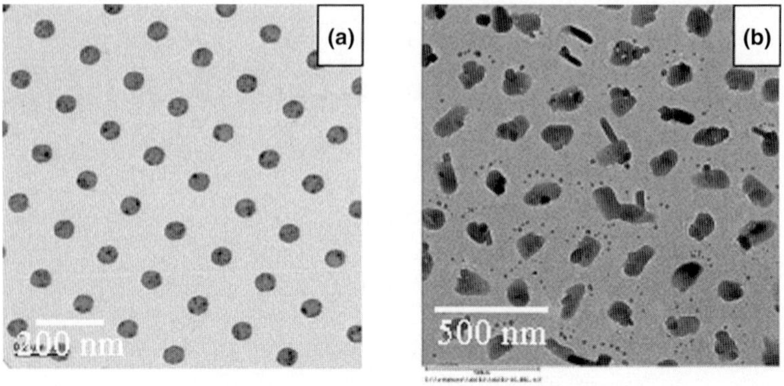

Fig. 4.30. Approximately 70-nm diameter Cu particles, 15-nm thick, with 200-nm interparticle spacing prepared by electron-beam lithography: (**a**) the as-prepared sample, and (**b**) after oxidation at 500°C in 1.1 mbar H_2 at 400°C. These images were taken with the in situ TEM by Stig Helveg at Haldor Topsøe A/S

diameter) migrating away from the larger Cu particles at an average distance of roughly 50 nm from the edge of the main particle. Following this crystallite migration in situ, in real time, allow Cu particle diffusion to be studied in great detail under relevant technical conditions. HR-TEM will allow the crystal structure of the individual crystallites to be determined. Moreover, it is possible by this approach to monitor surface structures that may be present only under operation conditions (high pressures and temperatures).

Apart from the obvious advantages of being able to study 2D diffusion and particle structures of well-defined model catalysts at high pressures and temperatures, there are other more subtle reasons why nanofabrication of planar model catalyst on TEM membranes are advantageous. Since the electron microscopy can be done on the very same samples that are used in the reactivity measurements (remember that the total model catalyst area is $2.4 \times 2.4\,\text{mm}^2$ in our case, while TEM is performed on a small section in the center of the membrane), the correlation of structure and dynamics with reactivity is greatly simplified. Moreover, the planar geometry ensures that the same active phase of the catalyst is studied in both cases. Finally, again we emphasize that the samples are well suited for a wide range of complementary surface science techniques, such as XPS, AES, LEIS, and SIMS.

4.5 Experimental Case Studies with Nanofabricated Model Catalysts: Catalytic Reactions and Reaction-Induced Restructuring

In this section, we present experimental results for each of the planar model catalysts presented above. We have chosen to use CO oxidation at atmospheric pressure as the main model reaction. There have been several studies using *simple* reactions, like CO oxidation [29] or ethylene hydrogenation [68], and no doubt there is a need to go beyond such model systems. There are, however, three main reasons for the choice of CO oxidation as a model reaction. First, the CO oxidation is a relatively simple reaction, and probably the most studied reaction in surface science. The elementary steps are well established under UHV conditions [137], and have been explored in detail also at atmospheric pressure conditions (see [29, 138], and references therein). Second, in view of the complexity of the lithographic fabrication procedure itself, it is desirable to reduce the complexity of the catalytic evaluation. Finally, the CO oxidation is a technologically important reaction; it is, e.g., one of the key reactions in automobile emission cleaning [1].

Studies of the CO oxidation over supported catalysts at atmospheric pressures (of which we present two examples below) impose several challenges compared to corresponding UHV studies. First, in the viscous flow regime the reactor and sample geometries are intimately coupled with the overall catalytic response. Second, at high conversion, significant chemical power is dissipated in the catalyst material, which in combination with a poor thermal conductivity of the support materials, can give rise to high local temperature gradients. Third, at high pressures, high surface coverages can be maintained also at high temperatures, which may cause significant adsorbate-induced surface restructuring, which prevents the use of (extrapolation to) UHV databases. Simultaneously, many surface-sensitive techniques are not applicable at high pressures. The first two points usually require significant modeling to extract the relevant contribution from the measured reaction kinetics. At the same time, the modeling introduces large uncertainties due to various simplifications/assumptions that have to be made in the analysis. The third point is the actual reason for dedicated model catalyst studies, and is the main subject here.

Obviously, it is desirable to devise experiments that simplify the analysis as much as possible and provide a framework for extracting quantitative information of the catalytic reaction of interest. Quite generally, it is desirable to perform steady-state experiments at constant pressures and temperatures to avoid rapidly varying pressure and temperature gradients, which are both difficult to measure and control. Steady-state experiments are, in general, time-consuming, which involve large experimental work plan matrices but are often the only alternative. In some cases, transient studies may be the subject of interest but detailed mechanistic analysis usually requires that such studies are complemented with steady-state experiments.

Here, we describe several approaches to study catalytic surface reactions (primarily CO oxidation) on lithographically fabricated model catalysts at atmospheric pressures, in UHV, and in electrocatalysis. In addition, we discuss reaction-induced restructuring of these lithographically fabricated samples.

4.5.1 Model Catalysts Fabricated by Electron-Beam Lithography

We give three examples of the use of EBL in catalysis research, the first two involving CO oxidation. In the first example, we use EBL to fabricate identical particle arrays of Pt/silica and Pt/ceria to obtain detailed information of the oxygen-spillover mechanism in Pt/ceria. This is a high-pressure viscous regime experiment. In the second example, performed under UHV conditions, we discuss arrays of Pd particles on SiO_2 where communication between facets on the particles could be measured. In addition, direct measurement of the O-mobility as a function of reactant mix and temperature is discussed. In the third example, we show how EBL can be used to systematically study reaction-induced particle restructuring. In particular, we demonstrate that the combination of novel TEM membranes and in situ HR-TEM gives unprecedented opportunities to study catalyst particle structure and dynamics under technically relevant conditions. Other reported catalytic case studies employing EBL include ethylene hydrogenation [30, 68] and cyclohexane dehydrogenation [139] over Pt/SiO_2 and Pt/Al_2O_3 model catalysts.

Before we proceed with the case studies, it is appropriate to comment on the special nature of EBL samples and its implications for catalysis studies. Due to the serial process involved in the EBL pattern writing, it is impractical with present technology to prepare large areas of small catalyst particle arrays on a routine basis. The small sample sizes are problematic to use in traditional reactor set-ups, and new micro-reactor concepts have to be developed. To elucidate the problem, consider a typical laboratory experiment, where a wet-impregnated monolith catalyst with a volume of $1\,cm^3$ is used. Taking typical values for the noble metal catalyst loading (1 wt%), and dispersion (20% fraction of exposed metal atoms), a total surface area of ca. $3 \times 10^5\,mm^2$ is obtained. In flow reactor studies, this normally implies flow rates in the range of $100–1{,}000\,ml\,min^{-1}$ to achieve acceptable total reactant conversion and signal-to-background ratio. In contrast, a typical EBL sample has a total metal surface area of the order of $1\,mm^2$ (corresponding to an array of ca. 10^9 disc-shaped particles, 10-nm high and 30 nm in diameter). This is $\sim 10^5$ times smaller total surface area than the wet-impregnated catalyst, and would thus require a reactant flow of the order of $0.001–0.01\,ml\,min^{-1}$, or ca. $0.02–0.2\,\mu l\,s^{-1}$. This is approximately the same amount of gas that flows through a capillary leak ($\sim 10^{-3}\,mbar\,L\,s^{-1}$) from 1 bar to a vacuum chamber at a pressure of $10^{-6}\,mbar$ with a typical pumping speed ($\sim 100\,L\,s^{-1}$). A mass spectrometer-based detection system is thus compatible with a microflow reactor system to study EBL samples. Further considerations leads to the conclusion that the reactor volume should be small ($1\,\mu L$) to have reasonable response times. The design of microreactors is currently an active research area in chemical reaction engineering, in particular, in rapid screening applications. We have previously described several microreactor concepts designed for catalysis research on micro- and nanofabricated samples [72, 85]. The microflow reactor we have used in the studies presented here is shown in Fig. 4.31 and has been described in detail elsewhere [29, 85]. Briefly, the microflow reactor acts as a quasi-batch reactor with a gas resident time of $\sim 100\,s$, and a short mixing time through diffusion inside the reactor (in the range of seconds). Thus, there is no significant gradient in the gas composition

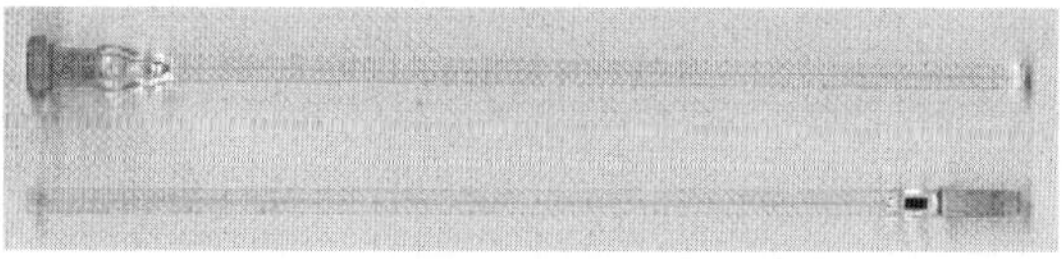

(a)

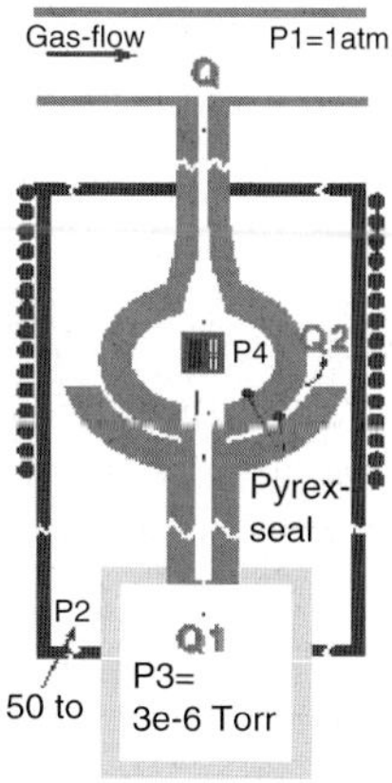

(b)

Fig. 4.31. (**a**) Photograph of a Pyrex-glass microflow reactor with a EBL model catalyst sample ($2 \times 2\,\mathrm{mm}^2$) inside the reactor and (**b**) schematic picture of the reactor with the different pressures (P1, P2, P3, P4) and flow directions indicated (from [85])

in the axial direction of the reactor (i.e., the gas flow direction). The response time is long for a step change in the gas feed (around 400 s) but moderate for depletion of reactants due to a fast reaction occurring on the catalyst sample inside the microflow reactor (tens of seconds).

CO Oxidation on Pt/silica and Pt/ceria

In this example, we use the well-known bistability regime in CO oxidation on Pt. Bistability and reaction hysteresis is exhibited at suitable combinations of absolute pressure, CO/O_2 mixing ratio and temperature. The characteristic of this bistability is that two kinetic regimes exist for the same, just mentioned, parameters; a low reaction rate regime, where the surface is dominated by CO coverage (self-poisoned state), and a high reaction rate regime, where the surface is dominated by oxygen coverage. Which state that is *occupied* depends on the prehistory. For example, if the bistability regime is entered from the oxygen-rich side, the high reactive regime is entered, while the low reactive regime is entered from the CO-rich regime. Scanning the CO/oxygen ratio in the gas phase back and forth, over the bistability regime, produces a hysteresis loop, where the reaction repeatedly switches between the high and low reactive (kinetic) states, and where the switch from the high reactivity state to the low reactivity one occurs at a different reactant mixture than the reverse switching.

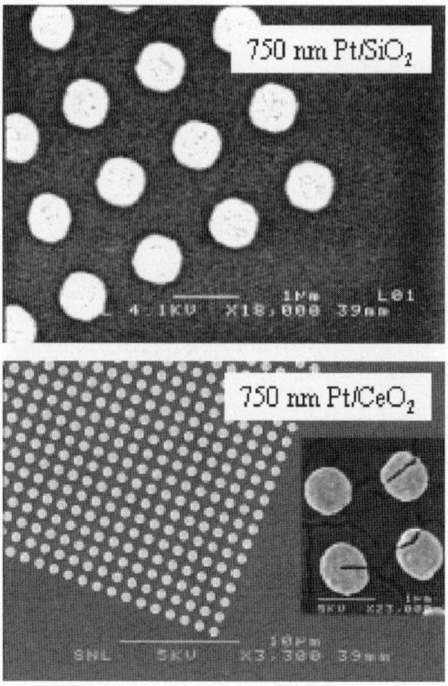

Fig. 4.32. SEM pictures of EBL-fabricated Pt/SiO$_2$ and Pt/CeO$_x$ samples. The particle size is 750 nm in diameter, 20 nm in height, and the center-to-center distance between particles is 1,400 nm. The inset shows a few particles from the larger array (from [29])

CO oxidation measurements were performed on EBL-fabricated, 750-nm (in diameter) Pt/CeO$_x$ and 750-nm Pt/SiO$_2$ model catalysts (Fig. 4.32). Fig. 4.33 shows the bistable reaction regions for these samples and their temperature dependences, as well as 200-nm Pt/CeO$_x$ model catalysts (not shown in Fig. 4.32). Within the bistability region, there exist two steady-state reaction rates for a given set of reaction conditions (reactant gas composition and temperature). These two steady-state reaction rates correspond to two branches, if plotted versus the fraction of CO in the total reactant flux ($\beta \equiv P_{CO}/(P_{CO}+P_{O_2})$), and at each end of the bistability region there are discontinuous transitions in the reaction rate connecting the two branches. The origin of the bistability can be formulated quite generally [138].

In the case of CO oxidation on Pt, it has its origin in the combination of a rapid Langmuir–Hinshelwood step, $CO_{(a)} + O_{(a)} \rightarrow CO_{2(g)}$, and the different adsorption kinetics for CO and O_2, respectively, where CO blocks the (dissociative) adsorption of oxygen but not vice versa. Simultaneously, the CO_2 desorption rate is rapid, so that no reaction products are present on the surface. Hence, in certain temperature and gas mixture intervals, a *bistable region* with two different stable reactive states (with different steady-state coverage of O and CO, respectively) can be reached depending on the preceding history. One state is reached from the oxygen excess regime (high-active state), and the other from the CO excess regime (low-active or CO-poisoned state). A bifurcation is thus observed in the reaction rate as the gas mixture is swept back and forth over the bistable region. Such a bifurcation occurs at constant temperature and should not be confused with catalytic ignition, for which a rapid increase in both reaction rate and temperature occurs when the Franck–Kamenetskii condition is fulfilled, i.e., the point where the chemical power generated by the exothermic reaction is no longer balanced by the heat conducted away from the Pt particle [140]. The bifurcation mentioned above is a kinetic phase transition and is due to sudden changes in the reaction kinetics on the surface only, and can occur even without an increase in temperature.

The observed shift of the measured bistability region of the Pt/ceria sample compared to the Pt/silica sample can qualitatively be explained by oxygen spillover between ceria and Pt. This can be understood by considering that for Pt/ceria O_2 may adsorb on ceria and diffuse to the Pt–ceria boundary and react with CO on the Pt particles, even if the Pt particle is covered by CO. The spillover channel thus extends the gas mixing (β) regime, where a high reaction rate can be maintained. The data shown in Fig. 4.33 is particularly

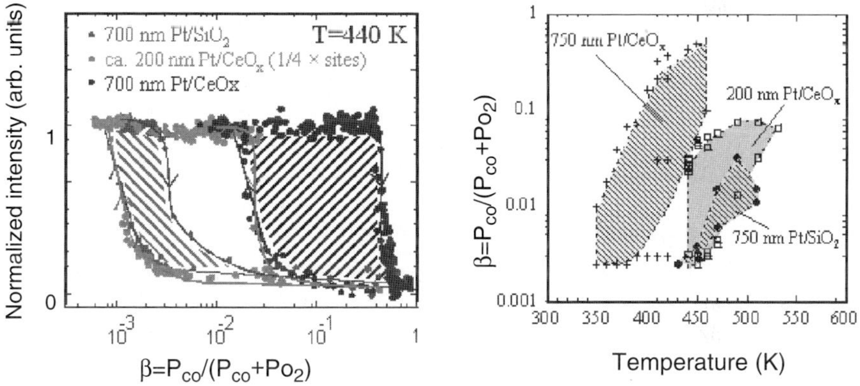

Fig. 4.33. Measured CO oxidation rate on EBL-fabricated model catalysts with different support materials and particle sizes: (**a**) The inlet gas mixture represented by the parameter $\beta = P_{\mathrm{CO}}/(P_{\mathrm{CO}} + P_{\mathrm{O}_2})$ has been scanned up/down at a constant temperature of 450 K, and the rate of CO_2 production has been monitored during the gas scan in β. This has been made for three different samples: (**a**) 750-nm Pt/SiO_2 (blue), (**b**) 750-nm Pt/CeO_x (black), and (**c**) Pt/CeO_x sample (200 nm, red, disintegrated particles). Results both from experiments (filled circles) and simulations (*solid lines*) are shown. The arrows indicate which reaction rate branch that has been observed while scanning up/down in β. A bistable region (hysteresis) is observed for all samples. (**b**) The bistability diagrams determined from a series of measurements as those shown in (**a**) at different temperatures (Pt/SiO_2, blue and open squares); Pt/CeO_x, black *hatch* marks and *crosses*; and the 200-nm Pt/CeO_x, red *hatched* area and open *squares*). The observed differences can be traced back to a pronounced O-spillover effect on ceria. Note the logarithmic scale for the β-value (from [29])

well-suited as input for microkinetic modeling, and allows for detailed information of the surface reaction kinetics to be extracted. Here, we outline briefly the main points of the results of the simulations, which have been presented in detail elsewhere [29]. First, an increased boundary length enhances the oxygen-spillover mechanism; smaller particles have larger interface length due to a larger surface-to-volume ratio than larger particles. This can be inferred from Fig. 4.33, since the number of Pt sites is different for the 200 nm particles (smaller by a factor of ca. 4). The enhancement due to the larger interface length is clear after normalization to the number of Pt sites (not shown). Second, the major effect of including a reactant supply via the support, compared to direct adsorption from the gas-phase on the Pt particles, is the much weaker coverage dependence of oxygen adsorption onto the support and subsequent (through diffusion) attachment to the Pt. Finally, to obtain an appreciable effect of the spillover mechanism, one finds that the surface coverage on the particles must be close to saturation. Disregarding the trivial case of O saturation, this means that finite CO coverages must be considered to achieve an effective spillover channel, which, in turn, implies that the CO + O reaction rate must be substantially lower than inferred from similar UHV studies. Also, the results for the Pt/silica system, which is here used as a reference system, indicate that the surface reaction step between CO and O is substantially lower than previously reported for UHV. A consequence of this is that the reaction rate on the low reaction rate branch may be just an order of magnitude lower than on the high reaction rate branch. This is an intriguing result, which suggests either a different CO oxidation mechanism in oxygen excess on Pt at high pressures and high temperatures ($400\,K < T < 600\,K$), and/or that the mean-field modeling employed here neglects essential (maybe local) ingredients. It is interesting to note that in a recent high-pressure STM study [141] it was observed that the Pt(110) surface underwent a significant restructuring during CO oxidation in oxygen excess under similar conditions as employed here, which was interpreted as Pt oxide island formation. In a later STM and DFT study, it was confirmed that the Pt(110) surface form a disordered surface oxide at high O coverages, which transforms into ordered PtO_2 islands above 400 K (under reaction conditions relevant to many CO oxidation experiments on Pt). The Pt surface oxides are, however, metastable and the chemisorption structure is the energetically preferred structure [142]. These findings suggest that different reaction kinetics should be considered on the low and high reaction branches, respectively [143].

CO Oxidation on Pd/SiO_2: Molecular Beam Studies

In the second examples, we present results of reactivity measurements performed on Pd/SiO_2 EBL samples in combination with in situ vapor deposited samples employing a combination of state-of-the-art UHV techniques. The EBL-fabricated model catalyst consisted of hexagonally arranged Pd *towers* with an aspect ratio of 1. These towers were ~500-nm high and ~500 nm in diameter, with 1,500-nm interparticle spacing. As described in detail by Libuda, and coworkers [144, 145], the reactivity measurements were performed in a UHV system, where molecular beams of CO and O_2 were generated with two high-intensity effusive beam sources and a supersonic beam source. The global reaction rates were detected with a nonline-of-site quadrapole mass spectrometer, and the local reaction rates with a rotatable mass spectrometer.

A well-defined bistability was found within a given temperature range (<440 K) and CO/O_2 mixing ratio (Fig. 4.34). This behavior is consistent with results for single-crystal surfaces under low-pressure or UHV conditions. The explanation for this bistability regime was given in the previous section. However, a well-defined bistable region was not seen for smaller-sized (2- and 6-nm Pd particles) model catalysts prepared with vapor deposition, as predicted theoretically [146]. This is shown in Fig. 4.35 [147], where a stable bistability region is seen on the large EBL-fabricated particles (model system I), but not at all on the very smallest particles (model system III), and only transiently seen on the 6-nm particles (model system II), while the kinetics reach a single global state over approximately 6 min. In addition, it is possible to switch between the two kinetically stable states on the EBL-fabricated particles by pulsing either CO or O_2, whereas this cannot be done on the smaller particles. This lack of hysteresis on the smaller particles cannot be explained (only) by the presence of defects, where CO is more weakly adsorbed. Instead, the influence of coverage fluctuations was invoked to explain the experimental observations. The amplitude of the coverage fluctuations of the adsorbed species increases with decreasing particle size, due to differences in surface diffusion effects on large and small catalyst areas, as also predicted in [146]. Thus, a dynamic equilibrium is established between the two-reactive states (CO- and O_2-rich in this case) as a result of fluctuation-induced spontaneous transitions between the two kinetic states, resulting in a monostable kinetic regime, as seen in this case for the 2- and 6-nm particles. This kinetic regime was modeled in detail with stochastic mean-field simulations [147]. These simulations showed that as the particle size decreased, the frequency of the spontaneous transitions between the CO- and O_2-rich states increases. This is due to rising amplitude of coverage fluctuations with smaller particle sizes and an increase of the width of the probability distributions over the particle surface in coverage space.

In addition, the combination of these EBL-fabricated Pd model catalysts and UHV molecular beam system has allowed the first experimental confirmation that surface diffusion over the whole particle must be taken into account to describe the global kinetics, and thus support the theoretically predicted communication effects between different facets on a nanoparticle [4, 146, 148, 149]. These experiments also allowed measurement of the surface mobility of oxygen under reaction conditions [83, 150]. On a given nanoparticle, regions with locally different adsorption or reaction properties (such as different crystallographic orientations, i.e., facets) can communicate,

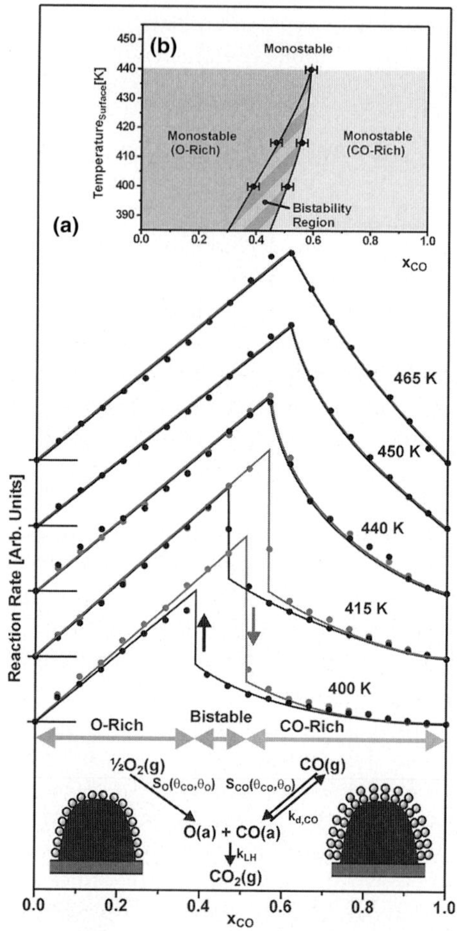

Fig. 4.34. Chemical bistability during CO oxidation on a Pd model catalyst fabricated by EBL (model system I). (**a**) At surface temperatures >440 K, there is a single well-defined transition point between two reaction regimes (at $\sim x_{CO} = 0.6$), which are characterized by high O_2 or CO coverage, respectively. Below 440 K, there is a region with two stable reactive states, which are reached by starting the reaction either from a CO-saturated or from an O-saturated surface. The arrows indicate the direction of the hysteresis. The total flux of CO and O_2 beams at the sample position was equivalent to a local pressure of 10^4 Pa, whereas the β-value or x_{CO} was varied systematically. (**b**) Bistability diagram for the CO oxidation. The origin of the bistable kinetics is related to the coverage dependencies of CO and O_2 sticking probabilities (see text). Above 440 K, the bistability vanishes because of the increasing CO desorption rate (from [147])

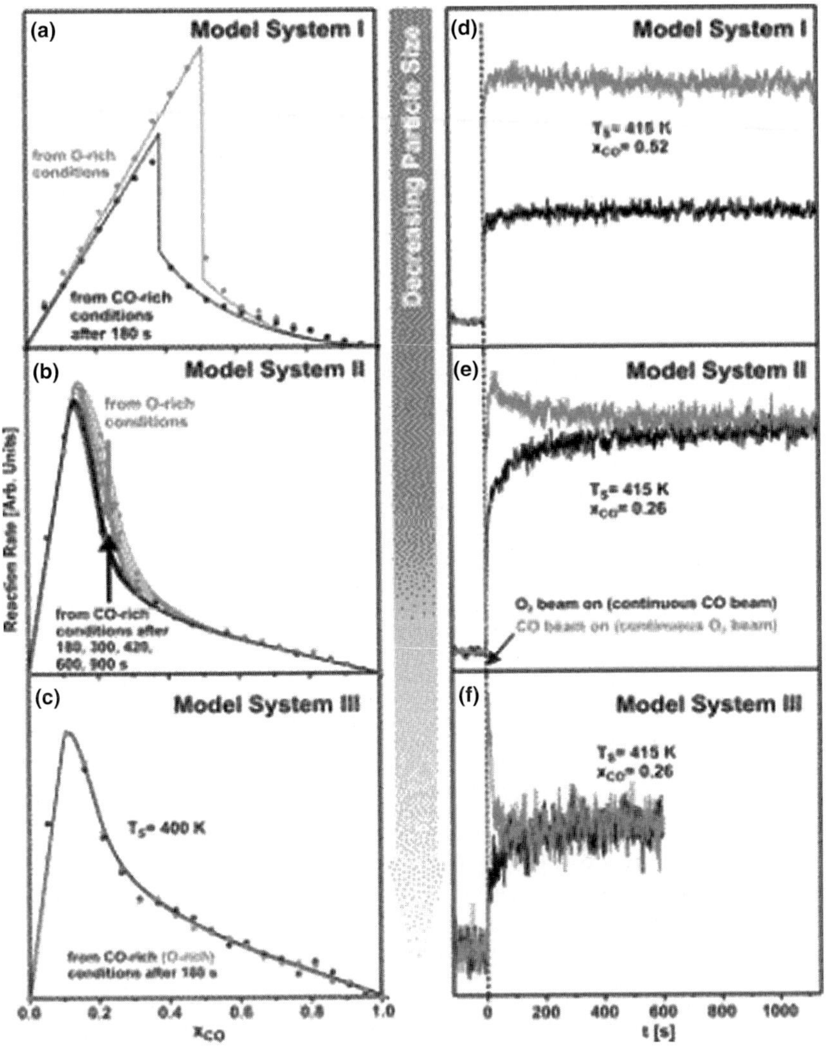

Fig. 4.35. Particle size-dependent bistability and hysteresis. On model system I (500-nm EBL-fabricated particles), the CO oxidation shows a perfectly stable bistability behavior. On the time scale accessible by the experiment ($>10^3$ s), we can arbitrarily switch between the two states by pulsing either pure CO or O_2 (**a** and **d**). For the model system II (6-nm particles), a very slow transition toward a single global state is observed in the transition region between the CO- and O-rich reaction regimes (**b** and **e**). This behavior is assigned to fluctuation-induced transitions, which are accelerated by the presence of defect sites. For the smallest particles of the model system III (1.8 nm), a globally monostable kinetics is rapidly established under all conditions (**c** and **f**). For all experiments, the total flux of CO and O_2 beams at the sample position was equivalent to a local pressure of 10^4 Pa. The surface temperature in (**a**–**c**) was 400 K and in (**d**–**f**) 415 K (from [147])

and thereby couple kinetically, through surface diffusion over facet boundaries. This can give rise to locally very different reaction rates on a nanoparticle, and a global (average) rate that cannot be predicted from data from single facets.

Finally, using angular resolved measurements of the distribution of CO_2 desorption from the above described array of nearly identical Pd nanoparticles, during CO reaction with oxygen, distinct variations of the angular distribution of CO_2 were observed as a function of O_2 beam angle (which can be explained by the weak interaction of CO_2 with Pt, leading to almost instantaneous desorption). A symmetric distribution of CO_2 under O_2-rich conditions ($x_{CO} = 0.34$), and an asymmetric distribution under CO-rich conditions ($x_{CO} = 0.67$) with an increase toward the O_2 flux was observed. This indicates complete equilibration of the O-coverage under O_2-rich conditions but not under CO-rich conditions. This can be explained by a large change in O-residence time (over 2 orders of magnitude) and in oxygen diffusion length, when switching between the two regimes. Through the use of a reaction-diffusion model, the diffusion activation energy, $E_{diff,O}$, for O on Pd during CO oxidation was calculated to be $55\,kJ\,mol^{-1}$ [83, 150].

Reaction-Induced Restructuring

Heat treatments of nanofabricated catalyst particles in different reactant mixtures can result in restructuring and eventually crystallization. Figure 4.36 shows two differently sized Pt/ceria samples made by EBL, before and after reaction in a lean $H_2 + O_2$ gas mixture for 10 h at 1,000 K. The smaller 130 nm particles reshaped and formed crystallites. In contrast, the larger 500-nm Pt particles (and 20 nm in height), were stable at 623 K, but were found to disintegrate and form clusters of smaller crystalline particles at 773 K [29]. The particle disintegration was also found to be dependent on the support material, occurring more readily on ceria than on silica [72, 77].

The disintegration is most certainly related to grain boundaries formed in the particles during deposition, where the formation of coherent final particles are preceded by nucleation and coalescence of many smaller particles (see discussion in [72, 77]). The stronger Pt–ceria interaction, compared to silica, promotes the formation of smaller initial grain sizes on ceria, which probably builds in higher stress in Pt/ceria, which subsequently is released during reaction. In addition, the surface of the ceria films was rougher than the oxidized Si wafers, because they are vapor-deposited from a CeO_2 pellet (as in colloidal lithography, see Sect. 4.3.2), while the SiO_2 film was grown on a smoother, polished Si wafer surface. The larger roughness probably contributes to a higher nucleation density in the Pt film.

The experiments described above were performed in O_2 excess, which is likely to promote Pt–O complex formation, thus weakening Pt–Pt bonding. It has frequently been suggested that oxygen acts as a surfactant for Pt metal

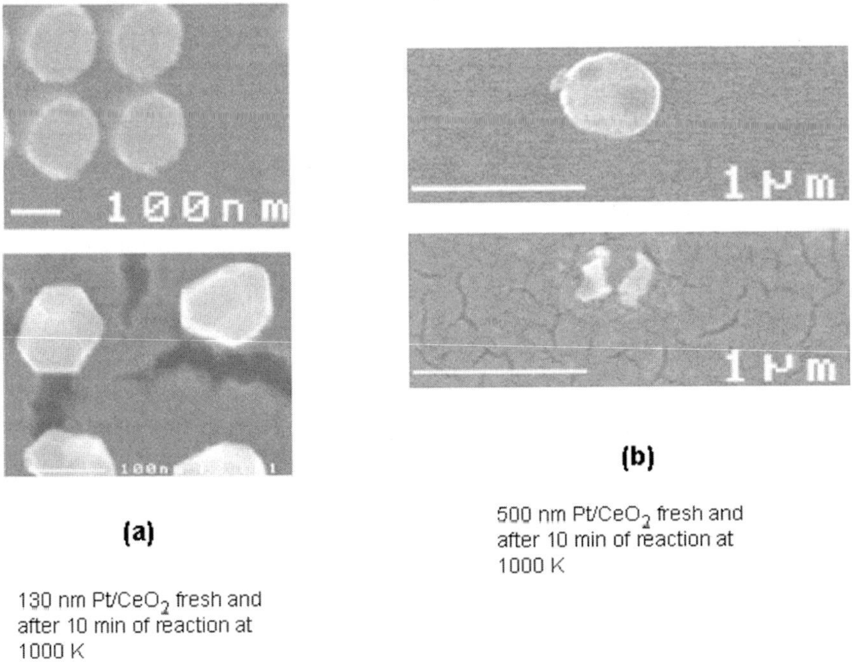

Fig. 4.36. Particle size effect of particle reshaping and disintegration for EBL-fabricated Pt/ceria: (**a**) As-prepared SEM images of 130-nm Pt particles supported on ceria (top) and after 10 min of H_2 oxidation ($P_{H_2}/(P_{H_2}+P_{O_2}) = 0.33$) at 1 atm in a flow reactor (bottom). (**b**) SEM images 500-nm Pt/ceria particles before (top) and after identical treatment as above (bottom) (from [82])

redistribution, thereby facilitating metal redistribution [135, 151]. A requirement for such a mechanism is that Pt is not fully oxidized, since this seems to inhibit restructuring. It is interesting to note that this restructuring behaviour is similar to the sintering behaviour of real, wet impregnated, porous alumina supported Pt catalysts, which showed accelerated sintering in lean $H_2 + O_2$ gas mixtures at approximately the same temperatures as where the morphology changes were observed with the above EBL samples [135].

The formation of mobile metal–oxygen species at high pressures and temperatures have recently been reported by in situ STM studies of the NO + CO reaction on Rh [133]. Avoiding Pt oxidation is accomplished by balancing the $H_2 + O_2$ reaction to near stoichiometric conditions, so that the fast reaction (H_2 oxidation) effectively cleans the surface from adsorbates and simultaneously prevents it from becoming (fully) oxidized. The exothermicity of the $H_2 + O_2$ reaction may further enhance the restructuring due to local high energy release [72].

A more systematic study was undertaken with Pt/SiO_2 EBL samples, where the diameters of the nanoparticles were varied between 50 and 300 nm

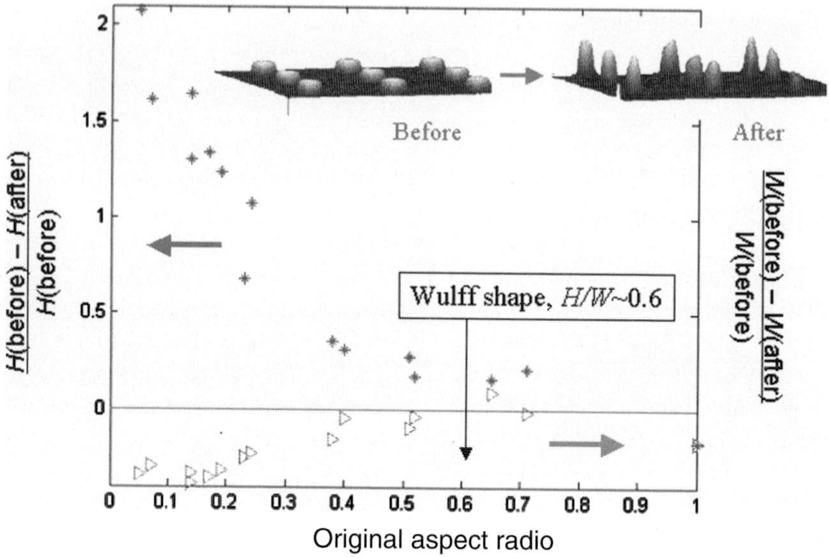

Fig. 4.37. Summary figure of the change in the height, H, and width, W, of EBL-fabricated Pt particles on SiO_2 after running H_2 oxidation ($P_{H_2}/(P_{H_2}+P_{O_2}) = 0.67$) at 800 K in a flow reactor at 1,200 ml min^{-1} for 1–10 h. The original aspect ratio is the width-to-height of the as-prepared particles. The width is measured from the projected diameter of the particles (average of ~50 particles) deduced from SEM and the height is measured from AFM images (average of ~10 particles). The inset shows AFM images taken before and after H_2 oxidation showing the increase in height for particles with a small initial aspect ratio ($W/H = XX$)

and the thickness was varied between 7 and 150 nm, resulting in aspect ratios, i.e., height-to-width, H/W, ratios varying from 0.04 to 1.0. After treatment at 800 K at a flow rate of 1,200 ml min^{-1} in a stoichiometric mixture of H_2 and O_2 (diluted in Ar), it was found that all nanofabricated particles with an $H/W < 0.1$ disintegrated and those with an $H/W \sim 0.6$ (as anticipated by thermodynamics) tended to show the least change in height and width (Fig. 4.37).

4.5.2 Catalytic Reaction Studies with Model Catalysts Made by Colloidal Lithography

CO Oxidation on Pt/alumina and Pt/ceria

All ex situ prepared catalysts must be activated prior to experiments to remove residual contamination from the preparation procedure and hydrocarbons from ambient air exposures. Such activations normally include an oxidation step followed by a reduction step. The pretreatments may be complemented with a mild O_2 plasma or UV-ozone pretreatment step as a final preparation step of the nanofabrication procedure. Experience from flow reactor studies with polycrystalline Pt samples also suggests that reaction runs in $H_2 + O_2$ mixtures are effective in removing contaminants from the catalyst surface [152]. With plasma or ozone treatments, sample oxidation may cause significant chemical modifications of the catalysts, and additional reduction steps may therefore have to be employed, such as H_2 plasma or high-temperature H_2 treatment (see "Characterization of Model Catalysts").

CO oxidation experiments were performed on Pt/alumina and Pt/ceria model catalysts, prepared by colloidal lithography. The samples were prepared without any of the additional plasma or UV-ozone pretreatments steps described in "Preparation Procedures. Figure 4.38 shows T_{50} (the temperature at which 50% of reactant conversion is reached) and E_a^* (the apparent activation energy) as a function of CO oxidation cycle (ramping up and down in temperature). It is seen that both T_{50} and $E^*{}_a$ initially shifts up during

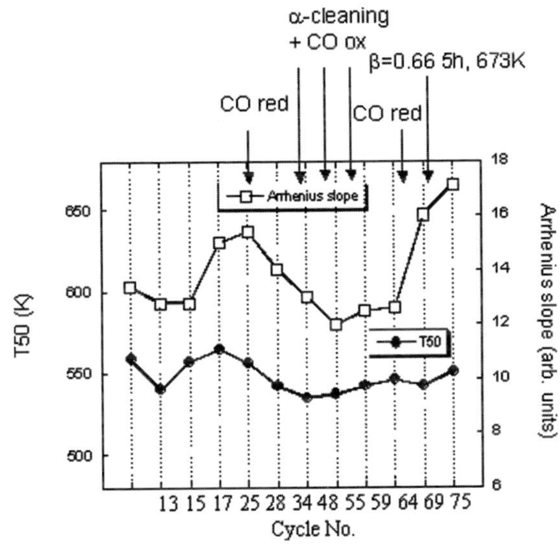

Fig. 4.38. The effects of various pretreatments (oxidative and reductive) on CO oxidation on a 40-nm Pt/ceria model catalyst prepared by colloidal lithography as measured by the temperature of 50% of CO conversion and the apparent activation energy from the Arrhenius plot. CO reduction was made in 0.5% CO for 1 h at 573 K, H_2 oxidation (α-treatment) was done at $\alpha \equiv P_{H_2}/(P_{H_2} + P_{O_2}) = 0.33$ at 573 K for 1 h, and finally $\beta =$ CO oxidation (β-treatment) was done in the O-rich regime (oxidative conditions), $\beta = P_{CO}/(P_{CO} + P_{O_2}) = 0.2$ with 0.3% CO and 1.2% O_2 at temperatures between 300 and 673 K. It is seen that reduction leads to a lower T_{50} and activation energy, while sustained CO oxidation leads to an increase of the activation energy, which is not recovered by reductive treatments. The latter is explained in terms of strong-metal–support interactions (SMSI) and particle reshaping (see text)

CO oxidation in O_2 excess on the fresh model catalyst. XPS analysis indicates stabilization of Pt oxide in the case of Pt/ceria [98]. In contrast, similar measurements on Pt/alumina show that these Pt particles are metallic and the chemical state does not change significantly after several CO oxidation cycles in O_2 excess [98]. The long-term lean CO oxidation experiments on Pt/ceria indicate that the catalyst activity gradually changes as a function of CO oxidation experiment time. Extensive CO reduction causes an up-shift of both T_{50} and $E^*{}_\mathrm{a}$, which is attributed to C deposition on the catalyst [9], probably via CO disproportionation at CO excess. A lower value of T_{50} and $E^*{}_\mathrm{a}$ can, however, be *restored* by running a H_2 oxidation over the catalyst (cycles 5–7) around stoichiometric conditions ($\alpha \equiv P_{\mathrm{H}_2}/(P_{\mathrm{O}_2} + P_{\mathrm{H}_2}) \approx 0.67$). The influence of the pretreatments on the catalytic activity decreases substantially as a function of cycle number. After extended CO oxidation in a stoichiometric gas mixture at 673 K, the shape of the conversion curve changed significantly, which is manifested in an increased $E^*{}_\mathrm{a}$ (Fig. 4.38). These results can be explained by a strong metal–support interaction (SMSI) occurring on Pt/ceria, which is particularly evident by the accompanying structural changes of the Pt/ceria particles. The reaction-induced restructuring is discussed further in "Reaction-Induced Restructuring on Pt/ceria". The Pt/alumina system is not sensitive to reduction treatments, as described above (no C deposition). Similarly, no *long-term* ageing effect is observed as for Pt/ceria. The overall results from the CO oxidation measurements on the Pt/alumina and Pt/ceria model catalysts are in good agreement with previous studies on wet-impregnated catalysts [20, 153, 154]. Ceria is shown to promote the low-temperature reactivity, which is manifested in a lower $T_{50,\mathrm{up}}$(ca. 50 K). Both observations can be explained by the well-known oxygen storage capacity of ceria and/or interfacial Pt–Ce–O species formation [9], which is also in agreement with the observed stabilization of PtO complexes.

CO Electrooxidation on Pt/C Electrocatalysts

By varying the particle size, shapes, separation, and support on planar model electrocatalysts, the influence of these properties on the electrocatalytic reactions, e.g., on fuel cell electrodes, can be evaluated systematically. Some new challenges arise, such as the adhesion of the catalyst particles on new types of support materials (e.g., glassy carbon). However, most of the procedures and concepts of preparation and characterization are the same as in heterogeneous catalysis and photocatalysis.

Here, we give an example where the electrochemical and electrocatalytic properties of nanostructured Pt catalysts supported on glassy carbon, prepared with colloidal lithography, were tested with oxidation of preadsorbed CO, so-called *CO-stripping*, as a test reaction [99]. The Pt nanoparticles were prepared as described in Fig. 4.11, except that the polystyrene particles were removed by UV-ozone treatment. The particles had an average diameter of 122 ± 11 nm and were about 40 nm high. There was no indication (by AFM

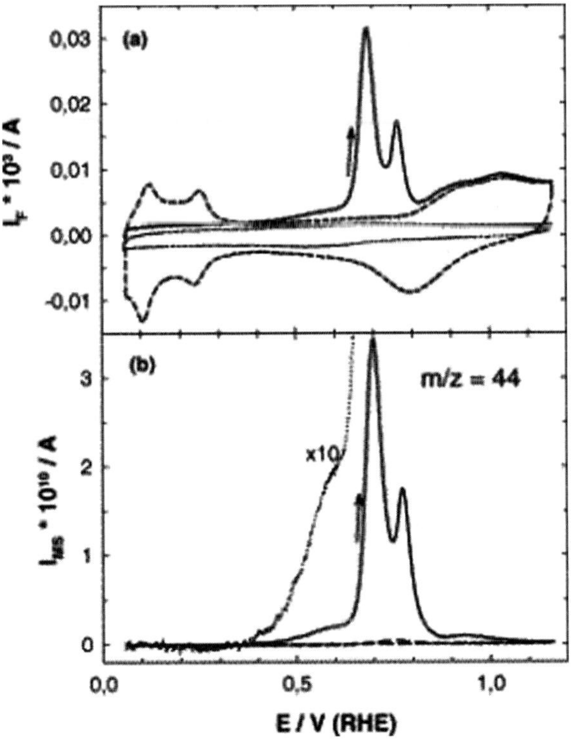

Fig. 4.39. Potentiodynamic CO stripping (*full line*) and subsequent base voltammogram (scan speed 10 mV s^{-1}, *dashed line*) on a nanostructured Pt/C model catalyst and, for comparison, base voltammogram recorded on a bare, Pt-free glassy carbon disk (*dotted line*) in 0.5 M H_2SO_4 solution. Prior to the measurement possible hydrocarbon contaminants had been removed by potential cycles in 0.5 M H_2SO_4 solution (multiple scans to 1.16 V), then the sample was saturated with CO at 0.06 V, and finally, the electrolyte was exchanged against CO_{free} base electrolyte. (**a**) Faradaic current and (**b**) mass spectrometric current at $m/z = 44$ (*dotted line*: signal multiplied by a factor of 10) (from [99])

or SEM imaging or XPS) that the electrochemical treatments induced any change in the morphology or chemical composition of the nanoparticles, as described in "Reaction-Induced Restructuring on Pt/ceria", with H_2/O_2 treatments. The potential was kept below 1.16 V during all cleaning and analytical potential sweeps in order to prevent electrochemical annealing. Figure 4.39 shows both the cyclic voltammogram from the electrochemical oxidation of a preadsorbed monolayer of CO and the accompanying CO_2 ($m/z = 44$) signal recorded by on-line mass spectrometry. The dashed line in Fig. 4.39a shows the base voltammogram from the Pt/C nanocatalyst without preadsorbed CO. The initial peaks at low potential are characteristic for the underpotential deposition of H from the electrolyte (0.5 M H_2SO_4). This H-uptake is completely suppressed by the preadsorbed CO indicating a saturated CO adlayer. The CO stripping doublet peak is centred at approximately 0.68 and 0.78 V, and is characteristic for polycrystalline Pt electrodes. This doublet can be explained by CO oxidation at different facets of low index planes found on the polycrystalline surface.

The mass spectrum in Fig. 4.39b reflects the rate of CO oxidation. The enlargement of the 0.3–0.6 V region indicates that the CO electrooxidation starts at about 0.3 V, and this can be explained by the initial oxidation of weakly bound CO (a) species, reacting with adsorbed OH species at defect sites. Further experiments with potentiostatic CO stripping transients also revealed behaviour similar to polycrystalline Pt surfaces. Thus, the conclusion was that Pt/C model electrocatalysts are viable for electrocatalytic studies. Further work is continuing on the electrochemical effects of Pt particle stability and morphology, and the effects on the electrochemical kinetic and transport behaviour with varying of particle size and particle separation. Particularly interesting is to explore smaller particles, where the relative number of edge atoms increase and where the particles' susceptibility to morphological changes during the electrochemical treatment may be larger.

Reaction-Induced Restructuring on Pt/ceria

Colloidal lithography has been used to study reaction-induced restructuring in $H_2 + O_2$ gas mixtures similar to those shown for EBL model catalysts discussed above. The principal advantage of the well-defined 2D geometry is demonstrated, which provides access of analytical tools (here microscopy) and give a simplified description of the two-phase system that can be connected to the macroscopic response (reported in "CO Oxidation on Pt/alumina and Pt/ceria"). In Fig. 4.40, the appearance of a sample with 107-nm Pt particles on ceria is shown, before and after reactions in a $H_2 + O_2$ mixture at 973 K for 2 h. As expected, the Pt particle shapes change and appears facetted. In addition, it is seen that adjacent particles merge into larger particles. In

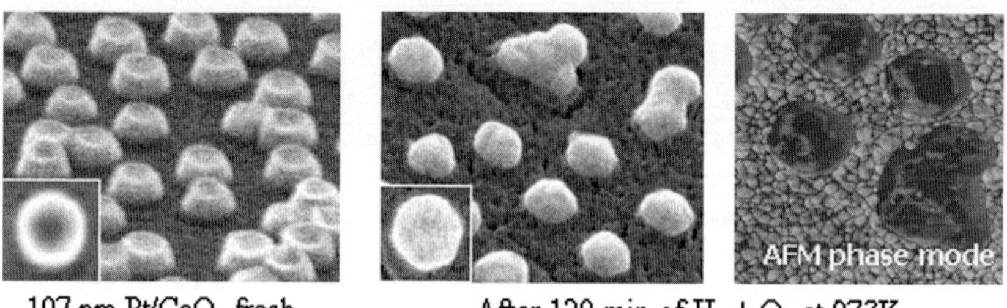

Fig. 4.40. 107-nm Pt/CeO$_x$ model catalysts prepared by colloidal lithography both before and after a H_2 oxidation reaction ($P_{H_2}/(P_{H_2} + P_{O_2}) = 0.67$) at 973 K for 2 h (nonflammable mixture in Ar)

view of the results presented below, it cannot be excluded that ceria suboxide species are present on the Pt particles after reaction. Furthermore, it is seen that the ceria support becomes granular, with cracks building up in the ceria layer. The granular ceria structure on the aged Pt/ceria catalyst is evident from the AFM measurements depicted in Fig. 4.40c. A similar change of ceria following catalytic reactions is evident on the EBL samples presented above.

The so-called SMSI is a name for an effect to explain why metal particles supported on certain reducible oxides (such as titania and ceria) show unique catalytic behavior unlike those seen with more inert supports [9,155]. This was initially though to originate from an electronic effect, whereby charge transfer from the oxide support to metal particle atoms led to a chemical modification of the metal atoms; hence the name SMSI. Studies on model oxides have played a crucial role in disclosing the nature of the SMSI effect. Today, there is a consensus that the SMSI is (at least in many cases) a rather inappropriate name, since the effect involves mobile suboxides that form on the support or at the particle–support boundaries augmented by chemical interfacial interactions [9]. These oxide species migrate onto the metal particles and lead to a decrease of the number of metal surface sites or even complete encapsulation of metal particles. Covering a metal with mobile oxide species is commonly associated with a decrease of catalyst activity, however rate enhancement may also occur, as in hydrogenation of C–O containing molecules. The latter has been inferred from model studies of *inverted* model catalysts, where an oxide is deposited on a metal surface [156].

Figure 4.41 shows SEM images of aged 40-nm Pt/ceria and Pt/alumina catalysts, used in the CO oxidation experiments described above. In Fig. 4.41a

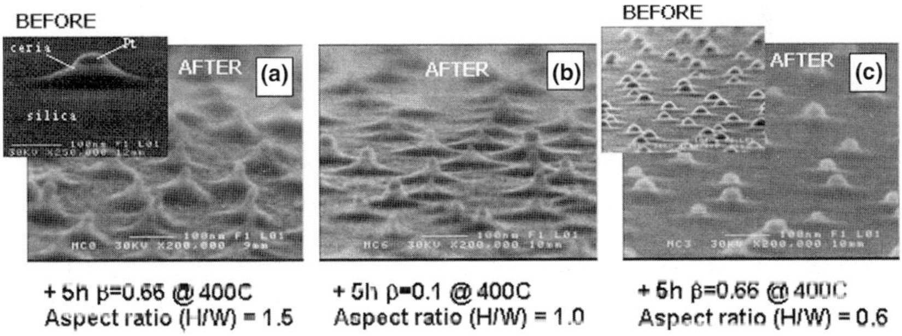

Fig. 4.41. EM images of the before (insets) and after reaction 40-nm Pt/ceria and Pt/alumina model catalysts. (**a**) Pt/ceria after 5 h in $\beta = P_{CO}/(P_{CO} + P_{O_2}) = 0.67$ (stoichiometric) at 673 K; (**b**) Pt/ceria after 5 h in $\beta = 0.1$ (O-rich) at 673 K; (**c**) 40-nm Pt/alumina after 5 h in $\beta = 0.66$ at 673 K. It is seen that on Pt/ceria the prolonged stoichiometric CO oxidation ($\beta = 0.67$) leads to pronounced restructuring, with a height-to-width ratio much larger than expected from thermodynamics (Wulff shape). On Pt/Al_2O_3, the reaction induces a reshaping that approaches the thermodynamically predicted H/W-ratio of 0.6 (from [98])

is shown the 40-nm Pt/ceria sample subjected to the history given in Fig. 4.38. The original hemispherical particles have reshaped into pillars, and the underlying ceria islands have also changed shape. The aspect ratio (height-to-width ratio, H/W) is $H/W \sim 1.5$ for the pillar structures as shown in Fig. 4.41a, compared to ~0.4 for the fresh 40-nm Pt/ceria sample (inset in Fig. 4.41a). In Fig. 4.41b, an SEM image is shown of a 40-nm Pt/ceria sample, which has been subjected to a similar history as the sample in Fig. 4.41a, except that in this case a nonstoichiometric gas mixture was used. The latter particles acquire the same qualitative form, but here H/W is only ~1.0. The Pt/alumina catalyst does not change the structure as dramatically as Pt/ceria (Fig. 4.41c). This is in good agreement with the reactivity results described in "CO Oxidation on Pt/alumina and Pt/ceria", where it was noted that the reactivity of a Pt/alumina model catalyst exposed to same reaction history as the Pt/ceria catalyst in Fig. 4.41a does not deteriorate significantly. A slight reshaping is observed with some sharp corners visible in SEM, suggesting faceting; the H/W ratio is increased from ~0.4 to ~0.6 compared to the as-prepared Pt/alumina catalyst.

It is well known that Pt interacts strongly with ceria (much stronger than with alumina) [9]. Previous studies have demonstrated an SMSI effect of Pt/ceria after H_2 reduction at high temperatures ($T \sim 1{,}000\,K$) whereby Pt–Ce–O species decorate the Pt particles [157]. The comparatively low temperatures employed in the present study along with the net oxidizing conditions suggest that SMSI would not be present. On the other hand, a decrease of the XPS Pt 4f peak intensity (Pt/Ce concentration ratio) on the aged sample compared to the fresh sample, suggests that the particles are encapsulated by, e.g., a ceria layer. Moreover, a small chemical shift between the fresh and the aged sample is also evident in XPS. Finally, as described in "CO Oxidation on Pt/alumina and Pt/ceria," $E^*{}_a$ increases substantially after the stoichiometric CO oxidation pretreatment, while the T_{50} value is not affected so much. This is in agreement with previous findings of CeO_2/Pt(111), where a fully covered Pt surface was found to maintain a high reaction rate, though with a higher $E^*{}_a$ compared to the bare Pt [158]. The form of the Pt/ceria particles must also include some Pt metal restructuring. The pillars are thus not due (solely) to ceria growing out from the particles. The Pt reshaping is promoted and accentuated by running the CO oxidation at stoichiometry. Similar to the arguments put forward in "Reaction-Induced Restructuring", it is believed that oxygen acts as a surfactant for Pt metal redistribution. Performing the reaction near stoichiometry means that the CO + O surface reaction effectively cleans the surface from adsorbates and simultaneously prevents it from becoming (fully) oxidized. In contrast, Pt/alumina particles exposed to the same reaction condition leads to crystallization where the particle shape is mainly governed by the substrate surface energy and the Pt-alumina interface energy. This is expected, since the interface energy is larger for Pt/ceria than for Pt/alumina. The overall form of the Pt/alumina particles agree with the Wulff relation, and an H/W ratio of ~0.6 is about twice the corresponding values observed for small (~3 nm) Pd/Al_2O_3 [159], suggesting a weaker interaction with the support, i.e., a larger effective surface energy, and hence a smaller work of adhesion.

4.6 Summary and Future Directions

The results presented here demonstrate that well-defined model catalysts with particle sizes in the 10–100-nm regime can be efficiently processed with lithographic methods. These model catalysts possess most of the desirable characteristics that are required in fundamental heterogeneous catalysis research. There are considerable advantages, compared to model catalysts prepared by, e.g., incipient wetness impregnation methods, which are enumerated:

1. Control of particle mass and dimensions (size and shape – especially in the case of EBL).
2. Control of interparticle distances.
3. Control over the lateral dimensions of an active support (capture zone, spillover distances).
4. Simplifications in the physical and chemical characterization; the 2D nature of the nanofabricated model catalysts make them easier to inspect with traditional surface science tools and electron microscopy.
5. Simplifications in kinetic modeling; with monodisperse particle sizes and separations, and the detailed characterization offered via [4]; more systematic studies of the influence of these parameters can be made, and the associated kinetic modeling is then simplified and becomes more accurate. Furthermore, the active catalyst phase is in direct contact with the reacting gas [e.g., no pore diffusion needs to be accounted for, which simplifies mass transport calculations (when needed)].
6. Studies of reaction-induced structural changes (sintering, dispersion changes, crystallization, particle-shape changes, etc.) are greatly simplified by combining nanofabricated model catalysts and prefabricated TEM windows. The latter even allows in situ real-time recording of such effects.

Having noted these advantages, it should also be said that the area is in an early stage. Improvements will occur through a combination of instrument development for fabrication (e.g., EBL) and characterization (e.g., small-spot AES, approaching 10 nm, and scanning probe techniques), and in the actual processing steps. The latter includes both the nanofabrication processing steps as such and the after-treatment (activation) of catalysts to make them to either attain their energetically most stable shapes (the Wulff shape), if this is desired, or alternatively to stabilize metastable shapes and structures. Another methodological improvement, necessary in the future, involves further development of the cleaning procedures to remove contaminants from the processing steps.

The model catalysts discussed here are still less well defined than the corresponding single crystal catalysts, both regarding chemical purity and structural properties. At the same time, they are compared with practical catalysts on porous supports, easier to characterize and can be varied more systematically with respect to chemical and geometric/structural properties. In this way, they bridge the gap between *real* catalysts and idealized model systems. The experimental case studies of reactivity measurements presented above show that qualitative and quantitative new information is obtained with such catalysts. It is thus tempting to say *yes* to the question raised in Fig. 4.1, i.e., that nanofabricated catalysts indeed can help to bridge the structure gap in heterogeneous catalysis. However, a drawback with methods that rely on elaborate pattern transfer processing with accompanying development, liftoff and cleaning steps, is the risk of embedding impurities in the final catalyst structure – we cannot stress this important aspect enough. Tedious pretreatment schemes and characterization procedures are instrumental to eliminate contamination-induced effects. While some standard pretreatment schemes apply in many cases, the different nanofabrication methods usually require their individual solutions, which make it hard to device general cleaning recipes. The most promising route in lithographic nanocatalysis has therefore been to adopt a parallel process development and reactivity measurement scheme with iterative feedback between the two. The obvious drawback in this scheme is that it is rather time consuming (accentuated by the heavy instrumentation that is required in the fabrication process), which, in turn, depends on persistent and long-term commitments. It is, however, expected that routine methods will be developed, that systematically cover both fabrication and system-specific aspects analogous to those that have been developed in UHV for sputtering and cleaning of single crystals or in microelectronic industry for etching, deposition, and cleaning steps.

Future developments in this field will undoubtedly very soon see methods and instrumentation improvements that will take down the attainable catalyst structures to the 1- to 10-nm range on a routine basis. This is the size most particles in (fresh) *real* catalysts have. Catalyst particles in this size range enter the regime, where the band structure picture of extended solids break down, and truly new size related phenomena/reactivities occur [6,21–23,160]. In this size regime, the lithographic samples will complement those made by, e.g., PVD [6,15,30], self-assembly methods [33,34], and cluster deposition methods [161]. Questions that will be addressed include relationships between reactivity and electronic structure for different particle sizes and shapes, the catalytic activity of different edge and perimeter (against the support) sites, spillover effects [28, 29] and interparticle communication [4, 148], the influence of size on kinetic phenomena, such as interfacet communication on a single particle [4, 148], bistability and kinetic phase transitions [4, 146, 148] and oscillatory, and chaotic phenomena versus particle size [77, 162]. In the latter area, the type of model catalysts described here can also be used to study synchronization [11,27] of oscillations on adjacent particles and particle arrays.

Beyond single component metal catalysts, the nanofabricated model catalysts can be used to study alloy catalysts, with compositions controlled by co-evaporation from two or more PVD sources. Alternatively, arrays of alternating particles or areas of two different materials can be made to study lateral communication between two types of catalysts at the nanoscale. For example, sequential reactions consisting of a first step on one type of catalyst and a second step on another catalyst particle could be studied systematically. The role of reactants and reaction intermediates as surfactants, affecting particle shape and morphology [163], will be possible to study in detail by in situ TEM studies in reactive environments.

All these and many other similar studies will help increasing the fundamental knowledge of catalytic phenomena at the nanoscale, and thereby ultimately lead to design of improved catalysts, according to the (iterative) loop depicted in Fig. 4.5. One type of study will involve refined studies of phenomena already known or suspected to occur on real catalysts, but which are difficult to nail down because of the structural and chemical heterogeneity of the real catalysts. On the other hand, completely new phenomena may also be discovered just because the model catalysts are simplified and do not obscure complex phenomena in the same way as real catalysts.

An important point to emphasize is that some of the methods used to make the model catalysts, like EBL, are so expensive that they are unlikely to be used to make real catalysts, at least in a foreseeable future, except possibly in high-value products, like catalytic sensors. The merit of the technique is that it allows detailed scrutiny of what type of catalyst is best for a particular application. This insight can then be used to develop cheaper or better catalyst with the same end result. The latter *translation* is much easier if one knows exactly, e.g., what composition or structure one is aiming at.

A prerequisite for the development indicated above to occur, is a parallel development in instrumentation to facilitate both physical and chemical characterization. TEM and SPM based methods will continue to play a central role in this development, since they possess the required nanometer (and subnanometer) spatial resolution. Optical spectroscopy using reflection adsorption infrared spectroscopy (RAIRS), polarization modulation infrared adsorption reflection spectroscopy (PM-IRRAS), second harmonic generation (SHG), sum frequency generation (SFG), various in situ X-ray absorption (XAFS) and X-ray diffraction spectroscopies (XRD), and maybe also surface enhanced Raman scattering (SERS), etc., will play an important role when characterizing adsorbates on catalyst surfaces under reaction conditions. Few other methods fulfill the requirements of being able to operate over a wide pressure gap (to several atmospheres) and to be nondestructive.

Reactivity measurements are in most cases possible with standard techniques (mass spectrometry, gas chromatography, FTIR, or other specific detectors), which are available today. New preparation methods will probably allow for scaling up possibilities to fabricate large enough active sample areas for these methods to be applicable in a standard manner. However, in the meantime, when this is not possible, better microreactor solutions than available today are needed. The coupled sample–microreactor system constitutes an interesting optimization problem; the larger area of active catalyst, that one can make on the nanofabricated sample, the smaller is the demand on the microreactor system and vice versa. Considering a sample consisting of an array of $1,000 \times 1,000$ particles, each with a width of 10-and 50-nm separation between the particles (corresponding to an active catalyst area of about $100\,\mu m^2$). A 10-nm particle has of order 10^3 surface atoms; hence the whole array amounts to 10^{12} sites. Assuming a TON of 10^{-2} site^{-1} s^{-1}, we would have a reaction product generation rate of order 10^7 molecules s^{-1}, which is not trivial but still possible to detect with fairly standard schemes. In addition to these requirements of detection sensitivity, such reactors put strong demands on handling very small gas flows and volumes, with accompanying impurity problems from the inlet gas and adsorbed molecules on reactor walls among other potential issues.

Finally, we comment briefly on extensions of the nanofabrication approach to photocatalysis. The techniques and strategies outlined above for making and evaluating nanofabricated, supported model catalysts are applicable, in almost every respect, to photocatalysis with optically active nanoparticles. One of the basic ideas in nano-photocatalysis is to use the unique, size- and shape-dependant optical properties of nanoparticles [108, 164–166] to effectively generate electron–hole (e–h) pairs, which can drive surface reactions on the nanoparticles via the interaction of hot photogenerated electrons (holes) with the LUMO (HOMO) orbitals of adsorbed reactants. The overall efficiency of the process is a combination of: (a) e–h pair generation, which might be a direct interband transition or amplified by nanoparticle plasmon resonances, (b) transport of hot electrons (holes) to the nanoparticle surface, and (c) attachment of the electron (hole) to the LUMO (HOMO) orbital of the adsorbate to be activated. It is obvious that optimization of this process requires systematic studies of the influence of particle size, shape, separation, geometric arrangement, and particle material and support. Note also that most (and in some cases all) of these properties affect the spectral response. In this context it is interesting to note that the nanoparticle plasmon resonances are not restricted to the most studied materials Au and Ag, but are also well developed in Pt and Pd [167].

Acknowledgments

The authors are grateful to many colleagues worldwide in the field of heterogeneous catalysis for exchange of ideas concerning the nanocatalysis approach and model catalyst fabrication. In particular, we are thankful to J. Libuda, H.-J. Freund and coworkers at the Fritz-Haber Institute in Berlin, Germany, who did the reactivity measurements on the EBL nanofabricated Pd/SiO_2

samples presented in "CO Oxidation on Pd/SiO_2: Molecular Beam Studies"; S. Helveg, B.S. Clausen and coworkers at Haldor Topsøe A/S in Lyngby, Denmark, who did HR-TEM imaging of the vapor-deposited Pt/SiO_2 and some of the EBL Cu/SiO_2 nanofabricated model catalysts on TEM windows presented in Sects. 4.2 and 4.3; J. Behm and coworkers at the University of Ulm, Germany, for the electrocatalytic evaluation of the nanofabricated Pt/C model catalysts presented in "CO-Electrooxidation on Pt/C Electrocatalysts". We acknowledge fruitful discussions with F. Besenbacher, G. Ertl, W. Goodman, C. Henry, J. Norskøv, H. Topsoe, G. Somorjai, and V.P. Zhdanov. Finally, we recognize that this work should not have been accomplished without diligent colleagues and students at the Department of Applied Physics and many others at Chalmers. L.Ö. and A.G. were financially supported by the Competence Center for Catalysis (KCK), Chalmers University of Technology, and the Swedish Defence Research Agency (FOI) in various parts of this work.

References

1. Ertl G, Knözinger H, Weitkamp J (eds.) (1997) Handbook of Heterogeneous Catalysis, vols. 1–4. Wiley, Weinheim, Germany
2. Somorjai GA (1994) Introduction to Surface Chemistry. Wiley, New York
3. Freund HJ, et al. (2001) Top. Catal. 15:201
4. Zhdanov VP, Kasemo B (2000) Surf. Sci. Rep. 39:29
5. Niemantsverdriet JW, Ribeiro FHE (2000) Top. Catal. 13:1
6. Henry CR (1998) Surf. Sci. Rep. 31:235
7. Österlund L, et al. (2001) Phys. Rev. Lett. 86:460
8. Verstergaard EK, et al. (2002) Phys. Rev. Lett. 88:259601/1
9. Trovarelli A (ed.) (2002) Catalysis by Ceria and Related Materials, vol. 2. Imperial College Press, London
10. Miyoshi N, et al. (August 1995) Paper presented at the SAE
11. Persson H, Thormahlen P, Zhdanov VP, Kasemo B (1999) J. Vac. Sci. Technol. A 17:1721
12. Geissler M, Xia YN (2004) Adv. Mater. 16:1249
13. Hammer B, Nørskov JK (2000) Adv. Catal. 45:71
14. Henrich VE, Cox PA (1994) The Surface Science of Metal Oxides. The Cambridge University Press, Cambridge
15. Campbell CT (1997) Surf. Sci. Rep. 30:227
16. Weiss W, Schlögl R (2000) Top. Catal. 13:75
17. Over H, Seitsonen AP (2002) Science 297:2003
18. Lauritsen JV, et al. (2001) J. Catal. 197:1
19. Ertl G (1990) Adv. Catal. 37:213
20. Zafiris GS, Gorte RJ (1993) J. Catal. 140:418
21. Haruta M, et al. (1993) J. Catal. 144:175
22. Chen MS, Goodman DW (2004) Science 306:252
23. Valden M, Lai X, Goodman DW (1998) Science 281:1647
24. Lopez N, et al. (2004) J. Catal. 223:232
25. Zanella R, Giorgio S, Shin CH, Henry CR, Louis C (2004) J. Catal. 222:357

26. St Sinclair TP, Goodman DW (2000) Top. Catal. 13:5
27. Zhdanov VP (2001) Phys. Rev. B 6419:193406
28. Zhdanov VP, Kasemo B (1998) Appl. Surf. Sci. 135:297
29. Johansson S, Österlund L, Kasemo B (2001) J. Catal. 201:275
30. Gunter PLJ Niemantsverdriet JW, Ribeiro FH, Somorjai GA (1997) Catal. Rev. Sci. Eng. 39:77
31. Freund HJ, Bäumer M, Kuhlenbeck H, (2000) Adv. Catal. 45:333
32. Henrich VE, Cox PA (1993) Appl. Surf. Sci. 72:277
33. Shah DO (ed.) (1998) Micelles Microemulsions and Monolayers: Science and Technology. Marcel Dekker, New York
34. Service RF, Szuromi P, Uppenbrink J (2002) Science 295:2395
35. Möller M, Spatz JP (1997) Curr. Opin. Colloid Interface Sci. 2:177
36. Spatz JP, Mössmer S, Hartmann C, Möller M (2000) Langmuir 16:407
37. Wei BQ, et al. (2002) Nature 416:495
38. Ingelsten HH, et al. (2002) Langmuir 18:1811
39. Moore GE (1965) Electronics 38:114
40. Krauth AC, Lee KH, Bernstein GH, Wolf EE (1994) Catal. Lett. 27:43
41. Zuburtikudis I, Saltsburg H (1992) Science 258:1337
42. Kielbassa S, Kinne M, Behm RJ (2004) Langmuir 20:6644
43. Smith HI, et al. (1996) Microelectron. Eng. 32:143
44. Schildenberger M, et al. (1998) Catal. Lett. 56:1
45. Schildenberger M, Bonetti Y, Gobrecht J, Prins R (2000) Top. Catal. 13:109
46. Greyson EC, Babayan Y, Odom TW (2004) Adv. Mater. 16:1348
47. Eigler DM, Schweizer EK (1990) Nature 344:524
48. Gimzewski JK, Joachim C (1999) Science 283:1683
49. McIntyre BJ, Salmeron M, Somorjai GA (1994) Science 265:1415
50. Maoz R, Cohen SR, Sagiv J (1999) Adv. Mater. 11:55
51. Brandow S, et al. (1997) J. Vac. Sci. Technol. B 15:1818
52. Klehn B, Kunze U (1999) J. Appl. Phys. 85:3897
53. Muller WT, et al. (1995) Science 268:272
54. Piner RD, Zhu J, Xu F, Hong S, Mirkin CA (1999) Science 283:661
55. Demers LM, et al. (2002) Science 296:1836
56. Hong S, Mirkin CA (2000) Science 288:1808
57. Wada Y (1998) Microelectr. J. 29:601
58. Chou SY, Keimel C, Gu J (2002) Nature 417:835
59. Odom TW, Love JC, Wolfe DB, Paul KE, Whitesides GM (2002) Langmuir 18:5314
60. Kuo CW, Shiu JY, Chen PL, Somorjai GA (2003) J. Phys. Chem. B 107:9950
61. Choi YK, Zhu J, Grunes J, Bokor J, Somorjai GA (2003) J. Phys. Chem. B 107:3340
62. Choi YK, et al. (2003) J. Vac. Technol. A 21:2951
63. Gambardella P, et al. (2001) Phys. Rev. Lett. 87:056103
64. Dahl S, et al. (1999) Phys. Rev. Lett. 83:1814
65. Kato N, Nakagaki K, Ina N (February 1996) Paper presented at the SAE
66. Jacobs PW, Ribeiro FH, Somorjai GA, Wind SJ (1996) Catal. Lett. 37:131
67. Yang MX, Gracias DH, Jacobs PW, Somorjai GA (1998) Langmuir 14:1458
68. Grunes J, Zhu J, Somorjai GA (2002) J. Phys. Chem. 106:11463
69. Grunes J, Zhu A, Somorjai GA (2003) Chem. Commun. :2257
70. Grunes J, Zhu J, Yang MC, Somorjai GA (2003) Catal. Lett. 86:157

71. Somorjai GA, Yang MC (2003) Top. Catal. 24:61
72. Wong K, Johansson S, Kasemo B (1996) Faraday Discuss. 105:237
73. Broers AN, Hoole ACF, Ryan JM (1996) Microelectronic Eng. 32:131
74. Tennant DM, Fullowan R, Takemura H, Isobe M, Nakagawa Y (2000) J. Vac. Sci. Technol. B 18:3089
75. Rai-Choudhury P (1997) In: Handbook of Microlithogarphy, Micromachining, and Microfabrication, vol. 1. SPIE Optical Engineering Press, Washington
76. Brodic I, Muray JJ (1992) In: The Physics of Micro/Nanofabrication. Plenum, New York
77. Kasemo B, Johansson S, Persson H, Thormahlen P, Zhdanov VP (2000) Top. Catal. 13:43
78. Ferry DK, et al. (1996) Semicond. Sci. Technol. 11:1552
79. Liu K, Avouris P, Bucchignano J, et al. (2002) Appl. Phys. Lett. 80(5):865
80. Tsirlin T, Zhu J, Grunes J, et al. (2002) Top. Catal. 19:165
81. Vieu C, Carcenac F, Pepin A, et al. (2000) Appl. Surf. Sci. 164:111.
82. Johansson S, Wong K, Zhdanov VP, Kasemo B (1999) J. Vac. Sci. Technol. A 17:297
83. Laurin M, et al. (2004)
84. Rinnemo M, et al. (1997) Surf. Sci. 376:297
85. Johansson S, Fridell E, Kasemo B (2000) J. Vac. Sci. Technol. A 18:1514
86. Eppler AS, Zhu J, Anderson EA, Somorjai GA (2000) Top. Catal. 13:33
87. Chen M, Wang T, Schmidt LD (1979) J. Catal. 60:356
88. Hanarp P, Sutherland D, Gold J, Kasemo B (1999) Nanostruct. Mater. 12:429
89. Hanarp P, Sutherland DS, Gold J, Kasemo B (2001) J. Colloid Interface Sci. 241:26
90. Hanarp P, Sutherland DS, Gold J, Kasemo B (2003) Colloid Surf. A 214:23
91. Werdinius C (2002) Ph.D. Thesis, Chalmers University of Technology and Göteborg University
92. Werdinius C, Österlund L, Kasemo B (2003) Langmuir 19:458
93. Burmeister F, et al. (1999) Appl. Surf. Sci. 145:461
94. Haginoya C, Ishibashi M, Koike K (1997) Appl. Phys. Lett. 71:2934
95. Hulteen JC, et al. (1999) J. Phys. Chem. B 103:3854
96. Trau M, Saville IA, Aksay IA (1996) Science 272:706
97. Dimitrov AS, Nagayama K (1996) Langmuir 12:1303
98. Österlund L, Kielbassa S, Werdinius C, Kasemo B (2003) J. Catal. 215:94
99. Gustavsson M, et al. (2004) J. Electroanal. Chem. 568:371
100. Feder J, Giaever I (1980) J. Colloid Interface Sci. 78:144
101. Adamczyk Z, Zembala M, Siwek B, Warszynski P (1990) J. Colloid Interface Sci. 140:123
102. Evans JW (1993) Rev. Mod. Phys. 65:1281
103. Talbot J, Tarjus G, Van Tassel PR, Viot P (2000) Colloids Surf. A 165:287
104. Senger B, Voegel JC, Shaaf PD (2000) Colloids Surf. A 165:255
105. Adamczyk Z, Weronski P (1998) J. Chem. Phys. 108:9851
106. Semmler M, Mann EK, Ricka J, Borkovec M (1998) Langmuir 14:5127
107. Sutherland D (unpublished results)
108. Hanarp P, Kall M, Sutherland DS (2003) J. Phys. Chem. B 107:5768
109. Whitney AV, Myers BD, Van Duyne RP (2004) Nano Lett. 4:1507
110. Ormonde AD, Hicks ECM, Castillo J, Van Duyne RP (2004) Langmuir 20:6927
111. Bouffard S, Balanzat E, Leroy C, Busnel JP, Guevelou G (1997) Nucl. Instrum. Meth. B 131:79

112. Kubota T, et al. (2004) Appl. Phys. Lett. 84:1555
113. Dubas ST, Schlenoff JB (1999) Macromolecules 32:8153
114. Maya L, Riester L, Thundat T, Yust CS (1998) J. Appl. Phys. 84:6382
115. Thomas JM, Terasaki O (2002) Top. Catal. 21:155
116. Hansen TW, et al. (2001) Science 294:1508
117. Hansen PL, et al. (2002) Science 295:2053
118. Poppa H (1993) Catal. Rev. Sci. Eng. 35:359
119. José-Yacamán M, Avalos-Borja M (1992) Catal. Rev. Sci. Eng. 34:55
120. Poppa H (2000) Top. Catal. 13:139
121. Thomas JM, Gal PL (2004) Adv. Catal. 48:171
122. Rupprechter G, Eppler AS, Somorjai GA (1998) In: Benavides EHAC, Yacamàn MJ (eds.), Electron Microscopy, vol. II. Institute of Physics Publishing, Bristol, p. 369
123. Enquist F, Spetz A (1986) Thin Solid Films 145:99
124. Grant AW, Hu QH, Kasemo B (2004) Nanotechnology 15:1175
125. Österlund L, Helvig S, Hu QH, Grant AW, Hansen PL, Clausen BS, Kasemo B (in preparation)
126. Madou MJ (2002) Fundamentals of Microfabrication: The Science of Miniturization, 2nd ed. CRC Press LLC, Boca Raton, FL, USA
127. Merlos A, Acero M, Bao MH, Baussell J (1993) Sens. Actuators A 37–38:737
128. Chen PH, Peng HY, Hsieh CM (2001) Sens. Actuators A 93:132
129. Shikida M, Masuda T, Uchikawa D, Sato K (2001) Sens. Actuators A 90:223
130. Wagner CD (1990) Auger and X-ray photoelectron spectroscopy. In: Seah MP (ed.), Practical Surface Analysis, vol. 1, 2nd ed. Wiley, West Sussix
131. Taylor TN, Butt DP, Pantano CG (1998) Surf. Interface Anal. 26:134
132. Grunwaldt J, Clausen BS (2002) Top. Catal. 18:37
133. Rider KB, Hwang KS, Salmeron M, Somorjai GA (2002) J. Am. Chem. Soc. 124:5588
134. Schmid AK, Bartelt NC, Hwang RQ (2000) Science 290:1561
135. Lööf P, Stenbom B, Norden H, Kasemo B (1993) J. Catal. 144:60
136. Horch S, Lorentsen HT, Helveg S (1999) Nature 398:134
137. Winterlin J, Volkening S, Janssens TVW, Zambelli T, Ertl G (1997) Science 278:1931
138. Zhdanov VP, Kasemo B (1994) Appl. Surf. Sci. 74:147
139. Zhu J, Somorjai GA (2001) Nano Lett. 1:8
140. Frank-Kamenetskii DA (1969) Diffusion and Heat Transfer in Chemical Kinetics, 2nd ed. Plenum, New York
141. Hendriksen BLM, Frenken JWM (2002) Phys. Rev. Lett. 89:046101/1
142. Li W, et al. (2004) Phys. Rev. Lett. 93:14104–1
143. Carlsson PA, Zhdanov VP, Kasemo B (2005) Appl. Surf. Sci. 239:424
144. Hoffmann J, et al. (2002) Chem. Phys. Lett. 354:403
145. Libuda J, Meusel I, Hartmann J, Freund HJ (2000) Rev. Sci. Instrum. 71:4395
146. Zhdanov VP, Kasemo B (2002) Surf. Sci. 496:251
147. Johanek V, et al. (2004) Science 304:1639
148. Zhdanov VP, Kasemo B (1997) J. Catal. 170:377
149. Pavlenko N, Imbihl R, Evans JW, Liu DJ (2003) Phys. Rev. E 68:12322
150. Johanek V, et al. (2004) Surf. Sci. 561:L218
151. Wynblatt P (1976) Acta Metal. 24:1175
152. Rinnemo M, Deutschmann O, Behrendt F, Kasemo B (1997) Combust. Flame 111:312

153. Serre C, Garin F, Belot G, Maire G (1993) J. Catal. 141:9
154. Törncrona A, Skoglundh M, Thormahlen P, Fridell E, Jobson E (1997) Appl. Catal. B: Environ. 14:131
155. Tauster SJ (1987) Acc. Chem. Res. 20:389
156. Boffa A, Lin C, Bell AT, Somorjai GA (1994) J. Catal. 149:149
157. Bernal S (1999) Catal. Today 50:175
158. Hardacre C, Ormerod RM, Lambert RM (1994) J. Phys. Chem. 98:10901
159. Hansen KH, et al. (1999) Phys. Rev. Lett. 83:4120
160. Freund HJ, Libuda J, Bäumer M, et al. (2003) Chem. Rec. 3:181
161. Heiz U, Schneider WD (2000) J. Phys. D 33:85
162. Zhdanov VP (2004) Catal. Lett. 93:135
163. Zhdanov VP, Kasemo B (1998) Phys. Rev. Lett. 81:2482
164. Haynes CL, McFarland AD, Zahao LL, et al. (2003) J. Phys. Chem. B 107:7337
165. Krenn JR, Dereux A, Weeber JC, et al. (1999) Phys. Rev. Lett. 82:2590
166. Gotschy W, Vonmetz K, Leitner A, et al. (1996) Appl. Phys. B: Lasers Opt. 63:381
167. Langhammer C, Yan Z, Zoric I, et al. (2006) Nano Lett. 6:833

5

Nanometer and Subnanometer Thin Oxide Films at Surfaces of Late Transition Metals

K. Reuter

5.1 Introduction

Confinement to nanoscale dimensions and the resulting quantization of electrons gives rise to two-dimensional quantum films, one-dimensional quantum wires, or zero-dimensional quantum dots. All three situations may exhibit properties and functionalities that make them different from the corresponding bulk material. In this respect, research on nanocatalysis is inspired by the expectation that due to such a reduced dimensionality of the active material new catalytic behavior can arise, which is not scalable from bulklike properties [1]. A salient target of such investigations is nanometer-size clusters of atoms, representing the three-dimensional confinement situation. While the potential of such clusters for catalytic applications is just being explored, it has already been recognized that they often exhibit the undesirable propensity to coarsen into larger aggregates on the support material. This problem could be less severe for the case of confinement in only one dimension, i.e., when striving for appealing properties of nanometer or subnanometer thin films. In fact, at the catalyst surface such films may even form automatically during the induction period, as a consequence of the exposure to the operating conditions. A prominent example of this is oxidation catalysis, where the oxygen-rich environment might oxidize the sample. Recent experimental and theoretical studies noted that particularly at late transition metals (TMs) few or even one atomic-layer thin, so-called surface oxide films are formed, when these are held in a realistic gas phase environment under catalytic or alike (T, p)-conditions. The vertical confinement makes these films distinct to surfaces of both bulk metals and bulk oxides, and their specific properties can have beneficial as well as detrimental effects on the desired functionality.

Appropriately tailoring and exploiting these functionalities requires an atomic-scale characterization and understanding of the oxidation process, as well as of the resulting surface composition and stoichiometry (under the steady state reaction conditions). Although oxide formation has been of technological and scientific interest since long, such an understanding is, however, only just emerging, and the underlying microscopic processes are unfortunately still largely unclear for everything that goes beyond submonolayer oxygen-adsorbate phases on low-index single crystal surfaces. This holds in particular for the transition from an oxygen-adsorbate layer to the surface oxide. A main reason behind this lack of understanding is that atomic-scale investigations of TM surfaces on the verge of oxide formation still pose a significant challenge for surface scientists: The need for rather high oxygen partial pressures (high compared to those possible in standard surface science experiments) and elevated temperatures required to initiate the process, a low degree of order at the surface, and often complex, large unit-cell geometries name but a few of the obstacles encountered. Since this often exceeds the analytical capabilities of one experimental technique alone, multi-method approaches (e.g., combining scanning, diffraction, and spectroscopic measurements) have become an important tool in this field. In addition, first-principles calculations, almost exclusively employing density-functional theory (DFT), have played a key role in this research area, in particular when suitably combined with concepts from thermodynamics or statistical mechanics. Although most of the arguments and understanding presented in this review are based on such calculations, I will not describe the methods here, but refer to excellent monographs (see for example [2, 3]) for DFT, to [4, 5] for applications to surfaces, and to [6] for the hybrid statistical mechanics methods.

Instead, this chapter will focus on recent progress in the modeling and understanding of oxide formation of late TMs, using particularly the late 4d series from Ru to Ag to discuss trends. Section 5.2 will highlight aspects leading to the formation of surface oxides, especially on-surface adsorption and accommodation of oxygen in the subsurface region. Key factors (also for the surface oxide stability) are the bulk-oxide heat of formation, diffusion energy barriers, elastic strain energies, and the nature of the surface-oxide–metal interface. Sometimes the nano- or subnanometer thickness of the formed oxide film is ruled by the thermodynamic conditions but often the thickness is determined by kinetics, e.g., by the diffusion of metal or oxygen atoms (between interface and surface) or by the ongoing oxidation reaction at the surface in a catalytic environment. Examples for both situations will be discussed in Sect. 5.3, where we will also see that such thin oxide films can have a different structure to what is known from the stable bulk oxides. Setting these results into a "catalytic perspective," I will conclude in Sect. 5.4, for example that the catalytic activity of Ru in high-pressure oxidation reactions is likely due to nanometer thin, but nevertheless bulklike oxide patches, whereas for Pd and Ag the catalysis is more likely actuated by novel, surface-confined subnanometer thin oxidic structures.

5.2 Initial Oxidation of Transition-Metal Surfaces

5.2.1 Formation of Adlayers

Oxide formation at metallic surfaces begins with the dissociation of the O_2 molecule. This process is still not fully understood (see for example [7] and references therein), but for the present review we do not need to discuss this in detail. As the end product of the dissociation atomic oxygen is chemisorbed at highly coordinated surface sites. I will mainly address the lowest energy surfaces of the late 4d TMs which have the hcp(0001) geometry for Ru and the fcc(111) geometry for Rh, Pd, and Ag. For these systems, the oxygen adatoms occupy threefold hollow sites, which are sketched in Fig. 5.1. These two adsorbate positions, named hcp and fcc sites, differ only with respect to the second substrate layer. Below the hcp hollow site there is a second-layer atom, and below the fcc hollow site there is none. In other words, these sites represent positions that atoms would occupy, if the crystals were continued in an hcp- or an fcc-stacking sequence. Interestingly, on the four 4d materials covered in this chapter, oxygen prefers initially those threefold hollow sites that correspond to the stacking type of the substrate, i.e., the hcp hollow sites on Ru and the fcc hollow sites on Rh, Pd, and Ag. With increasing number of oxygen adatoms repulsive lateral interactions between the adsorbates lead to the formation of ordered superstructures, typically with a (2×2) surface unit-cell at low coverages. Then, depending on the material and whether higher coverage on-surface structures are formed at all, this may be followed by either (2×1) or $(\sqrt{3} \times \sqrt{3})R30°$ periodic overlayers, with oxygen atoms still occupying either hcp or fcc hollow sites.

The quantity ruling the adsorption-site preference and the formation of ordered adlayers is the adatom binding energy:

$$E_{\mathrm{b}}(\Theta) = \frac{1}{N_{\mathrm{O}}}\left[E_{\mathrm{O@surf}} - E_{\mathrm{surf}} - \left(\frac{N_{\mathrm{O}}}{2}\right) E_{\mathrm{O_2}}\right] \tag{5.1}$$

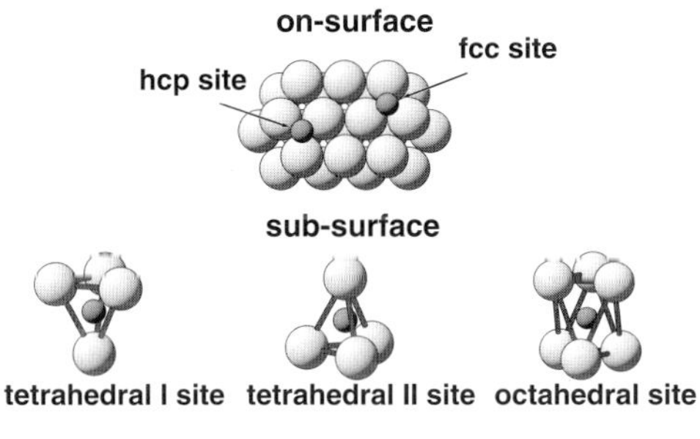

Fig. 5.1. Highly coordinated adatom sites at an fcc(111) or an hcp(0001) surface. The upper picture shows a top view of the surface, labeling the two threefold coordinated hollow sites. The lower row illustrates the local geometries of the three high-symmetry interstitial sites between the first and second substrate layers. Metal atoms are depicted as light, large *spheres*, and oxygen atoms as dark, small *spheres*

Here, N_O is the total number of O atoms in the unit-cell at the considered coverage Θ (measured in monolayers (ML), i.e., the fraction compared to the number of first layer metal atoms). $E_{O@surf}$, E_{surf}, and E_{O_2} are the total energies of the adsorbate system, of the corresponding clean surface before adsorption, and of the isolated oxygen molecule. A positive binding energy $E_b(\Theta)$ reflects that the dissociative adsorption of O_2 is exothermic. Equation (5.1) can be evaluated by DFT, and for big systems, this is in fact the only approach that is currently feasible and predictive. One has to stress, however, that for oxygen at TM surfaces such calculations are still demanding. Without describing the technical challenges of performing such large-scale DFT studies, I only mention the basic problem: Though DFT is in principle exact, in practice it is not, because the exchange-correlation (xc) functional is (and probably ever will be) only known approximately. For many systems, this is not a severe problem, but for the O_2 molecule even present-day gradient-corrected (GGA) functionals suffer from a binding energy error of about 0.5 eV per O atom [8–10]. Though some error cancellation will occur in (5.1), this O_2 problem calls for caution when judging on the exo- or endothermicity of oxygen adsorption or oxide formation. Binding energy differences between different adsorbate systems, for which the gas phase O_2 problem cancels out at least partially, appear fortunately often much less affected by the xc approximation. Still, caution is advisable, and if doubts exist, a regional xc correction, e.g., following the approach of Filippi et al. [11] may be necessary.

Evaluating (5.1) is most useful for identifying adsorption sites, lateral interactions, or the saturation coverage, and by such calculations the interplay between theory and experiment has indeed been most stimulating and synergetic in this field. The atomic-scale characterization of higher coverage O adsorbate layers forms already an intriguing example for this. Although not really fully addressed and understood, it seems that the sticking coefficient for O_2 drops either dramatically after a certain threshold coverage at the surface is reached (Ru, Rh, Pd) or is very low from the beginning (Ag). Particularly, the sharp drop after reaching a coverage of about 0.5 ML at Ru(0001) and Rh(111) leads to an apparent uptake saturation, when employing low gas exposures as typical for many early ultrahigh vacuum (UHV) surface science studies [12–14]. DFT calculations predicted, however, that much higher coverages should still be possible [15–17], and Fig. 5.2 shows such data for some selected adsorbate configurations spanning the whole coverage range up to 1 ML [18]. Although the decreasing binding energies with coverage indicate repulsive interactions between the adsorbed O atoms, they are still clearly exothermic up to the full ML for Ru (and also for Rh). As for the O/Ru case adsorption into other sites (e.g., below the surface) or other structures can be ruled out (see below), these results imply that energetically the formation of denser overlayers is possible, well beyond the apparent saturation at $\Theta \approx 0.5$ ML. And in fact, by using more oxidizing gases that create higher partial pressures (e.g., NO_2), orders of magnitude higher O_2 exposures (sometimes millions of Langmuirs) or the more novel, direct in situ studies at elevated pressures such overlayers

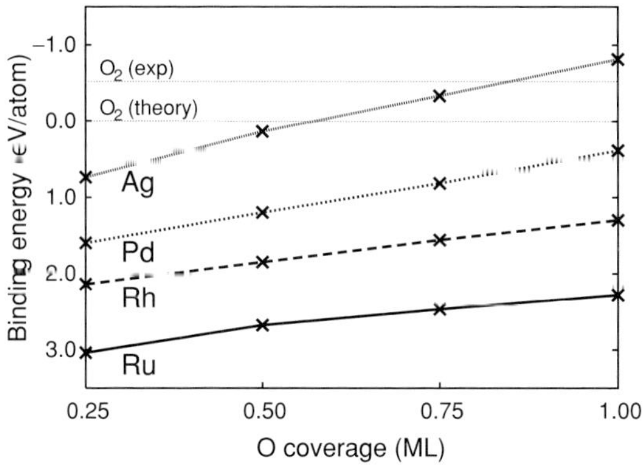

Fig. 5.2. Computed (DFT-GGA) binding energies $E_b(\Theta)$ for on-surface oxygen chemisorption into the most stable hollow sites of the basal surface of the late 4d TMs (Ru = hcp site, Rh/Pd/Ag = fcc site) and employing (2×2) surface unit-cells (the *lines* are only to guide the eye). The energies are given with respect to the theoretical O_2 binding energy, cf. (5.1), while the upper horizontal dotted *line* indicates the zero level, if the experimental O_2 binding energies were used (after [18])

(and configurations corresponding to even much higher O uptakes) are now routinely observed experimentally. One should stress, that on all four TM surfaces discussed in this chapter, such experiments have opened a new door to study the oxidation beyond submonolayer overlayers in a controlled fashion, significantly improving the scientific understanding over the last years.

Figure 5.2 also illustrates the mentioned O_2-molecule problem. Using in (5.1) the *experimental* O_2 total energy, E_{O_2}, instead of the theoretical one, would shift the energy zero to the upper dotted horizontal line. This is a too extreme alternative though, because some xc error also exists in $E_{O@surf}$ and E_{surf} Thus, the true energy zero is probably somewhere in between the two dotted lines. Obviously, in particular, for silver the inaccuracy in the molecular binding requires special attention, even when only discussing the sign of the binding energy [19]. What is, however, unambiguously evident from Fig. 5.2 is that the oxygen–metal bond strength decreases steadily from Ru to Ag, and this is in fact known qualitatively since long. The reason behind it is clearly visible in the projected density of states (PDOS), which is defined as [4]

$$N_\alpha(\varepsilon) = \sum_{i=1}^{\infty} |\langle\phi_\alpha|\varphi_i\rangle|^2 \delta(\varepsilon - \varepsilon_i). \qquad (5.2)$$

Here φ_i are the Kohn–Sham orbitals and ϕ_α is a properly chosen localized function, in the present case most suitably the Ru 4d and the O 2p atomic orbitals within a sphere around given atomic sites. According to the Anderson–Grimley–Newns model of chemisorption, the interaction of the O 2p level with the narrow 4d band of the TM surface gives rise to bonding states at the lower edge of the d band and antibonding states at the upper

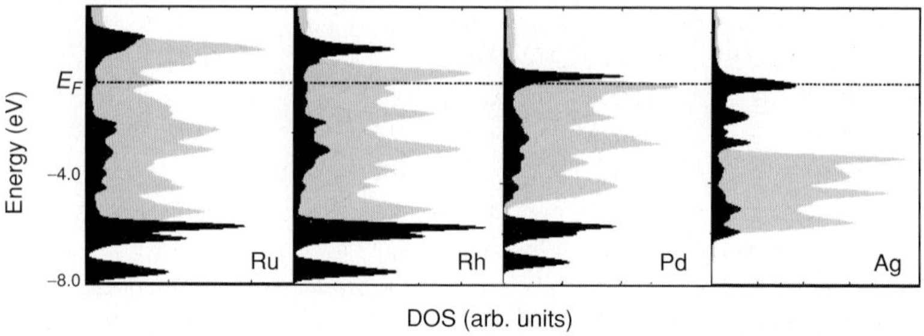

Fig. 5.3. Computed (DFT-GGA) projected density of states (PDOS), cf. (5.2), of the clean basal surfaces of the late 4d TMs (gray shading), and of the O(1 × 1) adlayer (black shading). Shown are the $N_\alpha(\varepsilon)$ contributions of the Ru 4d and O 2p states. The decrease of the oxygen PDOS bandwidth (from left to right) and the increased filling of antibonding states (the higher energy oxygen PDOS region) are clearly visible

edge of the d band. The corresponding peaks can be readily identified in Fig. 5.3, illustrating the PDOS of a high-coverage O adsorbate layer for the late 4d TM series. The Fermi level moves from the d-band center (for Ru it is already at a ≈ 70% filling) to about 3.5 eV above the top of the d band (for the noble metal Ag), and thus the antibonding oxygen–metal states become progressively more and more occupied [4, 20, 21]. As a consequence, the bond strength is weakened, explaining the steadily decreasing binding energies of Fig. 5.2. For the discussion below, it is finally worthwhile mentioning that despite these large variations of the binding energy with element (as well as with coverage), the oxygen–metal bond length does not change appreciably and stays always roughly at 2 Å.

5.2.2 Oxygen Accommodation Below the Top Metal Layer

It is often assumed that deposition of oxygen on metal surfaces initially gives rise to stable overlayers (either ordered or disordered). Apparently this is often correct, but there is no reason that it holds in general. In fact, oxygen atoms may well go below the surface, either staying in the near-surface region or dissolving into the bulk. Or, metal-oxide patches could form immediately. Experimentally, very little is unfortunately still known about this. For the basal surfaces of the late 4d TMs recent DFT-GGA calculations indicate that initially oxygen adatoms stay at the surface, and only when the coverage exceeds a certain value some adatoms will go subsurface [18]. They then remain preferentially close to the surface, i.e., they will not dissolve into the bulk [22, 23]. The interaction between these subsurface oxygen atoms is attractive, and this situation may well already be called a surface oxide or "stage 1" of oxide formation.

In principle, oxygen may occupy interstitial or substitutional sites in the crystal, but mostly due to the high cohesive energy of the late TMs, the latter are found to be energetically much less favorable [19]. Thus, they will not play a role in thermodynamic equilibrium, or close to it, and we may focus our discussion on interstitial sites. Directly below the topmost metal layer, i.e., in the immediate subsurface region, there are three high-symmetry possibilities for such interstitial sites (cf. Fig. 5.1). Namely, there is an octahedral site (henceforth called octa) and a tetrahedral site (tetra-I) directly below the on-surface fcc and hcp site, respectively. And there is a second tetrahedral site (tetra-II) directly below a first layer metal atom. The main trends in the energetics of on- versus subsurface oxygen are already discernible by comparing surface structures containing from one up to four O atoms in a model subset of (2×2) unit-cells. Of these up to four adatoms, consider that any one can be located either on or below the surface in any of the high-symmetry sites displayed in Fig. 5.1. Despite the ostensible simplicity of this model subset, the number of possible combinations of on- and subsurface oxygen atoms in various site and symmetry configurations is still quite large, as exemplified in Fig. 5.4 by all configurations where one out of a total of four oxygen atoms is located below the surface. Interestingly, from all these structural possibilities, the combination of on-surface O in fcc sites and subsurface O in tetra-I sites was in all four late 4d metals found to be energetically either most stable or very close to the most stable geometry [18, 19, 22, 24, 25].

Occupation of subsurface sites will commence at a certain total coverage Θ_c, when a structure with a nonzero fraction of the O atoms below the sur-

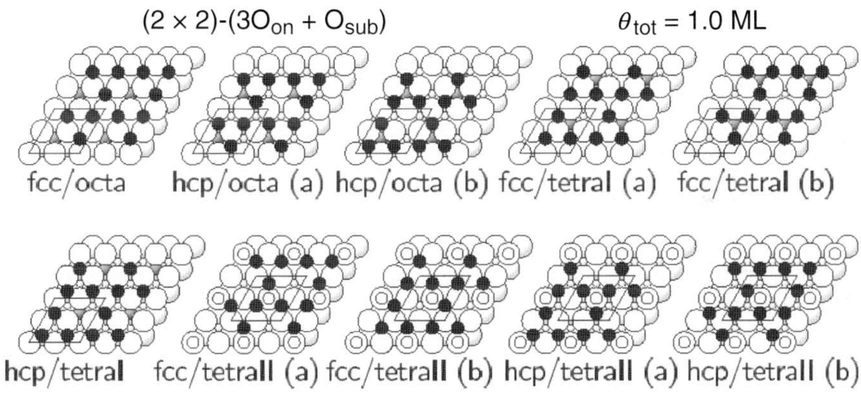

Fig. 5.4. Top view of all possible on-surface/subsurface site combinations at a total coverage $\Theta = 1$ ML with one oxygen atom per (2×2) surface unit-cell located below the surface, i.e., corresponding to a fraction of 25% below the surface (see Fig. 5.1 for the explanation of the different sites) The metal atoms are drawn as big *spheres* (white for the surface layer, gray for second layer), and the oxygen atoms are drawn as small *spheres* (black for on-surface, gray for subsurface). Oxygen atoms in tetra-II sites below the first layer metal atoms are invisible in this plot and are schematically indicated by small white *circles*. Symmetry inequivalent occupation of the same kind of on- and subsurface sites are denoted with (**a**) and (**b**), respectively

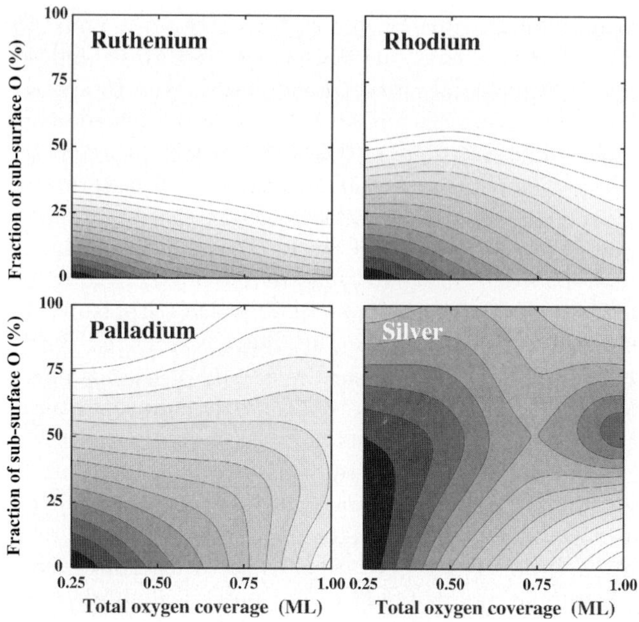

Fig. 5.5. Averaged binding energy E_b [using DFT-GGA and (5.1)] as a function of the total O coverage Θ of which a certain percentage is located below the basal surface (on-surface in fcc and subsurface in tetra-I sites). The highest E_b are always found for the pure on-surface chemisorption phase at $\Theta = 0.25$ ML (chosen as zero reference, black area), with each contour *line* (lighter gray areas) at 0.1 eV steps toward less stable E_b (from [18])

face is energetically more favorable than the pure on-surface adsorption (corresponding to a zero fraction of oxygen atoms below the surface). Figure 5.5 summarizes the energetics within the described model subset at the basal surface, and we see that for all four late 4d TMs on-surface adsorption is initially more favorable than subsurface incorporation. However, with increasing coverage this picture changes as the aforedescribed repulsive lateral interactions between the on-surface oxygen atoms progressively diminish the preference for adsorption into these sites. As can be deduced from Fig. 5.5, this threshold coverage Θ_c is beyond 1 ML for Ru and Rh, i.e., oxygen incorporation will only start after completion of a full $O(1 \times 1)$ overlayer, in nice agreement with experimental findings [26, 27]. Close to $\Theta = 1$ ML, the energy differences are very small in the case of Rh though. Hence, a finite concentration of subsurface oxygen might already be present at total coverages even slightly below 1 ML at elevated temperatures [24, 28]. For Pd, Θ_c lies around 3/4 ML, and for Ag subsurface incorporation starts almost immediately (cf. Fig. 5.5).

This trend for subsurface accommodation of oxygen atoms at the late 4d TMs can be understood by looking at the subsurface geometries. In all relaxed structures, the subsurface oxygen–metal bond lengths are slightly larger than

2 Å, i.e., they are similar to the on-surface O–metal distances. Comparing these bond lengths with the available geometrical space inside a frozen metal lattice, one finds them to be rather incompatible. For the unrelaxed tetra-I site of the late 4d metals, the bond length would range from 1.65 Å (Ru) to 1.80 Å (Ag). Thus, oxygen incorporation necessarily must go together with a substantial local expansion of the metal lattice. As shown by Todorova et al. [18], the bond strength of subsurface oxygen atoms may therefore roughly be divided into a stabilizing contribution due to the chemical binding and a destabilizing contribution from the lattice deformation cost. Going from Ru to Ag, the first component should follow the same d-band-filling trend of the oxygen–metal bond strength that was discussed in Sect. 5.2.1 for on-surface oxygen. And also the latter component follows this trend, i.e., it scales roughly as the bulk cohesive energy or the bulk modulus, which are largest for Ru and smallest for Ag as noted in Fig. 5.6. The two contributions (O–metal bond strength and metal–lattice distortion energy) have opposite sign and apparently roughly compensate each other, leading to a rather similar energetic stability of subsurface oxygen in all four elements [18].

Due to the lattice-deformation cost, subsurface oxygen turns out initially less favorable than on-surface adsorption, but with increasing coverage the repulsive lateral interactions for on-surface oxygen drive this preference down until eventually penetration of some oxygen adatoms becomes favorable. As the on-surface bond strength decreases from Ru to Ag, while the subsurface bond strength stays roughly constant, this crossover will occur progressively earlier, i.e., Θ_c is large for Ru and small for Ag.

If oxygen penetrates, the deformation cost also rationalizes why subsurface incorporation was found to be more favorable compared to bulk dissolution.

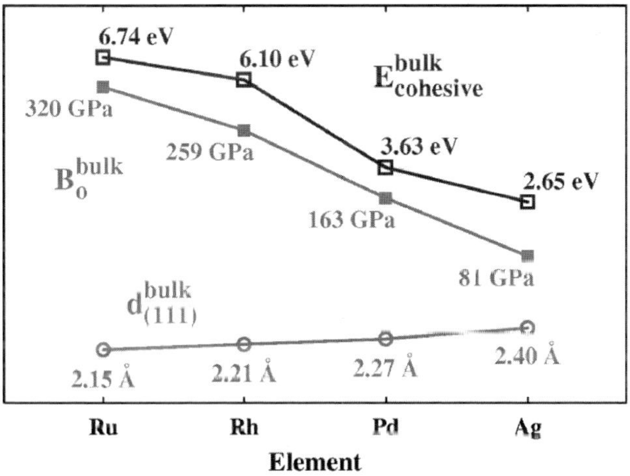

Fig. 5.6. Trend of DFT-GGA computed bulk modulus, cohesive energy, and layer distance in (111) direction for the late 4d TMs

Although the latter occurs of course always to some extent just on entropy grounds, there will at any rate be an enrichment of oxygen atoms in the easier to relax interstitial sites in the immediate near-surface fringe. And one could expect this deformation argument to hold even more for subsurface sites close to even lower coordinated atoms, i.e., interstitials in the vicinity of point defects, steps, or dislocations. The latter are in fact frequently believed to be the nucleation centers for surface oxide formation, and kinetic arguments like an easier O penetration are often put forward as explanation. Yet, we see that the thermodynamic deformation cost factor could also favor an initial oxygen accommodation close to such sites.

5.2.3 Oxygen Accumulation in the Surface Region and Surface Oxide Formation

Thermodynamically, the bulk oxide will start to form at the oxygen coverage, ${\Theta_c}^{thd}$, at which the average oxygen binding energy, cf. (5.1), equals the oxide heat of formation [29]. The word "coverage" is used here in a wider sense, i.e., meaning the total amount of oxygen atoms in the near-surface region. Interestingly, the just discussed Θ_c for the initial incorporation of oxygen at subsurface sites in the late 4d metals does not only follow ${\Theta_c}^{thd}$ in trend but also in approximate value [18]. This suggests subsurface oxygen as an intermediate, or a precursor, for the formation of the bulk oxide or other oxidic structures beyond on-surface adsorbate layers. Recent experimental studies seem to support this view, i.e., they indicate that thin oxidic structures, now mostly coined surface oxides, start to form roughly at coverages Θ_c [30–34].

For the "true" bulk oxide a higher oxygen concentration might be required, but for oxide nucleation and for the formation of oxidic islands such increased oxygen content is only needed locally. In this respect, it is remarkable that although for the on-surface adsorption at the late 4d TM basal surface the adatom–adatom interaction is repulsive (see the discussion of Fig. 5.2), an attractive interaction was found for subsurface oxygen atoms in Ru(0001) and Rh(111) [22, 24]. This is illustrated in Fig. 5.7, which shows the course of the average binding energy of O at Ru(0001) upon increasing the amount of oxygen atoms between the first and second substrate layer up to 1 ML subsurface coverage, i.e., 2 ML total coverage. As discussed in the Sect. 5.2.2, for Ru(0001) oxygen penetration starts only after the full on-surface monolayer coverage is reached. Up to $\Theta = 1$ ML, the binding energy decreases steadily due to the repulsive lateral interactions. For even higher coverages, i.e., upon the ensuing filling of subsurface sites, the binding energy curve exhibits an inflection. This implies overall attractive interactions that favor an accumulation of subsurface oxygen in dense two-dimensional islands with a local (1×1) periodicity [22].

While for Ru(0001) [22] and Rh(111) [24] the attractive interactions lead to dense oxygen islands with a local (1×1) ordering below the surface, cf. Fig. 5.7, at Ag(111) it already becomes more favorable to incorporate oxygen rather

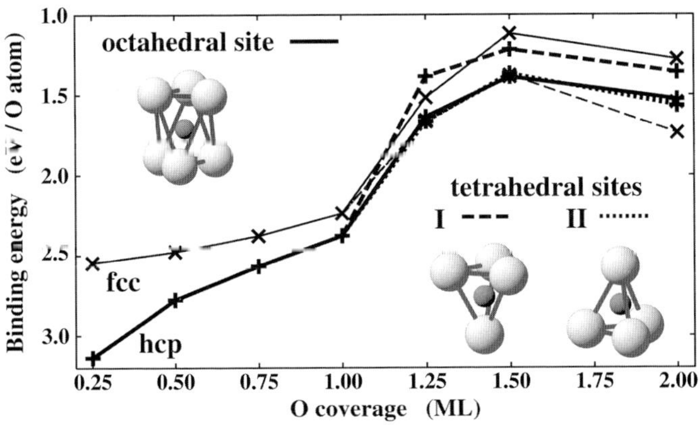

Fig. 5.7. Binding energies, calculated by DFT-GGA, for oxygen at Ru(0001). Coverages with $\Theta \leq 1\,\text{ML}$ correspond to pure on-surface adsorption (hcp or fcc site). For $\Theta > 1\,\text{ML}$ an $O(1 \times 1)$ arrangement is always present at the surface, cf. Fig. 5.5, while the remaining O is located in either of the three interstitial sites between the first and second substrate layer. From the six possible structures at each coverage, the three with on-surface O in hcp (fcc) sites are drawn with thicker (thinner) *lines* (from [22])

between deeper layers, when a local subsurface oxygen coverage of $\sim 1/4\,\text{ML}$ is exceeded [23]. Apparently, each metal is only able to sustain a certain amount of oxygen per layer, and the differences between Ru and Ag are significant. Whereas Ru(0001) can accommodate a full oxygen monolayer on the surface plus a full oxygen monolayer below the top Ru layer, for Ag(111) this is only 25% of a monolayer on and 25% below the top Ag layer. What seems to be a more general feature, however, is the O–metal–O trilayer type structure, that results from the accumulation of subsurface oxygen below the topmost, oxygen-adsorbate covered metal layer. This accumulation will eventually lead to the formation of an ordered structure, which as already mentioned above, one could denote as a surface oxide, or "stage 1" of oxide formation. This is to distinguish it from subsurface oxygen, which often conveys more the notion of a randomly incorporated O species in the near-surface fringe. Although we expect from the deformation cost argument an enrichment of the latter compared to bulk dissolved O, one should stress that an experimental identification of either (bulk or subsurface O) species will be very difficult. The concentration of bulk dissolved O is thermodynamically quite low in the late 4d TM crystal lattices [22,25] and will thus only yield small signatures (which as a note aside does not mean that the total amount of dissolved O in the whole crystal must be small though). And even though possibly enriched to a higher concentration, the signals from subsurface O will also be small, since now only a small volume of the crystal will contribute. In fact, subsurface oxygen has hitherto been quite elusive with only very few experimental studies claiming to have provided direct evidence for it. What is more frequent is that spectroscopic features that could not otherwise be explained have been interpreted as being due to a subsurface oxygen species.

Within the deformation cost picture, it appears in any case rather likely that (depending on the O exposure and kinetic barriers) a (lateral and vertical) aggregation of subsurface oxygen will directly lead to ordered surface oxide structures. We will see below, that these structures are indeed oxide precursors, which may often have a somewhat similar stoichiometry as the bulk oxides (e.g., RuO_2 and Ag_2O for the two mentioned trilayer structures, respectively), but the geometries can differ significantly. Experimentally it may be similarly involved, though not impossible, to identify such structures, because also an ensuing transition to other oxidic precursors or eventually the "true" bulk oxide (film) may proceed fast.

5.2.4 Formation of the Bulk Oxide

Continued accommodation of oxygen at the surface will eventually yield to the phase transition toward the "true" bulk oxide structure. For different materials, this happens at different oxygen load, and the pathway from the first formed surface oxide to the final bulk oxide may well comprise other oxidic precursor structures. In general, one has to admit that our atomic-scale knowledge specific to this step in oxide formation is again very shallow. Particularly for the more noble TMs, the low stability of the bulk oxides requires either very high oxygen pressures and/or rather low temperatures for the experimental preparation. At the lower temperatures, the growth is largely affected by kinetic limitations though, so that at Pd(111) at best small PdO clusters without apparent crystalline order were hitherto reported [35], while at Ag(111) there exists literally no atomic-scale knowledge on the oxidation process beyond the formation of an ordered $p(4 \times 4)$ surface oxide (see below). This situation is far better for Ru(0001) and Rh(111), where it is at least known that the oxidation process ends with the formation of crystalline rutile-structured $RuO_2(110)$ [31] and corundum-structured $Rh_2O_3(0001)$ films [32], respectively.

This detailed knowledge of both initial state (the aforedescribed O–Ru–O trilayer) and final state (the rutile structured $RuO_2(110)$ film) made it possible to elaborate for the first time on an atomistic pathway for the structural phase transition toward the bulk oxide [36]. If the oxygen load at the Ru(0001) surface is increased beyond the 2 ML coverage O–Ru–O trilayer, cf. Fig. 5.7, the surprising finding of the corresponding DFT-GGA study was that these additional oxygen atoms would also go preferentially between the first and second substrate layer, i.e., where already one full oxygen layer exists. The result is displayed in Fig. 5.8. It shows that a significant relaxation takes place such that the formed O–Ru–O trilayer is essentially floating on an O/Ru adsorbate system. With the trilayer decoupled, increasing the oxygen load further then gives rise to two O–Ru–O trilayers and so forth. Still, this is

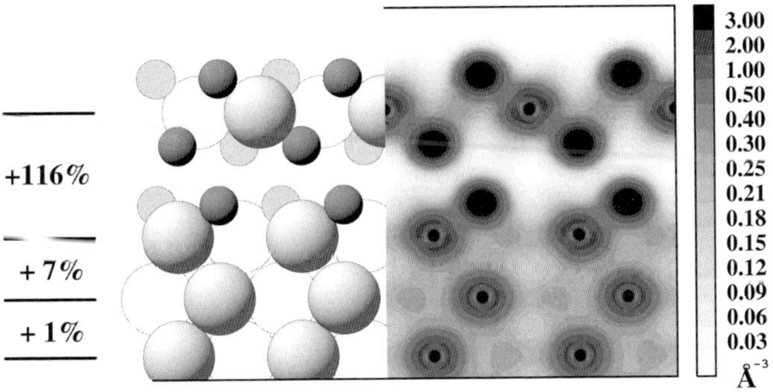

Fig. 5.8. Side view of the detached "floating trilayer" of O at Ru(0001), corresponding to the most stable geometry with $\Theta = 3\,\mathrm{ML}$. The left half of the figure shows the schematic geometry together with the relative layer expansions compared to the bulk distance, and the right half shows the calculated electron density. Ru = large *spheres*, O = small *spheres*, and Ru atoms not lying in the shown plane are whitened (from [36])

not yet a known Ru-oxide structure either with respect to the energy or the geometry.

As the trilayer on top has apparently only negligible influence on the underlying metal, a continued exposure to oxygen is likely to produce sequentially more and more of these trilayers, which one after the other will become decoupled from the metal substrate. This way a loosely coupled stack of trilayers will successively be formed. Comparing the geometry of these trilayers with the one of a rutile (110) layer, it can be noticed that the former can be transformed into the latter by a simple accordion-like lateral expansion of the trilayer as explained in Fig. 5.9. With this expansion, the self-contained trilayer unfolds into a more open geometry, in which every second Ru atom is undercoordinated compared to the ideal sixfold oxygen surrounding both offered in the trilayer and in bulk RuO_2. It is precisely these coordinatively unsaturated (cus) metal atoms that may eventually lead to a stabilization of the rutile structure for higher film thicknesses, as they can provide extra interlayer bonds that cannot be formed between the saturated trilayers. While for only one layer thick films, the more compact trilayer offers therefore a more favorable configuration, this preference was already found reversed at a two-layer thickness, eventually inducing the transition to rutile through the accordion-like expansion [36].

Admittedly, this idealized pathway for one specific surface can barely be able to represent the complex atomic rearrangements and stoichiometry changes one can expect more generally during this oxidation step. Still, it contains at least one aspect that appears quite crucial: The effect of strain. As already pointed out before, oxygen incorporation will always lead to a substantial local expansion of the crystal lattice of late 4d TMs, and known bulk oxide structures of the latter are accordingly much more open compared to the

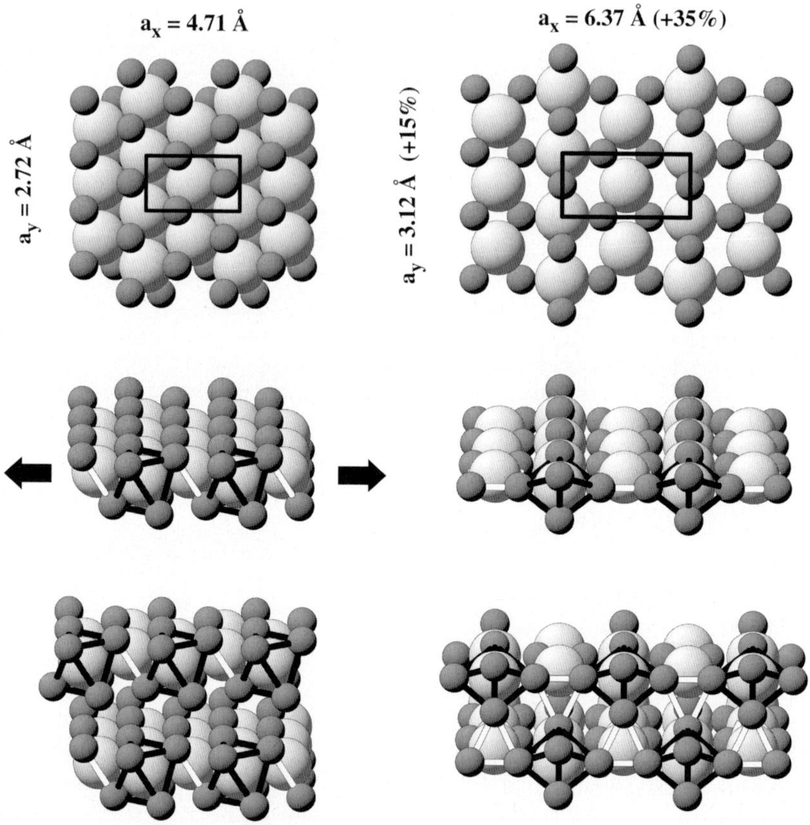

Fig. 5.9. Atomic geometries of the O–Ru–O trilayer shown in Fig. 5.8 (left-hand side) and of the final rutile $RuO_2(110)$ oxide structure (right-hand side). Top panel: Top view, indicating the largely different sizes of the surface unit-cells. Middle panel: Perspective view. The bulklike rutile structure is achieved by simply stretching the trilayer in direction of the arrows, keeping the length of the drawn O–O bonds (white and black *lines*) rigid. This way an alternative sequence of fully coordinated Ru atoms (inside the tilted O octahedra, black bonds) and coordinatively unsaturated (cus) Ru atoms (surrounded by only four O atoms, white bonds) is created. Bottom panel: Coupling of two trilayers versus that of rutile layers. Whereas the self-contained trilayers hardly bind, new bonds can be formed in the $RuO_2(110)$ structure. The bridging O atoms of the full octahedra (black bonds) form the apex atoms of the new octahedra (white bonds) around the former cus Ru atoms. Note, that the "bonds" drawn do not refer to chemical bonds but are merely used to guide the eye (from [36])

close-packed bulk metals. In the direction perpendicular to the surface, nothing opposes such an expansion and the corresponding vertical relaxations upon O incorporation can be quite substantial. Lateral relaxations, on the other hand, will conflict with the binding of the formed oxidic fringe to the underlying metal lattice. In the formation of (surface) oxides, the system therefore has to find an optimum energy compromise between elastic strain relaxation and good coupling to the underlying substrate. This may result in commensurable or incommensurable overlayers, and both types have already been observed experimentally on the late 4d basal surfaces [31–34]. In addition, strain may be easier relaxed at more open sites like steps, pointing again at the relevance of the latter for the oxidation process. Apart from the strain energy, another influential factor is the thermodynamic driving force to form an oxide. The latter will roughly scale with the bulk oxide heat of formation, which steadily decreases from RuO_2 to Ag_2O. The concomitant, lower stability of oxidic structures per se, together with easier to deform metal lattices, might then render the coupling to the underlying substrate more important toward the more noble metals, i.e., the nature of the oxide–metal interface plays an increasing role. In this respect it is intriguing to notice that particularly at Pd(111) and Ag(111) pseudomorphic surface oxide structures have been reported, with a structure that seems particularly suited to maximize the interaction with the metal lattice below.

5.3 Implications for Oxidation Catalysis

5.3.1 The Role of the Gas Phase

A salient feature emerging from studies of the kind briefly sketched in the first part of this review is that the initial oxidation of late 4d TMs proceeds via few atomic-layer thin oxidic structures, the electronic and geometric structure of which may deviate significantly from their known bulk oxide counterparts. This makes these surface oxide films distinct to surfaces of both bulk metals and bulk oxides, and one is led to wonder whether they exhibit specific new properties that can be exploited for a desired functionality like catalysis. In this respect, one has to recognize, however, that in such applications (and many experiments) the actual thermodynamic variable is not the oxygen accommodation at the surface but the partial pressures and temperature in the surrounding gas phase. Instead of looking at the change of the state of the surface as, e.g., a function of increased exposure (or uptake) as done in the preceding sections, the relevant question to ask is then rather what happens to the TM surface given certain environmental conditions?

Exposed to an unlimited supply of gas phase particles characterized by the applied pressures and temperature, the surface will adapt on time scales set by the kinetic limitations. Already these time scales could be sufficiently long to render corresponding metastable states interesting for applications. In fact, the classic example is a slow thickening of oxide films due to limitations in the diffusion of oxygen atoms from the surface to the oxide–metal interface or in the diffusion of metal atoms from the interface to the surface [37,38]. Directly at the surface a similar bottleneck can be the penetration of oxygen, which might be significantly facilitated at steps and defects and thus adds to their relevance particularly for the formation of surface oxides also from a kinetic point of view. Still, although such kinetic barriers might slow the oxidation process down beyond noticeable time scales, in a pure oxygen gas phase, the true thermodynamic ground state will eventually be reached. With respect to oxide formation this is often loosely equated with a completely oxidized sample, and it thus came as quite a surprise that for the more noble 4d TMs, DFT-GGA indicates that there are in fact (T, p)-conditions, for which a nanometer thin surface oxide represents the thermodynamically stable phase. This means that in such environments a thin film is automatically formed at the oxidizing surface, the growth of which is self-limited to the potentially particularly interesting nanometer or subnanometer width.

Reactive multicomponent gas phases, as typical for, e.g., catalysis, add another twist to this situation. In this case, we are dealing with an open system, in which the supply of reactant gases comes into contact with the solid surface, where a chemical reaction produces a new substance that is then transported away. The average surface structure and composition is in this case entirely determined by the kinetics of the various underlying atomistic processes like adsorption, desorption, diffusion, or reaction. Still, for given temperature and partial pressures of the reactants a steady state situation may be reached (or even aspired for a stable catalyst performance), but this steady state must not necessarily correspond to any thermodynamic state. Particularly for oxidation reactions, the O-rich environment might for example constantly oxidize the catalyst surface while the ongoing reaction continuously consumes oxygen in such a way that oxide formation never proceeds beyond a certain state. Similar to the thermodynamic situation described above, this could also lead to a self-limited thickness of the oxidized surface fringe.

5.3.2 Stability of Surface Oxides in an Oxygen Environment

If we initially focus on the effect of a pure oxygen gas phase, it is useful to first resort to a thermodynamic description. This allows assessing the various stages in the oxidation process when the surface is fully equilibrated with its environment and thus provides a sound basis for an ensuing discussion on how kinetic limitations could affect the picture. Particularly interesting in the context of this review is in this respect whether the thin surface oxides are mere kinetic precursors to bulk oxide growth or may actually represent a stable phase for certain gas phase conditions, as well as whether we can identify a trend over the late 4d TM series. For a surface in contact with a gas phase characterized by temperature T and pressure p_{O_2}, a thermodynamic description takes place in the grand canonical, constant (T, p)-ensemble and one needs to generalize (5.1) to evaluate the Gibbs free energy of adsorption [23, 39, 40]

$$\Delta G\left(\mu_{\mathrm{O}}\right) = -\frac{1}{A}\left(G_{\mathrm{O@surf}} - G_{\mathrm{surf}} - N_{\mathrm{O}}\mu_{\mathrm{O}} - \Delta N_{\mathrm{M}}\mu_{\mathrm{M}}\right). \tag{5.3}$$

Here, A is the area of the surface unit-cell, and $G_{\mathrm{O@surf}}$ and G_{surf} are the Gibbs free energies of the oxidized and clean surface, respectively. μ_{O} and μ_M are the chemical potentials of oxygen and metal atoms, and N_{O} is the number of oxygen atoms contained in the oxidized surface structure for which $\Delta G(\mu_{\mathrm{O}})$ is computed. Finally, ΔN_{M} is the difference in the number of metal atoms between the reference clean surface and the oxidized surface structural model.

In the difference between the two Gibbs free energies, the contributions due to vibrational free energy, configurational entropy and pV-term cancel to some extent, and for the oxidized late 4d TM surfaces discussed here replacing the Gibbs free energy difference by the difference of only the leading total energy terms appears to provide a reasonable approximation [41, 42]. This simplifies (5.3) to

$$\Delta G(\Delta\mu_{\mathrm{O}}) \approx \frac{1}{A}[N_{\mathrm{O}} E_{\mathrm{b}}(\mathrm{O@surf}) + N_{\mathrm{O}}\Delta\mu_{\mathrm{O}} + \Delta N_{\mathrm{M}}\mu_{\mathrm{M}}], \qquad (5.4)$$

where $E_{\mathrm{b}}(\mathrm{O@surf})$ is the average binding energy of oxygen in the oxidized surface, evaluated via (5.1), and where the oxygen chemical potential is now measured with respect to the total energy of an isolated O_2 molecule as reference zero, $\Delta\mu_{\mathrm{O}} = \mu_{\mathrm{O}} - 1/2\ E_{\mathrm{O}_2}$.

Equation (5.4) has a rather intuitive structure: Forming the oxidized surface by accomodating N_{O} oxygen atoms yields an energy gain of $N_{\mathrm{O}} E_{\mathrm{b}}$ (O@surf) that is opposed by the cost of taking these O atoms out of a reservoir, hence $N_{\mathrm{O}}\Delta\mu_{\mathrm{O}}$. The equivalent term $\Delta N_{\mathrm{M}}\mu_{\mathrm{M}}$ comes only into play for oxidized surfaces, where the total number of metal atoms is different to the one of the reference clean metal surface, and represents then the cost of transferring the corresponding number of metal atoms to or from a reservoir with chemical potential μ_{M}. For the metal atoms, this reservoir is for oxidized surfaces often conveniently chosen to be the metal bulk, with which the surface is assumed to be in equilibrium [23,39,42]. The oxygen atom reservoir, on the other hand, is given by the surrounding gas phase, allowing to relate $\Delta\mu_{\mathrm{O}}$ to temperature T and pressure p_{O_2} via ideal gas laws [42]. For a given $\Delta\mu_{\mathrm{O}}(T, p_{\mathrm{O}_2})$ in the environment, the thermodynamically stable phase maximizes $\Delta G(\Delta\mu_{\mathrm{O}})$, and (5.4) can be used to compare structural models for possible oxidation states of the surface, e.g., oxygen adlayers with different coverage versus surface or bulk oxides. Since $E_{\mathrm{b}}(\mathrm{O@surf})$ from (5.1) enters into (5.4), all the caveats about DFT calculations of the oxygen binding energy mentioned in the beginning hold here as well. This uncertainty is further aggravated by the additional approximation due to the neglected free energy contributions, and dictates utmost care in the interpretation of the obtained results.

When explicitly comparing a set of structures in the sketched atomistic thermodynamics approach, surfaces with higher O content will become more favorable, the more O-rich the environment. In the limit of low oxygen pressures, the clean surface will obviously be most stable [$\Delta\mu_{\mathrm{O}} \rightarrow -\infty$, cf. (5.4)], but this will eventually change with increasing chemical potential in the gas phase. The more oxygen is accommodated in a structure, the steeper the slope of the corresponding $\Delta G(\Delta\mu_O)$ curve with $\Delta\mu_O$. In the limiting case of the bulk oxide with its infinite number of O atoms, this will finally yield a vertical line at a value of the oxygen chemical potential that equals the bulk oxide heat of formation per O atom [40]. For any higher value of $\Delta\mu_O$, the bulk oxide represents then the thermodynamically stable phase.

This overall structure is nicely visible in Fig. 5.10, which compares several of the above discussed oxidation stages at the Ru(0001) surface. While for $\Delta\mu_O < -3.1\,\text{eV}$, the clean surface is most stable, four different O adsorbate phases with O in hcp sites and increasing coverages from 1/4 to 1 ML become progressively more stable for more O-rich environments. Above $\Delta\mu_O < -1.7\,\text{eV}$, bulk RuO_2 represents the most stable phase. The stability range of this bulk oxide is therewith so large that it in fact covers all realistically attainable pressures for temperatures below $T \sim 600\,\text{K}$, as illustrated in Fig. 5.10. In particular, one also notices that the O–Ru–O trilayer structures discussed above as possible intermediates before the bulk oxide formation are never thermodynamic phases, suggesting that at best they could only be stabilized by kinetic limitations in the ensuing phase transition to the bulk oxide.

A qualitatively similar picture has recently been reported for Rh(111) [32], where after adsorbate phases with O in fcc sites, bulk Rh_2O_3 becomes most stable at higher oxygen chemical potentials. Due to the already decreased heat of formation per O atom of Rh_2O_3 compared to RuO_2, the stability range of the bulk oxide is smaller, but starts in contrast to the situation at Ru(0001) nevertheless before higher coverage adsorbate phases corresponding to $\Theta \sim 0.5\,\text{ML}$ could become favorable. This limit is only approximate though, since one should keep in mind that in such atomistic thermodynamic plots only the stability of a given set of structures is compared, and there could well be yet uncharacterized, higher coverage structures with a sufficiently high average binding energy to show up as a new stable phase. Still, bulk oxide formation appears at this surface thermodynamically preferred already before a full ML $O(1 \times 1)$ adsorption phase, which is consistent with the trend toward easier O accommodation from Ru to Ag discussed in the first part of this chapter. Although an O–Rh–O trilayer structure equivalent to the one discussed for Ru(0001) above results similarly as only a kinetic precursor in DFT-GGA, it is interesting to note that at Rh(111) this structure could recently be experimentally prepared and characterized close to step edges [32].

Proceeding to Pd(111) one observes in most aspects a continuation of the identified trends. As apparent in Fig. 5.10, the again reduced thermal stability of the PdO bulk oxide shows up by a further decreased stability range, which commences nevertheless already at an O chemical potential below the one required to stabilize higher on-surface adsorbate coverages with $\Theta \sim 1/4\,\text{ML}$. On the other hand, there is also a qualitatively new feature, since now the experimentally observed, so-called "$(\sqrt{6} \times \sqrt{6})$" pseudo-commensurate surface oxide structure displayed in Fig.5.11 is not only a kinetic precursor but

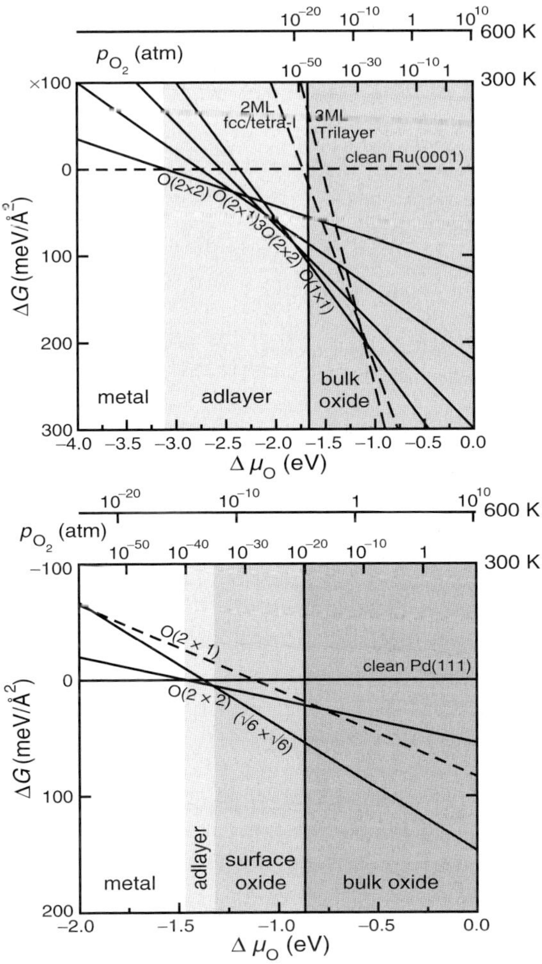

Fig. 5.10. Computed DFT-GGA Gibbs free energy of adsorption for several ordered on-surface adsorbate structures, surface oxides, and the bulk oxide. The dependence on $\Delta\mu_O$ is also translated into pressure scales at $T = 300$ and 600 K for clarity. In the bottom of the figure, the "material type" which is stable in the corresponding range of oxygen chemical potential is listed and indicated by the shaded regions. Upper panel: Ru(0001). Compared are $O(2\times2)$, $O(2\times1)$, $3O(2\times2)$, and $O(1\times1)$ adsorbate phases with O in hcp hollow sites, cf. Fig. 5.2, with the two O–Ru–O trilayer structures discussed in connection with Figs. 5.7 and 5.8. Lower panel: Pd(111). Compared are $O(2\times2)$ and $O(2\times1)$ adsorbate phases with O in fcc sites, cf. Fig. 5.2, with the so-called "$(\sqrt{6}\times\sqrt{6})$" surface oxide shown in Fig. 5.11 (the binding energies used to construct these graphs are taken from [33, 36, 40])

represents the stable phase for an intermediate range of $\Delta\mu_O$ [33]. This is therefore similar to the reported findings at Ag(111), of a $p(4\times4)$ surface oxide that is thermodynamically stable over a wide range of O chemical potentials,

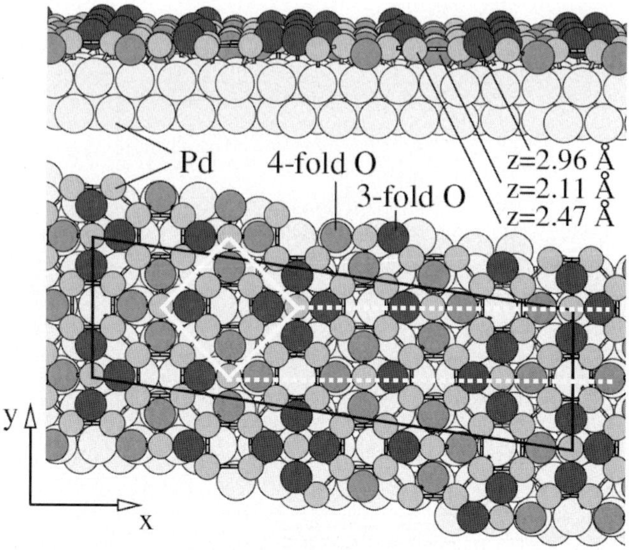

Fig. 5.11. Top- and side view of the pseudo-commensurate, so-called "($\sqrt{6} \times \sqrt{6}$)" surface oxide structure determined on Pd(111). The atomic positions do not seem to resemble the bulk PdO arrangement but exhibit roughly an O–Pd–O trilayer type structure (from [33])

bounded on the one end by the stability range of a low-coverage adsorbate phase and on the other end by the low-stability Ag_2O bulk oxide [23, 39]. Both the "($\sqrt{6} \times \sqrt{6}$)" surface oxide on Pd(111) and the p(4×4) surface oxide on Ag(111) seem to have a highly complex atomic structure that does not resemble the corresponding bulk oxides, cf. Fig. 5.11 [29, 32]. Instead, these structures appear particularly suited to achieve a good coupling to the underlying substrate, and it is this extra contribution that enhances their stability over a wide (T, p)-range beyond the one of the known bulk oxides.

Summarizing the trend, we therefore see a changing role of thin surface oxide films in the oxidation process over the late 4d TM series. For Ru and Rh, the present data suggests that such precursor structures to bulk oxide formation can only be stabilized by kinetic limitations. At the more noble metals Pd and Ag, they instead represent a thermodynamically stable phase over an increasing range of gas phase conditions. In corresponding oxygen environments, these nanometer thin films will eventually form on time scales set by possible kinetic limitations but never grow thicker. Due to this finite thickness, the coupling at the oxide–metal interface and an atomic structure that can be quite different to the one of the known bulk oxides, one might suspect new properties that are distinct to those of surfaces of both bulk metals and bulk oxides, and could thus be of potential interest for applications. For the specific functionality in oxidation catalysis, one needs, however, also to assess the role played by the second reactant.

5.3.3 Constrained Equilibrium

Although the average surface structure and composition are in catalysis even under steady state conditions determined by the kinetics of the ongoing elementary processes, it has been suggested that a thermodynamic approach with certain constraints could still provide a first insight into the effect of the reactive environment on the state of the catalyst surface [43,44]. A straightforward generalization of the above described atomistic thermodynamics concept to a multicomponent environment would be to account in (5.4) for a corresponding number of reservoirs described by chemical potentials $\Delta\mu_i$, where i denotes the different gas phase species. Modeling the catalyst surface in the reactive environment, each reactant is thus represented by a corresponding reservoir, with which the surface is in contact. However, in a full thermodynamic description all reservoirs would also be mutually in equilibrium. Under catalytically interesting environmental conditions, this yields directly the products as most stable environment, and is thus inadequate to treat the effect of exposure to the reactant gas phase.

Since it is precisely the high free energy barrier for the direct gas phase reaction that requires the use of the catalyst in the first place, it is appealing to instead consider the surface in contact with separate, independent reactant reservoirs. The surface would then be fully equilibrated with each reactant species, but the latter are not equilibrated with each other. Although neglecting the direct gas phase reaction is well justified, such a "constrained equilibrium" approach is still highly approximate, as it effectively also neglects the ongoing catalytic product formation at the surface itself, i.e., in other words exactly what one is in principle out to study. In this respect one should stress that the whole concept of a "constrained equilibrium" is not designed to yield precise answers to the microscopic state and reactivity of the catalyst surface. For that, the very dynamic behavior must be modeled by statistical mechanics, as will be illustrated below. Yet, with atomistic thermodynamics, the large scale behavior can be traced out first, possibly identifying those (T, p_i)-conditions where a more refined treatment explicitly accounting for kinetic effects is necessary.

How this works in practice is nicely illustrated by comparing the stability range of the bulk oxides of the late 4d TM series within the context of CO oxidation catalysis, i.e., in a gas phase formed of oxygen and CO [10]. The progressively decreasing heats of formation of the more noble metal oxides already discussed in the Sect. 5.3.2 are also clearly discernible in Fig. 5.12, which shows the oxide stability range as a function of the two chemical potentials in the gas phase. At very low CO chemical potential, the situation in a pure O_2 surrounding is recovered, and the stability range of RuO_2 and PdO coincides with the corresponding bulk oxide shaded regions in Fig. 5.10. With increasing CO content in the gas phase, this picture changes and the stability range gets eventually reduced due to CO-induced decomposition. Although uncertainties in the modeling of this decomposition translate

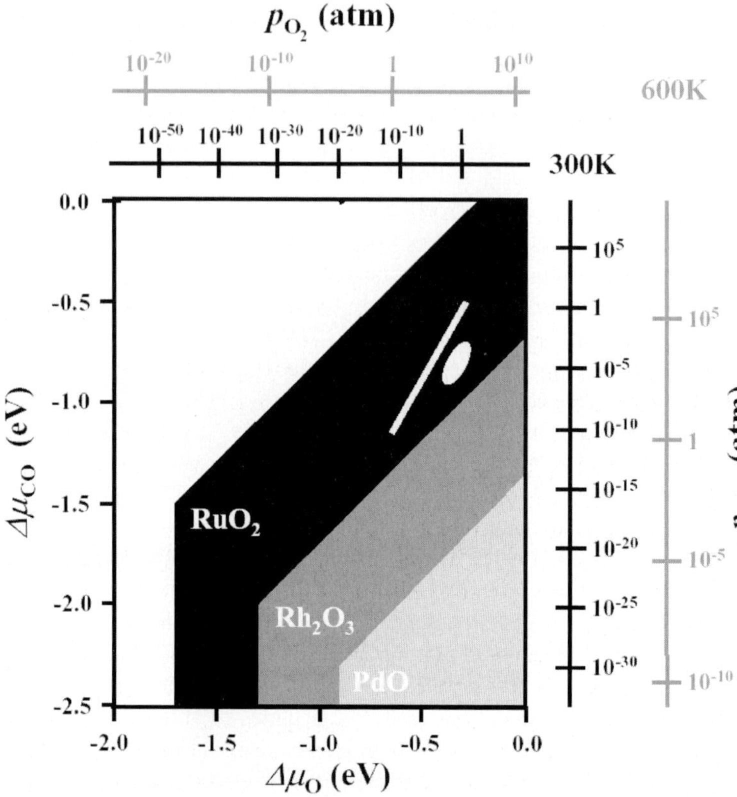

Fig. 5.12. Stability regions of bulk oxides of the late 4d TMs in $(\Delta\mu_O, \Delta\mu_{CO})$-space, as computed with DFT-GGA. Additionally, pressure scales are drawn for $T = 300$ and 600 K. The white *line* indicates environmental conditions corresponding to $p_{O_2} - p_{CO} = 1$ atm. and $300\,K \leq T \leq 600\,K$, and environments relevant for CO oxidation catalysis over these metals would correspond to the near vicinity of this *line*. More specifically, the small white area indicates the (T, p)-conditions employed in a recent experimental study by Hendriksen et al. on Pd(100) [45] (from [40])

presently into rather large error bars in the respective diagonal limits in Fig. 5.12 [40], a clear trend still emerges: Only the stability range of bulk RuO_2 extends well into environmental conditions representative of technological CO oxidation catalysis, roughly indicated in Fig. 5.12 as a white line. For Rh on the other hand the stability range of its oxide has already decreased significantly. Even if the real system was in a situation close to thermodynamic equilibrium, either metal or bulk oxide could prevail under ambient conditions, depending on whether the exact partial pressures and temperatures in the gas phase correspond to a situation above or below the stability line (which is at present impossible to assess due to the large uncertainty in the calculated position of this line). The bulk oxides of Pd and of Ag are finally already so unstable that they are unlikely to play a role in CO oxidation catalysis. The stability of Ag_2O in fact being so low, that it does not even show up anymore within the chemical potential range plotted in Fig. 5.12.

It is, however, important to realize that already quite thin oxide films at the surface would be sufficient to induce a changed catalytic functionality. In this respect, just considering the stability of bulk oxides as representatives of thick, bulklike oxide overlayers is not enough. Particularly at the more noble metals, we discussed already the extended stability range of the few atomic-layer thin surface oxide structures in a pure O_2 gas phase, which could similarly prevail in a reactive multicomponent environment. While the results of Fig. 5.12 indicate that thick, bulklike PdO and Ag_2O overlayers will hardly play a role in CO oxidation catalysis, the picture could be quite different for the nanometer thin surface oxide structures. Figure 5.13 exemplifies this with corresponding results for the case of a Pd(100) surface in constrained equilibrium with an O_2 and CO environment [46]. Similar to the aforediscussed situation at Pd(111), oxidation proceeds also at Pd(100) via a trilayer-structured O–Pd–O surface oxide film [35], that was recently identified as essentially a strained and rumpled layer of PdO(101) on top of the Pd(100) substrate [47]. Although the (101) orientation is not a low-energy surface of bulk PdO [48], it gets stabilized in the commensurate $(\sqrt{5} \times \sqrt{5})R27°$ arrangement of the thin oxide film due to a strong coupling to the underlying metal substrate. In line with our previous discussion, one would then expect an extended stability range compared to bulk PdO, which was indeed subsequently found both theoretically and experimentally for a pure O_2 gas phase [49]. The atomistic thermodynamics results summarized in Fig. 5.13 suggest furthermore that this would indeed hold equally for Pd(100) in an O_2 and CO surrounding. Comparing the stability of a large set of on-surface (co)adsorption, surface oxide and bulk oxide structures in constrained equilibrium with the reactive environment [46], several phases involving the $(\sqrt{5} \times \sqrt{5})R27°$ surface oxide are found to be most stable over a wide range of (T, p_{O_2}, p_{CO})-conditions that largely exceed the small stability range of bulk PdO.

The range extends in fact so much that it even just comprises the gas phase conditions typical for technological CO oxidation, i.e., partial pressures of the order of 1 atm. and temperatures around 300–600 K. Similar conditions were, e.g., also chosen in a recent in situ reactor scanning tunneling microscopy (STM) experiment of CO oxidation at Pd(100) by Hendriksen et al. [45], cf. Figs. 5.12 and 5.13. With increasing O_2:CO partial pressure ratio, the authors observed substantial changes in the surface morphology, which they assigned to the formation of a thin oxidic overlayer [45]. Inspecting Fig. 5.13, it is tempting to identify this overlayer as the already characterized $(\sqrt{5} \times \sqrt{5})R27°$ surface oxide, but this would be rushing beyond what can be safely concluded on

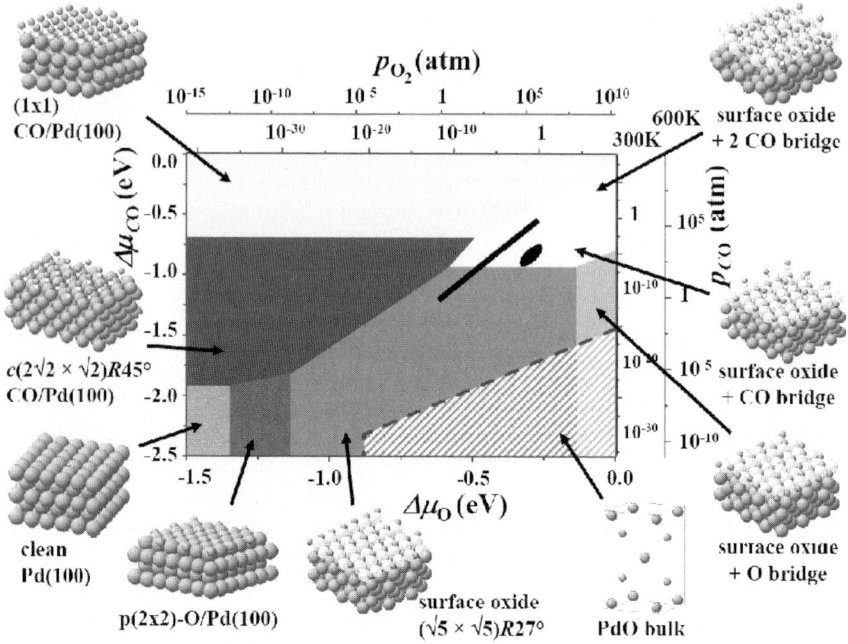

Fig. 5.13. DFT-GGA surface phase diagram of stable surface structures at a Pd(100) surface in "constrained equilibrium" with an O_2 and CO environment. The stability range of bulk PdO is the same as the one shown in Fig. 5.12. Phases involving the $(\sqrt{5}\times\sqrt{5})$R27° surface oxide exhibit a largely increased stability range, which now comprises gas phase conditions representative of technological CO oxidation catalysis ($p_{O_2} = p_{CO} = 1$ atm. and $300\,\text{K} \leq T \leq 600\,\text{K}$, marked by the black *line*, as in Fig. 5.12) (from [46])

the basis of the presently available DFT data: First, surface structures that are not yet explicitly compared in the atomistic thermodynamics approach could still change the surface phase diagram. Second, kinetic effects could significantly affect the oxidation state of the surface. This is the case, when the ongoing catalytic reaction consumes surface species faster than they can be replenished from the surrounding gas phase, thus yielding an average surface structure and composition that can be quite distinct to the predictions obtained within the constrained equilibrium concept. It could in particular be that a fast consumption of surface oxygen atoms reduces the stability range of the surface oxide well below the technologically interesting environmental conditions [which are even in constrained equilibrium quite close to the transition to the CO-covered Pd(100) surface phases]. At the moment, the only safe take-home message from the data presented in Fig. 5.13 is therefore that instead of the already ruled-out thick bulklike PdO film, the nanometer thin $(\sqrt{5} \times \sqrt{5})$R27° surface oxide appears as a likely candidate to actuate the increased catalytic activity connected with the morphology changes reported by Hendriksen et al. [45].

Still, the proximity of the technologically relevant high-pressure gas phase conditions to the phase transition between surface oxide and CO-covered Pd(100) in Fig. 5.13 is intriguing and stimulates further speculations. Depending on the exact partial pressures and temperatures, either the surface oxide or the Pd(100) metal surface could prevail. Even oscillations between the two phases are conceivable as a consequence of fluctuations, when operating under environmental conditions very close to the stability limit. At finite temperatures, configurational entropy may furthermore give rise to a complex phase coexistence at the surface [42, 44], which would then go in the direction of what was in the in situ reactor STM experiment interpreted as a continuous consumption and reformation of the surface oxide even under the employed stationary reaction conditions [45]. Instead of the one-time formation of a catalytically active oxide film in the case of Ru, this would point at a much more dynamic state of the Pd surface in the reactive CO oxidation environment: The catalytic activity could be intimately connected to a reversible (local) formation and reduction of the thin surface oxide, possibly even giving rise to spatiotemporal pattern formation phenomena as often reported for CO oxidation at Pt-group metals [45,50,51]. In such cases, also the domain boundaries between surface oxide and adlayers could be of particular importance for the overall reactivity.

5.3.4 Kinetically Limited Film Thickness

The constrained equilibrium description discussed up to now conveys the impression that a possible oxidation of the catalyst surface in the O-rich environments of oxidation catalysis would rather yield bulklike thick oxide films on Ru but thin surface oxide structures on the more noble 4d metals. This reflects the decreasing heat of formation of the bulk oxides over the late TM series, and seems to suggest that it is primarily at Pd and Ag where oxide formation in the reactive environment could be self-limited to nanometer or subnanometer thin overlayers. Particularly for the case of Ru, Fig. 5.12 shows that the gas phase conditions typical for technological CO oxidation catalysis fall deep inside the stability regime of the bulk oxide, indicating that thermodynamically nothing should prevent a continued growth of the once formed oxide film.

The formation of crystalline, bulklike $RuO_2(110)$ during high-pressure CO oxidation catalysis has indeed been observed experimentally at Ru(0001), but it is interesting to note that even after long operation times the film thicknesses never exceeded about 20 Å [30, 31, 52]. Generally, one could interpret this finding as reflecting kinetic limitations to a continued growth, presumably due to slow diffusion of either O or Ru atoms through the formed film [37,38]. While this could slow the thickening of the oxide overlayer down beyond hitherto measured time scales, it could also be that the kinetics of the ongoing catalytic reaction affect the surface populations in such a way that no further net diffusion through the film occurs and the nanometer thin oxide overlayer represents in fact a kinetically limited steady state in the given reactive environment. This would, e.g., be the case, if the oxidation reactions at the surface consume adsorbed oxygen atoms at such a pace that on average O penetration into the film occurs not more frequently than the filling of surface sites with oxygen from inside the film.

To explore this in more detail, one needs to go beyond the constrained equilibrium description and explicitly take kinetic effects into account in the theoretical modeling. This can for example be achieved via kinetic Monte Carlo (kMC) simulations, which can essentially be viewed as coarse-grained molecular dynamics simulations, following the time evolution of the system on the basis of the rates of all individual elementary processes occurring in the system [6,54,55]. Evaluating the statistical interplay between this manifold of atomic-scale processes, kMC simulations can easily cover sufficiently long time scales to arrive at proper averages of quantities like surface populations, even if the system is not in equilibrium, but only in a steady state as typical for catalysis. Such a modeling was recently performed for the CO oxidation over $RuO_2(110)$, using DFT-GGA based rates for 26 elementary surface processes, including adsorption, desorption, diffusion, and reaction events [53]. Of particular interest for the film thickness issue is the occupation of surface sites with O atoms in catalytically relevant gas environments. In this respect, the $RuO_2(110)$ surface offers two prominent adsorption sites, commonly denoted as bridge (br) and cus sites as explained in Fig. 5.14 [42].

Figure 5.15 shows the obtained steady state average surface populations at $T = 600\,K$ as a function of the gas phase partial pressures, together with the simultaneously computed TOFs (in units of formed CO_2 per cm^2 per s). The latter result from evaluating the average occurrence of the reaction events over

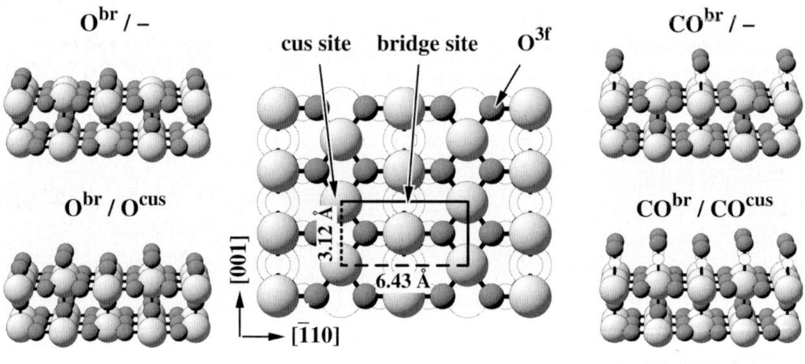

Fig. 5.14. Top view of the $RuO_2(110)$ oxide surface explaining the location of the two prominent adsorption sites (coordinatively unsaturated, cus and bridge, br). Also shown are perspective views of the four steady state adsorption phases present in Fig. 5.15 (Ru = light large *spheres*, O = dark medium *spheres*, C = white small *spheres*)

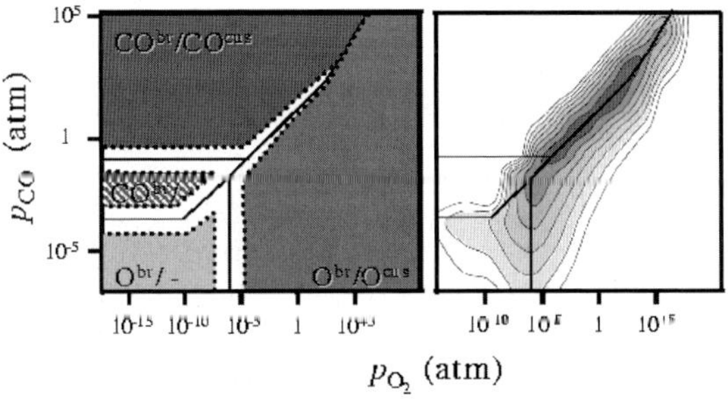

Fig. 5.15. Left panel: Steady-state surface structures of $RuO_2(110)$ in an O_2/CO environment obtained by first-principles kMC calculations at $T = 600\,K$. In all nonwhite areas, the average site occupation is dominated (> 90%) by one species, and the site nomenclature indicates the species at the two prominent adsorption sites offered by the surface: br = bridge sites, cus = coordinatively unsaturated sites, cf. Fig. 5.14. Right panel: Map of the corresponding catalytic CO oxidation activity measured as so-called turnover frequencies (TOFs), i.e., CO_2 conversion per cm^2 and second: White areas have a TOF $< 10^{11}\,cm^{-2}\,s^{-1}$, and each increasing gray level represents one order of magnitude higher activity. The highest catalytic activity (black region, TOF $> 10^{17}\,cm^{-2}\,s^{-1}$) is narrowly concentrated around the phase coexistence region in which O and CO compete for both br and cus sites (from [53])

long time periods as a measure of the catalytic activity and are intriguingly peaked around a narrow range of gas phase conditions. Contrary to the situation at most other partial pressures, the site occupation at br and cus sites is then not dominated by one species, i.e., either O or CO (or vacant). Instead, the kinetics builds up a surface population in which O and CO compete for either site type at the surface. This competition is in fact nicely reflected by the huge fluctuations in the surface populations with time apparent in Fig. 5.16. The dynamics at the surface in this "active state" is furthermore extremely fast, and the average time adsorbed O atoms stay in br or cus sites before desorbing or being reacted away is only of the order of fraction of milliseconds. Remarkably, it is particularly the concentration of O at the surface that turns out significantly lower than predicted within the constrained equilibrium approach [43, 44]. Even at the br sites, where O atoms bind very strongly [42], the O occupation is only 90% and not 100% as in the normally stable surface termination. This is quite surprising and indicates already how effectively the ongoing reaction consumes surface O species. It could therefore well be that the surface kinetics indeed limits a continued growth of the formed oxide film under this optimum catalytic performance.

5.3.5 Surface Oxidation and Sabatier Principle

The partial pressures and temperatures leading to this high-activity state of the $RuO_2(110)$ surface in the calculations, and even the absolute TOF

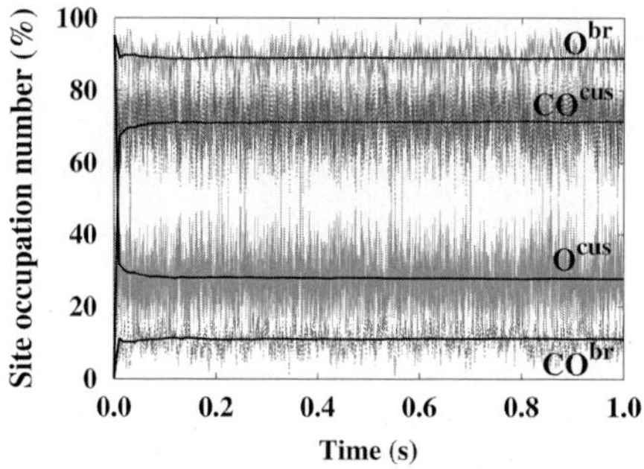

Fig. 5.16. Time evolution of the site occupation by O and CO of the two prominent adsorption sites at the $RuO_2(110)$ model catalyst surface shown in Fig. 5.14. The temperature and pressure conditions chosen ($T = 600$ K, $p_{CO} = 20$ atm., $p_{O_2} =$ 1 atm.) correspond to optimum catalytic performance, cf. Fig. 5.15. Under these conditions, kinetics builds up a steady state surface population in which O and CO compete for either site type at the surface, as reflected by the strong fluctuations in the site occupations (from [53])

values under these conditions, agree extremely well with detailed experimental studies measuring the steady state activity in the temperature range from 300 to 600 K and both at high pressures and in UHV [56–58]. This statement comprises already the emerging consensus that this Ru oxide also forms at the surface of Ru model catalysts in high-pressure environments [52, 53], i.e., the high activity that was originally ascribed to the Ru substrate was in fact due to RuO_2 [56,57]. With this understanding a puzzle is finally resolved that had bothered surface scientists for quite a while. For a long time the high CO oxidation reactivity of supported Ru catalysts [56], as well as of Ru(0001) single crystals [57] at high pressures stood in sharp contrast to exceedingly low turnover rates observed under UHV conditions [59], thus representing a nice example of a so-called "pressure gap" system [60].

Looking at the computed binding energies of oxygen at Ru(0001) shown in Fig. 5.2, the low catalytic activity of this surface appears in fact quite plausible. According to the early Sabatier principle [61], high activity would in general rather be expected for intermediate bond strengths at the surface (stable enough to form but weak enough to yield the final product). In this respect the computed medium binding energies at, e.g., Rh(111) and Pd(111) in Fig. 5.2 conform much better with the widespread use of these metals in catalytic car converters, whereas the oxygen bonding at Ru(0001) seems simply too strong. The correspondingly somehow enigmatic high catalytic activity of Ru(0001) at ambient pressures only became comprehensible when experimental studies reported the formation of bulklike $RuO_2(110)$ films at the surface in such reactive environments [30, 31]. The subsequently established [62] and above described high activity of this oxide surface then not only explained the pressure gap difference (Ru metal in UHV = low activity, RuO_2 at high pressures = high activity) but also reconciled this system with the Sabatier principle: The $RuO_2(110)$ surface indeed features oxygen species that are bound with intermediate bond strength comparable to the one of O at Rh(111) and Pd(111) [42,58].

An equivalent, though reverse explanation has also been suggested for the most noble of the late 4d metals [63,64]. According to Fig. 5.2, oxygen binds to Ag(111) only very weakly, which in view of the Sabatier principle is again difficult to combine with the known importance of Ag as catalyst for several oxidation reactions. However, the bond strength of oxygen in the p(4 × 4) surface oxide is substantially higher and would therefore well fall into the more favorable medium binding energy range [64]. If the surface oxide instead of Ag metal was the actually active surface in the high-pressure oxidation catalysis, the Sabatier principle would finally be reconciled with the complete trend of oxygen binding over the late 4d series apparent in Fig. 5.2. Both the too strong O binding on Ru and the too weak O binding on Ag is alleviated by the formation of oxides, which exhibit a similarly intermediate bond strength as oxygen at Rh or Pd.

It thus almost appears as if the interaction of oxygen with TM catalysts would serve to "tune" the O–metal bond strength for optimum reactivity through an appropriate amount of surface oxidation. Since the thermal stability of bulk Ag_2O is already too low and in line with our preceding extended stability discussion, this oxidation can at Ag at best correspond to the formation of a thin surface oxide though. Yet, compared to the surface oxides at Pd, even the latter is probably not stable enough to play an active role in CO oxidation. This can be different in the less reducing environments of partial oxidation reactions, consistent with the predominant use of Ag catalysts for ethylene epoxidation or formaldehyde synthesis [61].

These examples underline therefore already nicely the potential relevance of surface oxidation in the reactive surroundings of oxidation catalysis. In all cases discussed so far, this had a beneficial effect on the overall catalytic performance, but detrimental effects due to the formation of an inactive oxide film are of course equally well conceivable. Such a catalyst deactivation through oxidation has, e.g., been frequently suspected in the experimental literature concerning Rh surfaces [65–67]. In other cases, the catalytic activity of oxide and metal surface might be comparable, so that surface oxidation would likely remain unnoticed unless explicitly looked for by experimental techniques sensitive to the atomic scale.

5.4 Conclusions

Novel experimental and theoretical approaches addressing the effect of realistic environments on surface structure and composition have significantly increased the atomic-scale understanding of the oxide formation process of late TMs over the last years. Some identified key features have been reviewed for the late 4d sequence from Ru to Ag and are summarized schematically

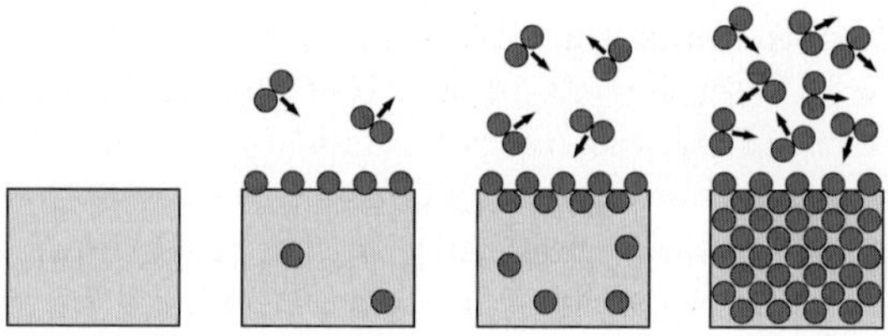

Fig. 5.17. Cartoon sideviews illustrating the effect of an increasingly oxygen-rich atmosphere on a late TM surface. Whereas the clean surface prevails in perfect vacuum (left), finite O_2 pressures in the environment first lead to oxygen adsorption phases. Apart from some bulk-dissolved oxygen, the lower deformation cost will at increasing pressures lead to a preferential accommodation of oxygen in the near-surface fringe. Thin surface oxide structures are the salient consequence before eventually thickening of the oxide film and formation of an ordered bulk compound set in. A thermodynamic or kinetic stabilization of the nanometer thin surface oxide could lead to novel functionalities, different from both bulk metal or bulk oxide surfaces

again in Fig. 5.17. A progressively more O-rich gas phase leads first to on-surface adsorption, followed by O penetration into the surface at increasing O accommodation. The incorporation takes place after the formation of dense oxygen overlayers in the case of Ru but already at lower on-surface coverages toward the more noble metals. This results from the opposing trends in decreasing O–metal bond strength and decreasing deformation cost of the metal lattice, the latter being a consequence of the substantial local expansion necessarily connected to O incorporation into the close-packed TM lattices. The deformation cost is also responsible for an enrichment of oxygen in the easier to relax immediate surface fringe, eventually inducing the formation of thin oxidic structures at the surface. Although some bulk dissolution will always occur on entropy grounds, there is thus primarily a propensity to form few atomic-layer thin surface oxides, which mostly goes hand in hand with substantial structural changes. The resulting geometry is then not only characterized by an optimum binding within the more open compound structure but can also be largely affected by the coupling to the underlying metal substrate (i.e., the nature of the oxide–metal interface) and residual strain. As a consequence, complex atomic arrangements not resembling surfaces of bulk oxides or bulk metals have been reported for this step in the oxidation process. Only after continued oxygen accommodation, the thickening oxide film will eventually switch to a bulk oxide structure and the continued film growth will then be more and more limited by the slow diffusion of O and/or metal atoms between surface and interface.

A most intriguing aspect of this surface oxidation is that it proceeds via the few atomic-layer thin surface oxide structures. This thickness can be ruled by kinetics, but at Pd and Ag it even seems that there are thermodynamic conditions for which the surface oxides represent the most stable phase. In corresponding environments such films are formed automatically, but the growth is self-limited to the potentially particularly interesting nanometer or subnanometer width. If stabilized in applications, the often quite distinct geometries of these films already suggest that new properties and functionalities might emerge that are not scalable from the known bulk materials. At present, very little to nothing is known about these properties, and one should recall that not even the corresponding bulk oxides of the late 4d TMs have hitherto received a lot of attention in atomic-scale studies. This opens up a wide new field for investigations of this new "material class" and its functionalities in various applications.

With respect to oxidation catalysis, oxide films may even form as a natural consequence of the reactive environment, and since maybe hitherto unnoticed, they might then actuate some of the catalytic functions that have traditionally been ascribed to the metal substrates. In this respect, the decreasing heat of formation leads to a clear trend for the bulk oxides. A bulklike RuO_2 film can even sustain the reducing environments of high-pressure total oxidation reactions, whereas the feeble Ag_2O will hardly prevail. This is, however, not enough to exclude the relevance of oxidized structures for oxidation catalysis over Pd and Ag catalysts. Also due to the coupling to the underlying substrate, the thin surface oxides exhibit a much extended stability range, and it seems at the moment at least feasible that for example CO oxidation over Pd is actuated by them, or their constant consumption and reformation. The present data suggest a similar role for surface-confined oxidic structures at Ag but only in less reducing environments, e.g., typical for partial oxidation. This would then also offer an intuitive explanation for the often reported high catalytic activity of noble metals for such reactions. Together with the formation of highly active, nanometer thin RuO_2 films at the surface of Ru model catalysts, this underlines the potential importance of surface oxidation in the reactive environments of oxidation catalysis over late TMs. Although often not yet fully conclusive, already the reviewed new insights into the oxidation behavior help sometimes to reconcile the data with more general concepts like the Sabatier principle. Much more often, however, they lead to the fall down of "established" rules and preconceptions, which is a prospect that makes working in this field exciting and all the worthwhile.

Acknowledgments

I gratefully acknowledge the valuable contributions and help from my (former) colleagues working on metal oxidation in the Theory Department of the Fritz-Haber-Institut: Jörg Behler, Veronica Ganduglia-Pirovano, Wei-Xue Li, Angelos Michaelides, Jutta Rogal, Cathy Stampfl, and Mira Todorova. Special thanks go to Matthias Scheffler for his continued support and for the insightful discussions on the intriguing facets of surface oxidation and catalysis. The EU is acknowledged for financial support under contract NMP3-CT-2003-505670 ($NANO_2$), and the DFG for support within the priority program SPP1091.

References

1. Haruta M (1997) Catal. Today 36:153
2. Dreizler RM, Gross EKU (1990) Density-Functional Theory. Springer, Berlin
3. Parr RG, Yang W (1994) Density-Functional Theory of Atoms and Molecules. Oxford University Press, Oxford
4. Scheffler M, Stampfl C (2000) Theory of Adsorption on Metal Substrates. In: Horn K, Scheffler M (eds.), Handbook of Surface Science. Vol 2: Electronic Structure. Elsevier, Amsterdam
5. Groß A (2002) Theoretical Surface Science—A Microscopic Perspective. Springer, Berlin
6. Reuter K, Stampfl C, Scheffler M (2005), Ab Initio Atomistic Thermodynamics and Statistical Mechanics of Surface Properties and Functions. In: Yip S (ed.), Handbook of Materials Modeling. Part A: Methods Springer, Berlin
7. Behler J, Delley B, Lorenz S, Reuter K, Scheffler M (2005) Phys. Rev. Lett. 94:036104
8. Perdew JP, Burke K, Ernzerhof M (1996) Phys. Rev. Lett. 77:3865
9. Zhang Y, Yang W (1998) Phys. Rev. Lett. 80:890
10. Perdew JP, Kurth S, Zupan A, Blaha P (1999) Phys. Rev. Lett. 82:2544
11. Filippi C, Healy SB, Kratzer P, Pehlke E, Scheffler M (2002) Phys. Rev. Lett. 89:166102
12. Surnev L, Rangelov G, Bliznakov G (1985) Surf. Sci. 159:299
13. Pfnür H, Held D, Lindroos M, Menzel D (1989) Surf. Sci. 220:43
14. Comelli G, Dhanak VR, Kiskinova M, Prince KC, Rosei R (1998) Surf. Sci. Rep. 32:165
15. Stampfl C, Scheffler M (1996) Phys. Rev. B 54:2868
16. Stampfl C, Kreuzer HJ, Payne SH, Pfnür H, Scheffler M (1999) Phys. Rev. Lett. 83:2993
17. Ganduglia-Pirovano MV, Scheffler M (1999) Phys. Rev. B 59:15533
18. Todorova M, Li WX, Ganduglia-Pirovano MV, Stampfl C, Reuter K, Scheffler M (2002) Phys. Rev. Lett. 89:096103
19. Li WX, Stampfl C, Scheffler M (2002) Phys. Rev. B 65:075407
20. Nørskov JK (1990) Rep. Prog. 53:1253
21. Hammer B, Nørskov JK (2001) Adv. Catal. 45:71
22. Reuter K, Ganduglia-Pirovano MV, Stampfl C, Scheffler M (2002) Phys. Rev. B 65:165403
23. Li WX, Stampfl C, Scheffler M (2003) Phys. Rev. B 67:045408
24. Ganduglia-Pirovano MV, Reuter K, Scheffler M (2002) Phys. Rev. B 65:245426
25. Todorova M, Reuter K, Scheffler M (2005) Phys. Rev. B 71:195403
26. Gibson KD, Viste M, Sanchez EC, Sibener SJ (1999) J. Chem. Phys. 110:2757
27. Stampfl C, Schwegmann S, Over H, Scheffler M, Ertl G (1996) Phys. Rev. Lett. 77:3371
28. Wider J, Greber T, Wetli E, Kreutz TJ, Schwaller P, Osterwalder J (1998) Surf. Sci. 417:301

29. Carlisle CI, Fujimoto T, Sim WS, King DA (2000) Surf. Sci. 470:15, Michaelides A, Reuter K, Scheffler M (2005) J. Vac. Sci. Technol. A 23:1487; Schnadt J, Michaelides A, Knudsen J, Vang RT, Reuter K, Laegsgaard E, Scheffler M, Besenbacher F (2006) Phys. Rev. Lett. 96:146101
30. Böttcher A, Niehus H, Schwegmann S, Over H, Ertl G (1997) J. Phys. Chem. B 101:11185
31. Over H, Kim YD, Seitsonen AP, Wendt S, Lundgren E, Schmid M, Varga P, Morgante A, Ertl G (2000) Science 287:1474
32. Gustafson J, Mikkelsen A, Borg M, Lundgren E, Köhler L, Kresse G, Schmid M, Varga P, Yuhara J, Torrelles X, Quiros C, Andersen JN (2004) Phys. Rev. Lett. 92:126102
33. Lundgren E, Kresse G, Klein C, Borg M, Andersen JN, De Santis M, Gauthier Y, Konvicka C, Schmid M, Varga P (2002) Phys. Rev. Lett. 88:246103
34. Carlisle CI, King DA, Boucquet ML, Cerda J, Sautet, P (2000) Phys. Rev. Lett. 84:3899
35. Zheng G, Altman EI (2000) Surf. Sci. 462:151
36. Reuter K, Stampfl C, Ganduglia-Pirovano MV, Scheffler M (2002) Chem. Phys. Lett. 352:311
37. Fromhold AT (1976) Theory of Metal Oxidation, Vols. I and II. North-Holland, Amsterdam
38. Roosendahl SJ, Vredenberg AM, Habraken FHPM (2000) Phys. Rev. Lett. 84:3366
39. Michaelides A, Bocquet ML, Sautet P, Alavi A, King DA (2003) Chem. Phys. Lett. 367:344
40. Reuter K, Scheffler M (2004) Appl. Phys. A 78:793
41. Xie J, de Gironcoli S, Baroni S, Scheffler M (1999) Phys. Rev. B 59:970
42. Reuter K, Scheffler M (2002) Phys. Rev. B 65:035406
43. Reuter K, Scheffler M (2003) Phys. Rev. Lett. 90:046103
44. Reuter K, Scheffler M (2003) Phys. Rev. B 68:045407
45. Hendriksen BLM, Bobaru SC, Frenken JWM (2004) Surf. Sci. 552:229; Hendriksen BLM (2003) Model Catalysts in Action: High-Pressure Scanning Tunneling Microscopy. Ph.D. Thesis, Universiteit Leiden
46. Rogal J, Reuter K, Scheffler M (in preparation)
47. Todorova M, Lundgren E, Blum V, Mikkelsen A, Gray S, Gustafson J, Borg M, Rogal J, Reuter K, Andersen JN, Scheffler M (2003) Surf. Sci. 541:101
48. Rogal J, Reuter K, Scheffler M (2004) Phys. Rev. B 69:075421
49. Lundgren E, Gustafson J, Mikkelsen A, Andersen JN, Stierle A, Dosch H, Todorova M, Rogal J, Reuter K, Scheffler M (2004) Phys. Rev. Lett. 92:046101
50. Bondzie VA, Kleban P, Dwyer DJ (1996) Surf. Sci. 347:319
51. Hendriksen BLM, Frenken JWM (2002) Phys. Rev. Lett. 89:046101
52. Over H, Muhler M (2003) Prog. Surf. Sci. 72:3
53. Reuter K, Frenkel D, Scheffler M (2004) Phys. Rev. Lett. 93:116105; Reuter K, Scheffler M (2006) Phys. Rev. B 73:045433
54. Landau DP, Binder K (2002) A Guide to Monte Carlo Simulations in Statistical Physics. Cambridge University Press, Cambridge
55. Kang HC, Weinberg WH (1995) Chem. Rev. 95:667
56. Cant NW, Hicks PC, Lennon BS (1978) J. Catal. 54:372
57. Peden CHF, Goodman DW (1986) J. Phys. Chem. 90:1360
58. Fan CY, Wang J, Jacobi K, Ertl G (2001) J. Chem. Phys. 114:10058

59. Madey TE, Engelhardt HA, Menzel D (1975) Surf. Sci. 48:304
60. Stampfl C, Ganduglia-Pirovano MV, Reuter K, Scheffler M (2002) Surf. Sci. 500:368
61. Ertl G, Knözinger H, Weitkamp J (eds.) (1997) Handbook on Heterogeneous Catalysis. Wiley, New York
62. Kim YD, Over H, Krabbes G, Ertl G (2001) Topics Catal. 14:95
63. Li WX, Stampfl C, Scheffler M (2003) Phys. Rev. Lett. 90:256102
64. Li WX, Stampfl C, Scheffler M (2003) Phys. Rev. B 68:165412
65. Oh SH, Carpenter JE (1983) J. Catal. 80:472
66. Kellog GL (1985) Phys. Rev. Lett. 54:82; Kellog GL (1986) Surf. Sci. 171:359
67. Peden CHF, Goodman DW, Blair DS, Berlowitz PJ, Fisher GB, Oh SH (1988) J. Phys. Chem. 92:1563

6

Catalytic Applications for Gold Nanotechnology

Sónia A.C. Carabineiro and David T. Thompson

6.1 Introduction

Gold is often considered the most inert of all metals. In fact, it is a widely held belief instilled in us from the earliest days in our school chemistry laboratories that gold is amongst the most stable and incorruptible of substances. For that reason, its chemical reactivity has not yet been as fully investigated as that of the platinum group metals (PGMs) and its remarkable catalytic properties were ignored until recently. Defining methods for preparing catalysts having nanoparticles of gold on oxide supports have opened up this new area of opportunity.

Although gold has been known and used in jewellery since ancient Egyptian times, it has to date found fewer practical applications than other precious metals. This is partly because gold is regarded as noble and therefore thought even by chemists to have limited potential for chemistry! This is in spite of the fact that there is much more gold in the world than PGMs [1, 2]. More gold could easily be made available for new industrial applications when increased demand materialises. Annual usage of gold is more than 20 times that for platinum, but jewellery accounts for close to 80% of this in the case of gold and only 50% with platinum: One of the indications that non-metallic uses are less well developed for gold than for the PGMs.

Recent increased interest in investigating the chemical properties of gold and that of its derivatives is, however, leading to a much richer picture emerging. Not least amongst these developments has been the demonstration that when gold catalysts are prepared in such a way that gold is dispersed as small nanometre-sized particles on transition metal oxides of the first row of the transition series in the periodic table, or alumina and some other oxides, then the resulting species become very active. For example, gold on iron oxide catalyses the oxidation of carbon monoxide at 197 K. The recent resurgence of interest in catalysis by gold is thus currently disturbing the balance of conventional wisdom amongst precious metal catalyst scientists and technologists.

In this chapter, we highlight the preparative methods, which have been used successfully to produce active, selective and durable nanogold catalysts and look at their properties and the reactions which they promote. We then consider the potential commercial applications for these new catalysts and look into the future.

6.2 Preparative Methods

For gold to be an active catalyst, use of a careful preparation procedure is crucial in order to obtain it as nanoparticles well dispersed on the oxide support. In this section we describe the methods devised for the preparation of these nanoparticulate gold catalysts. Catalysts are thought to be most active in the 2- to 5-nm gold particle range but both smaller and larger particles have also been shown to have activity and may play a significant role.

6.2.1 Naked Gold, Including Gold Single Crystals and Colloidal Gold

Supported nanogold catalysts, also called "real" catalysts, have received more study to date than naked gold, and they are thought to have more viable commercial applications. However, the study of gold catalysis using the surface science approach provides useful fundamental information relevant to reaction mechanisms and the nature of active centres of supported gold catalysts. Even for CO oxidation, the most studied reaction to date, the mechanistic pathways are still uncertain. Recent studies have, however, provided strong evidence that CO adsorbs on the gold particles themselves [3–8], probably on edge or step sites [6,7,9]. These findings make the CO–gold interaction an interesting topic for a model study on gold single crystal surfaces in ultra-high vacuum (UHV).

For preparation of a gold *single crystal surface*, the sample is cut from a single crystal ingot of gold by spark erosion and mechanically polished. The crystal is then mounted in a sample holder by means of tungsten wires [10], or tantalum wires or foil [11–14]. Passing current through the wires allows heating of the crystal resistively. The sample temperature can be monitored by means of a thermocouple attached to an edge of the crystal [11–13]. The sample is then cleaned in situ in a UHV unit by cycles of Ar^+ ion bombardment at room temperature and annealing at temperatures as high as 800 K. After the first few sputter and annealing cycles, a repolishing step may be necessary in order to get rid of pinning and faceting due to segregation of bulk contamination [15]. The surface cleanliness can be checked by Auger electron spectroscopy (AES), low-energy electron diffraction (LEED) and X-ray photoelectron spectroscopy (XPS).

Colloidal gold has a very long history. Colourful aqueous solutions of gold colloids date back to Roman times and were known to the medieval alchemists

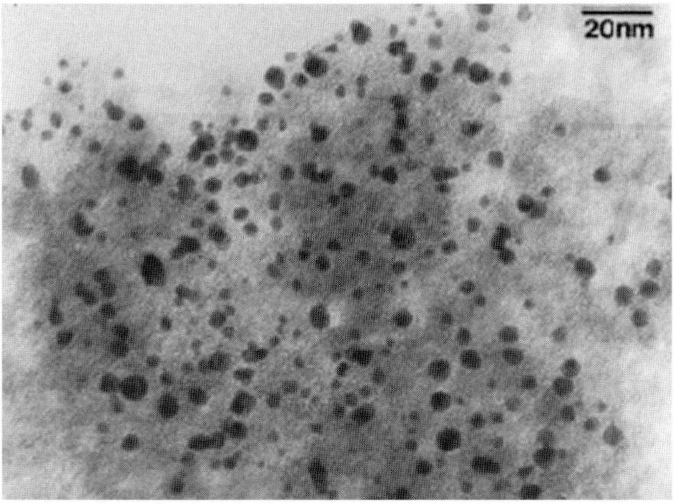

Fig. 6.1. HRTEM (High Resolution Transmission Electron Microscopy) of colloidal gold on carbon. Gold particles are viewed as dark points. Conditions: [Au] = 5×10^{-4} M, THPC/Au (mol mol^{-1}) = 2 (from [27])

as *aurum potabile* [16]. The notion that gold sols contain small metallic particles was first noted by Faraday as early as 1857 [17]. New recent developments in preparative and analytical methods have led to the preparation of much smaller particles and to the determination of intimate details of their structures [18].

The preparation of colloidal gold is generally accomplished by reduction of chloroauric acid ($HAuCl_4$) solution with sodium citrate [19–23], which is known to be influenced by several factors such as temperature and reactant concentrations [22]. Alternatively, other reducing agents can be used such as tetrakis(hydroxymethyl)phosphonium chloride (THPC) [24], borohydride [20, 25] and toluene [26].

The product is a hydrosol of gold of a specific colour (see Fig. 6.1). Inferences are drawn concerning particle size through comparison with the brilliant colours often shown by colloidal gold. Thermal decomposition of the chloroaurate ion on silica gives mauve products, but when its concentration is so high that reduction is needed to form the metallic state, the colour is dark red to brown. It was noted that colloidal gold is red when particles are smaller than 20 nm, and blue when they are larger. The observed intermediate colours correlate with XRD and TEM evidence on particle size. The colours shown appear to depend on the support used as well as on gold concentration, morphology, size and size distribution [18].

Control over the gold colloidal particle size, by simple variation of the relative surfactant concentration in the salt solution during reduction has been studied in detail for alkanethiols by Leff and co-workers [26].

6.2.2 Co-Precipitation

This is one of the simplest ways to prepare gold catalysts and it was one of the first to be used [28]. An aqueous solution of $HAuCl_4$ and water-soluble metal salts, such as a nitrate, is poured into an aqueous alkaline solution (Na_2CO_3 and/or NH_4OH) and agitated for a few minutes. The two hydroxides (or hydrated oxides) are precipitated simultaneously. After ageing for about 1 h, the precipitates are washed several times with water and filtered. The hydroxide and/or carbonate mixture is dried overnight and calcined to obtain powder catalysts [28–30]. In order to obtain a homogeneous dispersion of gold nanoparticles, the metal salt concentration has to be around 0.1–0.4 ML^{-1}, the pH range of 7–10 and the temperature between 320 and 360 K for precipitation and 550 and 670 K for calcination [28]. Then a reduction step is necessary. Many gold catalysts prepared by this method will, however, contain a significant concentration of sodium ion and chloride ion, if a metal chloride is used as precursor. Both can act as a catalyst poison [18, 31].

The applicability of this technique is limited to metal hydroxides or carbonates that can be co-precipitated with $Au(OH)_3$. Gold can be supported in the form of well-dispersed nanoparticles, by CP, on α-Fe_2O_3, Co_3O_4, NiO and ZnO, but not on TiO_2, Cr_2O_3, MnO_x and CdO [28].

6.2.3 Deposition Precipitation

This method is convenient and is used for producing commercial gold supported catalysts [7, 28]. It is also the one that can be applied to the widest range of different support materials [32].

The precursor to the active species is brought out of solution in the presence of a suspension of the support, usually by raising the pH in order to precipitate a hydroxide. The surface of the support acts as a nucleating agent and the method, if performed correctly, leads to the greater part of the active precursor being attached to the support. It is advisable to prevent the precipitation occurring away from the support by avoiding local high concentrations of alkali [18, 33, 34].

In the procedure described by Haruta, after the pH of an aqueous solution of $HAuCl_4$ is adjusted with NaOH, to a fixed point in the range of 6–10, a metal oxide support, in the form of a powder, bead, honeycomb, or thin film, is immersed in the solution. The partially hydrolysed species $[Au(OH)_nCl_{4-n}]^-$ $(n = 1–3)$ then react with the surface of the support. Ageing for about 1 h results in the deposition of $Au(OH)_3$ exclusively on the surface of the metal oxide support, if the concentration and temperature are properly chosen [18, 28, 31].

The dispersion and size of the particles and therefore their catalytic activity, depend critically upon the pH used (typically 6–10) and the amount of gold in the solution [8, 9, 31–33, 35–41].

Haruta emphasises that the influence of the pH on the particle size of Au is remarkable. Above pH 6, the main species of Au in solution are transformed from $AuCl_4^-$ to $[Au(OH)_nCl_{4-n}]^-$ ($n = 1$–3), and the mean particle diameters of Au in the calcined catalysts become smaller than 4 nm [31].

Consequently, it has been found that preparation at a pH ranging from 7 to 8 is preferable depending on the oxide support [8,9,32]. At this pH, the value of n is close to 3. At lower values of pH, there is less hydrolysis of the Au–Cl bond. Furthermore, at a pH below the isoelectric point of the support, the surface is positively charged and capable of adsorbing more of the negatively charged gold species. This results in not only a larger gold loading but also a high concentration of chloride on the surface. The presence of chloride increases the mobility of Au on the support, leading to large Au particles [9,42]. At pH values above the isoelectric point of the oxide, adsorption of the negatively charged $[Au(OH)_nCl_{4-n}]^-$ complex decreases rapidly, resulting in a lower gold loading. However, there will also be less chloride at the catalyst surface, so small gold particles can be formed. Consequently, there is a narrow range of pH where sufficient Au can be deposited onto the support with minimal chloride in the Au complex [9].

Bond and co-workers [43] showed that pH 9 was the optimum value to be reached during deposition precipitation for high activity of Au/TiO_2 catalysts used in CO oxidation. At this pH, the main species in solution were anionic Au complexes, from which most of the chlorine had been removed by hydrolysis. At lower pH, the Au complexes contained more chloride, Au particles were larger, and activities were lower [43].

Another effect of the pH of the solution during preparation has been identified. In addition to $[Au(OH)_nCl_{4-n}]^-$, Cl^- ions are adsorbed onto the oxide. The amount of chloride adsorbed decreases rapidly as the isoelectric point is approached, when the oxide surface is no longer positively charged [9]. Adsorbed Cl^- provokes an increase in particle size of gold during calcination, by the formation of Au–Cl–Au bridges in all directions, and therefore causes agglomeration of Au particles due to the proximity of Au atoms, such that the average Au particle size is larger in samples with higher residual chloride content [9,42,44,45]. The adsorbed chloride can be displaced by other anions.

Addition of magnesium citrate to the preparation solution can lead to an active catalyst [39,42,46]. It was suggested that citrate ions were adsorbed on the support, and they suppressed migration of Au particles during heat treatments [39]. Some authors, however, believe that the primary function of the citrate is to displace chloride from the support and possibly from the $[Au(OH)_nCl_{4-n}]^-$ complex [9,42,46]. As a step for reduction of gold complex to metallic gold is required for activating gold catalysts prepared by this method [47], chloride can thus be removed by reduction or calcination [9,47]. However, high-temperature calcinations or reduction cause agglomeration of Au particles. In contrast to $HAuCl_4$, $Au(OH)_3$ agglomerates only minimally during calcination [41].

An alternative method for adjusting the pH is to use urea [18]. This consists in the slow decomposition of urea in the solution. Hydroxyl ions are generated slowly and uniformly throughout the liquid phase, and their concentration is always low because they are consumed almost as soon as they are formed. The chemistry, as it applies to the active component, therefore, closely resembles that operating in CP [18].

Louis and co-workers [33, 48–50] have used the urea method to prepare Au/TiO_2 catalysts. These authors found that it gives the same gold particle sizes as those obtained by Haruta and co-workers with the deposition–precipitation (DP) method using NaOH, i.e., 2–3 nm, and no sodium poison is introduced. The main advantage of the urea method over that using sodium hydroxide for raising the pH is that all the gold present in solution is deposited on the TiO_2 surface as a gold(III) precipitate, which is not gold(III) hydroxide, but an amorphous compound containing nitrogen, oxygen and carbon [50]. This compound arises from a reaction between the gold precursor and the products of decomposition of urea. The metallic gold particles obtained after calcination have a decreasing size, when the time of urea addition increases. The progressive increase of pH due to urea decomposition at 353 K probably results in changes in the surface charge density of the gold precipitate particles and leads to fragmentation of the particles. Using this method, a higher gold loading can be achieved and the gold loading of the catalysts can be easily controlled [49, 50]. Venezia and co-workers [51] also used this method to prepare bimetallic Pd–Au catalysts supported on ASA (86% SiO_2, 13% Al_2O_3 and minor oxide components). Also, Nieuwenhuys and co-workers [38, 52–54] prepared multi-component gold catalysts, $Au/M^IO_x/M^{II}O_x/Al_2O_3$ (M^I, transition metal; M^{II}, alkaline earth metal) using this method. Deevi and co-workers [55, 56] also prepared Au/Fe_2O_3 catalysts, concluding that it produced a catalyst with higher final gold content than the CP method because it achieved more complete precipitation of the gold from solution. Gold catalysts supported on hydrotalcite (HT) with an Mg:Al molar ratio 4:2 were also prepared by Pitchon and co-workers [57].

In this slow DP method, the support is added to an aqueous solution containing $HAuCl_4$ (together with another metal precursor in the case of bimetallic catalysts) and urea. The initial pH is around 2. The suspension is thermostated at 353 K and vigorously stirred for several hours and urea decomposition generates OH^- ions, leading to a gradual rise in pH from 2 to 7. Centrifugation, washing, drying and calcination at 573 K follow [33, 48–51].

A constraint of the DP technique is that it is not applicable to activated carbon [7, 28, 59] or zeolites [60] due to their high isoelectric point. It is interesting that graphite substrates have often been used in photoemission experiments designed to explore the electronic properties of gold nanoparticles and that such particles appear to have similar electronic character to gold particles on oxides [41].

However, DP has the advantage over CP in that all of the active component remains on the surface of the support and none is buried within it [8, 18]. Also,

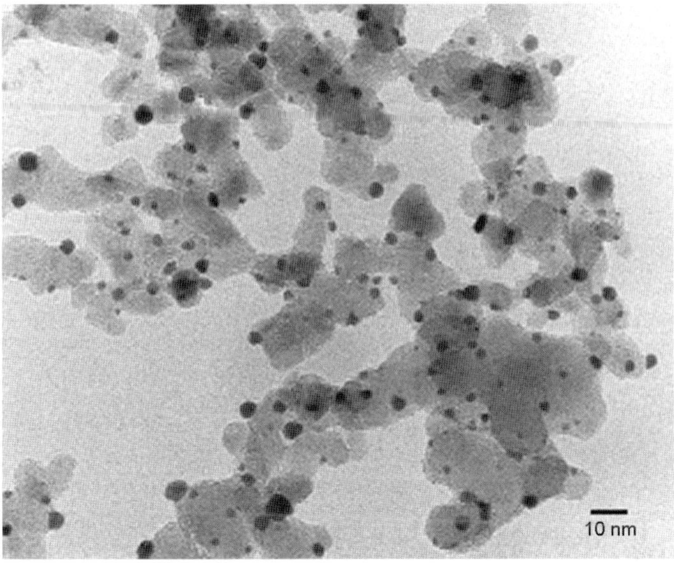

Fig. 6.2. TEM photograph of an Au/Al_2O_3 catalyst calcined in air at 623 K for 4 h prepared by DP with NaOH. Gold particles are seen as dark points (from [58])

it gives narrower particle size distributions, but it is recommended that the support should have a specific area of at least $50\,m^2\,g^{-1}$ [18]. Fig. 6.2. shows an Au catalyst prepared by DP.

6.2.4 Impregnation

This is the classical method used to make supported noble metal catalysts and consists of impregnating a support with a solution of a metal salt. This usually involves suspending the support in a larger volume of solution, from which the solvent is then removed. An alternative variation is the so-called incipient wetness (IW) technique, in which the pores of the support are filled with the solution [61–63].

Chloroauric acid ($HAuCl_4$) or auric chloride ($AuCl_3$ or Au_2Cl_6) is usually used but complex salts, such as potassium aurocyanide ($KAu(CN)_2$) and the ethylenediamine complex $[Au(en)_2]Cl_3$, are alternatives. Traditional supports are silica, alumina and magnesia, but titania, alumina, boehmite ($AlO(OH)$), ferric oxide (α–Fe_2O_3) and magnesium hydroxide can also be used [18].

After drying, the precursor has to be calcined at temperatures as high as 1,073 K and reduced. This reduction can be by hydrogen at 523 K, by aqueous oxalic acid at 313 K or by aqueous magnesium citrate [18].

Conventional impregnation techniques result in much larger gold particles than deposition or precipitation methods [5, 18, 38, 39, 64]. It is difficult to deposit Au as nanoparticles on metal oxide supports by this method and to obtain a high dispersion of gold (except for the few cases of very low metal loading), mainly because Au has a lower melting point and lower affinity to metal oxides than Pd and Pt. Also, during calcinations of $HAuCl_4$ crystallites, which are dispersed in the support surfaces, the gold particles experience severe agglomeration as $HAuCl_4$ interacts only weakly with the support [28, 31,39,41]. The early use of classical impregnation techniques was the principal reason why catalysis by gold was thought for so long to be inferior to that of the PGMs.

However, Datye and co-workers [65] reported a new impregnation method for supporting gold on alumina using $HAuCl_4$. Since impregnation under acidic conditions has been found to lead to poor dispersion of Au and the resulting catalysts are not as active as those prepared by deposition precipitation, a two-step procedure has been developed. In the first step, gold chloride was adsorbed on alumina from an acidified solution. After washing off the excess gold precursor, the solid was treated with strong base to convert the chloride to an adsorbed hydroxide. Drying and calcining at 673 K yielded a catalyst with gold particles having a number average diameter of 2.4 nm, with reactivity for CO oxidation at room temperature comparable with catalysts prepared by the DP method. These catalysts are stable to hydrothermal sintering, with an average particle size of around 4 nm after sintering in 10-mol% H_2O at 873 K for 100 h [65]. This new method is a successful impregnation preparative route for gold catalysts, allowing them to have durability at least until 873 K, since no chloride is present.

6.2.5 Vapour-Phase Methods and Grafting

These methods are similar, differing only in whether or not a solvent is present. In the *vapour-phase method* [sometimes called chemical vapour deposition (CVD)], a stream of a volatile compound of gold is transported onto a high area support by an inert gas and it reacts chemically with the surface of the support to form a precursor to the active species [18,31]. The most widely used gold precursor is $AuCl_3$ or $HAuCl_4$, but other substances have also been used, mainly to prevent chloride contamination [5]. Dimethyl-gold(III) β-diketone has been applied to a number of supports, including alumina and titania and to acidic supports, such as silica and MCM-41, for which DP would not work. Gold acetylacetonate has also been used to prepare Au/SiO_2 catalysts [5,18] and Au/Al_2O_3 [47]. Also, organometallic compounds of gold can be used for CVD [18,47,66,67]. The adsorbed organic gold compound is pyrolised in air to decompose it into small gold particles [31].

In the *grafting* method, a gold complex in solution reacts with the surface of a support, forming species convertible to a catalytically active form. Thus, the compounds $[Au(PPh_3)]NO_3$ and $[Au_9(PPh_3)_8](NO_3)_3$ have been grafted onto the surface for a number of freshly precipitated wet hydroxides [18,68–73].

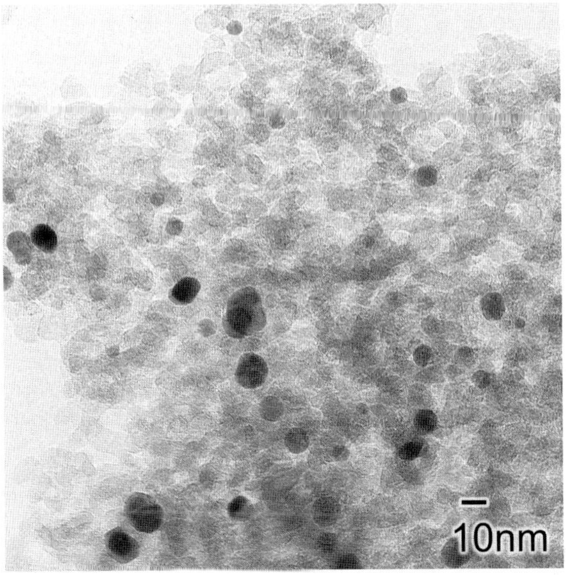

Fig. 6.3. TEM image of Au/SiO_2 catalyst prepared by gas-phase grafting and calcined in air at 673 K for 4 h. Black points are Au particles (from [75])

These supports have many surface OH groups reactive to the Au complexes. Vacuum-drying at room temperature and temperature-programmed calcination in air follows. This step causes simultaneous transformation of both precursors to gold particles and oxides, respectively, under their chemical interactions by temperature-programmed calcination [69, 71, 74]. Au–phosphine complexes are good candidates for metal precursors because they are thermally decomposed to Au metal in a temperature range similar to that used for the transformation of wet metal hydroxides to oxides and the phosphine ligands are expected to retard the growth to large Au metallic particles [68, 69, 71, 74].

Gold can be deposited on SiO_2 (see Fig. 6.3), MCM-41, SiO_2–Al_2O_3, activated carbon as nanoparticles with high dispersion by the gas-phase grafting of an acetylacetonate complex of gold, while liquid-phase preparation methods are not that effective with these supports [75]. Only gas-phase grafting is effective for depositing Au nanoparticles on activated carbon; however, the sizes of Au particles are relatively large (around 10 nm) [28].

There have been two reports of direct deposition of gold nanoparticles onto oxide or carbon supports [76, 77]. This is an important development from the viewpoint of avoiding contamination of the catalyst with chloride and gives rise to durable catalysts (see Sect. 6.3.3 later)

6.2.6 Ion-Exchange

The process of cation exchange is that by which protons and other cations on the surface or within the structure of the support are replaced by cations of the active metal and this leads, with the noble elements of Groups 8–10, first to atomically dispersed species and then, after an optimal calcination and careful reduction with hydrogen, to extremely small metal particles [18]. The procedure is especially effective with zeolites. But the introduction of catalytically active species into the cavities of these materials, as opposed to placing them on their external surface, presents certain difficulties, namely the lack of suitable cations or cationic complexes [18]. Nevertheless, different kinds of Au/zeolite systems have been prepared this way [78–89].

A new method to place gold-based nanoparticles on Y, NaY and HY zeolites, used by Fraissard and co-workers [82, 85, 87, 89] is based on introducing the metal by cation exchange using the $[Au(en)_2]^{3+}$ (en = ethylenediamine = $H_2N–CH_2–CH_2–NH_2$) complex ion. The gold particle size was found to be between 2 and 4 nm, depending on the zeolite type. The particles were located inside the zeolite supercages as proved by TEM microtomy and ^{129}Xe NMR studies [85, 87, 89]. Nevertheless, these gold catalysts were contaminated by residual chloride. An additional high-temperature pretreatment was required to activate the catalyst, but this resulted in sintering of the gold particles.

Kang and co-workers [79, 80, 84] prepared Au/NaY and Au–Fe/NaY catalysts by ion exchange from a solution of $HAuCl_4$. When only gold was present, they obtained gold particles of 10 to 40 nm diameter, while in the presence of iron, the mean diameter decreased to 5–8 nm. These results suggest that the presence of iron retards the sintering of metallic gold.

Chao and co-workers [90–92] have described an ion-exchange method to prepare highly dispersed gold particles in *mesoporous silica*. To overcome the problem of depositing negatively charged solution species on the negatively charged silica surface, the host was functionalised with a cationic organosilane to generate a monolayer of positively charged groups on the pore surface. This facilitated a uniform distribution of negatively charged metal species. The $[AuCl_4]^-$ species were subsequently incorporated into the channel system via ion exchange. Upon reduction with $NaBH_4$, highly dispersed nanoparticles of gold were formed in the channels of the mesoporous host. It was shown that the gold particles are to some extent stabilised by the pore system, but the interaction is relatively weak and sintering was observed even at 373 K [90]. Compared with a previously reported material obtained by the CVD method, which is the most appropriate for a silica support, the reaction rate for CO oxidation is still lower by about one order of magnitude. However, the straightforward preparation method reported by Chao and co-workers still yields more active catalysts than any other Au/SiO_2 system made by a solution technique, which usually show almost no activity at room temperature.

Pitchon and co-workers have developed a novel method for preparing supported gold catalysts based on the direct anionic exchange (DAE) of the gold species with the hydroxyl groups of an *alumina* support [45] and on a HT support with an Mg:Al molar ratio of 4:2 [57]. An aqueous solution of $HAuCl_4$ is added to the support, heated to 343 K and aged for 1 h. The slurry is then filtered, washed with warm water, dried in an oven at 393 K overnight and calcined in air at 573 K for 4 h. In order to completely remove the chloride ions, these authors take a fraction of the catalyst after drying and wash it with a concentrated solution of ammonia [45, 57]. However, this may be a dangerous procedure, since gold oxide and ammonia sometimes produce fulminating gold, which is explosive [93].

6.2.7 Sol–Gel Method

In the method described by several researchers [30, 94–101], a gold precursor (e.g., gold acetate, hydrogen tetranitratoaurate, chloroauric acid) is added to a sol solution of the desired support (such as alumina, titania, etc.). This sol is previously made by the mixing of a support precursor (e.g., aluminium tri-sec-butoxide or aluminium isopropoxide for alumina, tetrabutoxy-titanium for titania, tetra-ethyl-*ortho*-silicate for silica) with water and ethanol or methanol and/or nitric acid.

After vigorously stirring for a variable time, gel formation begins. The resulting gel is then dried at a temperature around 373–473 K for 12–24 h. A calcination step follows.

The sol–gel method produces mean particle sizes below 6 nm range, involving materials derived from soluble precursors which form a three-dimensional network on adding base [59].

Zhu and co-workers [102] have combined sol–gel processing with a liquid crystalline templating approach in order to synthesise meso-structured silica–titania mixed oxide monoliths, in which the nanocrystalline titania was well dispersed inside the silicate framework. It was shown that the mixed oxide monoliths were excellent support materials for Au catalysts.

6.2.8 Gold Alloy Catalysts

Since gold is now established as a catalyst in its own right, more advances are expected when other metals are added to it, in order to increase its activity/selectivity [103, 104].

It is convenient to define an alloy as a system containing two or more metals, irrespective of their intimacy of mixing or precise manner of mixing [105]. Alloys are usually formed by the extremely fast cooling of a melt of two or more components [18, 106]

Since bulk alloys have a small surface area, they are of limited use as catalysts and prone to recrystallise [18]. However, the product of the oxidative decomposition can be a supported metal catalyst. Thus, a gold–zirconium alloy affords an Au/ZrO_2 catalyst [107], and alloys with the Au_5FeZr_{14} and

Au_5AgZr_{14} give the corresponding zirconia-supported gold, promoted by the second element [108].

In preparing Au/ZrO catalysts, chemical routes are commonly used but they are restricted to low concentrations of noble metals because the processing temperature required induces particle growth [109]. The oxidation of glassy Zr–Au [107] and Zr–Au–Ag [108] alloys has been shown to result in high surface area catalysts.

There are a growing number of examples, where alloy nanoparticles containing gold have increased activities compared with gold alone or the other metal [7,51,103,110–123], as will become apparent from the reaction sections of this chapter.

Cortie and co-workers [124] have reported a new method for alloying gold with aluminium, to form a gold-containing inter-metallic precursor. Thus, $AuAl_2$ was produced by melting gold and aluminium together under a protective atmosphere. The aluminium is then dissolved from the $AuAl_2$ precursor using sodium hydroxide solution, to produce a highly porous gold structure. These authors also describe a method for incorporating transition metal oxides with the porous gold to form an active heterogeneous catalyst. Promoter elements can also be melted with $AuAl_2$ to improve catalytic activity. Alloying has the advantage that the promoter element can be in contact on the atomic level with gold [124].

Njoki and co-workers [121] have described an effective synthetic route for the preparation of water-soluble Au–Pt nanoparticles with bimetallic compositions ranging from ≈4% to 90% Au and particle sizes ranging from 2 to 8 nm. Hydrogen was used as reducing agent for the reduction of Pt(II), whereas acrylate was used a reducing agent for the Au(III). The latter reaction was found to be catalysed by the formation of Pt from the reduction of Pt(II). The size and composition of the Au–Pt alloy nanoparticles can be controlled by controlling the synthetic feed ratios and metal precursor concentrations. One possible area of application for these Au–Pt alloy nanoparticles is the bifunctional catalysis, in which Pt provides the main oxidation sites, whereas Au modifies the surface adsorption/desorption properties for reaction intermediates or products.

6.3 Properties of Nanoparticulate Gold Catalysts

Properties of the supported Au nanocatalysts have been found to depend on the method of preparation, and the pretreatment and calcination procedures used [18,31,44,125]. This section discusses the most important properties of catalytic gold nanoparticles.

6.3.1 Activity

Although, the reasons for the catalytic activity of gold are not as yet fully understood, the presence of gold as well-dispersed *nanoparticles* (<10 nm) is one of the best established requirements for preparing active catalysts [5–9, 18, 31, 32, 37, 38, 40–42, 47, 52–54, 59, 74, 75, 90, 126–137]. One example is shown in Fig. 6.4.

Nevertheless, unsupported powdered gold (mean diameter, 76 nm) proved to be active for CO oxidation [138, 139], and so did naked gold [140] and gold nanotubes [141] under mild conditions. In addition, investigations by Flytzani-Stephanopoulos and co-workers [142–144] on the *water-gas shift* (WGS) reaction (see Sect. 6.4.1) have challenged the idea that gold nanoparticles are essential, but these investigators have not exhaustively proved the complete absence of Au(0).

There are suggestions that there is an optimal Au particle size between 2 and 5 nm, although under the right circumstances particles as large as 10 nm, or even more, can exhibit high activity [6, 7, 41, 145], and smaller particles, not always visible under the electron microscope, could also be effective. Recent results have shown that there can be a significant fraction of Au particles smaller than 2 nm in the active catalyst [146]. The smallest supported gold particles (< 1 nm) reported to date were generated by DP on $Mg(OH)_2$ [147] and by the method of amine functionalisation that produces highly dispersed Au in the mesoporous silica [148]. The essence of this methodology lies in the combination of the complex-mediated growth of nanoparticles and surfactant-template synthesis. The formation of nanoparticles was controlled by the complexing ligand and the unique micellar media [149, 150]. Dai and co-workers [151] also reported the successful preparation of ultra-small gold nanoparticles (0.8–1 nm) deposited on titania-modified SBA-15 via a DP method.

Although some investigators claim that it is the size of gold nanoparticles that is most important [47], the presence of small gold particles alone cannot explain the catalytic behaviour of gold. Some studies made with similar sized gold particles, on different supports, have shown differences in catalytic activity. This is because the effect of the support must also be considered [6–8, 18, 34, 38, 41, 74, 90, 152, 153]. Among macroscopic factors, morphology and the specific area of catalyst supports have an influence on performance of heterogeneous catalysts [74]. Also, Andreeva and co-workers [153] showed that the nature of support and, in particular, its redox properties, play a decisive role in the reactivity of the catalysts.

The most usual supports are transition metal oxides [7, 38, 53, 54, 154, 155]. The role of these is still a matter of discussion and two different classes of oxides can be identified [90, 152]. The first class comprises *active supports*. These are materials that can be relatively easily reduced such as NiO, Fe_2O_3 or TiO_2. The second class is formed by inert, so-called *inactive supports* such as MgO, Al_2O_3 and SiO_2. These supports can, in some cases, lead to catalysts as active as those with reducible supports [32], but it is very important to ensure a high dispersion of very small particles of gold on the oxide [47, 90].

Alumina is the support most commonly used for PGM industrial catalysts [8, 36, 47], but not all investigators regard Al_2O_3 as a good support for gold catalysts. Schubert and co-workers found that Al_2O_3 was not as effective

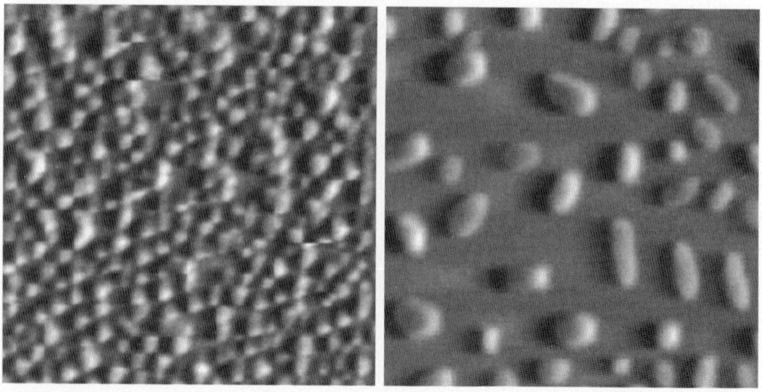

Fig. 6.4. Gold particles on a MICA substrate imaged by atomic force microscopy. Left: Active gold particles (average particle size below 10 nm) formed by evaporation in Ar atmosphere and annealing at 773 K. Right: Inactive gold particles (average particle size of 1 μm) formed by evaporation in vacuum and annealing at 773 K (adapted from [134, 156])

as reducible oxides such as FeO_x, NiO_x and CoO_x [152]. When gold was supported on alumina, which is believed to behave as neutral in the reaction process, only moderately active catalysts were obtained. Au/SiO_2 was much less active.

Au nanocatalysts supported on reducible transition metal oxides exhibited very high activities. Also, some composite oxide supports, such as $ZnFe_2O_4$, $NiFe_2O_4$ and $MgFe_2O_4$, have been investigated [8]. Composite oxides supports were employed to prepare $Au/MO_x/Al_2O_3$ (M = Fe, Co, Mn, Cu) catalysts [8, 38, 52, 120, 127, 157–160]. The latter proved to be more active than the former. Some authors believe that this result comes from synergistic effects between the gold nanoparticles and the support material [4,5,8,38,74,152,161].

Haruta also states that strong contact is necessary between gold and the underlying support [6, 7], and that this may be achieved via the calcination procedure. Since calcination at high temperatures generally leads to agglomeration of particles, which is undesirable [36, 41], calcinations at mild temperatures (373–573 K) have been preferred by several authors [9, 29, 111, 162, 163]. Consequently, the challenge consists in selecting a temperature for calcining the sample to produce good contact between the gold particles and the support, without producing too much sintering of particles. Some investigators report highest activities for uncalcined catalysts [43].

6.3.2 Selectivity

Gold nanoparticles show extraordinary selectivity in a large number of reactions. One remarkable example is in the *oxidation of propene* to the corresponding epoxide. A notable selectivity of above 99% was reported from early investigations over an Au/TiO_2 catalyst, using a combination of H_2 and O_2 in the gas stream, and the yield of propene oxide (PO) can be increased by optimising the support composition [28,164–166]. Another interesting feature is that only when gold is deposited as particles larger than 2 nm in diameter by DP onto certain titania-support materials, such as anatase, titania-modified silica, Ti-MCM-41 or TS-1, is high selectivity to propene oxide achieved [167, 168]. Gold particles smaller than 2 nm in diameter, produce a shift in selectivity from PO to propane [7,169–171].

Selective *dehydrochlorination of chlorofluorocarbons* (CFCs) is a very important environmental issue, and the need to replace these detrimental, ozone-depleting compounds by benign hydrochlorofluorocarbons (HCFCs) and/or hydrofluorocarbons (HFCs) has stimulated intensive work on the subject [172–182]. Palladium has been the most extensively investigated catalytic metal in this reaction, but the moderate selectivity for CH_2F_2 exhibited by Pd/SiO_2 (40%) can be significantly increased, up to 95%, with a 20–40 at%. Au addition [180], and, in Pd/C, from 70% to 90% with Au addition [181,182].

In contrast to PGM catalysts, gold nanocatalysts are generally more active for CO oxidation than for H_2 oxidation. In addition, gold catalysts are almost insensitive to CO_2, and their activity is enhanced by moisture, the products of the two reactions. Gold catalysts are therefore strong candidates for the *selective oxidation of CO in H_2-rich gases* [2, 35, 38, 46, 52, 53, 101, 157, 159, 183–200], a reaction needed to produce pure hydrogen for fuel cells.

Partly because gold catalysts are active under mild conditions, they have been investigated for other *selective oxidation* and for *selective hydrogenation reactions*. Details are given in Sect. 6.2.4.

6.3.3 Durability

From Tammann temperature considerations, gold would be expected to sinter at around 773–823 K (half of the melting point of metallic gold). Small Au nanoparticles might well sinter at much lower temperatures than this, as their melting temperature would be expected to decrease drastically with decreasing particle size (therefore, a particle of around 2 nm would be expected to melt at 573 K [5, 38, 134, 201], a significantly lower temperature than the melting point of bulk gold at 1,337 K). However, there is increasing evidence that gold catalysts are much more durable than would be expected from these considerations [104].

One example is a catalyst consisting of gold on cobalt oxide particles supported on a mechanical mixture of zirconia-stabilised ceria, zirconia and titania, that survived 773 K for 157 h, with some deactivation [145]. Grisel and Nieuwenhuys [38, 127] have shown that the addition of transition metal oxides to form Au/MO_x/Al_2O_3 catalysts, massively suppresses Au particle sintering in methane oxidation tests up to 973 K. Also, Seker and Gulari [30] found that Au/Al_2O_3 catalysts survive rigorous pre-treatments of 873 K in air for 24 h, followed by several cycles of 423–773 K; and they were then kept

Table 6.1. Thermal durability of composite gold oxide catalysts used in lean-burn NO_x reduction (adapted from [202, 203])

Catalyst composition		T at 50% C_3H_6 conversion[a] (K)	
Chemical formula	Au content (wt%)	Initial	After durability test[b]
$Au_2Sr_5O_8$	0.4	618	619
$Au_2Sr_5O_8$	0.4	613	628
$La_2Au_{0.5}Li_{0.5}O_4$	0.2	614	621
$La_2Au_{0.5}Li_{0.5}O_4$	0.2	617	618
Au/Al_2O_3	2	651	706

[a] Evaluation conditions: CO 1,000 ppm, C_3H_6 670 ppm C, NO 250 ppm, O_2 7.3%, CO_2 6.7%, H_2O 5%, balance N_2 at 150,000 h^{-1}

[b] Durability test conditions: CO 1,000 ppm, C_3H_6 670 ppm C, NO 500 ppm, O_2 6.5%, CO_2 10%, H_2O 10%, balance N_2 at gas temperature of 800°C for 5 h

at 773 K overnight. The catalysts gave some of the highest recorded activities (90–100% for NO conversion). Datye and co-workers [65] have shown that Au/Al_2O_3 is hydrothermally stable at 873 K for 96 h. Since 873 K is higher than the Tamman temperature for gold, these results point to the fact that the active gold species are produced by interaction between the gold and the support. These catalytically active gold oxide species are much less likely to sinter than metallic gold and these results, therefore, indicate a much higher potential for practical applications than if the catalyst was metallic gold itself [202].

Another example of thermal durability has been patented by the Toyota Motor Company and some of the results are given in Table 6.1 [202, 203]. Complex gold oxides, such as $Au_2Sr_5O_8$, in which the gold is entirely ionic and trapped in the oxide lattice. The claim is that the highly dispersed ionic gold exhibits good catalytic activity and is inhibited from sintering and therefore has improved durability. Results at 1,073 K, given in Table 6.1, bear this out and show that a standard Au/Al_2O_3 catalyst under the same conditions suffers significant degradation.

Consequently, we conclude that there is scope to improve the thermal durability of gold-based catalysts through inhibition of gold-particle sintering, by strong interaction between Au and the supporting metal oxides or even formation of stabilised ionic Au. It should, therefore, be possible for Au catalysts to at least be stabilised for the moderate temperatures required in diesel exhaust systems. In an European urban driving cycle, temperatures average 353–453 K (maximum 503 K), while averages of 453–503 K (maximum 713 K) are experienced in the extra-urban part of the cycle [202, 204].

Stability at lower temperatures is easier to achieve. Some examples are 350 h for an Au/ZnO catalyst and 500 h for an Au–Pt/ZnO catalyst, at 353 K, for selective catalytic removal of CO by the oxidation method in a hydrogen-rich system [191].

Residual *chloride* can act as a catalyst poison (see Sect. 6.2 above), and supported Au catalysts are generally prepared from chloride-containing Au precursors, where their properties are highly sensitive to preparation procedures. The residual chloride has been found to affect the activities in two different ways: It facilitates the agglomeration of Au particles during heat treatment; and it inhibits the catalytic activity by poisoning the active site [9, 41, 42, 44, 133]. Haruta and co-workers also reported agglomeration of Au particles after they impregnated an Au/Ti-MCM-41 catalyst with CsCl [205].

The fact that residual chloride can affect the catalytic properties of supported Au catalysts in more than one way is a possible reason for the reported performances of Au catalysts being very different amongst the various research groups [125], and why apparently inconsequential changes in synthesis variables sometimes have deleterious effects [42]. Understanding the chloride effect will enable a better control of the variables, which are critical to active catalyst preparation.

Pitchon and co-workers [45] strongly improved catalytic activity of their Au catalysts prepared by DAE on an alumina support by the complete removal of chloride using an ammonia washing procedure (see Sect. 6.6). These catalysts were resistant to an oxidative ageing treatment at 873 K and to the presence of water.

As discussed in Sect. 6.4, Datye and co-workers [65] reported a new impregnation method for supporting gold on alumina using $HAuCl_4$. In this method, the acidified gold solution is contacted with alumina to adsorb the gold chloride on it. After washing off the excess gold precursor, the solid is treated with ammonia to convert the chloride to an adsorbed hydroxide. Catalysts prepared by this method were shown to be stable up to 873 K for 100 h [65].

6.3.4 Poison Resistance

One feature of particular interest is the resistance that gold-based systems might display against *sulphur* poisoning. Only a little work has been carried out in this area to date, but gold on titania catalyses the Claus reaction and is 5–10 times more effective than titania itself, indicating that gold can be sulphur-tolerant:

$$SO_2 + 2H_2S \rightarrow 2H_2O + 3S_{solid} . \quad (6.1)$$

In addition, the reduction of SO_2 by CO using Au/TiO_2 is also 5–10 times more active than with pure titania [206].

Also, Marsh and co-workers [145] showed that gold on cobalt oxide particles, supported on a mechanical mixture of zirconia-stabilised ceria, zirconia and titania remains active in a gas stream containing 15 ppm SO_2. Haruta and co-workers [207] found that although the low-temperature CO oxidation activity of TiO_2-supported Au can be inhibited by exposure to SO_2, the effect on the activity for the oxidation of H_2 or propane is quite small. Venezia and co-workers [208] reported that bimetallic Pd–Au catalysts supported on silica/alumina are resistant to sulphur poisoning (up to 113 ppm S in the form of dibenzothiophene) in the simultaneous hydrogenation of toluene and naphthalene at 523 K.

The reaction of photo-induced sulphur desorption from the surfaces of the metal oxide-supported (rutile and anatase TiO_2, $SrTiO_3$, ZnO, Fe_2O_3 and SnO_2) Au nanoparticles in water at room temperature has also been studied [209]. It was found to be driven by an upward shift of the Fermi energy of the metal oxide-loaded Au nanoparticles with irradiation. It has also been demonstrated that this phenomenon is applicable to the low-temperature cleaning of sulphur-poisoned metal catalysts.

These examples offer an encouraging sign that gold may not be irreversibly poisoned by sulphur. Further work is required to investigate this poison resistance in more detail [202]. In a recent Chevron patent [210], it is claimed that an Au catalyst may be used for both high- and low-temperature WGS reactions: 0.1–3.0 wt% Au/sulphated ZrO_2 (0.02–2.5 wt% sulphur were used and, for example, 98.2% conversion was obtained at 20,000 GHSV over 350 h at 473 K and 30 psig). This patent provides evidence of advantages for the presence of sulphur in catalysis by gold [104].

6.4 Reactions Catalysed by Nanocatalytic Gold and Gold Alloys

6.4.1 Water-Gas Shift

The WGS reaction is of interest, both for the production of pure hydrogen for fuel cells and heavy chemical processing [143, 144, 211–218]. Moreover, the WGS is also one of the key steps involved in automobile exhaust processes, converting CO and water to hydrogen and CO_2, and the presence of the hydrogen provides a very effective reductant for NO_x removal [211, 212, 215].

Water-gas was first obtained via the reaction of steam with coke. Catalysts are used to promote the following equilibrium in the direction of hydrogen [18, 132]:

$$CO + H_2O \rightarrow CO_2 + H_2. \tag{6.2}$$

Commercial catalysts for the WGS are based on Ni and Cu supported on zinc oxide and alumina [18, 132, 143, 144, 217]. In comparison with these Ni- or Cu-based catalysts that operate at 900 K or at 600 K, respectively, supported Au catalysts appear to be advantageous for operation at a temperature as low as 473 K [6, 202]. This is a significant difference because lowering the temperature of the WGS improves the equilibrium concentration of CO and hence lowers the CO content of the resulting reformate gas. Potential system cost and size savings for on-board vehicle fuel processing might therefore be possible by employing gold catalysts, thereby removing the need for an additional on-board CO clean-up system [202, 219–225].

Moreover, the presence of 1% CO in the hydrogen causes major problems in Pt-based fuel cells because Pt is poisoned by CO at the operating temperatures of the fuel cell. Current commercial fuel processing systems rely on a subsequent CO clean-up system based on the reaction:

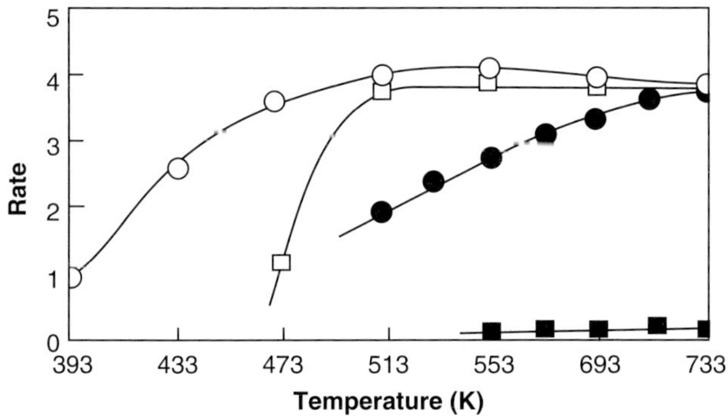

Fig. 6.5. Temperature dependence of the catalytic activity of supported gold and base metal oxide catalysts in the WGS. Starting reaction gas mixture was 4.88 vol% carbon monoxide in argon; water vapour partial pressure 223 Torr; SV 4,000 h^{-1}; 1 atm. Au/α-Fe_2O_3 (○); CuO/ZnO/Al_2O_3 (□); α-Fe_2O_3 (•); Au/Al_2O_3 (■). Rates are expressed in mol $m^{-2}\,h^{-1} \times 10^{-2}$ (adapted from [219])

$$2CO + O_2 \rightarrow 2CO_2. \tag{6.3}$$

Another advantage for gold catalysts is that they can be used to selectively catalyse this oxidation, since the selectivity for the competing hydrogen/oxygen reaction is significantly lower at ambient temperatures (see Sect. 6.4.4).

High catalytic activity for gold catalysts in the WGS was found by Andreeva and co-workers using Fe_2O_3 as a support [219, 220, 222]. Over this catalyst, CO conversion started at 393 K (see Fig. 6.5), and its activity exceeded that of the most efficient industrial WGS catalyst known at that time.

A strong dependence of the activity on the gold dispersion and the preparation method has been observed [220], showing that the support plays an important role in the WGS activity. It could therefore be presumed that gold deposited on readily reducible metal oxides, such as Co_3O_4, would be active, while that deposited on hardly reducible oxides, such as TiO_2 and ZrO_2, would not. The following order of catalytic activity was established experimentally [223, 226]:

$$Au/TiO_2 \approx Au/Fe_2O_3 \approx Au/ZrO_2 > Au/Co_3O_4. \tag{6.4}$$

The activity found for Au on titania and zirconia supports was therefore surprisingly high [223, 226, 227]. Only the Co_3O_4 catalyst was very rapidly reactivated. Haruta and co-workers have also reported a higher catalytic activity for Au/TiO_2 catalysts in the WGS in comparison with Cu/ZnO/Al_2O_3, Au/ZnO, Au/Fe_2O_3 and Au/Al_2O_3 [6, 224].

In the search for novel, more stable and efficient supports for gold catalysts for the WGS, mixed oxide supports (Fe_2O_3–ZnO and Fe_2O_3–ZrO_2), as well as supports of a different crystalline state—amorphous zirconia and titania—have been used also by Andreeva and co-workers (Fig. 6.6) [228]. Gold deposited on well-crystallised metal oxides was found to have higher activity than that on amorphous oxides. The same is true for the mixed oxide supports [223, 228].

It has been found that gold deposited on ceria (CeO_2) is very effective in the WGS [143, 144, 211, 212, 214, 215, 217, 218, 223, 229–231]. These results are of particular interest since CeO_2 is extensively employed as a component in automotive, three-way emission-control catalysts, mainly due to its capacity to undergo a relatively rapid change in oxidation state upon changes in the redox potential of the exhaust gases. Owing to its high oxygen storage capacity, CeO_2 acts as an active support for the WGS [144, 202, 211, 212, 218, 223, 230, 232].

As was observed for other gold catalysed reactions, the WGS activity of gold-based catalysts is sensitive to the size of the gold nanoparticles [38, 53, 202, 215, 223, 233]. Previous studies on Fe_2O_3, TiO_2 and ZrO_2 supports indicated that the gold particle size was 2–5 nm [220, 228], while the average size of gold particles for CeO_2 support was about 4–5 nm [211, 212]. Also, HRTEM (High Resolution Transmission Electron Microscopy) and EDS (Energy Dispersive Spectroscopy) have shown presence of very highly dispersed gold clusters (diameter of about 1 nm) on the surface of the Au/ceria catalysts prepared by DP [215, 218].

One of the problems in using gold catalysts for WGS is the activity loss during catalysis. This problem is connected with the stability of the gold dispersion, because the main reason for the fast deactivation of these catalysts is the agglomeration of the gold particles. The studies on Au/CeO_2 catalysts have, however, demonstrated their very stable catalytic activity for WGS (between 393 and 633 K), which can be related to the high and stable dispersion of gold on ceria [211, 212, 215, 223, 230].

The role of the support is very important due to its ability, depending on its nature, to modify the electronic properties of the small gold particles. The interactions between the metallic gold nanoparticles and ceria appear to be of a crucial importance in explaining the remarkably high stability and activity of Au/CeO_2 catalyst [212, 223]. Andreeva's group showed that the sample with 3 wt% Au exhibits higher activity and stability than the samples with 1 and 5 wt% [212]. This means that the support plays an active role in the reaction, because of the need for enough free support surface area. This could be indirect evidence for a mechanism involving participation of the support [212, 223].

However, Flytzani-Stephanopoulos and co-workers have suggested that gold nanoparticles are not the active species for the WGS catalysed by Au/CeO_2 [142–144]. The authors found that using a cyanide leaching process almost all the metallic gold could be removed from Au/CeO_2 that had been prepared by deposition precipitation or co-precipitation, leaving behind a highly dispersed gold species. These leached catalysts exhibited comparable activity to the as-prepared catalysts leading the authors to conclude that metallic gold may not contribute to the catalyst's reactivity. These authors SUPPORT a model

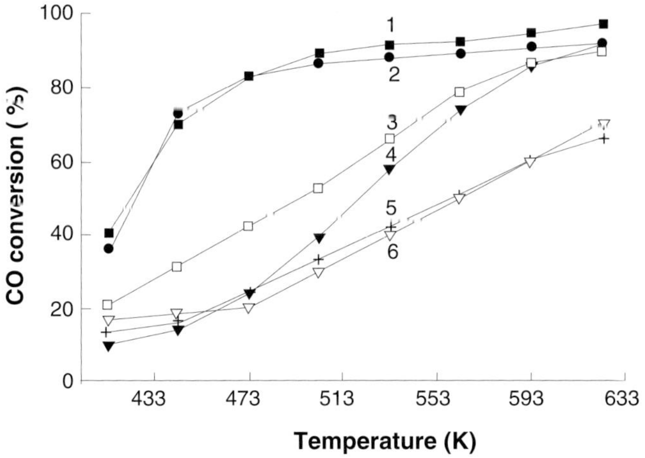

Fig. 6.6. Temperature dependence of the catalytic activity of WGS (degree of CO conversion) of the samples: (1) Au/Fe_2O_3; (2) Au/ZrO_2; (3) Au/Fe_2O_3–ZnO; (4) Au/Fe_2O_3–ZrO_2; (5) Au/ZrO_2; (6) Au/ZnO (adapted from [228])

in which Au is integrated into the ceria support. They suggest that the role of Au is to increase the amount of surface oxygen in ceria [143, 230].

Recent results showed that Au/CeO_2 catalysts are subject to deactivation by poisoning or blockage of the active sites by either carbonates, formates or hydrocarbons [217, 234, 235]. These species appeared to be formed by CO and H_2, and their formation was facilitated by oxygen-deficient sites on ceria [217]. However, regeneration of the WGS activity can be achieved by calcination: 95% of the initial activity is recovered by heating in air at 673 K [217, 234, 235].

A recent Chevron patent [210] claims that an Au catalyst may be used for both high- and low-temperature WGS reactions: 0.1- to 3.0-wt% Au/sulphated ZrO_2 (0.02- to 2.5-wt% sulphur were used and, for example, 98.2% conversion was obtained at 20,000 GHSV over 350 h at 473 K and 30 psig). This shows that there are advantages in having sulphur present in gold catalysis [104].

Scurrell and co-workers have studied bimetallic Au–Ru catalysts supported on Fe_2O_3 [225]. When compared with Au/Fe_2O_3 and Ru/Fe_2O_3 catalysts, the activity of the bimetallic system was found to be higher over all the reaction temperatures studied (373–513 K). TPR analysis indicated a shift in reduction temperature of the bimetallic system, when compared with the mono-metallics.

Su and co-workers [236] have evaluated titanium oxide nanotubes (TNTs), with diameters of 8–10 nm and lengths of several tens to several hundreds of nanometres, for the first time, as supports for the preparation of nano-sized Au-based catalysts. Catalysts were synthesised via the reaction of TiO_2 crystalline powders of either anatase or rutile phase and an aqueous solution of sodium hydroxide. Authors studied their application as active supports of Au particles prepared by DP. It was observed that nanosized Au particles

Table 6.2. Yields of VAM in Dupont process

Catalyst	VAM STY (g l^{-1} h^{-1})	VAM selectivity (%)
Au/Pd/KOAc	764	93.6
Pd/KOAc	100	95.4
Au/Pd	594	91.6
Pd	124	94.7

Fixed bed performance of Au/Pd catalysts after 40 h on stream. Test conditions: 165°C, 115 psig, with feed consisting of ethene, acetic acid, oxygen and nitrogen (from [237])

could well fit into the tubes. The catalytic activity of Au/TNTs was evaluated, for the first time, in the WGS reaction over a wide temperature range (413–573 K). The catalytic behaviour of the Au-supported TNTs correlates with the structural characteristics and nature of the TNTs. It may be possible to improve the catalytic performance of Au/TNTs by increasing the diameters of the nanotubes or modifying the crystalline structure of the nanotubes to the anatase phase [236].

6.4.2 Vinyl Acetate Synthesis

Vinyl acetate monomer (VAM) is an important chemical intermediate used in the production of several types of paints, adhesives as wallpaper paste and wood glue, and surface coatings such as the protective laminate films in automotive safety glass [2, 104, 237, 238].

The commercial process for vinyl acetate production has evolved over the years. Early in the 1930s, Wacker developed a process based upon the gas-phase conversion of acetylene and acetic acid over a zinc acetate catalyst supported on activated carbon. Later, in 1960s, a more economically favourable gas-phase process was introduced involving the acetoxylation of ethene over a Pd-based silica supported catalyst. Ethene, acetic acid and oxygen reacted to form vinyl acetate and water [122, 237–242]:

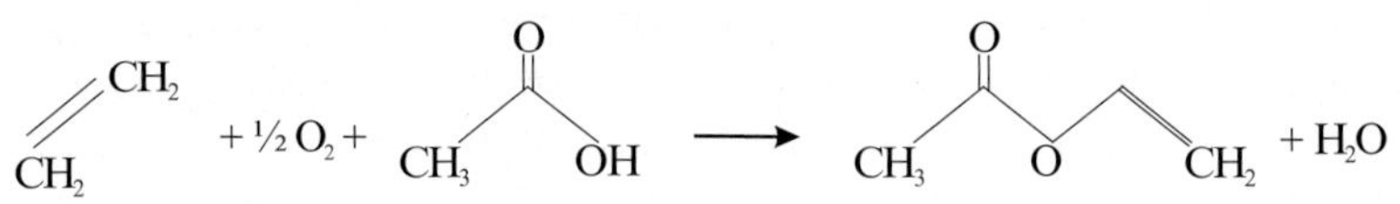

Most of current production of vinyl actetate is derived from this ethene-based process, which has been shown to be facilitated by many promoters. Gold and potassium acetate (KOAc) the most notable of these [237, 241–243].

In consequence, Pd–Au silica-supported catalysts promoted with KOAc were also studied. The results obtained showed that addition of Au to Pd catalysts significantly improves productivity [237, 238, 243], whilst also improving the intrinsic selectivity to vinyl acetate (Table 6.2) [237, 244]:

However, little work has been done to address the role of gold in these catalysts. Lerou and co-workers [237] assigned the promoting influence of Au in Pd/SiO_2-based catalysts to an electronic interaction, modifying the adsorption strength of various species on the metal surface and enhancing the desorption rate of vinyl acetate. Single crystal studies showed that the added Au had significant effects on the surface chemistry of the Pd. It promoted the formation of adsorbed mono-dentate acetate and favoured desorption of VAM as opposed to its decomposition [238,245]. Goodman and co-workers [242] evaluated Pd–Au alloys on SiO_2 for vinyl acetate synthesis, and concluded that the enhanced capacity of the Pd–Au surface for oxygen and/or the increased mobility of adsorbed oxygen under the reaction conditions are probably responsible for the unusually high reactivity of these Pd–Au alloy catalysts.

During the lifetime of a commercial fixed-bed Pd–Au–K/SiO_2 VAM synthesis catalyst, a progressive loss in activity occurs, restricting the useful lifetime typically to the order of a few years, but the corresponding selectivity (as high as 96%) remains essentially constant throughout, so that this is not generally an important issue [238, 239].

Deactivation processes have not been studied to any significant extent. Abel and co-workers [246], studying a Pd–Cd–K/SiO_2 catalyst, found that deactivation was mainly caused by sintering, due to the formation of mobile palladium acetate species. In a recent paper, Lambert and co-workers [238] have reported on the characterisation of fresh and industrially aged commercial Pd–Au–K/SiO_2 catalysts (5 to 6 mm diameter spherical SiO_2 pellets). Catalyst ageing resulted in an increase in the average particle size from 5 nm for the fresh sample to 12 nm for its deactivated analogue. However, a very large proportion of the palladium content of these catalysts was present in a highly dispersed form. Also, the support component did not sinter extensively and was therefore able to maintain a higher metal dispersion. Consequently, no significant change in alloy composition was observed with usage, which has significant implications in relation to the mechanism of sintering. If sintering of these alloy particles were to occur via an Ostwald-ripening process involving migration of palladium acetate species, as proposed by Abel and co-workers [246], a considerable enrichment in the palladium content within the alloy particles would be expected on ageing. This was not observed, however, and Lambert and co-workers [238] suggest that the sintering, observed by XRD, involves migration/coalescence of individual alloy particles. Also, TEM images obtained from the fresh catalyst indicate a tendency for alloy particles to form clusters. The sintering is thought to be due to the coalescence of each cluster into a single large alloy particle.

However, earlier findings [247] showed that very highly dispersed palladium catalysts had very low activity for VA formation. This was attributed to the very small particles being inaccessible to the ethene feed due to their being completely embedded within the acetic acid/acetate liquid layer (ethene has very low solubility in acetic acid). Vinyl acetate formation may therefore be restricted to the larger Pd–Au alloy particles accessible to gaseous ethene. This would also explain why ageing has little effect on selectivity. The catalytically active alloy particles grow but do not change composition, therefore, activity declines but selectivity remains (approximately) constant.

Goodman and co-workers [122] have also performed comparative kinetic studies of vinyl acetate synthesis in Pd/SiO_2 and $Pd\text{–}Au/SiO_2$ catalysts. They concluded by TEM–EDS that Au surface enrichment took place for the Pd–Au alloy, and the interaction between Au and Pd lead to a formation of more active Pd ensembles such as $PdAu_5$ and $PdAu_6$. The enhanced capacity of the Pd–Au surface for oxygen and/or the enhanced mobility of adsorbed oxygen under the reaction conditions are likely responsible for the unusually high reactivity of these Pd–Au alloy catalysts [122].

For the latest commercial developments in fluid bed production of vinyl acetate manufacture, see Sect. 6.5.1.

6.4.3 Hydrochlorination of Ethyne

The heterogeneously catalysed conversion of ethyne (acetylene) into vinyl chloride (chloroethene) by the addition of hydrogen chloride is a reaction important for the manufacture of polyvinyl chloride (PVC) [18, 213, 248, 249]. Mercuric chloride, supported on activated carbon, is often used as a catalyst commercially [18, 213, 248–252] but, unfortunately, this catalyst rapidly deactivates. The high rate of deactivation, together with the toxicity associated with mercury compounds, has prompted a search for alternative catalysts [213, 248, 249, 251, 252].

It was shown that the rate-determining step is the addition of HCl to a surface ethyne complex [18, 213, 248, 253]:

$$\mathrm{MCl}_n + \mathrm{HC} \equiv \mathrm{CH} + \mathrm{HCl} \rightarrow \mathrm{MCl}_n.\mathrm{HC} = \mathrm{CH.HCl}. \qquad (6.5)$$

Active catalysts should therefore be able to form surface metal-ethyne and metal·HCl complexes. One of the most extensive studies of metal chloride catalysts was carried out by Shinoda [254]: 20 metal chlorides supported on carbon were investigated for ethyne hydrochlorination and it was proposed that a correlation existed between the catalytic activity and the electron affinity of the metal cation, divided by the metal valence. The correlation consists of two straight lines and, for this reason, it cannot be used predictively. However, electron affinity is, necessarily, a one-electron process, whereas the hydrochlorination of ethyne is more likely to be a two-electron process, involving 2π electrons of ethyne. Many of the metal cations investigated in the original study of Shinoda [254] are divalent and, consequently, the standard electrode potential was proposed as a more suitable correlation parameter.

On the basis of this correlation, Hutchings predicted that gold would be the most active catalyst for ethyne hydrochlorination [255], and his subsequent research confirmed this prediction [252, 253, 256–259]. Gold catalysts were found to be about three times more active than the commercial mercuric chloride catalysts (see Fig. 6.7).

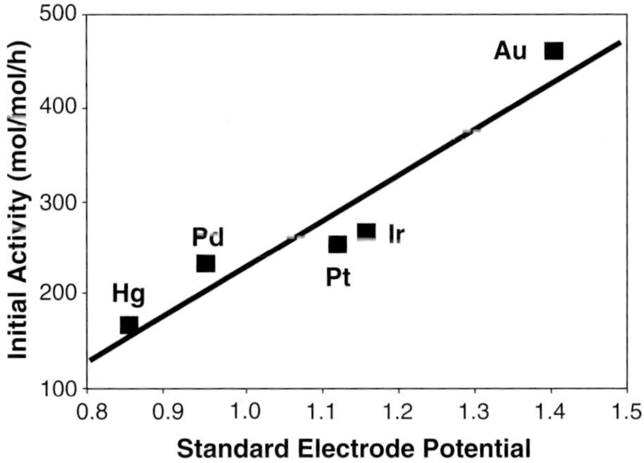

Fig. 6.7. Correlation of hydrochlorination activity of metal chlorides supported on carbon (453 K, GHSV 1,140 h^{-1}) with the standard electrode potential. Catalysts contain 0.0005-mol metal per 100 g_{cat} (adapted from [253])

Indeed, under appropriate conditions, supported catalysts are considerably more active than mercuric chloride and the best catalyst is gold on activated carbon, obtained using an IW method using $HAuCl_4$ dissolved in aqua regia [252, 258, 259].

One unfortunate observation for all the supported metal chloride catalysts is that they all deactivate with time on stream when used in a standard fixed bed reactor [18, 248, 249]. However, supported Au catalysts deactivate much less rapidly than other supported metal catalysts and if high loadings of gold are used ($\geq$1 wt%) then deactivation is minimised [18, 248, 249].

Interestingly, the deactivation with Au catalysts for ethyne hydrochlorination was found to be temperature dependent, and high rates of deactivation were observed at both high and low temperatures. The low-temperature deactivation was caused by coke deposition [252,258], probably the result of surface polymerisation reactions of vinyl chloride and ethyne. The high-temperature deactivation was probed using ^{197}Au Mössbauer spectroscopy [252, 258] and it was found that this was due to the reduction of Au^{3+} to Au.

Although deactivation was a problem with these Au catalysts, in common with other supported metal chloride catalysts, it was found that Au catalysts could be reactivated by treatment off-line with HCl or Cl_2 [259]. However, perhaps the most important observation was that NO could significantly enhance and restore the catalyst activity when co-fed with the reactants. Indeed, co-feeding NO with the reactants from the start of the reaction showed that deactivation could be virtually eliminated. Most important of all, no effect on vinyl chloride selectivity was observed during the NO co-feeding experiment [259]. This represents an important example of on-line reactivation of a heterogeneous catalyst.

During use, the commercial carbon-supported mercuric chloride deactivates primarily as the result of loss of mercury from the reactor [250]. This is not the case for gold catalysts since, under the reaction conditions used, no gold is lost from the catalyst. In this case, deactivation, as a result of the reduction of Au^{3+}, is readily overcome by co-feeding NO. Hence, supported gold nanocatalysts should be the catalysts of choice for this reaction.

Consequently, gold on carbon has considerable potential as a commercial catalyst for the manufacture of vinyl chloride (see Sect. 6.5.2).

6.4.4 Carbon Monoxide Oxidation

The oxidation of CO to form CO_2 is the simplest reaction catalysed by nanogold and the one which received most initial publicity [8, 104]. It is also the most intensively studied gold catalysed reaction and catalysts which promote this reaction at ambient temperatures will find practical applications [6,7,18,31,64,71,110,125,126,132,145,251,260–264], that include use in air-purification devices for indoor space (see Sect. 6.5.5), including gas masks, and for cleaning automotive exhaust gases (see Sect. 6.5.6). CO removal is also required in CO_2 lasers (which require recombination of the CO and O_2 produced in the laser to maintain a high power output, since CO_2 is decomposed in the laser discharge, causing a build-up of O_2 that decreases the laser efficiency) [261]. CO gas sensors are another application for gold catalysts (see Sect. 6.5.8).

The industrial catalysts currently used for CO oxidation are Hopcalite (mainly MnO_x and CuO), but these catalysts are not stable in a water environment, and do not exhibit catalytic activity at around room temperature and deactivate rapidly, being thus unsuitable for long-term use [183,251,263,265].

Haruta and co-workers have intensively investigated the ability of gold to catalyse CO oxidation [6,7,31,75,110,111,138,139,147,155,183,185,207,261, 262,265–275]. At the time of their early work in the 1980s, where the activity for gold-supported catalysts came as a surprise [183, 265], but it is now well known that supported gold nanocatalysts are effective for CO oxidation at very low temperatures, with activity as low as 203 K. Activities do, however, depend on choice of metal oxide support (see Fig. 6.8).

Significant progress has been made and a number of review articles written [6,7,9,18,74,110,252,276]. It has been shown that Au catalysts have advantages arising from their improved activity at low temperatures and stability in the presence of water [6, 7, 159]. Supported Pt, Pd, Ru and Rh catalysts exhibited appreciable activity only at temperatures higher than 423 K [277]. Gardner and co-workers have shown that Ag-, Ru- or Pd-based catalysts are less active and deactivate faster than Au supported on MnO_x and CeO_x [278].

Although unsupported powdered gold (mean diameter 76 nm) is active for CO oxidation [138,139], most of the studies made by Haruta and other groups have been on supported gold catalysts, often using TiO_2 which is considered a very *active support* [3,5,31,32,34,43,44,49,62,71,75,99,110,111,137–139,155,

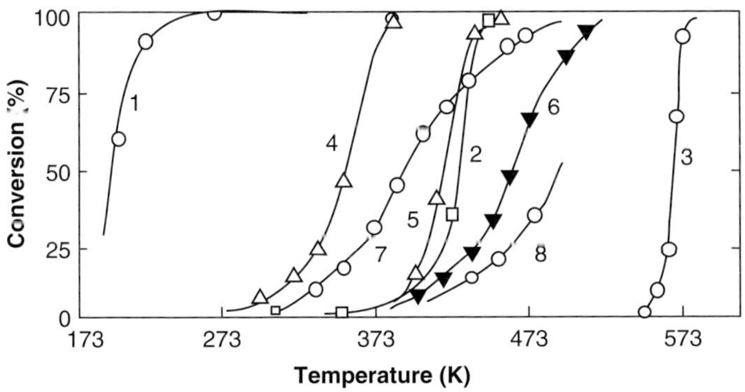

Fig. 6.8. Oxidation efficiencies for carbon monoxide oxidation for Au/α-Fe_2O_3 in relation to catalyst temperature in comparison with other catalysts. (1) Au/α-Fe_2O_3 (Au/Fe = 1:19), prepared by CP; (2) 0.5 wt% Pd/γ-Al_2O_3, prepared by IMP; (3) gold fine powder; (4) Co_3O_4, ex-carbonate; (5) NiO; (6) Au/α-Fe_2O_3; (7) 5 wt% Au/α-Fe_2O_3, prepared by IMP; (8) 5 wt% Au/γ-Al_2O_3, prepared by IMP (adapted from [183])

185, 207, 262, 264, 266, 269–271, 273–275, 279–297]. However, Au supported on Fe_2O_3 can have advantages over Au/TiO_2 for commercial applications arising from its lower cost, greater availability and improved catalytic performance, not only for CO oxidation but also for other reactions [70, 72, 73, 110, 163, 183, 261, 262, 266, 285, 298–301].

Other metal oxides, such as NiO [183], Co_3O_4 [183, 261, 266, 302], CuO [126, 183], ZnO [126, 263, 292], ZrO_2 [99, 283, 286, 287, 290, 292], MgO [196, 303], CeO_x [278, 304, 305], MnO_x [29, 260, 278], CeO_2 [306], Y_2O_3 [306, 307], have also been employed. This last support, considered "inadequate" for Au catalysts, was transformed into a highly active material if Au was deposited on nanocrystalline Y_2O_3 particles instead of being prepared by a conventional precipitation method, and use of zeolites (Mordenite, β, Y and HY) have also been reported [60, 74, 79, 87, 88, 308]. Metal hydroxides, such as $Be(OH)_2$ and $Mg(OH)_2$, are excellent choices for demonstrating high activity at a temperature as low as 196 K [68, 147]. Au on $Fe(OH)_x$ has also been shown to be a good choice at ambient concentrations (with a good lifetime and high resistance to moisture and high concentration of CO_2) [309, 310]. Among supported noble metal catalysts, Au supported on $Mg(OH)_2$ is the most active for CO oxidation at 196 K [147, 268, 271, 272].

"Inert" supports, such as SiO_2 [4, 5, 9, 75, 90, 99, 113, 122, 185, 273, 275, 279, 280, 292, 296, 311–313], Al_2O_3 [4, 5, 9, 28, 39, 42, 45, 64, 65, 75, 77, 99, 133, 137, 155, 185, 273, 275, 285, 290, 296, 313, 314], WO_3 [185], and non-metal-oxide supports, such as activated carbon (AC) [185] and activated carbon fibres (ACF) [64, 315], have also been used. Also, Au–Ag alloy nanoparticles supported on mesoporous aluminosilicate shown exceptionally high activity for CO oxidation at low temperatures (such as room temperature). The best

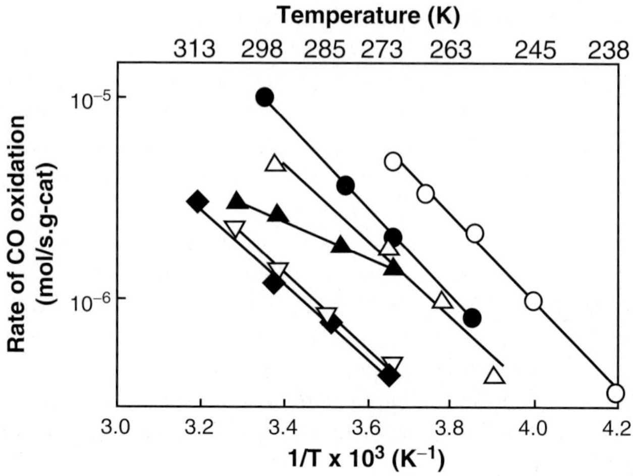

Fig. 6.9. Logarithmic reaction rate of CO oxidation over Au/Al_2O_3, Au/SiO_2 and Au/TiO_2 as a function of reciprocal temperature. (△) 5.3 wt% Au/Al_2O_3 by GG; (♦) 0.96 wt% Au/Al_2O_3 by DP; (▽) 4.2 wt% Au/Al_2O_3 by CP; (▲) 6.6 wt% Au/SiO_2 by GG; (•) 4.7 wt% Au/TiO_2 by GG; (○) 2.0 wt% Au/TiO_2 by DP (adapted from [75])

results were obtained with an Au/Ag ratio of 3:1 [123]. Another example of the use of alloy particles was reported by Goodman and co-workers [122] using Pd–Au alloys supported on silica. Results revealed that the onset temperature for CO oxidation shifted from 553 K for the Pd-only catalyst to 483 K for the Pd–Au catalyst. Moreover, CO oxidation was purported to occur exclusively on Pd in a Pd-rich alloy surface, since Au/SiO_2 shows no reactivity [113]. In a study of the electrochemical oxidation of CO, Gossner and co-workers [316] suggested that the high rate on Pd–Au alloy may be ascribed to the migration of O_{ads} from Pd to Au, which presumably would enhance O_2 adsorption.

Mesoporous silica SBA-15 was uniformly covered with one-layer titanium oxide by a hydrolytic surface sol–gel process by Dai and co-workers [151] and ultra-small gold nanoparticles (0.8–1 nm) were deposited on the resulting material via a DP technique. This catalyst has also shown high catalytic activity for CO oxidation.

It has been shown than the activity of Au supported on Al_2O_3, SiO_2 or TiO_2 (prepared by gas-phase grafting and CVD) at room temperature is nearly the same (see Fig. 6.9), indicating that the contributions of these metal oxide supports are more or less similar and that Au is deposited as nanoparticles with strong metal–support interaction [75, 185]. However, Au deposited onto the acidic supports, such as silica–alumina and activated carbon, exhibits a much lower CO oxidation catalytic activity than for active Au catalysts such as Au/TiO_2 [75].

Highly dispersed gold with *particle sizes of 2–5 nm* seem to be an absolute requirement for this low-temperature activity [6, 7, 9, 18, 38, 41, 53, 74, 77, 110, 159, 202]. Nørskov and co-workers [317] have suggested that the main effect of decreasing the gold particle size is to increase the concentration of low-coordinated Au atoms. However, the Au–Ag alloy nanoparticles supported on mesoporous aluminosilicate, referred above, inspite of the large particle size (of about 20–30 nm) were surprisingly very active at room temperature [123].

The activity of gold supported on TiO_2 strongly depends on the size of the gold particles with a maximum of activity usually observed for a particle size of about 3 nm [6, 7, 318–320]. With TiO_2, Fe_2O_3 and Co_3O_4 supports, for example, Au particles in the range of 2–10 nm in diameter exhibit high activity, but only gold clusters smaller than 1 nm [268] exhibit high activity with $Be(OH)_2$ and $Mg(OH)_2$ supports. Also, a sharp increase in the activity of Au/TiO_2 catalysts was obtained with a decrease in the particle size from 4 nm [266]. Later Haruta's group also observed that the rate of the CO oxidation reaction markedly increased with a decrease in the diameter of the Au particles [111]. However, it has been shown that the presence of small gold particles alone is not sufficient to achieve high activity in CO oxidation, but the support also plays an important role [305, 309, 313, 317]. For example, Schubert and co-workers [152] have argued that Au particles of 12–30 nm on "active" metal oxide supports (Fe_2O_3, TiO_2, NiO_x, CoO_x), enable a good supply of reactive oxygen, and these can show activities as high as those for small Au particles on "inert" supports (Al_2O_3, MgO, SiO_2). The mechanism of supply of oxygen may be via hydroxyl groups which may be renewed on the surface by the presence of water [321]. Also, it has been shown that the catalytic activity of the newly developed $Au/Ti(OH)_4^*$ catalyst is much higher than that of the conventional Au/TiO_2 [322]. Despite the fact that the conventional Au/TiO_2 catalyst has a higher surface area, as it has a micro-porous structure, the resistance to pore diffusion is higher. Consequently, the catalytic CO oxidation reaction predominates on the external surface area and the internal surface area would not be utilised efficiently [322].

Composite oxides Au/MO_x (M = Li, Rb, Mg, Co, Mn, Ce, La, Ti, Fe) on SiO_2 [4, 5] and Al_2O_3 [4, 5, 8, 38, 52–54, 159, 323] have also been investigated. Silica supported samples were less active than the alumina ones, probably due to the presence of large gold particles [4, 5]. The presence of a partly reducible oxide was beneficial to the activity [8, 38, 53]. For the $Au/CeO_2/Al_2O_3$ case, full conversion was already achieved around 333 K [323].

Nieuwenhuys and co-workers also investigated CO oxidation on multi-component gold catalysts, $Au/M^IO_x/M^{II}O_x/Al_2O_3$ (M^I, transition metal; M^{II}, alkaline earth metal) [38, 52–54]. It was suggested that M^IO_x plays an important role in O_2 activation via a Mars and van Krevelen-type mechanism, and $M^{II}O_x$ stabilises the gold particle size against sintering [38, 52–54]. The most active catalyst in CO oxidation was $Au/MgO/MnO_x/Al_2O_3$ with MgO being a stabiliser for the Au particle size and MnO_x being the co-catalyst [38,52–54].

This catalyst also exhibits good performance in selective oxidation of CO in a hydrogen atmosphere (see below).

To evaluate the effect of preparation procedure, Haruta and co-workers used Au/TiO_2 catalysts prepared by a conventional impregnation mechanism and noted that they were less active than Pt prepared in a similar way, while those prepared by DP methods had a very different behaviour [269]. The DP method yielded metal particles strongly attached to the TiO_2 support (with average particle size of 3 nm), while IMP methods yielded particles which were simply loaded on the TiO_2 support and were much larger (around 10 nm). Over Pt/TiO_2, CO oxidation took place preferentially on the Pt surfaces and the metal oxide support was less involved in the reaction. This explains why the different methods of preparation do not make much difference in Pt catalysts. However, the activity of Au/TiO_2 markedly depends on the preparation procedure and the activity of Au particles exceeded that of Pt by one order of magnitude [207, 269]. In these experiments, Haruta's group used the deposition–precipitation with NaOH, but Louis and co-workers have developed an alternative preparative procedure for Au/TiO_2 catalysts using deposition–precipitation with urea, achieving good results in CO oxidation at 278 K [33, 48, 49].

Soares and Bowker have also found that Au/TiO_2 catalysts prepared by DP are much more active at room temperature than those prepared by IMP which generally only show conversion above 673 K [62] or by IW [289] that usually do not show steady-state conversion until temperatures above 393 K. The poor performance of IMP and IW catalysts, when compared with DP catalysts, is ascribed to the presence of *residual chloride*, which gives vulnerability to sintering to a larger particle size [62, 289]. Soares and Bowker also showed that $Au(OH)_3$ and Au_2O_3 mixed with TiO_2 is catalytically active [62]. Consequently, it is clear that if chloride is essentially removed, active catalysts can be prepared via impregnation, as shown by some authors for Au/TiO_2 [279–281] and Au/Al_2O_3 [39, 65] catalysts. Moreau and Bond [43] showed that pH 9 was the optimum value for high activity of Au/TiO_2 catalysts prepared by DP, since at this pH the main species in solution were anionic Au complexes, from which most of the chlorine had been removed by hydrolysis. At lower pH, the gold complexes contained more chlorine, Au particles were larger, and activities were lower [43].

Ageing of the precursor solution at the optimum pH to hydrolyse the $HAuCl_4$ enabled the formation of small gold particles on alumina by impregnation [39], and Datye and co-workers have developed a two-stage impregnation procedure [65], which gives 2.4 nm particles without chloride (see Sect. 6.4). These results above are in agreement with the conclusions of Haruta's group that the contact structure between Au particles and the support is the most critical factor in supported Au catalysts and that the reaction probably takes place at the perimeter interface around the Au particles [6, 7].

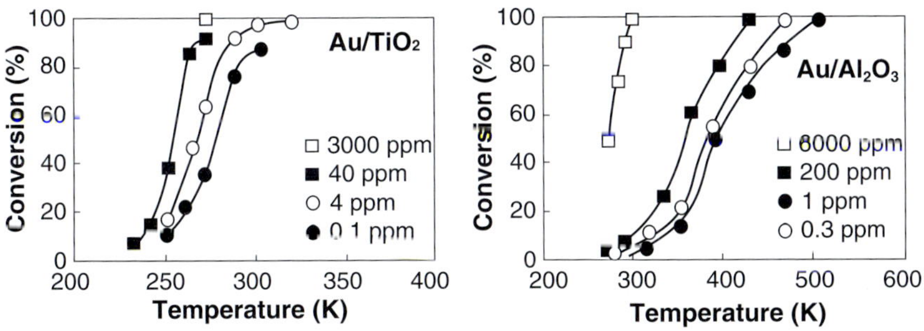

Fig. 6.10. Temperature dependence of CO conversions over Au/TiO_2 (*left*) and Au/Al_2O_3 (*right*) for various concentrations of H_2O, measured at the space velocity of 80,000 and 20,000 ml g_{cat}^{-1} h^{-1}, respectively (adapted from [275])

Since residual *water* is often present in CO streams at atmospheric pressure, it is important to know how moisture affects the catalytic activities, from the viewpoint of industrial applications. Wang and co-workers [263] found that Au/ZnO catalysts exhibit good catalytic activity and stability for CO oxidation at room temperature in humid circumstances. Chen and co-workers [309] also showed that catalysts consisting of Au supported on iron hydroxide are resistant to moisture. Haruta's group studied Au/Fe_2O_3, Au/Co_3O_4 and Au/NiO catalysts, showing that their activity was appreciable even at temperatures as low as 343 K, and that they were able to oxidise CO completely at 303 K even under relative humidity of 76% [183]. They also studied the moisture effect for the CO oxidation over Au/TiO_2, Au/Al_2O_3 (see Fig. 6.10) and Au/SiO_2 [273, 275, 296]. It has been found that the catalytic activity is generally enhanced by moisture in the reactant gas (unlike traditional metal oxide catalysts such as hopcalite). The enhancement is as large as two orders of magnitude. The moisture effect is reversible and the activity is mainly determined by the amount of moisture adsorbed by the catalyst. Conversion curves versus temperature drastically change for Au/SiO_2, indicating that the reaction over Au/SiO_2 does not take place at low temperatures without moisture. The moisture effect may be attributed to the decomposition of carbonate species accumulated during the reaction, while the –OH group might directly participate in the reaction pathways as a promoter [273,275,296]. Jiang and co-workers [324] prepared Au/MCO_3 (M = Ca, Sr, Ba) catalysts and achieved full conversions of CO at ambient reaction temperature in the presence of moisture. These researchers believe that water's positive influence might be due to its direct participation in the CO oxidation reaction [324].

Another important aspect of gold nanocatalysts is the *calcination temperature* [111,163,263,266,270,274]. Haruta's experiments using a mechanical mixture of Au colloids of about 5 nm with TiO_2 powder clearly showed that higher calcination temperatures lead to higher activity, although Au particles grew to larger particles [270]. They also studied an Au/TiO_2 catalyst calcined at

Table 6.3. Mean diameter of gold particles (estimated by TEM) and temperature for 50% conversion of CO over Au/TiO_2 catalysts calcined at various temperatures (from [274])

Calcination temperature (K)	Mean diameter of Au particles (nm)	Temperature for 50% conversion (K)
473	2.4	245
573	3.5	235
673	3.4	250
873	7.9	300

different temperatures, 473, 573, 673 and 873 K, and they found the best performance on CO oxidation for the catalysts calcined at 473 and 573 K [111,274] as shown in Table 6.3. Chen and co-workers [309] also obtained better results using low calcination temperatures (453 K) for $Au/Fe(OH)_x$ catalysts. It should be noted, however, that Bond and co-workers [18,325] and Galvagno and co-workers [326] obtained high activities for uncalcined catalysts.

The higher catalytic activity at these temperatures was related to the higher concentration of step sites over the Au surfaces and to a larger concentration of step sites at the borderline with the support. Calcination at 873 K promotes the coagulation of Au particles in forming larger particles with diameters above 10 nm, but at the same time with stronger contact leading to much higher catalytic activity than at 573 K [111]. While larger number of steps, edges, corners in smaller Au particles increases the catalytic activity for CO oxidation, contact with the support contributes to an additional increase in activity, at least by one order of magnitude [139]. These authors suggest that calcination at temperatures above 573 K is necessary to reduce oxidic gold species in metallic gold [266, 274].

It is claimed that the effect of the oxidation state of gold is also very important [137,285,315], however, the oxidation state of the gold in active catalysts is unclear. Several authors have suggested that part of the gold supported in Fe_2O_3 and Co_3O_4 is non-metallic, and that this may be responsible for the low-temperature activity [261, 327]. Other reports indicate that oxidised or partially oxidised gold species are the most active sites for gold supported on TiO_2, Al_2O_3, Fe_2O_3, Co_3O_4 and MgO [133, 137, 266, 285, 302, 303, 328]. In contrast, many other authors suggest that metallic gold is the active species in Au/TiO_2, $Au/Ti(OH)_3$, Au/ZrO_2 or Au/ACF [3, 7, 71, 74, 161, 286, 315]. Chen and Goodman [329] suggest that the active site at least for low-temperature CO oxidation, involves gold atoms that are nearly electrically neutral and bound to the surface via Au–Au covalent bonds and Au–Ti bonds. The authors find that the gold film completely covers the oxide and that reactants are sterically hindered from chemical bonding directly with the underlying Ti_2O_3 [329, 330].

Although metallic Au particles seem to be indispensable, a question arises as to why the periphery of Au particles can activate O_2 molecules at low temperatures. It is likely [125] that the perimeter interfaces contain hydroxyl groups bonded to Au atoms under usual conditions wherever H_2O is present at concentrations above 1 ppm. These hydroxides might be stabilised and reversibly formed and decomposed by the aid of the metal oxide supports and the presence of water.

Despite the increasing efforts in the investigation of supported Au catalysts for CO oxidation, the nature of the active sites and the corresponding mechanism for this reaction remain unconfirmed [9, 44, 125, 133, 137, 252, 314]. The mode of activation of oxygen still needs to be defined and spectroscopic evidence for the reaction intermediates obtained. Goodman and co-workers [318] using a combination of STM and spectroscopy considered that the unusual reactivity could be due to quantum size effects of the very small gold particles. Boyen and co-workers [331] concluded that Au particles containing 55 atoms (which were 1.4 nm in diameter), were extraordinarily stable and these Au^{55} particles could be active sites for CO oxidation. Models of the active site involving an ensemble of metallic Au atoms and Au^{+}–OH^{-} has been suggested [125, 133]. There is increasing support that the active site involving the perimeter of the Au particles and the Au–support interface [6, 7, 38, 111, 157, 266, 332] and for a role played by the presence of Au cations [162, 163, 319, 333], although these are matters for much current discussion [38, 157] and experimental proof of the details is still awaited.

It has been shown that the rate of CO oxidation over Au/TiO_2, Au/Fe_2O_3 and Au/Co_3O_4 is independent of the concentration of CO and is slightly dependent on the concentration of O_2 (on the order of 0–0.25) [266]. This is surprising since the conversion of CO increased with a decrease in the concentrations of CO and O_2 [139]. This independence suggests that CO and O_2 are adsorbed on the catalyst surfaces to near saturation and that the reaction of the two adsorbed species is the rate-determining step [6, 7].

The *oxidation of CO in an H_2-rich environment*, also called *preferential oxidation* (*PROX*) has long been of considerable technical interest for purification of hydrogen feed gas [101, 186, 188, 189, 191, 192, 199, 200], e.g., for its use as a supply in fuel cell technology (see Sect. 6.4.7). Particularly, proton exchange membrane (PEM) fuel cells have been attracting much attention as a method for improving the fuel efficiency and cleanliness of automobiles [101, 159, 188, 189, 191, 195, 197, 334–336]. This is due to low temperature of operation, fast cold start, perfect CO_2 tolerance by the electrolyte and a combination of high-power density and high-energy conversion efficiency.

Hydrogen is the most promising fuel for these fuel cells, since it simplifies system integration, maximises system efficiency and provides zero emission [101, 189, 335]. Currently, the supply of hydrogen needed for operation can be produced from methanol or other fuels, via partial oxidation, steam reforming and/or WGS reactions, but all these methods produce a large amount of CO_x and some water as by-products along with hydrogen [38, 52, 101, 184, 186, 189, 191, 192, 334–340].

The products coming out of the steam reformer containing ca. 10% CO (depending on the feedstock and conditions employed) are passed into WGSs, where CO is reacted with water to form CO_2 and hydrogen, and the CO content is reduced to a few ppm in the PROX reactor [189,334]. The hydrogen can be introduced in the fuel cell only after CO is eliminated from the stream [38, 46, 52, 186, 189].

The PROX reaction has been extensively investigated on Pt catalysts supported on Al_2O_3 [341–343] and zeolites [344]. Oh and Sinkevitch [341] compared the efficiency of several noble metals over alumina; the CO conversion was found to decrease in the following order: Ru > Rh > Pt > Pd (metal loading 0.5 wt% for all catalysts). Other systems, which have been investigated for the PROX reaction include metal oxides [345], oxide-supported noble metal catalysts [346] and bimetallic [347] catalysts.

The nanogold catalysts are also extremely promising as, in contrast to the Pt metal group catalysts, they are intrinsically more active for CO oxidation than for H_2 oxidation, and also the catalytic activity of Au is enhanced by moisture and almost insensitive to CO_2 [35, 38, 52, 183–187, 189, 191, 195]. Haruta and co-workers observed more than 95% conversion for CO (1% CO, 1% O_2, balance H_2) in a temperature range between 323 and 353 K on gold supported on manganese oxides [184]. This clearly is very promising, as the normal operation temperature of a PEM cell is about 353 K.

Alumina supported nanogold catalysts have also been effectively employed as PROX catalysts [46,200,346]. The conversion and selectivity for CO oxidation was further enhanced by the addition of MgO and MnO_x to the Au/Al_2O_3 catalysts [38,51,53,157,159] (see Fig. 6.11). The gold particle size was around 5 nm on the mixed oxides. The beneficial effect of MgO can be ascribed to stabilisation of small Au particles, which are intrinsically more active in CO and H_2 oxidation. MnO_x is thought to be able to supply the active oxygen needed for CO oxidation. These catalysts showed selectivities greater than 90% for the PROX reaction at temperatures lower than 373 K. At temperatures above 323 K, the thermal desorption rate of CO was large enough for H_2 oxidation to commence, implying a decrease in selectivity. Water was also found to have a pronounced effect on the CO oxidation rate even at room temperature. The effect of water is ascribed to a beneficial role of surface OH groups in CO oxidation [38, 52, 53, 157, 321].

It has been demonstrated that catalysts modified by Mn and Fe have higher activity, selectivity and stability in preferential CO oxidation, than the unmodified Au/MgO catalysts [196]. The results indicate that the addition of these modifiers changes the type of "metal ion–metal nanocluster" active sites involved in CO activation, and that the new type of sites are more active above room temperature.

Catalyst comparison studies for the Au/Fe_2O_3 [186–188,192,336], with Pt- and Cu-based systems, for the PROX reaction, revealed the former to be a better catalyst since it has a higher CO oxidation activity at lower temperature (353 K) and lower hydrogen consumption. CuO–CeO_2 showed to be effective

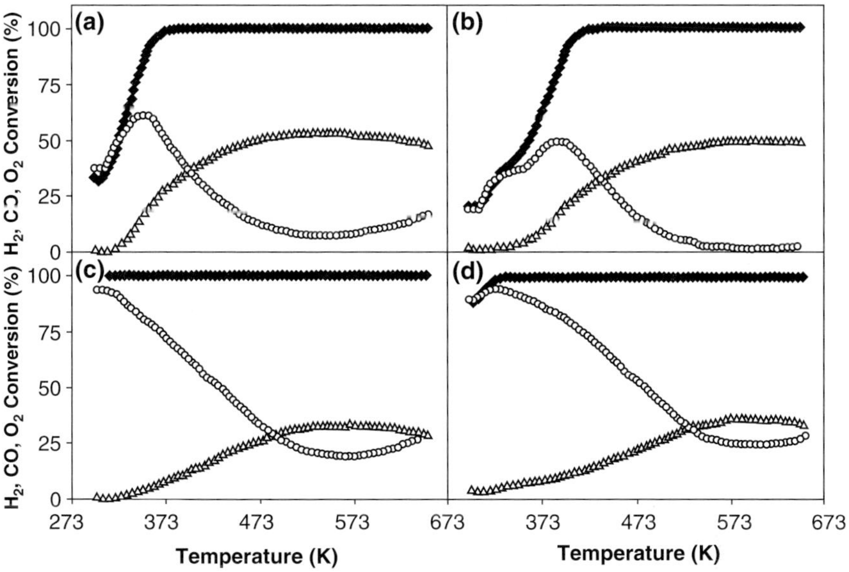

Fig. 6.11. Conversion of CO (○), H_2 (△) and O_2 (♦) versus temperature (K) over (**a**) Au/Al_2O_3, (**b**) $Au/MnO_x/Al_2O_3$, (**c**) $Au/MgO/Al_2O_3$ and (**d**) $Au/MnO_x/MgO/Al_2O_3$ when H_2 is present in the reactant flow (adapted from [38,52])

at higher temperatures (above 410 K) [194,348] but not as active as gold. Also, ferric hydroxide supported nanogold catalysts have been reported to be one of the most effective catalysts so far for PROX reaction at lower temperatures [190]. Rousset and co-workers [200] found that the reactivity order for CO oxidation to be Al_2O_3, ZrO_2 and TiO_2 for these supports, but in the presence of H_2, the reaction rates become rather similar on all three systems.

However, Pt catalysts are more stable than Au catalysts. In general, nano-Au catalysts are known to undergo rapid deactivation during CO oxidation [18,188,196,283,284,305,309,315,349]. Extensive efforts have been undertaken to study the deactivation mechanism [18,283,349–353]; STM/STS studies on model Au catalysts have shown that the deactivation is induced by oxygen. So, in order to employ Au as PROX catalysts, it is essential to synthesise nano-Au catalysts with high stability.

Goodman and co-workers obtained highly active and fairly stable catalysts by a temperature-programmed reduction–oxidation treatment of an Au–phosphine complex on TiO_2 [354]. Time-on-stream CO oxidation studies revealed a slow initial deactivation followed by stable activity for several hours. Moreover, the catalyst could be completely regenerated by a reduction–oxidation treatment. Behm and co-workers [198] performed isotope labelling studies also using Au/TiO_2 catalysts showing that H_2 affects the CO oxidation, most probably by competing with hydrogen adsorption on the Au nanoparticles and reaction with oxygen, and this results in a significantly higher CO reaction order.

Au/CeO_2 catalysts were found to be quite stable in the deactivation test [101, 144]. The stability of low-content gold–ceria catalysts for the PROX reaction in a realistic fuel gas mixture containing 1% CO–0.5% O_2–50% H_2–10% H_2O–15% CO_2–He was excellent. No drop in activity or selectivity was found in cyclic operation up to 423 K [144].

Some investigators [191, 197] believe that Au/ZnO catalysts could be the best candidates for the removal of CO from hydrogen-rich fuel gases. During 500 h of continuous experimental investigation for CO selective removal at 353 K, it was shown that the Au/ZnO catalyst had good stability, but its performance was slightly decreased when the time-on-stream exceeded 350 h. However, when small amounts of Pt were added to Au/ZnO, its stability was improved, as shown at the reaction temperature of 353 K during 500 h [191].

Rousset and co-workers [199] prepared Au/ZrO_2 catalysts simply via the oxidation of a $Zr_{0.5}Au_{0.5}$ single-phase alloy in air at 293 K. Authors think that this solvent-free method, which requires neither costly post-synthetic heat treatments nor the costly set-ups involved in the physical routes of preparation, is a good and alternative route in terms of cost and environment and could be suitable method for the preparation of active gold catalysts on a large scale [199].

Hutchings and co-workers [355] reported the most practical solution to this PROX requirement. They used Au/Fe_2O_3 catalyst prepared using a two-stage calcination procedure to achieve a commercially acceptable competitive oxidation of dilute CO in the presence of moist excess H_2 and CO_2. The catalyst was prepared by co-precipitation of Au^{3+} and Fe_2O_3. Subsequent calcination converts the cationic gold, which catalyses the reverse WGS reaction, to metallic gold. The calcination temperature has a crucial influence on the selective CO oxidation activity of the catalyst. Calcining the catalyst twice at 400 and 823 K gives a catalyst which removes more than 99.5% of the CO in the presence of moist hydrogen and CO_2 at 353 K, the operating temperature of the fuel cell.

6.4.5 Selective Oxidation

Selective oxidation is used in the activation of raw materials to provide useful products and chemical intermediates. Catalytic methods for the oxidation of organic molecules of this type are of growing interest with regard to eco-sustainable chemical processes [18, 356]. Heterogeneous catalysts may be used under mild conditions with gaseous oxygen or air in aqueous dispersion to promote these transformations effectively.

Rossi and co-workers have studied the *oxidation of organic substrates in the liquid phase*, and shown that gold on carbon is the preferred catalyst [27, 357–364]. One of the best preparation methods is the immobilisation of colloidal particles [27, 140, 357–360, 362–364], which can readily be prepared with a narrow nanoparticle size range. The microporous nature of carbon supports provides a shield for small gold particles and a limit for their mean diameter. Compared with other metals, appropriate gold catalysts show better resistance to deactivation, providing a convenient method for use in industry on several types of feedstock [213]. The advantage of using gold catalysts lies in their high resistance to poisoning, which is the limiting factor in the scaling-up of liquid-phase oxidation using dioxygen [59, 360, 363].

The oxidation of *aldehydes* with O_2 as the oxidant may be performed in the liquid phase using homogeneous catalysis based on salts of copper, iron, cobalt, manganese, etc. [362], but heterogeneous systems are advantageous when metal recovery is being considered. From the industrial point of view, gas-phase oxidation is more attractive as the absence of solvent facilitates product collection, but limitations due to high boiling points or thermal instability sometimes preclude this choice.

Liquid-phase catalytic oxidation can normally be carried out under milder conditions than in the gas phase and the choice of solvent helps to determine the reaction mechanism, with organic solvents favouring a radical pathway [362] and water an ionic pathway [365]. Moreover, water is a safe and environmentally compatible solvent and is favoured from the industrial point of view. However, in liquid aqueous phase oxidation, the classical heterogeneous catalysts based on platinum-group metals suffer deactivation when O_2 is the oxidant and catalyst lifetime is compromised [365]. In an attempt to meet both environmental (liquid aqueous phase and heterogeneous system) and industrial requirements (safety and high productivity), Rossi and co-workers have shown that gold on carbon has good activity in the liquid-phase oxidation of aldehydes [362]. It is also a good alternative to homogeneous systems that have a high environmental impact and to heterogeneous Pt/C that deactivates rapidly (Fig. 6.12).

The catalytic oxidation of vicinal diols in water is a valuable target, for example, in the synthesis of glycolic acid and lactic acid via the

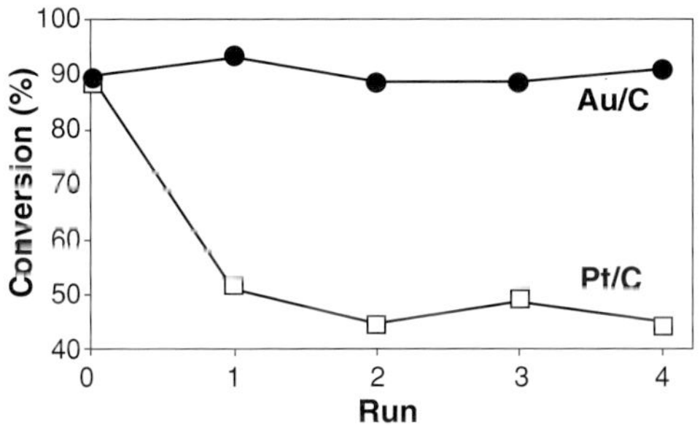

Fig. 6.12. Conversion on recycling of n-propanal in water. Reaction conditions: [n-propanal]=0.23 M in water; T=363 K; $p(O_2)$=300 kPa; t=2 h. After the first run, the catalyst was filtered-off, thoroughly washed with water and used in the next run with a freshly prepared solution of aldehyde (adapted from [362])

monooxygenation of *ethane-1,2-diol* (*ethylene glycol*) and *propane-1,2-diol* [27, 356, 357, 363, 364]:

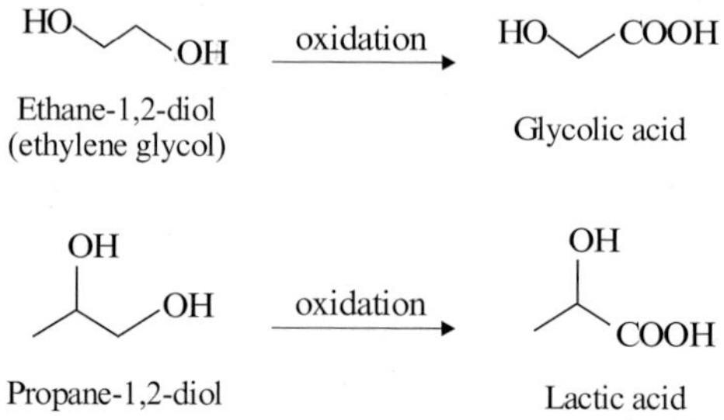

Current chemical methods involve the use of toxic or corrosive reagents as well as high-pressure equipment. An alternative fermentation process, available for lactic acid production, is affected by low productivity and severe purification problems.

Studies by Rossi and co-workers [27, 59, 356–360, 363, 364, 366] have shown that nanoparticulate gold on carbon can be conveniently employed as a catalyst for the *selective oxidation of 1,2-diols to α-hydroxyacids* under mild conditions, and have demonstrated superiority over mono-, di- and trimetallic-, Pd-, Pt- and Bi-based catalytic systems, since the gold catalyst is more selective and poison resistant. However, strong alkaline conditions were required for gold to speed up the reaction and, in the other cases, this also enhanced selectivity. Moreover, gold on carbon shows the best stability in recycling tests, where neither deactivation nor metal leaching was observed. Particle size ranged from 7 to 12 nm [357]. Ethane-1,2-diol and propane-1,2-diol has successfully been transformed into glycolate and lactate with selectivities exceeding 98% [27, 356, 357, 363] using gold catalysts.

In the selective liquid-phase oxidation of *ethylene glycol* to glycolate, colloidal Au solutions were immobilised by Rossi and co-workers not only on activated carbon [27, 359, 363, 364, 366] but also on Al_2O_3 [27], SiO_2 [27] and TiO_2 [359]. Smaller gold particles on alumina result in higher reactivity than larger ones, but they observed an opposite trend on carbon until a mean dimension of 7.5 nm was reached [27, 358]. Rossi and co-workers have explained this anomalous trend (Fig. 6.13) by considering that with decreasing dimensions, the gold nanoparticles can easily be fixed on the internal surface of carbon, consequently, being less accessible to reagents than gold particles on the external surface.

For larger particles than 7.5 nm, the reactivity follows the normal trend, in line with the decreasing area of active metal surface. This group also used *N*-dodecyl-*N*,*N*-dimethyl-3-amino-1-propane sulphonate in water for stabilising gold nanoparticles on activated carbon [363, 367]. Characterisation by XPS revealed a high-atomic percentage of superficial gold compared with other Au/C catalysts prepared by the immobilisation of different sols. The activity of these stabilised Au/C catalysts was found to be superior to that observed

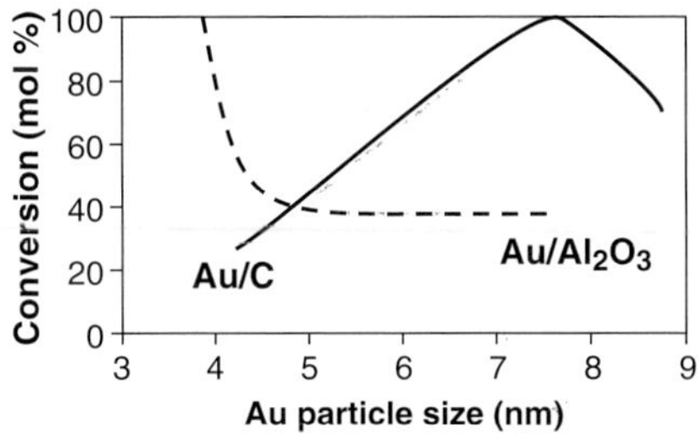

Fig. 6.13. Dependence of the conversion for glycol oxidation over Au/Al_2O_3 and Au/C catalysts on the Au particle size (adapted from [358])

in the liquid-phase oxidation of ethylene glycol using other gold-on-carbon catalysts [363, 367].

Martin and co-workers [368] obtained a selectivity to glycolic acid from ethylene glycol of higher than 95% for low and highly dispersed Au/Al_2O_3 catalysts. They could clearly demonstrate that the selective oxidation of ethylene glycol to glycolic acid is not a structure-sensitive reaction, i.e., this reaction is not restricted to a defined gold particle size, as has been found for the epoxidation of propylene over Au/TiO_2 (see below). However, Vorlop and co-workers [369] showed that glycolic acid formation from ethylene glycol increases steadily, as the Au surface area increases.

Rossi and co-workers have also found that *phenylethane-1,2-diol*, a cheap starting material, can be oxidised under mild conditions using 1% Au on carbon powder, prepared from a colloidal sol, where the average gold particle diameter was 7 nm. Gold on carbon thus represents a promising alternative to present methods for the synthesis of mandelic acid, an important pharmaceutical intermediate [356, 363].

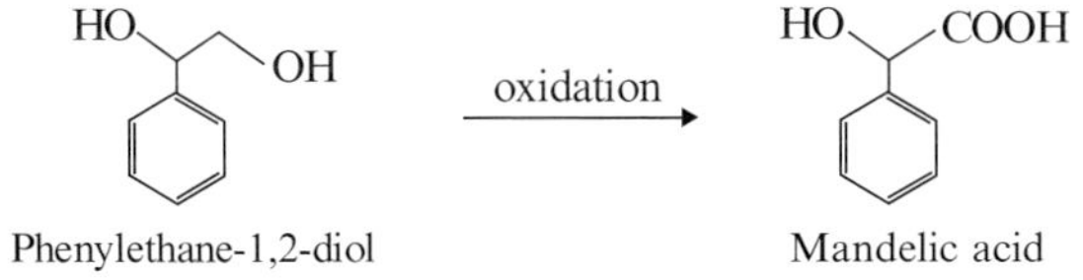

Another example of selectivity is that aminoalcohols can be oxidised to aminoacids in slightly alkaline solution. In this application, gold is remarkably superior to other metals because the free amino group has a strong interaction with other catalytic metals such as Pd and Pt [361, 363].

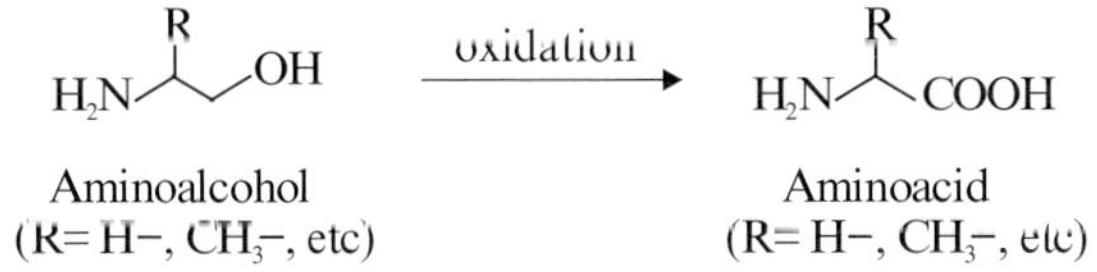

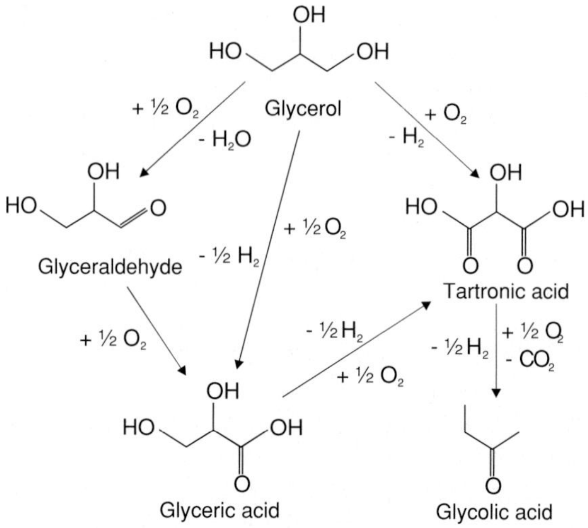

Fig. 6.14. Selected reactions of network of the glycerol oxidation (adapted from [363, 372])

Glycerol is a highly functionalised molecule that is readily available from bio-sustainable sources such as rapeseed and sunflower crops [252, 370–373]. A large number of products can be formed from glycerol oxidation (see Fig. 6.14 for examples) and selectivity to individual products is important for their economic use as chemical intermediates in the fine chemicals industry [252, 370–372, 374]. Given its high boiling point, the selective oxidation of glycerol with air/oxygen is preferably carried out in the liquid phase using water as the solvent [375].

Glyceraldehyde has been obtained in selectivities of 70–80% using Pt catalysts [376], but Hutchings and co-workers [370, 371] have shown that glycerol can be oxidised to glyceric acid with 100% selectivity at high conversion, using either 1% Au/charcoal or 1% Au/graphite catalyst under mild reaction conditions (333 K, 3 h, using water as solvent).

Claus and co-workers [372] studied the liquid-phase oxidation of glycerol over nanosized Au particles (2–45 nm) on a number of supports. When only the support was used, no glycerol conversion could be detected, whereas Au nanoparticles supported on different carbons (carbon black, activated carbon and graphite) and oxides (TiO_2, MgO and Al_2O_3) were active. The carbon black support gave a more active catalyst than activated carbon or graphite.

Moreover, comparing the catalytic behaviour of Au/C catalysts with different mean Au particle sizes in the range from 2.7 to 42 nm, it was shown that glycerol oxidation is structure-sensitive. In fact, the selectivity to glyceric acid was increased to 75% with smaller gold particle size until the optimum value of 3.7 nm was reached. With smaller mean gold particles (2.7 nm) the selectivity to glyceric acid decreased to 40% whilst that for glycolic acid increased from 15 to 36%. Consequently, the metal particle size of the Au/C catalyst plays a major role in the oxidation of glyceric acid.

Porta and Prati [363, 375] also studied the same reactions using Au on carbon as the catalyst. Two types of catalysts were recognised: First, characterised by well-dispersed nanoparticles with a mean diameter centred at 6 nm that did not maintain the initial selectivity of the oxidation through full conversion; second, characterised by larger particles (>20 nm) that, on the contrary, maintained constant selectivity throughout the reaction period. It thus became apparent that it is not only particle dimension that fulfils an important role but also the preparation method, with sol immobilisation being better than the impregnation or IW methods. Also, temperature plays an important role in determining product stability, increasing the temperature promotes the oxidation of glyceric acid to tartronic acid. Ninety-two percent of selectivity was achieved at full conversion [363, 375].

Prati and co-workers studied bimetallic Au catalysts supported on carbon [374] and graphite [373] using a sol method. It was shown that, in terms of activity either at 303 or 323 K, bimetallic catalysts are more active than monometallic catalysts, indicating that a synergistic effect exists between Au and Pd or Pt [373,374]. This effect was especially significant in the case of Pt as the mono-metallic catalyst was poisoned before reaching full conversion. In terms of selectivity to glyceric acid, Au–Pd catalysts showed better selectivity than Au–Pt catalysts, with Pd mainly to promoting the formation of tartronic acid and Pt, glycolic acid. The overall selectivity to glyceric acid increased using bimetallic (Au–Pd)/C catalysts with respect to mono-metallics. Using Au–Pd/C at 303 K, authors obtained the best selectivity to glyceric acid never obtained, i.e., 69% at 90% conversion [374].

Mixed Au–PGM/C activity enhancement compared with the activities found for mono-metallic catalysts has also been observed in the selective oxidation of D-*sorbitol* to gluconic and gulonic acids [363, 377, 378]. Bimetallic catalysts showed a resistance to poisoning by dioxygen and by reaction products as high as Au mono-metallic catalyst but also an improved selectivity toward mono-oxidation with respect to mono-metallic. Adding Au after the prior addition and reduction of the Pd or Pt metal resulted in the most active catalytic system irrespective of the reaction conditions used [378]. With regard to selectivity, Au and Au/Pt showed the best selectivity (60% and 62%, respectively) to mono-carboxylate (gluconate/gulonate) [363, 377, 378].

Interest in the selective oxidation of *glyoxal* (ethanedial) into glyoxalic acid (also called glyoxylic and oxo-acetic acid) is stimulated by the fact that it is an important intermediate in the cosmetic and food industries, e.g., for the synthesis of vanillin [379]. Vorlop and co-workers [369] showed that when glyoxal is oxidised rupture of the C–C bond is the predominant reaction, producing formic acid; and the desired product, glyoxalic acid, is observed in minor amount only (see Fig. 6.15).

Hermans and Devillers [379] have shown, however, that Au can have a clear promoting effect for the activity of Pd in the reaction of glyoxal selective

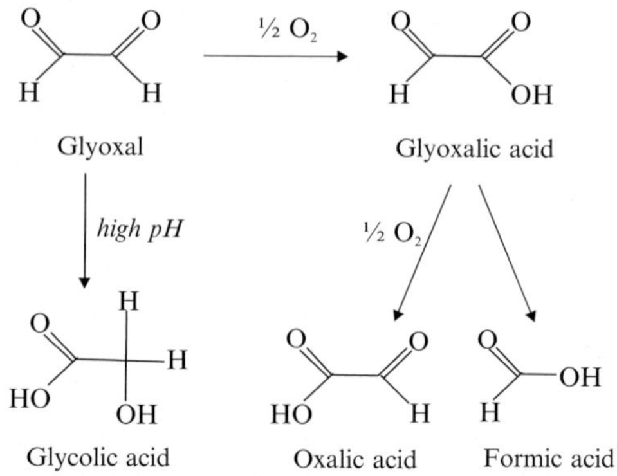

Fig. 6.15. Possible by-products in the catalytic oxidation of glyoxal (adapted from [379]). *Note:* The production of glycolic acid depends only on the pH by Cannizzaro dismutation and is thus truly homogeneous

oxidation carried out in water. The mono-metallic Au/C catalysts were inactive, but the bimetallic Pd–Au/C catalysts were at least twice as active as the best Pd/C materials. Moreover, the Pd–Au catalysts were more active than the Pd–Bi/C or Pd–Pb/C catalysts that gave the best performance for this reaction in studies carried out previously by the same authors [380, 381]. In addition, no metal leaching could be detected in the reaction mixtures (unlike with the previous catalysts), meaning that the Pd–Au catalysts are remarkably stable in very demanding conditions (in the presence of water and oxygen, with reactants that possess chelating properties), a fact that was the limitation with the Pd–Bi(Pb) systems. These Pd–Au/C catalysts represent thus a viable alternative for this reaction and the promoting ability of Au might be extendable to other liquid-phase oxidations [379].

Au/Fe_2O_3 catalysts prepared by CP [382] and Au catalysts prepared by IMP on Fe_2O_3, ZnO, CaO and Al_2O_3 [383] have been shown to catalyse the oxidation of *o-hydroxybenzyl alcohol* (salicylic alcohol) under mild conditions. Salicylic aldehyde is the main reaction product with a selectivity >90% at 90% conversion [383]. Sodium carbonate treatment of the catalyst probably removed most of the chloride which would otherwise be a poison.

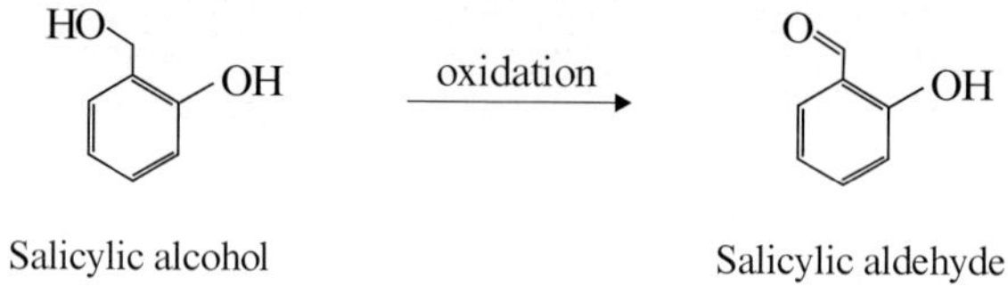

Mirescu and co-workers [384] have reported on the selective oxidation of *lactose* and *maltose* with Au/TiO_2 catalysts. The initial activity for maltose was more than twice that for lactose but both reactions gave 100% selectivity to maltobionic acid and lactobionic acid, respectively. The two acids have potential uses in the pharmaceutical and detergent industries [104].

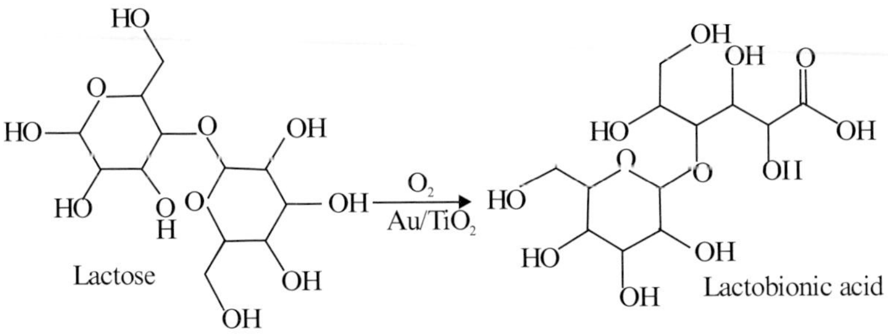

The oxidation of *glucose* or glucose-containing raw materials is commercially interesting, since gluconic acid and its salts are important industrial products used as water-soluble cleansing agents or as additives in food and beverages [360, 361, 364].

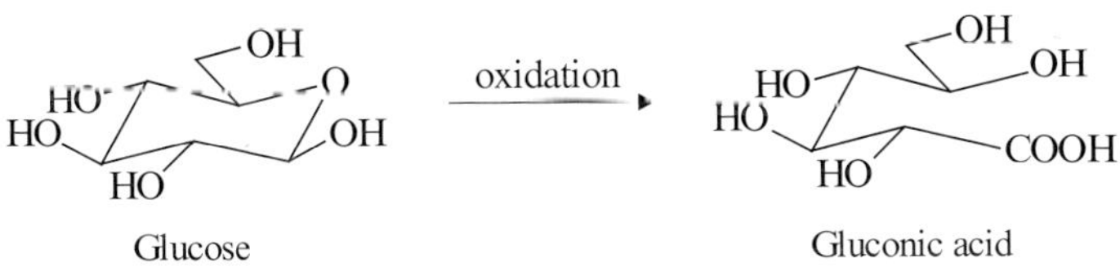

A fermentation process is currently the preferred industrial route, despite problems related to microbe separation, control of by-products and the disposal of wastewater [360, 361]. Only the development of catalysts with excellent activity, selectivity and durability is expected to compete with this.

Rossi and co-workers have investigated the use of gold on carbon as a catalyst for this reaction [140, 360, 361, 364, 385]. The average size of the gold particles was 2–5 nm and experiments were conducted at a pH of 7–9.5 in order to avoid alkali-promoted reactions [360]. It was shown that carrying out the oxidation at 323 K, without controlling the pH, led to the rapid poisoning of Pd and Pt catalysts, which was ascribed to strong adsorption of the reaction product on the catalyst [360, 363]. When they tested the behaviour of gold at neutral pH, however, they achieved 100% conversion. In uncontrolled pH conditions, gold showed a lower reaction rate (than under pH-controlled conditions), but it still enabled a quantitative transformation [360, 361]. Also, experiments on catalyst recycling have shown a better performance for the gold catalyst.

Thus, it was shown that gold on carbon was much better than the more complicated multi-metallic Pt/Pd/Bi catalysts evaluated previously [365, 386–392] for the liquid-phase oxidation of D-glucose, giving 100% selectivity to gluconic acid [360, 361]. Claus and co-workers also studied a series of Au/C catalysts prepared by the sol–gel method and got conversions and selectivities of up to 90% for the conversion of glucose to gluconic acid (323 K, pH 9–10, 60 min) [393].

This glucose to gluconic oxidation reaction was also investigated by Rossi and co-workers using Au, Ag, Pd and Pt colloidal sols [140,394]. It was found that gold nanoparticles are much more active than the others, reaching 21% conversion in 200 s, whereas 4% conversion for Pd and lower values for other metals were obtained after 1,000 s (see Fig. 6.16). However, Ag, Pd and Pt sols were stable for several hours, while the gold sol collapsed due to the formation of larger particles (from 3.5 to 10 nm), leaving a colourless and almost inactive solution. However, the activity observed for gold during the first 200 s is still surprisingly high when compared with supported gold particles in previous studies [361]. Although the short life of the gold sol does not allow its practical use as a catalytic system, its activity is notable.

It was however observed, in the same study, that by adding carbon and titania, the activity of gold was prolonged for a long time allowing the total conversion of glucose in the first case and more than 90% conversion in the second case, after 2,000 s. Gold particles immobilised on carbon and added to the reacting sol are very stable with their dimensions unchanged after the catalytic cycle. The activity of the original particles in the first 200 s was not enhanced by the addition of supports, showing that supported and unsupported gold particles are similarly efficient during this time and, therefore, metal–support interactions, if present, are negligible in terms of activity [140]. This is a surprising result in view of the importance of the metal–support interactions thought to be involved in most of the oxide-supported gold catalysts.

The *selective oxidation of cyclohexane to cyclohexanol and cyclohexanone* has been studied by Suo and co-workers [395, 396]. This reaction is a key process in the chemical industry, for the oxidation products of cyclohexane, via cyclohexanol and cyclohexanone, are important intermediates in the manufacture of nylon-6 and nylon-66 polymers and are also used as solvents for lacquers, shellacs and varnishes as well as stabilisers and homogenisers for soaps and synthetic detergent emulsions.

The first studies were performed over a calcined Au/ZSM-5 molecular sieve catalyst, with oxygen as oxidant, in a solvent-free system [396]. They found out that this calcined catalyst was very efficient and it would be recycled twice without any obvious loss of activity. The yield of both products appears to decrease with temperature (from 413 to 453 K; at the latter value, the total selectivity of cyclohexanol and cyclohexanone decreases) [396].

The same authors also studied the same reaction over an Au/MCM-41 catalyst in a solvent-free system, where oxygen was the only oxidant and the reaction conditions were very moderate [395]. They showed the conversion of cyclohexane oxidation and the selectivity of cyclohexanone were up to ≈16% and ≈76%, respectively. Authors claim this was the first report of such an excellent conversion and selectivity for cyclohexane oxidation under such reaction system. It was also shown that this novel environmentally friendly catalyst can be used repeatedly for at least three times without obviously

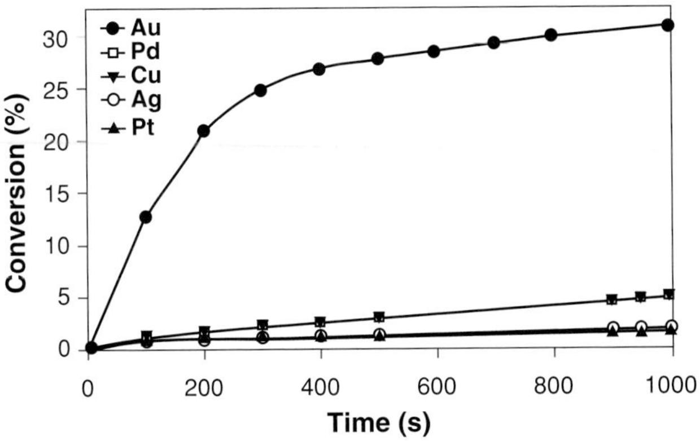

Fig. 6.16. Glucose oxidation using Au, Ag, Cu, Pd and Pt particles in the form of colloidal sols (adapted from [140, 394])

conversion and selectivity losing [395]. A Solutia Inc., USA patent, describing similar technology was published [397] and Corma have shown advantages in using Au/CeO_2 as a catalyst [398]. Au/C can also be used to promote this conversion in the presence of tertiary butyl hydroperoxide as initiator [399]. Improved nanoparticulate gold on mesoporous silica catalysts have also been reported for this reaction [400].

The *selective oxidation of hydrocarbons* is one of the most important chemical transformations in petroleum-based industrial processes, as the oxygenated products are used as key intermediates in organic synthesis [332]. The oxirane products formed from olefins are versatile products that easily undergo ring-opening reactions to form bifunctional compounds.

Propene oxide (PO, also known as propylene oxide, methyl oxirane and 1,2-epoxypropane) is an important building block for the manufacture of polyurethanes, polyethers, diols and numerous other valuable chemical products ranging from propylene glycol in toothpaste, cosmetics and deodorants to the structural units of block copolymers [2, 40, 104, 169–171, 251, 332, 401–405]. PO is currently produced using two commercial processes. These are the chlorohydrin process and the hydroperoxide process [2, 33, 40, 104, 169–171, 403, 404]. The chlorohydrin process is not environmentally friendly, due to the formation of $CaCl_2$ by product and hazardous organic chlorinated compounds. The hydroperoxide process has a demand–supply gap for the co-products. Consequently, there is a growing interest in developing an innovative process for the direct gas-phase synthesis of PO similar to that used for the epoxidation of ethene [40, 169, 171, 402]. Examples are the supported Ag catalysts containing organic/inorganic chloride promoter and potassium promoter [406, 407], and Pd and Pd–Pt catalysts supported titanosilicates such as TS-1 [406, 408, 409]. However, there has so far been no alternative process developed, probably due to low yields, low raw-material conversion, low productivities, numerous by-products, and complex catalyst systems [40, 169, 171, 402].

Table 6.4. Propene to propene oxide, reported by Dow [418]

T (K)	PP conv. (mol%)	PO sel. (mol%)	H_2O/PO (mole ratio)	gPO $kg_{cat}^{-1}h^{-1}$
433	0.36	92	2.04/1	8.3
Feed stream: 20% propene, 10% H_2, 10% O_2+He				

2 g_{cat}, flow rate: 160 $cm^3\,min^{-1}$, olefin GHSV, 480 h^{-1}, atm pressure. Catalyst 0.5 wt% TiO–SiO: calcined at 823 K

Haruta's research work on the catalysis of gold has initiated a new possibility for the direct *epoxidation of propene* using hydrogen and oxygen [7, 169]:

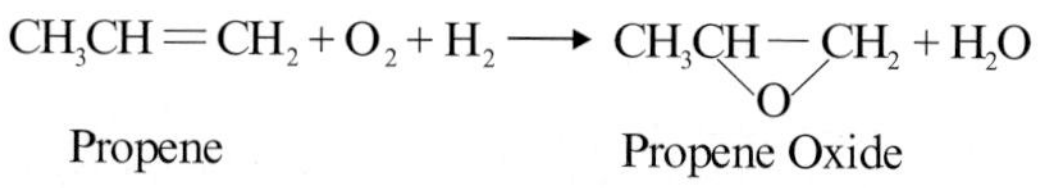

Thus, a number of papers have been published on the vapour-phase epoxidation of propene over highly dispersed nanosized Au particles supported on TiO_2 [164–169, 405, 410]. An amazing selectivity above 99% was reported over these supports [164–166]. Studies with TiO_2/SiO_2 [165, 166, 169, 171, 401, 411] and titanosilicates, such as TS-1, TS-2, Ti-β, Ti-MCM-41 and Ti-MCM-48 [28, 40, 167, 170, 205, 296, 332, 401–405, 412–415] and Au/Al_2O_3, and multi-component Au/MO_x/Al_2O_3 (M = Ce, Mn, Co, Fe) catalysts prepared by DP with urea [120, 160] were also reported. In the last case, it was concluded that the most active catalyst was Au/CeO_x/Al_2O_3. The average particle size was 3 nm.

Patents claiming use of gold catalysts for propene oxidation have been appearing in the literature [2, 104, 416, 417], indicating significant industrial interest in this application and pilot plants are understood to be operating within the industry. Bayer researchers have claimed an 8% yield of propene oxide [416], with 95% selectivity. Results from a Dow patent are given in Table 6.4.

Industrial application of the Au catalysts would be feasible if 10% conversion of propene could be achieved. However, there still remains a number of serious problems; low PO yields (<2%), low selectivities at high reaction temperature, deactivation with time-on-stream and low H_2 efficiency (<30%). The first two problems could be somewhat improved when titania-modified silica and titanosilicates were used as supports for Au nanoparticles. The initial conversion of propene was increased to ca. 5% from 2% or below, when the reaction temperature was raised to 423 K from 353 K or below without considerable loss in PO selectivity [40, 402]. To date, only small improvements in the deactivation of Au catalysts with time-on-stream have been reported,

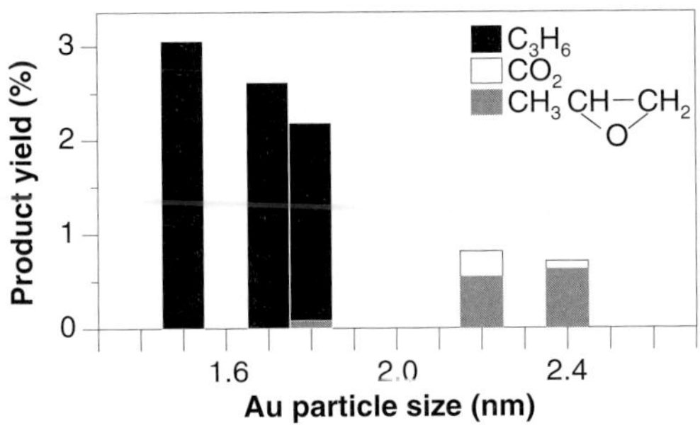

Fig. 6.17. Product yields of the reaction of propene, oxygen and hydrogen over Au/TiO_2 as a function of mean diameters of Au particles. Feed gas: $C_3H_6/O_2/H_2/Ar = 1/1/1/7$, SV $4{,}000\,\mathrm{ml\,g_{cat}^{-1}h^{-1}}$, $T = 353\,\mathrm{K}$ (adapted from [6,31,169])

although Au/Ti–SiO_2 catalysts were less deactivated than Au/TiO_2 [28] and this is thought to be because the product adheres less strongly to these supports. Hydrogen consumption could be greatly suppressed when a promoter, such as CsCl, is mixed with Au catalysts [205,403].

Haruta and co workers [296,403] have demonstrated that highly dispersed gold nanoparticles supported on 3D mesoporous, silylated titanosilicates with large (>7 nm) pores, in the presence of $Ba(NO_3)_2$ as a promoter, are very efficient catalysts for this reaction. This research group is carrying out further work on catalyst preparation, silylation optimisation, metal salt promoters, and optimisation of the balance between synergistic Au and Ti sites, which could result in the design of a commercially viable catalyst for propene epoxidation. Furthermore, these 3D mesoporous titanosilicates with a high titanium content (up to 8 mol%) can be expected to be highly active in other oxidation and epoxidation reactions [296].

The formation of PO over Au-based catalysts is a structure-sensitive reaction. Only hemispherical Au particles with a suitable size (2–5 nm) will selectively produce PO [167,168,403] and 2.2 to 2.4 nm particle size seemed to be optimum in the early experiments [31]. The most effective type of Au nanoparticles is prepared by the DP technique, which brings them in strong contact with the support. Gold particles smaller than 2 nm show a shift in selectivity from PO to propane [7,169–171,403]. This switch of epoxidation to hydrogenation for particles under 2 nm size indicates that small Au clusters exhibit different behaviour in surface properties from that of metallic Au [171] (see Fig. 6.17).

The crystalline and pore structure of TiO_2 and Ti–SiO_2 (Ti is dispersed as in isolated species in the matrix of SiO_2) supports also greatly affects PO formation. Only anatase TiO_2 support, not rutile and amorphous TiO_2, makes Au selective for PO [401, 403]. The activity of Au supported on Ti-MCM-48 is higher than that of Au supported on Ti-MCM-41, indicating that three dimensional, branched-pore structure is preferable [40]. The accumulation of

PO and other oxygenates on the catalyst surfaces is considered to be a main reason for lower PO yield and deactivation [40, 165, 166, 419].

FTIR results reported by Moulijn and co-workers [165] showed that there was an irreversible adsorption of PO on the TiO_2 and $Ti-SiO_2$ surfaces by the formation of propoxy species, which are linked to acidic Ti–OH and Si–OH sites. The internal and external surfaces of $Ti-SiO_2$, Ti–OH and Si–OH groups yields a relatively acidic and hydrophilic environment, especially in the presence of water. Modification of the hydrophilicity/hydrophobicity of $Ti-SiO_2$ supports may adjust the adsorption and desorption of reactants and products during reaction. The same authors showed that a limitation of the PO yield at around 2% over Au/TiO_2-based catalysts was caused by consecutive reaction over Ti–O–Ti containing units, or the existence of a PO adsorption–desorption equilibrium over active epoxidation centres [410]. A catalyst development or a temperature/pressure swing process could circumvent this yield limitation. Catalyst development has to be focused on an increase in the hydrophobicity of the catalyst. Selective removal of PO via membrane technology might open an option for further investigations.

Haruta and co-workers showed that silylation of the Au/Ti-MCM-48 catalysts prevents them from getting deactivated faster and also helps to improve PO selectivity and decrease H_2 consumption [28, 40, 170, 403]. Physical mixing of CsCl with Au supported on Ti-MCM-41 reduces H_2 consumption by about 90% and improves propene oxide (PO) selectivity up to 97% at a propene conversion of 1.7% [205, 403].

Styrene oxide is an important organic intermediate in the synthesis of fine chemicals and pharmaceuticals. It is traditionally produced by epoxidation of styrene using stoichiometric amounts of peracids as an oxidising agent [420–422]. However, peracids are very expensive, hazardous to handle, non-selective for the epoxide formation and also lead to formation of undesirable products, creating a lot of wastes [421–424]. In addition, the epoxidation of a terminal olefin, such as styrene, is relatively difficult and hence requires prolonged reaction time [421, 422, 424, 425].

The use of easily separable solid catalysts, such as TS-1 [426], $Ti-SiO_2$ [427], and TBS-2 and TS-2 [428], has been reported for the epoxidation of styrene. However, these catalysts showed poor selectivity when H_2O_2 was used as the oxidising agent. High styrene oxide selectivity (>80%) was achieved using organic hydroperoxide [429] or urea–H_2O_2 adduct [430] instead of aqueous H_2O_2, but at low styrene conversion (9.8% and 17.7%, respectively). Sharpless and co-workers [431] have reported pyridine ligand-accelerated methyl trioxorhenium (MTO)-catalysed olefin epoxidations with excellent yields (>90%) to epoxides. However, catalyst reusability was not reported and the MTO is very expensive. Hence, there is a great practical need to develop a novel reusable solid catalyst showing both high activity and high selectivity in the epoxidation.

Choudary and co-workers reported for the first time, the use of Au nanoparticles (5.7–8.9 nm) deposited on MgO, CaO, SrO and BaO [422] and transition metal oxides (viz TiO_2, Cr_2O_3, MnO_2, Fe_2O_3, Co_3O_4, NiO, CuO, ZnO, Y_2O_3, ZrO_2, La_2O_3 and U_3O_8) [423, 424] and Group IIIA metal (viz Al, Ga, In and Ti) oxide-supported [421] novel highly active/selective and reusable catalysts in the epoxidation of styrene to styrene oxide (SO) by anhydrous *t*-butyl hydroperoxide. Catalysts were prepared by DP using urea as the precursor for precipitating agent (ammonium hydroxide), using a similar method to that used by Louis and co-workers [33].

Catalysts prepared by usual DP had slightly poorer performance (in terms of conversion and selectivity) when compared with the ones prepared by homogeneous DP [421,422]. The Au loading, Au particle size and performance in the epoxidation of the supported Au catalysts are found to be strongly influenced by the transition metal oxide support used in the catalyst (prepared by HDP method). Generally, the catalysts containing more basic support showed better performance in the epoxidation. Au/In_2O_3 showed not only the best performance and also excellent reusability in the epoxidation (in the third reuse of the catalyst, the conversion and styrene oxide selectivity were found to be 51% and 48%, respectively) [421].

Arai and co-workers [432] reported also a direct route for the preparation of cyclic styrene carbonate from styrene, which avoids the preliminary synthesis and isolation of styrene oxide. A catalyst system consisting of Au/SiO_2, zinc bromide and tetrabutylammonium bromide (Bu_4NBr) was applied to the one-pot synthesis of styrene carbonate from styrene, organic peroxide and CO_2. Au/SiO_2 is active for the epoxidation of styrene, and zinc bromide and Bu_4NBr cooperatively catalyse the subsequent CO_2 cycloaddition to epoxide.

The *selective catalytic oxidation* (*SCO*) *of ammonia* has also been studied by Nieuwenhuys and co-workers [120, 158, 159] over various Al_2O_3-supported-Au-based catalysts. Mixed metal oxides, $MO_x{-}Al_2O_3$, (M = Cu, Fe, Ce, Li, Ti) were used as supports of 5% Au catalysts. The presence of Au enhanced the SCO activity. The following order was found for the temperature range 473–673 K: $Au/Ti{-}Al_2O_3 < Au/Al_2O_3 \approx Au/TiO_2 < Au/Li{-}Al_2O_3 < Au/FeO_x \approx Au/Fe{-}Al_2O_3 < Cu{-}Al_2O_3 \approx Au/Ce{-}Al_2O_3 < Au/Li{-}Ce - Al_2O_3 \leq Au/Cu{-}Al_2O_3$. The most active catalyst, $Au/Cu{-}Al_2O_3$, shows a conversion of 50% at 483 K and a N_2 selectivity over 90%. On the other hand, $Au/Li{-}Ce{-}Al_2O_3$, which has only a slightly lower activity than $Au/Cu/Al_2O_3$ is very selective to N_2O.

6.4.6 Selective Hydrogenation

Nanoparticulate gold supported on oxides (SiO_2, Al_2O_3, TiO_2, ZnO, ZrO_2 and Fe_2O_3) has been used for the selective catalytic hydrogenation of several organics such as α, β *unsaturated aldehydes* [98, 99, 249, 433–440], α,β-*unsaturated ketones* [58, 326, 441] and *unsaturated hydrocarbons* [18, 437, 442–445].

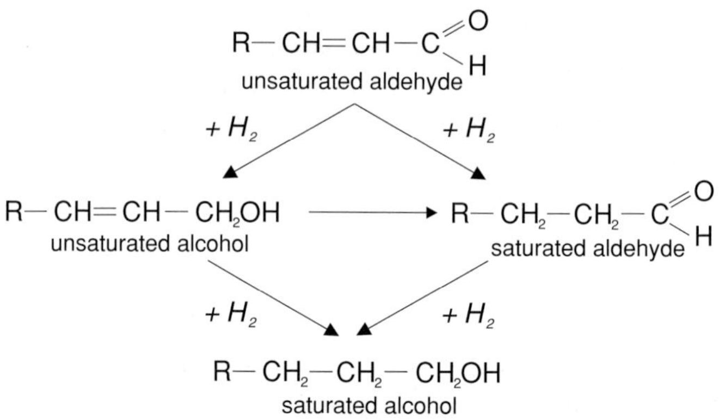

Fig. 6.18. General scheme for the hydrogenation of α, β-unsaturated aldehydes as acrolein (R: H) or crotonaldehyde (R: CH_3) (adapted from [99])

The hydrogenation of α, β-*unsaturated aldehydes* to give allylic alcohols is of considerable interest, since these reactions are an important step in the industrial synthesis of fine chemicals, particularly, of pharmaceuticals and cosmetics [98, 99, 249, 433–440, 446]. A simplified reaction scheme is shown in Fig. 6.18.

The desired products are the unsaturated alcohols, but these are difficult to obtain because the thermodynamics favours the hydrogenation of the C = C over the C = O bond by about 35 kJ mol^{-1}; and for kinetics reasons, the reactivity of the C = C bond is higher than that of the C = O bond [48,98,99,438,440,446]. In addition to the reactions shown above, further reactions, such as the decarbonylation of aldehydes, dehydration of alcohols and di/oligomerisation of the unsaturated aldehydes are possible. As expected, conventional hydrogenation catalysts mainly provide saturated aldehyde or trigger secondary hydrogenation into saturated alcohol. Hence, conventional hydrogenation catalysts mainly produce the saturated aldehyde, so special catalysts are required to control the intra-molecular selectivity, which is more desirable than high activity for the purpose of obtaining the olefinically unsaturated carbonyl compound as product. Steric and electronic effects and the structure and morphology of metal particle surfaces are some of the most effective factors controlling intra-molecular selectivity [98, 99, 438, 440, 446–448].

Allylic alcohols were hydrogenated using 1 to 3 nm diameter bimetallic Pd–Au dendrimer-encapsulated catalysts (DECs) [118]. Both alloy and core/shell Pd–Au nanoparticles were prepared. The catalytic hydrogenation of allyl alcohol was significantly enhanced in the presence of the alloy and core/shell Pd–Au nanoparticles as compared to mixtures of single-metal nanoparticles [118].

Some time ago, selective Pt/Y zeolite catalysts were developed for the liquid-phase hydrogenation of cinnamaldehyde into cinnamyl alcohol (selectivity >96% [449]) and an organometallic derived Rh[Sn-(n-C_4H_9)$_2$]SiO_2 catalyst for the conversion of citral to the corresponding α, β-unsaturated alcohols, geraniol and nerol (selectivity 96% at complete conversion [450]). Galvagno and co-workers [436] have now shown that gold supported on iron oxide can be used to hydrogenate citral to geraniol/nerol with selectivity higher than 95% and a conversion of 90%:

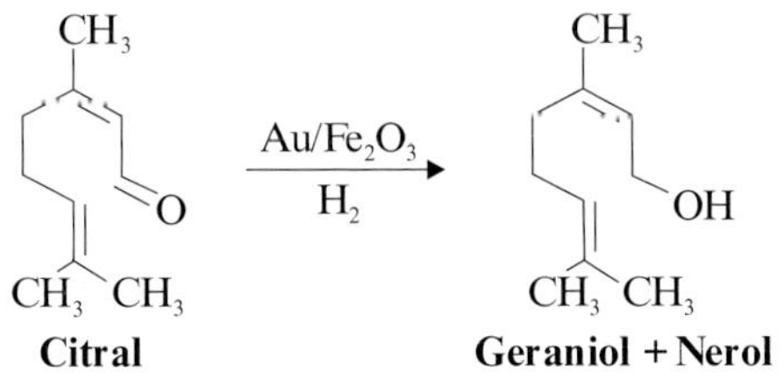

The selective hydrogenation of C_3 and C_4 α,β-unsaturated aldehydes, such as *crotonaldehyde* (but-2-enal) ($R = CH_3$ in Fig. 6.19) and *acrolein* ($R = H$ in Fig. 6.19), to the corresponding allylic alcohols is even more challenging [98, 99, 438].

Crotonaldehyde has been hydrogenated using special bimetallic Rh–Sn catalysts [447] and Ag catalysts [447, 451]. Interestingly, it has also been shown by Hutchings and co-workers [452, 453] that thiophene addition to Cu/Al_2O_3 catalyst promoted the catalytic activity for the synthesis of crotyl alcohol from crotonaldehyde hydrogenation. The same authors [434] also prepared Au/ZnO and Au/ZrO_2 catalysts by CP and Au/SiO_2 by IMP. They found that Au/ZnO was more selective for $C = O$ bond hydrogenation and that

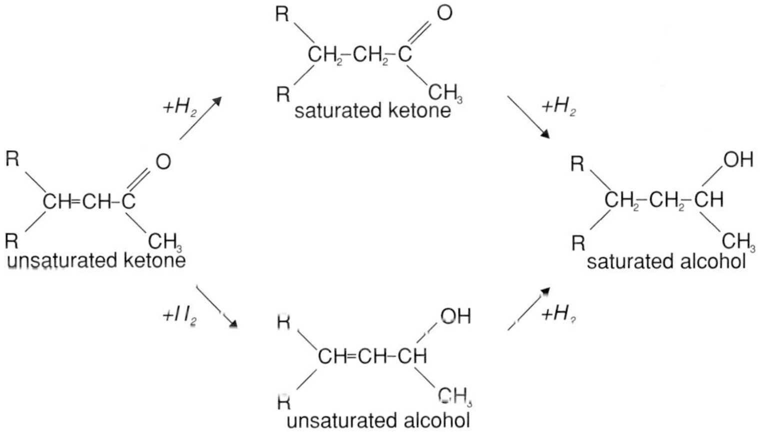

Fig. 6.19. General scheme for the hydrogenation of α,β-unsaturated ketones. The formation of the saturated ketone and of the unsaturated alcohol occurs through parallel reactions, whereas the saturated alcohol is mainly obtained by the further hydrogenation of the $C = O$ bond of the saturated ketone (adapted from [326])

the addition of thiophene acted as a promoter for selective hydrogenation. No crotyl alcohol was detected during crotonaldehyde hydrogenation for the Au/SiO_2 catalyst [434]. It was shown that the saturated aldehyde was the main product (selectivity >97%, at 523 K) and the conversion increased from 11 to 22% during the time-on-stream of 180 min. Improved crotyl alcohol selectivities were achieved with 5 wt% Au/ZnO (52% at 9% conversion) and 5 wt% Au/ZrO_2 (38% at 12% conversion) at 527 K. By comparison with Au/ZnO and Au/ZrO_2, it was concluded that interfacial sites responsible for C = O group adsorption were absent in the case of SiO_2 and the adjacent large Au particles. Nevertheless, they reported that the crotyl alcohol selectivity further increased up to 82% on Au/ZnO by addition of thiophene [434].

In a subsequent series of experiments [435], these investigators have shown that selectivities to crotyl alcohol ≥80% can be achieved using 5 wt% Au/ZnO when it is pre-treated in hydrogen at high temperatures (673 K). ZnO in the absence of Au is not active under these conditions. Interestingly, in this case, thiophene pre-treatment does not significantly enhance the catalyst performance. Using TEM, it was shown that high selectivity can be associated with large Au particles (≥10 nm) [435]. This is in distinct contrast to previous studies involving supported Au catalysts for which high activity for total oxidation is associated with smaller Au particles, typically in the range of 2–5 nm [249] (see previous oxidation sections).

Haruta and co-workers [437] have studied the same reaction using Au/Al_2O_3 and Au/TiO_2 catalysts prepared by DP with NaOH and gold particles sizes of 4.6 and 3.4 nm, respectively, and Au/SiO_2 (5.3 nm) prepared by grafting of gold acetyl acetonate. They found that the selectivity depends on the nature of the support. The highest selectivity to crotyl alcohol was obtained over Au/TiO_2(~25% instead of <10% for the other samples).

Louis and co-workers [48] have investigated the dependence of the catalytic properties with gold particle size in Au/TiO_2 catalysts, prepared by DP with urea for the same hydrogenation reaction [33]. The results obtained were compared with gold particles prepared by the classical DP method of preparation developed by Haruta using sodium hydroxide [28, 31] and found that there were no differences in activity and selectivity between those samples, as both exhibited similar gold particle size distribution (1.7–8.7 nm), although the gold loading in the samples prepared with urea (~8 wt%) was much higher than in those prepared using sodium hydroxide (~3 wt%).

The selectivity to carbonyl bond hydrogenation (to give 60–70% of crotyl alcohol) was constantly twice as high as for the olefinic bond hydrogenation (to give butanal). Selectivity was not sensitive to the reduction temperature, as was the case for platinum catalysts whose selectivity to crotyl alcohol was generally significantly lower (<50%) than for gold. The structure sensitivity of the Au/TiO_2 catalysts was clearly indicated by the increase of activity and turnover frequency for ca. 2 nm gold particles. It is thought that the dissociation of hydrogen on edges and corners sites of gold particles with low coordination numbers is the rate-determining step for this reaction.

Acrolein, because of its lack of space-filling substituents at the C = C group, is considered as the α, β-unsaturated aldehyde most difficult to hydrogenate to unsaturated alcohol [99, 438, 440, 447, 448, 454–456]. Nevertheless, Claus and co-workers [438, 440] have shown higher selectivities to the desired allyl alcohol using Au catalysts than are obtained using conventional hydrogenation catalysts such as Pt/ZrO_2. A comparison of the catalytic properties of Au/SiO_2 with Au/ZrO_2 showed that the interface or contact area between the support material and nanodispersed metal particles plays a special role in respect of the adsorption and activation of the C = O group [438, 440].

These same researchers also studied the effect of the particle size in the same reaction using Au catalysts deposited on titania, zirconia and alumina [98, 99, 440]. They concluded that the activity and selectivity to allyl alcohol increased with increasing gold particle size [98, 99, 440]. This was attributed mainly to two effects, a geometric and an electronic one. They conclude that adsorption of the C = O group of the α, β-unsaturated aldehyde is favoured by face atoms, so that most likely the increased fraction of dense (111) planes of the larger gold particles will give higher formation rates for allyl alcohol than the lower coordinated gold atoms of smaller-sized gold particles [98, 99]. A similar correlation has also been found by Hutchings and co-workers in their work on the hydrogenation of crotonaldehyde on Au/ZnO catalysts [435].

In a subsequent study, Claus and co-workers [439] observed in an Au/ZrO_2 series, prepared by DP with NaOH, that the activity decreased and the selectivity to allyl alcohol increased when the particle size increased from 4.0 to 7.7 nm. They also showed that the reaction occurred under the conditions used even on bulk-like gold powder, although the selectivities to the desired product, allyl alcohol, are much lower than for gold nanoparticles on supports [439]. Thus, the activation process on gold is neither exclusively limited to quantum size effects nor is it induced by the supports or a special design of the gold-support interface as assumed for oxidation reactions on gold [31] (see above). Claus's group suggested that a higher selectivity to the desired allyl alcohol seems to be restricted to smaller gold particles, that is, to special surface sites accessible, to a large extent, only on these particles [439, 440]. A concomitant HRTEM study led them to conclude that the changes in the catalytic properties were not induced by the support or by gold-support interfacial sites, but by changes in the proportion of the various surface sites (corners, edges and faces), due to changes in the morphology of the gold particles (round-shape, facetted, multiply twinned particles). Claus and co workers [440, 457] have reported that a ZnO-supported bimetallic Au–In catalyst produced allyl alcohol with a selectivity of 63%.

The liquid-phase reduction of α, β-*unsaturated ketones* [*trans*-4-phenyl-3 buten-2-one (benzalacetone $C_6H_5CH = CHCOCH_3$), 4-methyl-3-penten-2-one $(CH_3)_2C = CHCOCH_3$ and 3-penten-2-one $CH_3CH = CHCOCH_3$] by dihydrogen, to the corresponding α, β-unsaturated alcohols has been investigated on gold-supported catalysts by Galvagno and co-workers [44, 326].

These authors showed that Au supported on Fe_2O_3 can be used to hydrogenate the *trans*-4-phenyl-3-buten-2-one (benzalacetone) to the corresponding unsaturated alcohol with selectivity higher than 60% at 90% conversion [441]. They also prepared Au/Fe_2O_3 and Au/Al_2O_3 catalysts by CP and DP methods and compared the catalytic behaviour of these "homemade" samples with an Au/Fe_2O_3 "gold reference catalyst" supplied by the World Gold Council [326]. Particle size ranged from 2.8 nm (reference catalyst) to 6.6 nm. In the hydrogenation of benzalacetone and 4-methyl-3-penten-2-one on the "homemade" Au/Fe_2O_3 catalysts, the unsaturated alcohol is the main reaction product. Chemoselectivity higher than 60% was achieved. On Au/Al_2O_3, the selectivity is 10%. It is noteworthy that within the gold supported on Fe_2O_3, the reference catalyst shows the lowest selectivity toward the formation of the unsaturated alcohol. In the hydrogenation of 3-penten-2-one on the homemade Au/Fe_2O_3, the saturated ketone is the main reaction product and the selectivity toward the formation of the unsaturated alcohol is 15% at conversion >90%.

Galvagno and co-workers then concluded that gold is an intrinsically selective catalyst for the hydrogenation of the $C = O$ bond and that iron oxide favours the activation of this bond. It is likely that the absence of bulky substituents on the conjugated $C = C$ double bond favours its adsorption on the catalytic sites, leading to the formation of the saturated carbonyl compounds as the main reaction product [326].

The *selective hydrogenation of ethyne to ethene* is regarded as one of the very important industrial processes [58, 458–460]. Pd is, so far, regarded as a unique metal which has commercial ability to selectively hydrogenate acetylene. However, this reaction over a supported palladium catalyst is not 100% selective and a part of the ethylene product is further hydrogenated to ethane (as will be discussed later). To solve this problem, bimetallic palladium catalysts or addition of surface modifiers, such as lead acetate, have been investigated [458–461]. Jia and co-workers [58] have shown that Au/Al_2O_3 catalysts, prepared by DP, are able to catalyse the hydrogenation of ethyne to ethene with 100% selectivity, in the temperature range 313–523 K. The hydrogenation of ethene on the catalyst occurred only at much higher temperatures, over 573 K. The activity depends on the size of the ultra-fine gold particles, which showed a maximum at about 3 nm in diameter [58].

The selective *semi-hydrogenation of ethyne in ethene* is also an industrial process of vital importance, used in both laboratory practice and relevant to fine chemicals and polymer production [445, 462]. The reaction is generally performed over low loaded Pd catalysts. Systematic investigations have been performed over Pd–Ag, Pd–Cu and Pd–Au supported catalysts that are superior to mono-metallic Pd catalysts [445,461,463]. The presence of Au decreases the carbon coverage and improves the ethene selectivity [445].

MgO-supported mono-nuclear gold complex catalysts have been evaluated for *ethene (or ethylene) hydrogenation* at 353 K [464]. The results provided evidence of the stability of the metal complex, complex-support interactions and reactant-derived ligands on the metal during the catalysed hydrogenation. Au(III) complexes were identified as the catalytically active species.

Propyne hydrogenation over TiO_2- and Fe_2O_3-supported Au catalysts has also been investigated [465]. The Au/TiO_2 catalyst (mean particle size 3.7 ± 1.50 nm) prepared by DP was 100% selective towards propene production, but was characterised by deactivation on increased pulsing of the reaction mixture and substantial propyne retention presumably by the support material. The Au/Fe_2O_3 catalyst (mean particle size 3.7 ± 0.93 nm) prepared by CP, initially exhibited broadly similar behaviour to the titania-supported catalyst (100% propene selectivity, deactivation, propyne retention by the support material). However, the catalytic properties of the sample changed in time (4 months), where a significant reduction in propene selectivity and the absence of deactivation were observed. Higher reaction temperatures were necessary to recover the propene selectivity. However, pre-treatment at the higher temperature of 663 K ensured 100% selectivity at low and high reaction temperatures. The two gold catalysts exhibited optimum propene formation at two different reaction temperatures: 423 K for Au/TiO_2 and 523 K for Au/Fe_2O_3 [465].

Supported gold catalysts have been found to have high selectivity for the *hydrogenation of butadiene to butenes* [18]. The first study on 1,3-butadiene hydrogenation using supported Au catalysts was made by Bond and co-workers [442–444], who found that the Au/SiO_2 catalyst was selective to partial hydrogenation to form butenes. However, these Au catalysts prepared by impregnation were significantly less active than the Pd and Pt catalysts [442, 456]. Due to the low-melting point and poor affinity towards oxygen, Au could not be highly dispersed over the metal oxide supports by the IMP method. The selective removal of dienes and alkynes from the olefin streams used for large scale polymerisations in the petrochemicals industry is an important objective to prevent polymerisation catalyst poisoning and deactivation.

Haruta and co-workers [437] have investigated the *selective hydrogenation of 1,3-butadiene* to butenes over Au/Al_2O_3, Au/SiO_2 and Au/TiO_2 prepared by DP, GG and LG methods. All the Au catalysts were 100% selective to butenes, with 65–75% of selectivity to 1-butene. They found that the reaction was almost insensitive to the size of the Au particles (2.5–7 nm) and to the nature of the oxide support.

Claus and co-workers [99] also studied this reaction using silica-, titania-, zirconia- and alumina-supported 1 to 6 nm gold nanoparticles, prepared by various synthetic routes (sol–gel, DP, CVD, IMP). The hydrogenation of 1,3-butadiene over them gave butenes selectively. The amount of 1 butene was always higher than of the isomers *trans*- and *cis*-2-butene, in agreement with Haruta's results, but the selectivity to *trans*-butene was always higher than *cis*-butene. The selectivity to the thermodynamically favoured product *n* butane was always below 5%. Within the temperature range 493–533 K, conversions of butadiene between 10% and 30% were achieved.

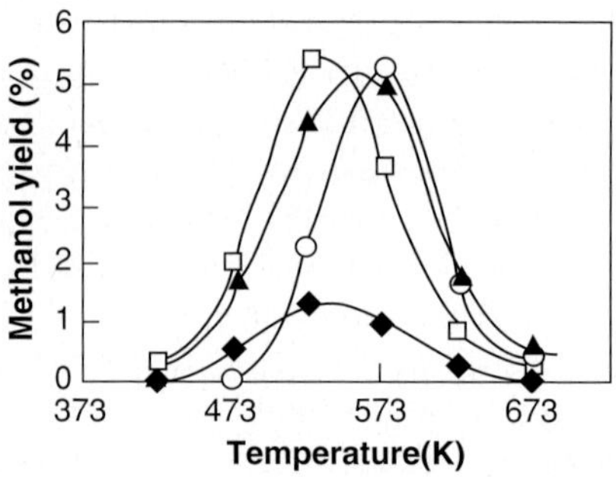

Fig. 6.20. Methanol yield as a function of reaction temperature in the hydrogenation of carbon dioxide over supported gold catalysts. Au/α−Fe_2O_3(□); Cu/ZnO (▲); Au/ZnO (○); Au/TiO_2 (♦): Au/M (M = Fe, Zn, Ti)= 1:19. CO_2: H_2:Ar = 23:67:10; 50 atm pressure; SV = 3,000 ml g_{cat}^{-1} h^{-1} (adapted from [31])

The *hydrogenation of carbon dioxide* to produce methanol is a very important commercial process [466], which could be used to reduce the greenhouse effect [467]. Haruta showed that gold supported on titanium, iron, or zinc oxides exhibits an appreciably high activity for carbon dioxide (Fig. 6.20) and carbon monoxide hydrogenation, between 423 and 673 K, when very fine gold particles are deposited on oxide supports by either CP or DP methods [31, 468, 469].

The particle size of gold was less than 4 nm. Product selectivity is greatly affected by the support oxides. Au/Fe_2O_3 and Au/TiO_2 were shown to be more active for hydrocarbon formation and for both forward and reverse water-gas shift reactions. Over all the catalysts tested by Haruta's group, carbon dioxide produced methanol at lower temperatures and produced methane more selectively than carbon monoxide [469].

The catalytic activity of a series of M/(3ZnO·ZrO_2) catalysts (where M = Cu, Ag, Au) was evaluated for the synthesis of methanol from CO_2 [467]. The catalysts containing Au and Ag were shown to have these metals distributed more uniformly. The catalytic activity was higher for the Cu catalysts, but Au catalysts were shown to be slightly more selective to methanol than the others.

6.4.7 Hydrogen Peroxide Formation

Hydrogen peroxide (H_2O_2) is a strong oxidising agent. It is often used: to make bleaching agents for detergents, textiles and paper; to prepare peroxides, particularly organic ones; in polymerisation reactions; to restore the original hues of paints containing lead; to combat odours in sewers and as an antiseptic in homes and hospitals (usually as a "20 volume" solution, so-called as 1 volume of this solution produces 20 volumes of oxygen gas and contains about 6% hydrogen peroxide) [2, 104, 466].

The demand for hydrogen peroxide has increased during recent years, especially due to its increased use as a bleaching agent in the paper industry [470]. There is also much interest in the design of new heterogeneous catalysts for selective oxidation under ambient conditions and these often use H_2O_2 as the oxidant [112]. As a consequence, much work is being performed in industry as well as in universities in order to improve the efficiency of the reaction. One indication of this is the large number of new patents [471–484]. It would be particularly advantageous to be able to make hydrogen peroxide, where it is required, in order to avoid the heavy costs of transportation caused by its hazardous nature.

Hydrogen peroxide is currently manufactured by the sequential hydrogenation and oxidation of alkyl anthraquinone and global production is about 1.9×10^6 tonnes per annum (tpa) [103, 112, 485]. This process is currently only economic on a large scale ($4 - 6 \times 10^4$ tpa) and H_2O_2 is often required on a much smaller scale. In view of this, there is considerable interest in the direct manufacture of hydrogen peroxide from the catalysed reaction between hydrogen and oxygen [482, 483, 485], so that it could be made on site in the quantities needed wherever and when required. At present, some success has been achieved using Pd as a catalyst, especially when halides are used as promoters [473, 484, 485].

As already discussed in Sects. 6.4.5 and 6.4.6, recent developments have stimulated interest in using supported Au catalysts as active oxidation and hydrogenation agents, respectively. H_2/O_2 mixtures have also been used for the oxidation of propene to propene oxide using Au catalysts [169], and it is considered that a surface hydroperoxy species may be formed as the oxidant (see Sect. 6.4.5) [41, 112, 470]. Theoretical calculations [470] predicted that formation of H_2O_2 should be promoted by gold surfaces.

Amongst the few reported studies to date on the use of supported Au catalysts for the direct synthesis of hydrogen peroxide from H_2/O_2, Hutchings and co-workers have shown that oxide-supported nanoparticulate gold can be very effective for this process and, furthermore, the rate of hydrogen peroxide formation can be significantly enhanced by using a supported Au/Pd alloy [112]. These authors used ZnO and Al_2O_3 as the support for catalysts.

Results showed that Au/ZnO and Au/Pd/ZnO catalysts exhibit some hydrogen peroxide synthesis at 308 K, although at a low rate. This is an improvement on use of the Pd catalyst that only generated water as a product (see Fig. 6.21).

However, experiments conducted at lower temperature (275 K) indicated that hydrogen peroxide can be formed at a high rate with the supported Au catalyst (see Table 6.5). The selectivity for H_2O_2 for the Au/Al_2O_3 catalyst was 53%. The supported Au catalyst produces more hydrogen peroxide than the supported Pd catalyst. However, more interestingly, the supported Au/Pd (1:1 wt/wt) catalyst produces significantly more than the pure Au catalyst. This indicated a synergistic effect of Pd acting as a promoter for the Au catalyst. These authors showed that the metal nanoparticles are in fact Au/Pd alloys [112].

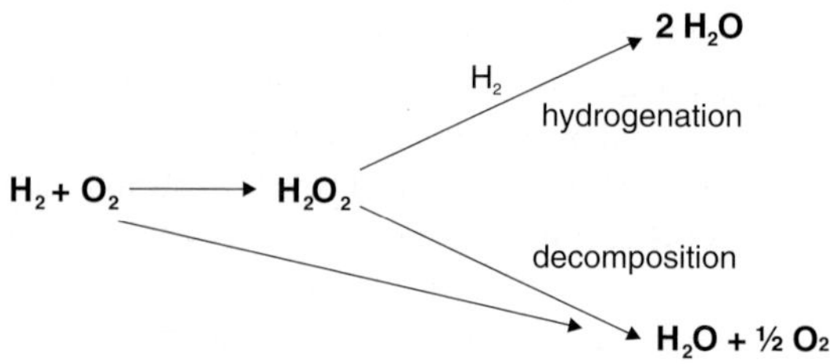

Fig. 6.21. H_2O_2 formation, hydrogenation and decomposition (adapted from [112])

Table 6.5. Formation of H_2O_2 from the reaction of H_2/O_2 over Au and Pd catalysts in methanol under 3.7 MPa pressure (adapted from [112])

Catalyst	T (K)	O_2/H_2 (mol ratio)	H_2O_2 (mmol g_{cat}^{-1} h^{-1})
Au/Al_2O_3	275	1.2	1,530
$Au:Pd(1:1)Al_2O_3$	275	1.2	4,460
Pd/Al_2O_3	275	1.2	370

This reaction was also studied on Au/SiO_2 catalyst without the addition of a halogen compound [485]. It was found that this catalyst promotes high activity to H_2O_2 formation by direct oxidation of H_2 by O_2. The effects of additives on Au/SiO_2 catalyst were further studied and it was found that the addition of a small amount of Pd is effective for increasing the H_2O_2 formation rate, due to the improved H_2 activation capability. However, the H_2O_2 formation rate decreased when an excess amount of Pd was added, since this increased the decomposition activity. The formation rate and the selectivity for H_2O_2 at initial 2 h were achieved to a value of 3.2 µmol h^{-1} and 30% under optimised conditions for Au–Pd/SiO_2 (1:1 wt%) catalyst. The amount of H_2O_2 accumulated was 3.5 mmol l^{-1} over 6 h [485].

As for other aspects of gold catalysis, the mechanism of hydrogen peroxide formation is not yet verified. The use of D_2 induced an isotope effect, suggesting that the adsorption (or reaction) of hydrogen is the rate-limiting step [168]. In a recent review, Haruta speculated that after adsorption of oxygen on gold, the particles become electron deficient due to the electronegativity of oxygen and thereby adopt an electronic structure similar to platinum, which allows for the facile dissociation of H_2 [486].

It can be concluded, therefore, that supported Au catalysts are effective for the direct synthesis of hydrogen peroxide from hydrogen and oxygen, and they may provide a significant improvement over the Pd catalysts that have been investigated previously [473, 482–485].

6.4.8 Reduction of NO_x with Propene, Carbon Monoxide or Hydrogen

The oxides of nitrogen (collectively known as NO_x) are primary pollutants of the atmosphere. These oxides are emitted into the air during the combustion of fossil fuels. Power stations, oil refineries, factories and car exhausts all contribute to this type of pollution [466, 487, 488]. The nitrogen oxides can dissolve in water to form nitrous and nitric acids. These are therefore secondary pollutants which together with sulphurous and sulphuric acid cause acid rain. Nitrogen oxides can also combine with hydrocarbons (also emitted during fossil fuel combustion) to produce photochemical smog, which can often be seen as a haze, arising from particles in the smog, above cities during warm weather [466].

About 90–95% of total NO_x emissions from combustion sources are of nitrogen monoxide (NO, also known as nitric oxide or nitrogen oxide). This is a colourless, water insoluble gas, with a strong characteristic smell [466, 487]. Nitrogen dioxide (NO_2) is also formed but in smaller quantities. This is a brown, extremely poisonous water-soluble gas. NO_2 is rarely detected, since it is only formed during cooling of exhaust gases via the following reaction [466, 487]:

$$2NO_{(g)} + O_{2(g)} \rightarrow 2NO_{2(g)} \quad (6.6)$$

The most practical and convenient method for removing NO_x gases is to reduce them to N_2, using unburnt CO, H_2, propene or other hydrocarbons in the exhaust gas. This can be achieved using conventional PGM catalysts, using a "three-way catalyst" (TWC) system, under stoichiometric conditions, where the amount of air supplied is controlled to a level just sufficient for the complete combustion of the fuel [132, 489] (see Fig. 6.22).

Rhodium, palladium and platinum have been employed to date as the key catalytic metals, due to their durability under both the oxidising and reducing parts of the cycles. These noble metals are usually supported on oxides with oxygen storage capacity such as CeO_2 deposited on Al_2O_3. Severe oxygen inhibition of the CO + NO reaction, over, e.g., Au/Al_2O_3, unfortunately foils the promise of very low-temperature TWC. Nevertheless, Marsh and co-workers [145] have demonstrated a gold-based TWC as shown in Sect. 6.5.7.

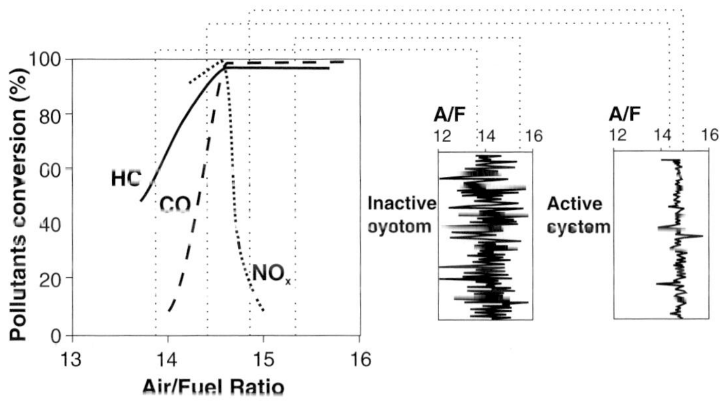

Fig. 6.22. Effect of air/fuel (A/F) ratio on the conversion efficiency of TWCs (adapted from [490])

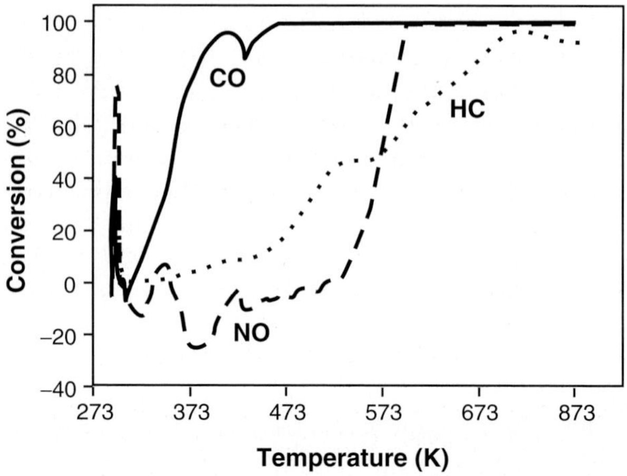

Fig. 6.23. AARLs gold-based TWC under reducing gasoline conditions. Gas composition is CO 1%, O_2 0.6%, CO_2 10%, $C_3H_8$350 ppm, C_3H_6 350 ppm, NO 1,000 ppm, SO_2 15 ppm, H_2O 2%, balance N_2. GHSV $= 60,000\,h^{-1}$ (redrawn after [145, 202])

The catalyst, developed at Anglo American Research Laboratories (AARL), consists of 1 wt% Au supported on zirconia-stabilised CeO_2, ZrO_2 and TiO_2, and contains as promoters 1 wt% CoO_x, 0.1 wt% Rh, 2 wt% ZnO and 2 wt% BaO. The catalytically active gold on cobalt oxide clusters are in the range of 40–140 nm (see Fig. 6.23).

The catalyst was evaluated under simulated gasoline engine emissions for oxidising (0.9% O_2) and reducing (0.6% O_2) conditions close to stoichiometric. Under oxidising conditions, the catalyst displayed a CO T_{50} of 348–358 K, an HC T_{50} of 513 K and as expected no NO_x reduction. The CO T_{50} remained unaffected under reducing conditions, whereas the HC T_{50} increased to 583 K. Complete conversion of NO_x occurred at approximately 603 K, with a T_{50} of 573 K under reducing conditions. Selectivity for reduction to N_2 over N_2O is not known.

However, diesel engines and some gasoline engines are operated under lean-burn conditions, where the oxygen is fed in excess, i.e., 10–20% more than is required to meet the stoichiometry for combustion of the fuel [132, 489]. Gold catalysts have therefore been examined for their potential in low-temperature activity to combat cold-start emission problems and removal of NO_x from lean-burn engines [202].

The most important strategies [490, 491] to effect NO_x reduction under lean-burn (i.e., oxidising, oxygen-rich) conditions include NO_x-storage-reduction (NSR) catalysts and selective catalytic reduction (SCR) of NO_x by a reducing agent. NSR catalysts (typically Pt/BaO/Al_2O_3 based) behave as "NO_x adsorbers", since they are used to store NO_x under oxidising conditions as adsorbed "nitrate" species, which are then released and reduced on a traditional TWC by temporarily running the engine under rich conditions.

Studies with Au-based catalysts have focused on SCR of NO by *propene* [61,96,489,492,493], *carbon monoxide* [78,489,494] and *hydrogen* [4,5,83,489]. Urea [97], methane [495] and other hydrocarbons [489,496,497] have also been used. SCR using the first three reagents will be discussed here.

Haruta and co-workers first reported the effectiveness of supported Au catalysts for SCR of NO using *propene* (Fig. 6.24a) [492]. Gold supported on ZnO, α-Fe_2O_3 and ZrO_2 is active at lower temperatures for this reaction. This is especially notable at 523 K, where the Au/ZnO catalyst gives a maximum of 49% in the conversion to N_2 and, in addition, 16% conversion to N_2O:

$$NO \xrightarrow{O_2} NO_2 \xrightarrow{C_3H_6} N_2 \qquad (6.7)$$

The low-temperature activity of Au/ZnO at ~523 K matches that for typical PGM-based catalysts [498]. Gold supported on MgO and TiO_2 gives the middle temperature conversion of NO. Kung's work on the higher temperature Au/Al_2O_3 catalysts [496, 499] suggests a gold particle size effect with maximum NO_x activity in the 15 to 30 nm particle range [496]. Smaller Au particles appear to favour the combustion of propene, lowering the NO_x activity [499].

Gold loading also appears to be important. Haruta and co-workers [489, 492] found that 0.17 wt% Au/Al_2O_3 is more active than 0.82 wt%, while Seker and Gulari [30] have shown that 0.8 wt% from a series ranging from 0.2 to 1.7 wt% Au is the most active on Au/Al_2O_3. The latter catalysts were prepared by sol–gel methods and the former by deposition–precipitation. As is the case of other reactions, the preparation method used for gold catalysts is very important in NO reduction and the size of the gold nanoparticles is one of the controlling factors for catalytic activity [489, 492].

SCR using propene over Au/Al_2O_3 is tolerant to the presence of water, which may even promote the reduction to a small extent [202, 489, 492]. The enhancing effect of moisture has also been observed in CO oxidation over other supported gold catalysts (see Sect. 6.4.4) and this provides gold catalysts with a significant potential advantage in applications for combustion of exhaust gases, since these usually contain water at a concentration higher than 10 vol% [489]. The catalysts which contain the oxides of base transition metals, such as Cu, Ag and Co, as catalytically active species are appreciably deactivated by moisture [489,500–502]. In the case of Pt supported on metal oxides, NO conversion is maintained even in the presence of high concentrations of water, but this is accompanied by enhanced formation of N_2O [489, 503–505]. The presence of CO has a positive effect on NO reduction by propene over Au/Al_2O_3, giving higher yields of N_2 (60–75%, at 673 K) [489].

Ueda and Haruta has shown that NO conversion is very low (about 8%) in the absence of oxygen [489]. The addition of oxygen in the reactant stream appreciably enhanced the reduction of NO up to a conversion of 69% (at an oxygen concentration of 4 vol%). The conversion to N_2 remains almost constant in the oxygen concentration range of 6–20 vol%. Therefore, oxygen is indispensable for the reduction of NO by propene. The steady conversion

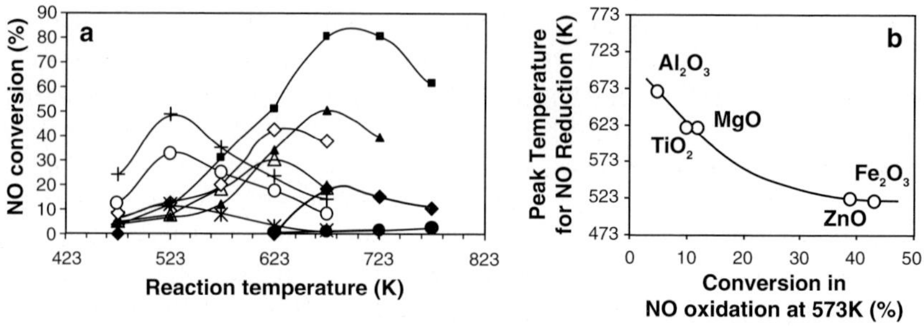

Fig. 6.24. (**a**) Temperature dependence of NO reduction by propene over 1.2 wt% Au/ZnO (+); 0.79 wt% Au/Fe_2O_3 (♦); 0.84 wt% Au/ZrO_2 (○); 0.85 wt% Au/MgO (◊); 0.76 wt% Au/TiO_2 (△); 0.82 wt% Au/Al_2O_3 (▲); 0.17 wt% Au/Al_2O_3 (■) 1.1 wt% Au/Al_2O_3 (♦) by impregnation; and Al_2O_3 (•). (**b**) Correlation of peak temperature for NO reduction with NO oxidation activity of the supported Au catalysts. Reaction gas is NO 1,000 ppm, propene 500 ppm, O_2 5%, H_2O 1.8%, balance He for NO reduction and NO 1,000 ppm, O_2 5%, H_2O 1.8%, balance He for NO oxidation. SV $= 20,000\,h^{-1}$ (redrawn after [492] and using data in [489])

of NO on Au/Al_2O_3 catalysts in the high oxygen concentration range is advantageous for applications with exhaust gases under lean-burn conditions, which usually contain oxygen at a concentration from 1 to 15 vol% [489].

Other hydrocarbons, such as *ethene, ethane and propane*, are also effective [489,496,497]. However, the presence of more than 5% oxygen is necessary for maximum NO reduction. Reduction of NO_2 does not require oxygen and this together with the observation that the Au-based catalyst catalyses NO oxidation to NO_2 suggests that NO_2 is a reaction intermediate [492]. Interestingly, there is a correlation between NO oxidation activity and NO reduction activity on supported gold catalysts (Fig. 6.24b). Results from Bamwenda and co-workers [506] suggest that formation of NO_2 as an intermediate acts as a reaction initiator or O_2 transfer agent. The importance of NO_2 is further supported by the observation of a significant enhancement in NO reduction, when Mn_2O_3 is mechanically mixed with Au/Al_2O_3. The mixed catalyst significantly increases the NO to NO_2 oxidation activity [489, 493].

Another advantage of Au-based catalysts for the SCR with propene is their apparent greater selectivity for $N_2(S_{N_2})$ over N_2O formation. On Au/Al_2O_3 S_{N_2} of ~100% have been observed [30, 489, 492, 496, 499], but this may be decreased in the presence of water [30]. With hydrogen [489] as reductant over Au/Al_2O_3 and Au/TiO_2 in 10% water, ~100% S_{N_2} is found at peak NO_x reduction activity at 473 K. However, at the lower temperatures needed for propene-SCR over Au/ZnO, significant N_2O formation is observed [489,492]. The results obtained thus far indicate that S_{N_2} over Au-based catalysts might be better than those obtained over PGM-based catalysts, which are typically <30% (but the catalysts are active at lower temperatures). Amiridis and co-workers [61] showed that for bimetallic Au–Pt/SiO_2

catalysts, if there is a strong Au–Pt interaction (catalyst prepared from a Pt_2Au_4-organometallic precursor) the temperature of peak NO_x reduction activity is significantly increased (~423 K), and there is a concomitant increase in the S_{N_2} from ~50% to 70–80%.

Preliminary tests conducted, using AARL catalysts, catalyst Marsh and co-workers [145] (see above) under simulated diesel engine conditions with oxygen content of 7.5%, showed significant activity for NO_x reduction. Favourable results showed a CO T_{50} of 333 K and an HC T_{50} of 443 K. An NO_x conversion window was observed at temperatures between 493 and 623 K, with a T_{50} of 523 K and a maximum conversion of 66% at 563 K.

Roberts and co-workers [507] prepared a new class of spongy Au catalysts, that are unsupported powdered Au that consist of micrometre particles (≈5μ) that have nanometre scale (≈10 nm) internal skeletal structures. These catalysts were shown to be very active in the SCR of propene under lean-burn conditions. Alloying these catalysts with Pd leads to a significant widening of the temperature window of activity.

Reduction of nitric oxide by *carbon monoxide* is one of the reactions occurring in catalytic converters for the purification of the engine exhaust gases [489]:

$$2NO + 2CO \rightarrow N_2 + 2CO_2 \tag{6.8}$$

Although there are some differences in the literature concerning the relative orders of activity reported for the PGMs, a well-established order is Rh > Pd > Pt [489].

Over Au/Al_2O_3, the reaction of NO with CO, in the absence of oxygen, takes place even at a temperature as low as 323 K, producing nitrous oxide (N_2O):

$$2NO + CO \rightarrow N_2O + CO_2 \tag{6.9}$$

Over rhodium-supported on Al_2O_3, the reaction takes place at temperatures above 400 K [508, 509], showing that under oxygen-free conditions, gold-based catalysts easily outperform those based on the PGMs for this reaction.

The catalytic activity for NO reduction by CO over supported gold catalysts changes drastically depending on the kind of metal oxide supports employed. It is especially noteworthy that oxides containing iron provide the highest catalytic activity for the NO + CO reaction. For Au catalysts, such as $Au/NiFe_2O_4$ and $Au/MnFe_2O_4$, the reduction to N_2O (at ambient temperature) or N_2 (at 423 K) is near complete [489]. These results show that gold catalysts are superior in their low-temperature catalytic activity to Rh-modified [510], Pt-modified [511] and Pd-modified catalysts [512, 513], where temperatures above 403 K are required over the most active Rh/Al_2O_3 catalyst and significantly higher temperatures of at least 573 K are needed for N_2 formation. This extraordinarily high activity of gold catalysts can be maintained in the presence of moisture [489].

However, as expected, reduction of NO to N_2 under the more commercially interesting lean-burn conditions is more difficult with gold catalysts, since gold catalysts are not as active as the PGM-based ones at the lower temperatures [132]. Ueda and Haruta [489] showed that over Au/Al_2O_3, the NO + CO reaction was strongly inhibited by the presence of oxygen, giving a maximum conversion of NO to N_2, of 5%. The O_2+CO reaction takes place in preference to the NO + CO reaction. On the other hand, a mechanical mixture of Mn_2O_3 with Au/Al_2O_3 gave a higher NO conversion to N_2 (21%) at 623 K. It is likely that Mn_2O_3 catalyses the reaction of NO with O_2 to form nitrogen dioxide (NO_2), which then reacts with CO adsorbed on the surface of gold particles supported on Al_2O_3:

$$2NO_2 + 4CO \rightarrow N_2 + 4CO_2 \tag{6.10}$$

Consequently, the severe oxygen inhibition of the CO + NO reaction, over Au/Al_2O_3, described earlier, unfortunately foils the promise of very low-temperature TWC. Nevertheless, Marsh and co-workers [145] have demonstrated on their gold-based TWC described above under simulated gasoline engine emissions for oxidising (0.9% O_2) and reducing (0.6% O_2) conditions close to stoichiometric.

In spite of the drawbacks, gold has still some advantages, being more selective for N_2 than the PGM catalysts and less susceptible to the deleterious effects of moisture than most ordinary transition metal oxide catalysts [132].

Reduction of nitric oxide can also be achieved by *hydrogen*, and this is also present in the exhaust gases from combustion of hydrocarbons as it can be generated either by the water-gas shift reaction (see Sect. 6.4.1) or the steam reforming reaction:

$$C_3H_8 + 6H_2O \rightarrow 3CO_2 + 10H_2 \tag{6.11}$$

It has been shown that the relative catalytic activity sequence for reduction of NO by H_2 is Pt > Rh > Pd > Au, Ru > Ir [489]. This order can be explained by the fact that Au gives the middle activity amongst the precious metals for both NO decomposition and the $O_2 + H_2$ reaction. The combination of NO oxidation by oxygen and NO_2 reduction by hydrogen may also give further improvement of NO_x conversion to N_2 as was the case with the $NO+CO+O_2$ reaction [489].

It has been shown that Au catalysts supported on Al_2O_3 and TiO_2 are active in NO reduction with H_2 [5]. The first showed activity already at room temperature. Au/SiO_2 proved to be inactive unless pre-oxidised and excess H_2 was present in the reaction. Then it was moderately active above 523 K [5]. It was also shown that Au/MO_x catalysts (M = Co, La, Ce) are active for this reaction. These type of catalysts supported on silica were less active than supported on alumina, due to the sintering of gold particles on silica. Nevertheless, the presence of a metal oxide was beneficial to the activity over Au/SiO_2. Opposite behaviour was observed with the NO reduction over Au/Al_2O_3, where the metal oxide might cover active species. Addition of CoO_x and LaO_x improved the selectivity towards N_2 formation [4, 5].

As discussed above, *nitrous oxide* (N_2O, also known as dinitrogen oxide) is one of the main products of NO reduction. It is regarded as an undesirable and

Table 6.6. Catalytic activity of gold-supported catalysts for N_2O/H_2 reaction with a reactant ratio $N_2O/H_2 = 0.5$ (from [54])

Catalysts	T_{50} (K)	T_{95} (K)
Au/Al_2O_3	384	538
$Au/CeO_x/Al_2O_3$	329	478
$Au/Li_2O/Al_2O_3$	348	418
$Au/Li_2O/CeO_x/Al_2O_3$	325	395
$Au/CoO_x/Al_2O_3$	330	478
$Au/CoO_x/LiO_2/Al_2O_3$	337	391
$Au/Rb_2O/Al_2O_3$	433	538
$Au/Rb_2O/CeO_x/Al_2O_3$	323	388
$Au/MgO/Al_2O_3$	376	498
$Au/MnO_x/Al_2O_3$	413	538
$Au/MnO_x/MgO/Al_2O_3$	342	418
$Au/TiO_2/Al_2O_3$	355	433

harmful gas in automotive exhaust gases [54]. It is a more harmful greenhouse gas than CO_2 and which may contribute to ozone depletion in the upper atmosphere [487,497]. In spite of this, there are still only a few papers dealing with N_2O decomposition and reduction, when compared with the attention given to the NO. Au/TiO_2 and $Au/CoO_x/TiO_2$ are promising catalysts for the CO/N_2O reaction [494], and Nieuwenhuys and co-workers [5,54,159] have also studied the N_2O/H_2 and N_2O/CO reactions, using Au catalysts supported on TiO_2, Al_2O_3 and mixed oxides and obtained a synergistic effect for mixtures of metal oxide supports (Table 6.6).

It was suggested that the role of the partly reducible metal oxide additive is to contribute to the formation of new active sites and increase N_2O dissociation. On the other hand, alkali and alkaline earth metal oxides stabilise gold particles against sintering.

6.4.9 Oxidative Decomposition of Dioxins and VOCs

Dioxins are a family of heterocyclic hydrocarbons, mostly poisonous chemical by-products of the manufacture of certain herbicides and bactericides. The most toxic are polychlorinated dibenzo-*p*-dioxins (CDDs) and dibenzofurans (CDFs, also called dioxins) [514, 515]. Amongst the most dangerous are the isomers of the 2,3,7,8-tetrachlorodibenzo *para*-dioxin (TCDD) and 2,3,7,8-tetrachlorodibenzo-furan (TCDF). The former occurs in small amounts in some herbicides and defoliants, including the so-called "Agent Orange" (a highly toxic herbicide sprayed as a defoliant in chemical warfare) [466].

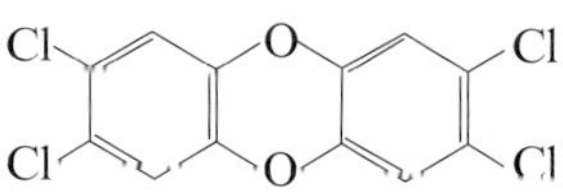

2,3,7,8 tetrachlorodibenzo-para-dioxin

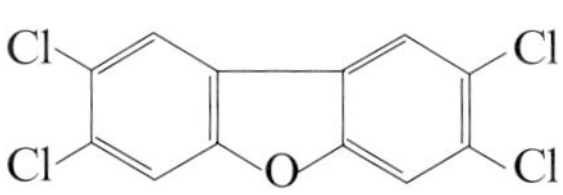

2,3,7,8 tetrachlorodibenzofuran

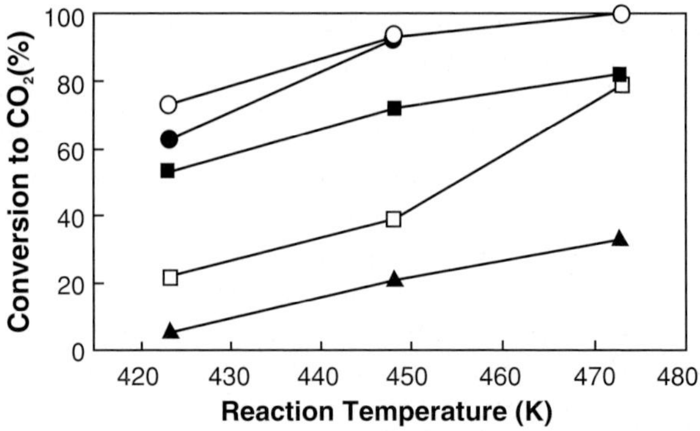

Fig. 6.25. Yield of CO_2 as a function of catalyst temperature in the oxidation of trimethylamine over single- and multi-component noble metal catalysts. (▲) 2 atm.% Ir/La_2O_3; (□) 2 atm.% Pt/SnO_2; (■) 2 atm.% Au/Fe_2O_3; (○) 2 atm.% Au/Fe_2O_3–2 atm.% Ir/La_2O_3; (•) mechanical mixture of 2 wt% Pt/SnO_2 and 2 wt% Ir/La_2O_3; reactant gas, trimethylamine 100 ppm in air; SV $= 10,000\,\text{ml}\,g_{cat}^{-1}\,h^{-1}$ (adapted from [516])

Consequently, the control of dioxin emissions from incinerators is one of the most urgent needs in environmental protection today [466, 516]. There are several treatment techniques presently in use, including the raising of the waste gas temperature to cause catalytic oxidative decomposition [516]. However, when the initial investment and mechanical complexity is taken into account, this method is not practicable for small-scale incinerators. A novel solution for this task would be to decompose dioxin and its derivatives at the dust filter using catalytic oxidation, preferably below 473 K. This particular temperature is relevant since the waste gases are usually reduced to this temperature at the dust filter after efficient heat recovery.

Since Au is stable in the presence of halogens, some supported Au catalysts have been reported to be more active and stable than other noble metal catalysts for reactions involving them such as the oxidative decompositions of CFCs and dioxins [6, 7, 253, 259]. Gold on Fe_2O_3–La_2O_3 is reported to have very good activity for the decomposition of several PCDDs and PCDFs at 413 K, by catalytic oxidation [517]. Other authors have used Pt, Pd and Ir catalysts and their oxides supported on silica–boria–alumina composites oxides and zeolites as a first catalyst and supported Au and some of other metals (such as Ag, Cu, Fe, Sb, Se, Te, Ta and their oxides) as a second catalyst [518]. The final catalyst eliminated dioxins and furans in the gases by oxidative decomposition, while suppressing the formation of secondary pollutants from gaseous precursors [518].

Haruta and co-workers have developed multi-component catalysts containing various supported noble metal systems [516]. These catalysts were found to interact and work in synergy for several catalytic reactions. Although Ir/La_2O_3 exhibits low activity in the decomposition of dioxin derivatives, its presence enhances the catalytic activity of nanoparticulate Au/α–Fe_2O_3. The promotion effect of Ir catalysts helped to achieve 95% decomposition of dioxin from the outlet gases of incinerators even at 423 K [28, 516]. This multi-component noble metal catalyst is a promising material for purifying the typical dioxin-containing exhaust gases emanating from incinerators.

Incinerator exhaust gases do, however, contain a variety of pollutants – not only dioxin derivatives but also other compounds such as odours, volatile organic compounds (VOCs) and the reaction intermediates of dioxins [516]. *VOCs* are a major contributor to air pollution because of their toxic and malodorous nature and their contribution to ozone and smog formation [131]. They are emitted from a wide range of industrial processes and transportation activities [131, 162, 519–521]. Some representative VOCs are methanol, ethanol, 2-propanol, acetone and toluene.

Catalytic combustion is the most promising technology for the destruction of VOCs [131, 162, 519–522]. This approach is preferred to a thermal one due to the lower temperature required and its greater selectivity, which implies a considerable saving of energy. Other advantages are its definitive character and the fact that it can operate with dilute effluent streams ($<1\%$ VOCs) [131, 162, 519–522]. For this reason, supported noble metals (Pt, Pd, Rh) or metal oxides (Cu, Cr, Mn) were the conventional catalysts used [523–529].

Au/Fe_2O_3 [162, 519, 520], Au/Al_2O_3 [131] and Au/CeO_2 [521] catalysts have been found to have high activities for oxidation of VOCs. The average size of gold particles varied from 3.8 to 8.2 nm [131,521]. The high activity of these Au-based catalysts has been explained on the basis of the capacity of gold nanoparticles to increase the mobility of the Fe_2O_3 lattice oxygen [519, 520] and to weaken the surface Ce–O bonds adjacent to Au atoms, thus enhancing the reactivity of the CeO_2 surface capping oxygen [521], which is involved in the oxidation via a Mars-van Krevelen reaction mechanism [519–521].

The addition of Au to Pd by simultaneous deposition–precipitation on CeO_2 increases the catalytic activity for methanol decomposition to CO and H_2 at 450 K [115]. No significant promotional effect was observed by the addition of Rh or Ir to Pd.

Zhong and co-workers [530] described recent results of an investigation of the electrocatalytic oxidation of methanol using carbon-supported Au and Au–Pt nanoparticle catalysts. The exploration of the bimetallic composition on carbon black support was aimed at modifying the catalytic properties for the methanol oxidation reaction at the anode in direct methanol fuel cells (DMFCs). Au and Au–Pt nanoparticles of 2–3 nm sizes encapsulated in an organic monolayer were prepared, assembled on carbon black materials and treated thermally. The results have revealed that these Au–Pt nanoparticles catalysts are potentially viable candidates for use in fuel cells under a number of conditions [530].

Also, Bowker and co-workers [116] prepared Au/TiO_2 catalysts by IMP and evaluated them for room temperature photocatalytic reforming of methanol in aqueous solution to produce hydrogen. These catalysts proved to be one the best metal support combinations for this reaction under anaerobic conditions. However, that rate is quite low because only photons of energy greater than the band gap of titania are utilised (i.e., those shorter than ∼360 nm). To make this a viable technology another type of support, photostable in aqueous media, needs to be found.

6.4.10 Catalytic Combustion of Hydrocarbons

The detoxification of hydrocarbon pollutants is one of the global environmental challenges [531–533]. The complete catalytic oxidation of hydrocarbons to carbon dioxide and water has received much attention in order to reduce their emission from motor vehicles and processing plants [531, 533, 534].

The promoting effect of nanogold particles in catalysts used for the complete oxidation of various saturated and unsaturated aliphatic hydrocarbons has already been reported [7, 31, 45]. Gold deposited on Co_3O_4 (which is the most active among base metal oxides) exhibits the highest catalytic activity amongst supported metal catalysts for the complete oxidation (combustion) of hydrocarbons [7, 31, 302, 535, 536]. Based on the temperature for 50% conversion for the oxidation of methane, propane and propene, the activity for 10 wt% Au catalysts decreases in the following order [535]:

$$Au/Co_3O_4 > Au/NiFe_2O_4 > Au/ZnFe_2O_4 > Au/Fe_2O_3 \tag{6.12}$$

Methane is the hydrocarbon most difficult to oxidise, and it is often used as a model for activity tests. In addition, methane itself is a powerful greenhouse gas [531]. Miao and Deng showed that the catalytic activity of 2.5 wt% Au/Co_3O_4 towards methane combustion could be enhanced when a small amount of Pt, e.g., 0.2 wt% was added and the temperature for 100% conversion of methane could be decreased by 323 K [536]. Liu and Flytzani-Stephanopoulos prepared gold/ceria-based catalysts that had synergistic effects in catalytic activity for both methane and CO oxidation [537]. The metallic gold particle size was around 8 nm average and the particles were evenly distributed in the cerium oxide matrix and were in intimate contact with the cerium oxide itself [537]. Nieuwenhuys and co-workers studied methane oxidation over Au nanocatalysts supported on Al_2O_3 [37, 38, 538] and MO_x/Al_2O_3 (M = Cr, Mn, Fe, Co, Ni, Cu and Zn) [38, 127, 538] and concluded that methane oxidation activity of Au/Al_2O_3 is improved upon addition of MnO_x, FeO_x, CoO_x and to a lesser extent NiO_x. Measured activities in methane oxidation over $Au/MO_x/Al_2O_3$ decreased in the following order for various MO_x:

$$CuO_x > MnO_x > CrO_x > FeO_x > CoO_x > NiO_x > ZnO_x \tag{6.13}$$

The high activities observed for CuO_x and CrO_x were assigned to the intrinsically high methane oxidation capability of these oxides themselves [38, 127].

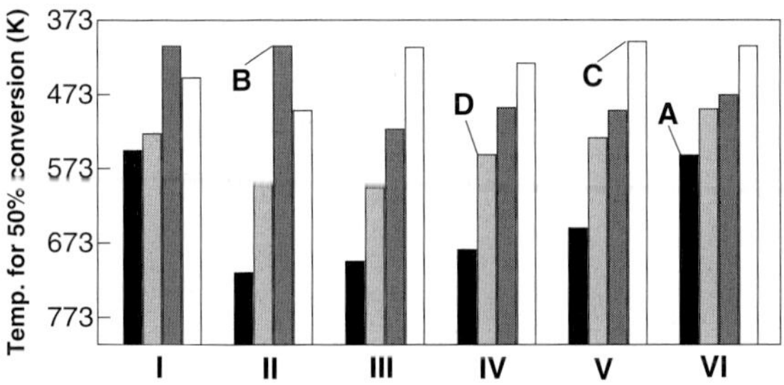

Fig. 6.26. Temperature for 50% conversion in the catalytic combustion of hydrocarbons; 1 wt% Pd/Al_2O_3 and Pt/Al_2O_3 were prepared by IMP. 1:19 Au/support catalysts were prepared by CP and calcined in air at 673 K. (I) 1 wt% Pd/Al_2O_3; (II) 1 wt% Pt/Al_2O_3; (III) 10 wt% Au/Fe_2O_3; (IV) 10 wt% Au/$ZnFe_2O_4$; (V) 10 wt% Au/$NiFe_2O_4$; (VI) 10 wt% Au/Co_3O_4. (A) Methane; (B) propene; (C) trimethylamine; (D) propane. The concentrations of reactant gases in air was CH_4: 0.25 vol%; C_3H_8 and C_3H_6: 0.1 vol%; $(CH_3)_3N$: 0.05 vol%; balanced with air to 1 atm; SV 2×10^4 ml g_{cat}^{-1} h^{-1} (based on [535])

Au is inferior to Pd and Pt catalysts in the combustion of *unsaturated hydrocarbons*, such as propene, but it is at least as active for *saturated hydrocarbons* such as CH_4 and C_3H_8. This is probably because olefins can adsorb on the defect sites of small gold particles a little too strongly and may retard the migration of oxygen adsorbed at the perimeter interface. In the case of Pd and Pt, they adsorb olefins much more strongly than Au; however, probably this may take the dissociation of C–C bonding in olefins faster than in alkanes to give higher catalytic activities for olefins [31].

Andreeva and co-workers have studied gold and vanadia supported on titania and zirconia for the complete oxidation of *benzene* [532–534, 539]. They observed a strong synergistic effect between gold and vanadia when molecular oxygen was used as an oxidising agent. This effect was more pronounced for titania than for the zirconia support [223, 532, 533]. Upon oxidation by molecular oxygen the oxidant activation takes place on the nanosize gold particles, while the vanadium oxide surface species are responsible for the activation of benzene. This is consistent with there being a synergistic effect between the finely dispersed gold and the surface vanadium structures [532]. The average particle size of the supported gold catalysts was 2–3 nm for those on titania and 3–4 nm for those on zirconia [532, 534].

Andreeva and co-workers also prepared gold–vanadia–ceria [539–542] and gold–vanadia–ceria–alumina [542] catalysts for the same reaction. The average size of gold particles was 3.5–6.0 nm. A high and stable catalytic activity was established. The presence of gold enhances the reducibility and the reactivity of VO_x and ceria surface layers, which results in the lowering of the reaction temperature [539–541]. The observed strong synergistic effect between gold and vanadia, when they are present simultaneously on ceria, could be related to the specific interaction between gold, vanadia and ceria and the possibility of obtaining nanosize gold and ceria particles, producing more active oxygen species [541]. The presence of alumina prevents the gold and ceria agglomerating [542]. The high stability of this catalyst at the extreme working conditions renders it promising for practical applications in complete oxidation of hydrocarbons [542].

6.4.11 Ozone Decomposition

The presence of ozone (O_3) in the upper atmosphere is beneficial since it absorbs ultraviolet radiation. However, the presence of ozone, which can be emitted by equipment, such as photocopiers and laser printers, close to the ground is harmful, as it causes respiratory illness and enhances photochemical pollution [2, 543]. The allowable concentration in a working environment and regulation threshold level for allowable exposure is 0.1 ppm [543].

Ozone is useful since it is a powerful oxidising agent, but the molecule itself is toxic to both animal and plant life and its release into the environment must be avoided. The problem of ozone decomposition is therefore technologically important. For example, ozone needs to be reduced in airplane cabins, submarines and office environments. Discharges from sterilisation, odour removal and waste water treatment units must also have their ozone levels lowered [466, 544].

Although the decomposition of ozone to dioxygen is a thermodynamically favoured process [545], ozone is thermally stable up to 523 K and catalysts are needed to decompose it at lower temperatures. In most cases, when used to treat air for ventilation, the catalysts are required to operate at ambient temperatures, in the presence of water vapour and high space velocity. Most of the work on ozone decomposition has reported some development of catalytic materials. It was reported that some adsorbents were used for the decomposition reaction. A limited number of reports of solid catalysts are available. The active components of the catalysts are mainly metals, such as Pt, Pd, Rh, and oxides of metals including Mn, Co, Cu, Fe, Ni and Ag. Catalyst supports used are γ-Al_2O_3, SiO_2, ZrO_2, TiO_2 and activated carbon [544, 546].

Activated carbon and zeolite have been used in adsorbent filters and need regeneration. Catalytic decomposition has no such disadvantages and, therefore, has attracted great interest. Most of the catalysts are reported in patent literature [547–549]. Research carried out in a variety of catalysts showed insufficient activity and stability when they are used under severe conditions such as a high ozone concentration, large space velocity and in the presence of moisture.

Supported silver and silver-containing catalysts have a high activity for ozone decomposition but they have low stability under severe reaction conditions. Au catalyst is similar to Ag in the activation mechanism [550], but has superior catalytic performance such as resistance to moisture [543]. Results obtained by Liang and co-workers [543] showed that gold on Fe_2O_3 is highly effective for ozone decomposition at low temperatures, being superior

Table 6.7. Decomposition of ozone on supported gold catalysts[a]

Catalyst	O_3 decomposition conversion at 273 K (%)	T (K)	Active time (h)
Au/NiO[b]	95	< 300	> 15[c]
Au/Fe_2O_3[b]	98	< 300	> 15[c]
Ag_2O/γ-Al_2O_3	77	< 300	3.1
NiO/γ-Al_2O_3	72	< 300	4.5

[a] Space velocity: $10.000\,h^{-1}$; $O_3 :\sim 3,000\,ppm(O_3 + O_2)$.
[b] Catalyst: 1 at.%.
[c] Subjective stop (from [543])

to Ag and Ni catalysts in both catalytic activity and stability. Gold particle size was 8 nm. Table 6.7 shows the activities of composite oxide catalysts on the decomposition of ozone as expressed by the time for which the ozone decomposition is near complete and also by percentage ozone decomposition at 273 K, respectively.

This group also used gold catalysts for the simultaneous removal of ozone and CO. Petrov did the same and reported that gold and a transition metal-oxide supported on ceria and titania achieved 100% ozone decomposition [551].

Andreeva and co-workers have reported that 3 wt% Au/V_2O_5 on titania and zirconia, having 5 nm gold nanoparticles, is very active for the decomposition of ozone at 293 and 303 K, and the resulting active oxygen species have been used to oxidise benzene at 313 K [552]. Using ozone as the oxidising agent, an additional lowering of the reaction temperature for benzene oxidation was achieved [532].

Engelhard Co. (USA) markets a base metal catalyst system, which converts ozone into oxygen when coated onto a car radiator at the moderate temperatures generated by the radiator [553]. From the limited amount of information on the properties of gold catalysts currently available, inclusion of gold in such a catalyst system could make them really effective [543].

6.4.12 SO_2 Removal

Nowadays, huge quantities of poisonous SO_2 are formed by the combustion of fossil-derived fuels and the incineration of solid waste in factories, power plants, houses, automobiles and industrial boilers and by volcanic activity [466,487,554–557]. Sulphur oxides (SO_2 and SO_3) are, together with nitrogen oxides, a primary and substantial source of acid rain [466,487,554–557,560, 561,564–567]. The atmospheric deposition of these acidic components has been known to be the main cause of damage to plants, fish and all biological systems due to the resulting reduction of pH of the ground waters, streams, rivers and lakes and the impoverishment of agricultural soils due to lixiviation of plant nutrients. They also have harmful effects on human health, and degradation of buildings and monuments in urban areas is caused by the resulting acid rain. Sulphur dioxide is responsible for the depletion of the ozone layer, and this has been linked to several illnesses and respiratory diseases [466,487,554, 556,557,563,564,567]. Consequently, the destruction of SO_2 ($DeSO_x$) is a very important problem in environmental chemistry [487,555,563–569]. Thus, new environmental regulations emphasise the need for more efficient technologies to destroy the SO_2 formed in combustion processes [558,562,564,565,568,570, 571].

The bond between adsorbed SO_2 and bulk metallic gold is very weak (bonding energy <10 kcal mol^{-1}) and the molecule does not dissociate [565]. This can be contrasted with the behaviour of SO_2 on most transition metal surfaces [553,559,560,563,569], where the molecule adsorbs strongly and readily decomposes at temperatures below 300 K. The only known exceptions to this trend are surfaces of silver, which adsorb the molecule reversibly [553,572]. In any case, Ag–SO_2 bonds are still stronger than Au–SO_2 bonds. In the presence of SO_2, the surface of bulk gold behaves as an ideal "noble metal" with a negligible reactivity.

Titania is the most common catalyst used in the chemical industry and oil refineries for the removal of SO_2 through the Claus reaction [563,565,566] and the reduction of SO_2 by CO [563,565]:

$$2H_2S + SO_2 \rightarrow 2H_2O + 3S_{solid} \qquad (6.14)$$

$$SO_2 + 2CO \rightarrow 2CO_2 + S_{solid} \qquad (6.15)$$

The main products of the adsorption of SO_2 on TiO_2 are SO_3 and SO_4 species, and no dissociation of the molecule is achieved [206, 565, 566].

The addition of gold to TiO_2 produces desulphurisation catalysts with a high efficiency for the cleavage of S–O bonds. Rodriguez and co-workers showed that a nanoparticulate (2–5 nm) Au/TiO_2 system is much more chemically active than either pure titania or metallic gold [206, 565]. Whereas, the surface of bulk metallic gold does not react with SO_2, the supported Au nanoparticles on titania have an extraordinary ability to adsorb and dissociate the molecule with an activity comparable to that mentioned in literature for surfaces of Ru, Rh, Ni, Pd, and Pt [555,559,560,562,569]. Also, Au/TiO_2 is 5–10 times more active than pure TiO_2 for the reduction of SO_2 by CO and for the Claus reaction above [206, 565]. Rodriguez and co-workers also tested the $DeSO_x$ activity of MgO, since this oxide is frequently used in scrubbers for SO_2 [559, 564]. As for TiO_2, adsorption of SO_2 on MgO also leads to the formation of SO_3 and SO_4 surface groups, without dissociation of the molecule [564, 566]. However, Au nanoparticles supported on MgO are still more reactive towards SO_2 than pure metallic gold.

The results show clearly that the Au/TiO_2 system is much more chemically active than the Au/MgO (see Fig. 6.27). It appears that MgO plays a minor role in the chemical activity of the supported Au nanoparticles. The dissociation of SO_2 on Au/MgO is very limited due to weak Au/MgO interactions [559]. In contrast, Au/TiO_2 interactions are complex and simultaneously

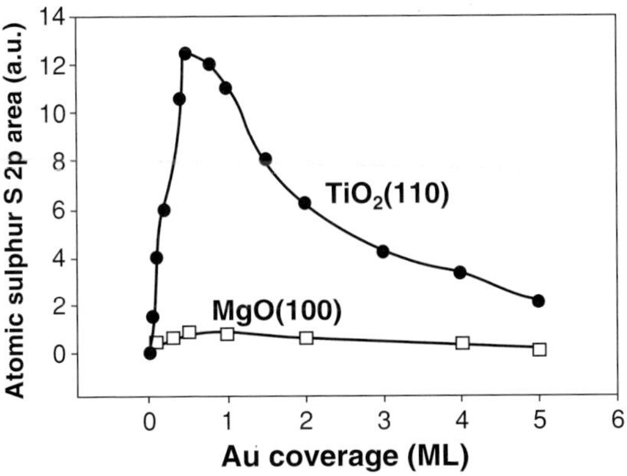

Fig. 6.27. Relative amounts of atomic S formed by the full dissociation of SO_2 on Au/MgO(100) and Au/TiO_2(110) surfaces at 300 K. Each surface was exposed to 5 L of SO_2. The S coverage is assumed to be proportional to the area under the corresponding S 2p XPS features, which were measured with Mg Kα radiation (from [566])

enhance the DeSO_x activity of both gold and titania. Once Au bonds to titania, its ability to adsorb and dissociate SO_2 increases significantly. In addition, Au adatoms modify the rate of exchange of oxygen vacancies between the bulk and surface of titania, enhancing in this way the chemical activity of the oxide [206, 565]. The addition of gold to the support proves to be essential, as neither MgO nor TiO_2 are able to dissociate SO_2 on their own [559, 565, 566].

6.4.13 Heck Reaction

Researchers have described a new catalyst for the Heck reaction [573]. The 1:1:1 Au/Ag/Pd tri-metallic nanoparticles are more efficient than the traditional palladium complex catalysts for the following coupling reaction.

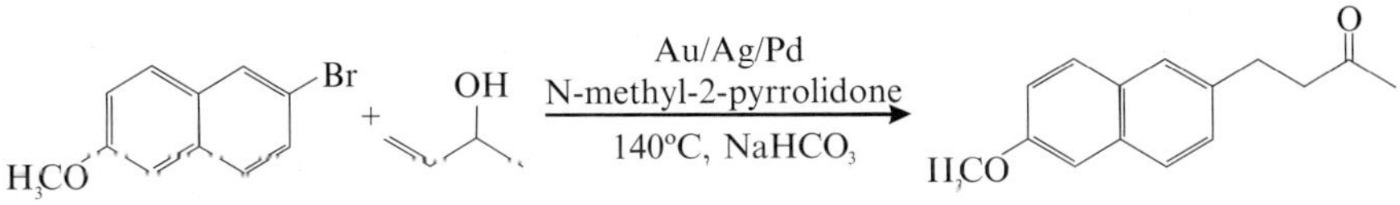

Using a developed method, laser irradiation of colloidal mixtures of the three metals, these authors dispersed Au/Ag/Pd nanoparticles with average diameters $4.4 \pm 1.5\,nm^2$. These tri-metallic nanoparticles were assessed for activity in the synthesis of nabumetone [4-(6-methoxy-2-naphthalenyl)-2-butanone], a non-steroidal anti-inflammatory drug that has greater activity than aspirin and is comparable to naproxen and indomethacin. The 0.45-mol% Au/Ag/Pd was sufficient to catalyse the coupling of the 3-buten-2-ol

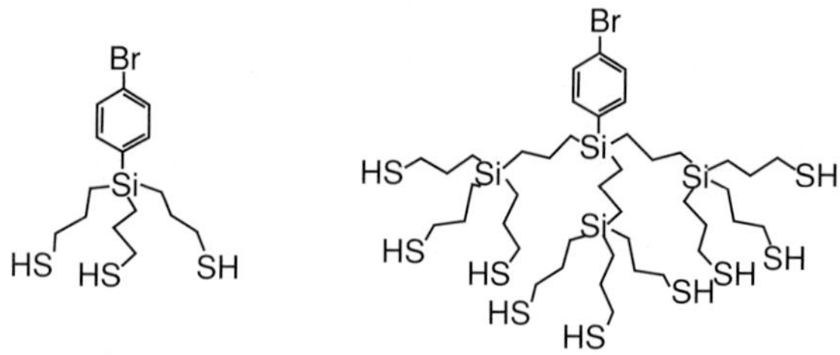

Fig. 6.28. Dendritic molecules contained 3 (left) or 9 (right) thiol groups at the periphery and a bromophenyl group at the focal point (from [574])

to 2-bromo-6-methoxynaphathalene with a product yield of at least 90%. Following identical procedures, use of a traditional Heck reaction catalyst, i.e., $[(PPh_3)_2PdCl_2]$, resulted in a yield of only 64% of nabumetone. Further optimisation studies on the catalytic cycle and immobilising the Au/Ag/Pd particles on a solid support are continuing.

Yam and co-workers [574] have prepared multi-dentate thiolate films by solution phase deposition of dendritic oligothiols on gold surfaces. The dendritic molecules contained 3 or 9 thiol groups at the periphery and a functional group (bromophenyl) at the focal point (Fig. 6.28). The high stability of the films allows for the palladium-catalysed Heck reaction on thiolate films. Results indicated that the presence of about 20 and 28% of the terminated S atoms were unbound to the gold surfaces and about 76 and 68% of Br atoms were consumed during the Heck reaction with 4-fluorostyrene. It was speculated that the Br atoms in the dendron molecule containing an unbound SH group were inactive toward the Heck reaction. About 60–70% of the reacted Br atoms were replaced by the 4-fluorostyryl groups via the Heck reaction. Prolonging the reaction time did not consume the remaining Br atoms, probably due to unfavourable orientation of the bromophenyl groups and/or the presence of adjacent thiol groups that may deactivate the catalyst in the film assembly.

6.4.14 CO_2 Activation

A remarkable advance was reported in this area by Deng and co-workers [575]. These investigators and findings show that nanogold can be a highly effective catalyst for cyclisation of epoxides (Fig. 6.29) or carbonylation of aliphatic amines (Fig. 6.30) with CO_2 to yield cyclic carbonate and disubstituted ureas, respectively. Polymers were used as the support to immobilise the nanogold catalysts because the polymer immobilised catalysts were closer to homogeneous catalysts in chemical character and were attractive for technological applications. Two types of polymers were used (which were described as Poly1 and Poly2). Poly1 functional group was quaternary ammonium, which may stabilise the nanogold particles, while the functional group in Poly2 was sodium sulphonate, which could not.

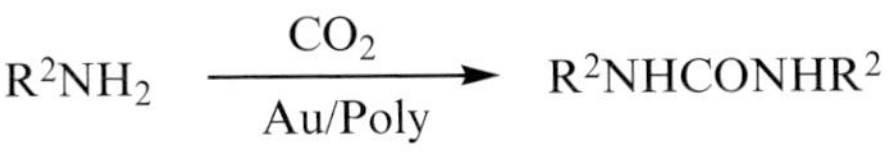

R^2 = benzyl, chlorohexyl

Fig. 6.29. Cyclisation of epoxides with CO_2 over nanogold catalysts (adapted from [575])

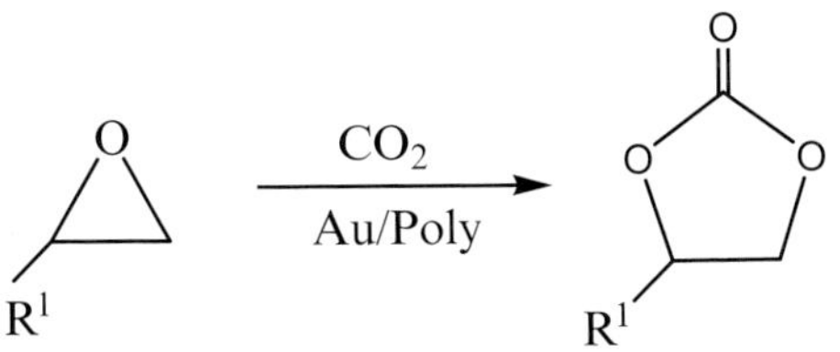

R^1 = CH_3-, $(CH_3)_2CHOCH_2$-, $PhOCH_2$- and $ClCH_2$

Fig. 6.30. Formation of carbamate from amines and carbon dioxide (adapted from [575])

The reaction of epoxide and carbon dioxide can be conducted with unprecedentedly high turnover frequencies (TOF), ca.>50,000 mol mol^{-1} h^{-1} (for Poly1), which is significantly superior to those obtained by a classical acid-basic catalyst system (up to ca. 5,000 mol mol^{-1} h^{-1}). All of the selectivities for the desired product nearly reached 100%, so the yield could be used to indicate the catalytic performances. The catalyst with 0.05 wt% gold loading exhibited the highest yield (71.2%), and lower yields were obtained over catalysts with higher gold loadings due to the formation of larger gold particles. For 0.05 wt% Au/Poly2, poor results were obtained, indicating that the properties of the polymer support had a great impact on the catalytic activity of nanogold, and the reason for such a tremendous difference in the activities could be attributed to the different functional groups of the polymers.

Poor results were obtained over catalysts Pd/Poly1 and Rh/Poly1 with 3.1% and 1.1% yields, respectively, although Pd and Rh complexes were usually regarded as more effective catalysts in comparison with gold complexes for carbonylation reactions. Consequently, these Au catalysts had exclusive catalytic activity for the carbonylation of epoxides with CO_2, i.e., activation of CO_2. To achieve a higher yield, a longer reaction time (10 h) was adopted over catalyst 0.05 wt% Au/Poly1, and an 89.5% yield was obtained.

The result of catalyst 0.05 wt% Au/Poly1, reused for a fifth time, indicated that the reusability of the catalyst system was possible, although relatively lower yield (66.4%) was given, which could be attributed to the loss of gold species during reaction (only 0.03 wt% gold loading remained after the fifth use)

The polymer-immobilised nanogold is also an effective catalyst for carbonylation of amines with CO_2 to yield corresponding disubstituted ureas

$$2R^2NH_2 + CO_2 \xrightarrow{\text{fast}} [R^2N(H)CO_2]\,[R^2NH_3]$$

Fig. 6.31. Carbonylation of amines with CO_2 over nanogold catalysts (adapted from [575])

(Fig. 6.31). The isolated yields of dicyclohexyl urea and dibenzyl urea were 85% and 83%. These results suggested that the activation of CO_2 over polymer-supported nanogold catalysts could be universal.

It was shown that nanogold particles were assembled on the surface of the polymer support, which ensured the high catalytic activity with lower gold content, and some interactions occurred between the nanogold particles and the cationic terminal group of the polymer. TEM characterisation indicated the formation of nanogold particles, and the average particle sizes of the nanogold were between 3 and 12 nm. Thus, catalytic activity of the polymer-supported nanogold catalysts was regulated by the particle size of nanogold particle, that is, higher catalytic activity was exhibited over smaller nanogold particles. XRD characterisation confirmed the formation of crystalline gold species when the gold loadings were high enough.

The reaction mechanism, it is not clear as yet, but the activation of carbon dioxide at the nanogold particle could be the key step, and the synergism between the nanogold species and the peculiar microenvironment of the polymer surface should be indispensable, and this may be completely different from previously reported processes for carbon dioxide activation with acid–base catalysts [575].

6.4.15 Other Reactions

Nanogold catalysts have proved to be successful in several other reactions. Chujo and co-workers report in their research on a new system for the *oligomerisation of aminoacids* using gold nanoparticles as nanosized supports [576]. In this system, they used L-*cysteine*, which is water soluble and can readily bind to the surface of gold nanoparticles via a thiolate linkage. The amino group and carboxyl group, which are polymerisable groups, are located on the outer surfaces of the gold nanoparticles. The surface layer of covalently bound L-cysteine molecules is densely packed. Thus, oligomerisation takes place effectively on the surfaces of the gold nanoparticles in water. After reaction, the gold nanoparticles were re-oxidised with iodine and suspended in water again. The resulting products were simply isolated by filtration.

This is not yet a route to oligomerisation in the conventional polymer chemistry sense, but gold particles should be considered as centres for gathering polymerisable molecules together. This proves once again that gold nanoparticles are indeed very reactive species.

The *water-splitting reaction* (dissociation of water to produce hydrogen and oxygen) is an alternative for the production of hydrogen, which is a clean

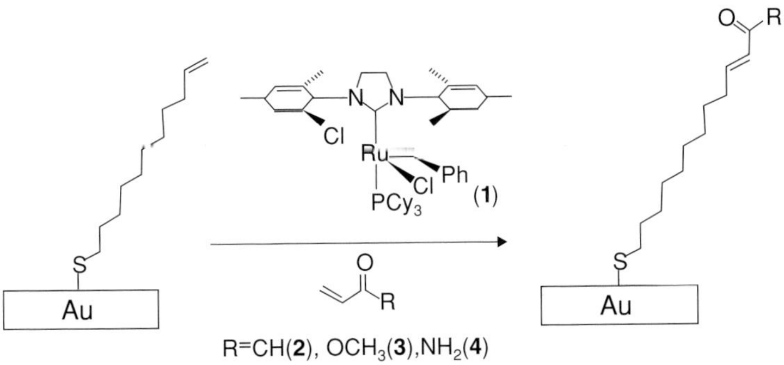

Fig. 6.32. Schematic description of the olefin CM procedure (from [579])

and renewable combustion fuel that can be used for fuel cells (see Sect. 6.5.7). The dissociation of water under sunlight without using fossil energy is attractive [577]. Several catalysts have been studied in the past for this reaction but only recently has Au been used. Wan and co-workers [577] using a perovskite titanate substrate ($K_2La2Ti_3O_{10}$), found that the gold catalyst prepared from an incipient wetness IMP process possessed a better activity for water splitting than that prepared from a DP process. This is because a better crystallinity of $K_2La_2Ti_3O_{10}$ was preserved in the former case. Moreover, the existence of gold on $K_2La_2Ti_3O_{10}$ can enhance the activity of this support for photocatalytic water splitting: The activity can be increased significantly, after the reduction of gold ions on $K_2La_2Ti_3O_{10}$ to nanogold metal (around 50-nm crystal size). When compared with the best metal–titanate catalyst reported in the literature ($Ni/K_2La_2Ti_3O_{10}$) [578], $Au/K_2La_2Ti_3O_{10}$ possessed a lower hydrogen production rate in the UV region and in the visible region. This may be because there is absorption from plasma resonance onto the nanogold surface of $Au/K_2La_2Ti_3O_{10}$ in the visible region, rather than onto the surface of $Ni/K_2La_2Ti_3O_{10}$ [577].

It has been reported that *cross-metathesis reactions* can be performed on *olefins* using nanoparticulate gold catalysts [579]. The introduction of various organic functional groups onto organic thin films is a first step in their application in sensors, catalysis and nanotechnology in general.

Choi and co-workers [579] studied the reactivity of vinyl-terminated self-assembled monolayers (SAMs) of undec-10-ene-1-thiol on gold (see Fig. 6.32) toward olefin cross-metathesis (CM). Vinyl groups on SAMs were successfully converted into α, β-unsaturated carbonyl groups by CM with acrylic acid, methyl acrylate, and acrylamide. Result shows that various useful functional groups can be introduced to SAMs on gold (and other solid surfaces) by olefin CM and suggests an alternative to the synthesis of desired molecules in solution [579].

The *hydrosilylation of C–C multiple bonds* provides a very convenient route to the preparation of organo-silicon compounds, which have found industrial application as photo resistors, semiconductors, adhesives, binders

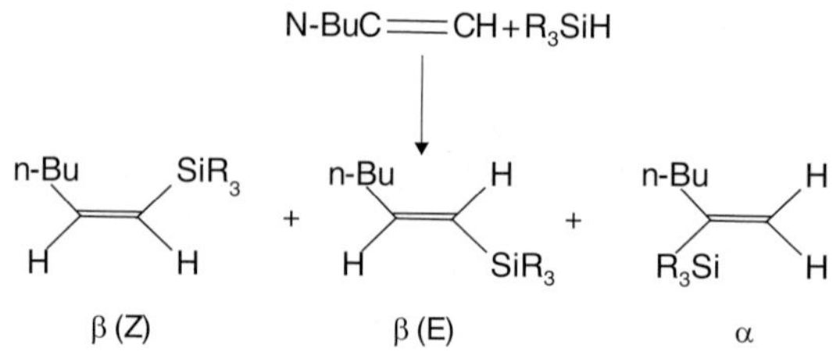

Fig. 6.33. Possible isomers from the hydrosilylation of 1-hexyne (adapted from [581])

and represent an important class of reagents for organic synthesis [580, 581]. Salvadori and co-workers [581] used supported Au nanoparticles, prepared by deposition of acetone solvated Au atoms on carbon and γ-Al_2O_3 supports, as catalysts for the *regioselective hydrosilylation of 1-hexyne* with different silanes. Au particle average size was 4 nm using both supports. The peculiar catalytic behaviour of the Au catalyst appeared independent on the type of support. Au/γ-Al_2O_3 catalyst exhibited exactly the same order of reactivity detected for Au/C, i.e., the best specific activities. Also, the Au-promoted reaction was shown to be highly regio- and stereo-selective. The hydrosilylation of a terminal acetylene, such as 1-hexyne, can yield three isomeric products: β-(Z), β-(E) and α (Fig. 6.33). The Au catalysts, in both supports, clearly favoured the formation of the *trans* isomer β-(E) which is almost the only product obtained [581].

The *hydrodechlorination of* CFCs is a very important environmental issue. These compounds have industrial significance as refrigerants, blowing agents in foams, aerosol propellants and solvents, due to their unique properties like chemical and thermal stability and non-flammability [172, 173, 176, 177]. They are also extremely stable and therefore when emitted into the atmosphere, they do not react before reaching the stratosphere. Once there, the CFCs photodissociate, producing Cl radicals that participate in ozone depletion reactions [172–178, 181]. They also make a relatively large contribution to the greenhouse effect, estimated to be up to 25% [173, 174, 176, 177].

The development of CFC alternatives has focused on non-chlorinated compounds, such as HFCs, which have similar physical properties, but will break down in the lower atmosphere [582, 583]. Consequently, hydrodechlorination is a simple method for the production of HFCs and also for destruction of CFCs [178–182]. Palladium was found to be unique among the catalytic metals for its activity and selectivity in this process [172–178, 584, 585].

Recent results by Karpinski and co-workers, however, describe the use of Pd–Au alloys using a direct redox method for catalyst preparation that involves reductive deposition of gold onto pre-reduced palladium [179–182]. This method results in a higher degree of Pd–Au alloying than impregnation procedures. The moderate selectivity for CH_2F_2 from CCl_2F_2 (CFC-12) exhibited by Pd/SiO_2 (40%) is significantly increased, up to 95%, with 20–40 at.% Au addition [180] and, with Pd/C, from 70% to 90% with Au addition [181, 182].

For the oxidative decomposition of *nitrogen-containing organic molecules*, ferric oxides and nickel ferrites have the highest level of catalytic activity owing to their strong affinities for nitrogen [7]. The oxidative decomposition of trimethylamine, which is a typical odour-producing compound, proceeds over $Au/NiFe_2O_4$ at temperatures below 373 K (see Fig. 6.26), yielding mostly N_2 and CO_2, while N_2O is mostly produced over Pd and Pt catalysts even at higher temperatures [7, 31]. Haruta and co-workers showed that the catalytic activity for the oxidation of trimethylamine of Au/Fe_2O_3 could be greatly improved by the addition of Ir/La_2O_3, creating a multi-component catalyst [516].

Chlorinated hydrocarbons (CHCs) are widely used in industry but bring both environmental and health risks [466, 586]. Catalytic oxidation is a low cost method for CHC destruction. The most active catalysts for the oxidation of hydrocarbons are noble metals supported on alumina, but a high reaction temperature is needed to increase the reaction rate and to overcome chloride poisoning [587]. At higher temperatures, where rates are more favourable, however, HCl attacks the alumina support.

The use of other supports that are active CHC oxidation catalysts at lower temperatures, such as Co_3O_4 and Cr_2O_3/Al_2O_3, have been evaluated by Wang and co-workers [586]. They found out that the addition of Au significantly improved the activity of Co_3O_4, increasing the rate by a factor of ca. 25. The Au/Co_3O_4 catalysts were more active than other catalysts; the reaction rate over 5% Au/Co_3O_4 was 10 times higher than over Cr_2O_3/Al_2O_3, 70 times higher than over 0.5% Pt/Al_2O_3, and 560 times higher than over 0.5% Pd/Al_2O_3, at 573 K. Moreover, Au/Co_3O_4 catalysts were stable in the presence of CH_2Cl_2, and the reaction product (HCl) and selectively converted CH_2Cl_2 to CO and HCl. No detectable by-products (other CHCs) or partial oxidation products (Cl_2, $COCl_2$, HCHO or CO) were formed at 623 K [586].

Remarkable hydrodechlorination activity has been reported over silica supported Ni–Au catalysts [588]. In gas-phase catalytic 2,4-dichlorophenol hydrodechlorination, a co-impregnated, thermally treated $Au–Ni/SiO_2$ catalyst was significantly more active than the equivalent Ni catalyst.

Hydrodesulphurisation (HDS) is the removal of sulphur from petroleum feedstock [51, 114, 571, 589]. The currently used HDS catalysts, i.e., alumina-supported Mo and Co oxides have a tendency to deactivate by coke formation and are not able to lower the sulphur content down to the limit of 50 ppm imposed by new legislation [51, 114, 590]. The inefficiency of these catalysts is more severe in the presence of polyaromatic compounds, which are hard to desulphurise because of the extra stabilisation of the aromatic rings. In addition to the sulphur problem, health hazards associated with aromatics [591], especially in the case of diesel fuels, are causing new limitations to their introduction.

White and co-workers [589] have deposited Mo nanoparticles onto an Au(111) surface. AES and TPD studies showed that bare Mo nanoparticles were very reactive and could cause complete dissociation of hydrogen sulphide, methyl mercaptan, and thiophene. However, the presence of Au atoms on the Mo nanoparticles modified their reactivity. In the case of H_2S and CH_3SH, the overall activity for desulphurisation was unaffected by Au encapsulation; but the selectivity to form methane from CH_3SH increased from 20% on bare Mo particles to 60% on Au-covered Mo particles. In contrast, Au-encapsulated Mo nanoparticles are relatively inert towards the dissociation of thiophene.

Palladium is a well-known hydrogenation catalyst, but only recently it has been shown that Pd(111) can directly activate thiophene decomposition, resulting in the deposition of sulphur and the formation of C_4 species on the surface [592]. Relevant to the present research on HDS catalysts is the strong affinity of gold for sulphur compounds, currently exploited in gas sensors for the detection of H_2S and also in molecular electronics using the specific bonding between Au- and sulphur-containing functional groups [18, 51, 593]. The addition of gold to palladium catalysts has been found to improve the activity and selectivity in hydrogenation reactions [594] and in hydrodechlorination (see above).

Venezia and co-workers [51, 114, 595] prepared bimetallic Pd–Au catalysts supported on SiO_2 [114, 595] and ASA (86% SiO_2, 13% Al_2O_3 and minor oxide components) [51] for use in the HDS of thiophene and dibenzothiophene. For the former reaction, the promoting effect of gold on the palladium activity was confirmed and, for the latter, pure gold catalyst exhibited larger activity and different selectivity compared with those of the palladium catalyst [51, 114, 595]. Au_xPd_y alloy particles are characterised by stronger resistance to sulphur poisoning compared with the pure palladium catalyst and by synergistic influence between the two metals in close contact [51, 114, 595]. The ASA-supported Au_xPd_y catalysts prepared by DP [51] contained only gold-enriched alloy particles of rather large size (13–45 nm). Preparation using colloidal dispersions of Pd and Au metals protected by the organic polymer polyvinylpyrrolidone (PVP), supported on SiO_2, yielded smaller (4–8 nm) alloyed particles [114].

Results consistent with these have been reported by Fierro and co-workers [117, 119] studying the simultaneous hydrogenation of naphthalene and toluene in the presence of dibenzothiophene. The supported $Au/Pd/Al_2O_3$ catalyst had a higher activity than Pd/Al_2O_3 and was more resistant to sulphur poisoning [117]. They also prepared Au, Pt and bimetallic Au–Pt catalysts supported on γ-Al_2O_3 and SiO_2 [119]. The silica support yielded a gold catalyst more active and more selective to decalin with respect to its homologue on alumina support.

Venezia and co-workers also used bimetallic Pd–Au catalysts supported on ASA for the *hydrogenation of aromatic compounds* [208]. The simultaneous hydrogenation of toluene and naphthalene in the presence of dibenzothiophene was studied. This type of catalyst was shown to be resistant to sulphur poisoning (at 523 K, until 113 ppm S in the form of dibenzothiophene).

The examples given above are taken from the literature published until June 2005. We believe that many other reactions will be studied using Au nanocatalysts and some of them will undoubtedly give new examples of catalysis by gold!

6.5 Potential Commercial Applications for Gold Nanocatalysts

6.5.1 Vinyl Acetate Synthesis

Vinyl acetate is being produced industrially from acetic acid, ethene and oxygen using Pd–Au catalysts and the reaction proceeds with selectivities as high as 96% [103, 237–239, 243, 244] (see Sect. 6.4.2):

$$C_2H_4 + CH_3COOH + 0.5O_2 \rightarrow C_2H_3OOCCH_3 + H_2O \qquad (6.16)$$

At the end of 2001, BP commissioned the brand new Leap Plant in Hull, UK. This is the world's first fluidised bed process for VAM, while 80% of today's VAM plants worldwide are more than 20 years old and use the old fixed bed process [596].

BP Chemicals have developed this cost saving VAM fluidised bed route that allowed process simplification and intensification, requiring only a single reactor compared with the two reactors usually needed in the fixed bed process. Hull in UK was an optimum location for the commercial scale plant. Acetic acid was already produced there and the ethylene supply from Teesside was approved for construction along with an air separation unit to produce oxygen. In a fixed bed reactor, the catalyst which promotes the reaction was in the form of spheres which are packed into tubes. The reaction gases pass through the tubes and around the catalyst particles in the spheres, between the spheres, without moving them.

In a fluidised bed reactor, the catalyst is in the form of a fine powder, similar to talcum powder, and as the reaction gases flow upwards through the reactor they blow the fine catalyst around, rather like the balls in a bingo blower. This gives much better mixing and contact between the gases and the catalyst, improving heat transfer and allowing the catalyst to be removed and replenished without having to shut down the reactor. Also, fluidised beds are cheaper and easier to build (this decision to go to a fluidised bed process saved 30% in capital costs).

Moving from a fixed to a fluidised operation also required a new catalyst. The selected VAM catalyst was an Au/Pd mix in the form of very fine spheres, so miniscule that they seem to flow almost as liquid, prepared in collaboration with the leading catalyst manufacturer Johnson Matthey. Hence, gold-based nanocatalysts are already being used for this process on an industrial scale. Fixed bed processes based on Au/Pd catalysts have been operated for some time by companies such as Celanese and Dupont.

6.5.2 Methyl Glycolate Synthesis

Nippon Shokubai has developed a gold-based catalyst that enables the methyl ester of glycolic acid to be made directly from ethylene glycol [597]:

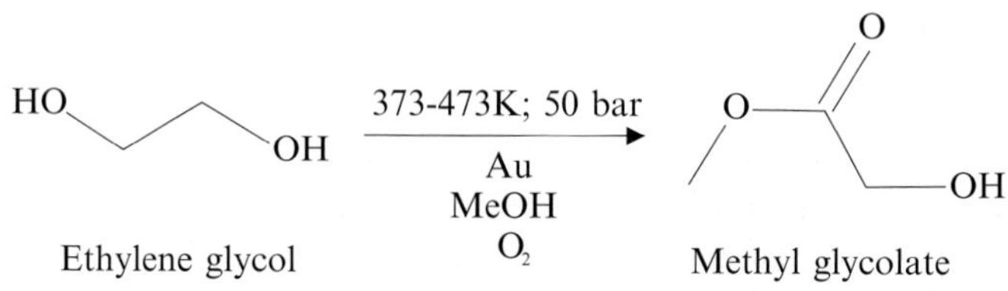

A 50 M tonne per annum pilot plant has been commissioned and there are plans to build a larger scale plant in about three years time. The proprietary catalyst is a nano-sized dispersion of activated gold supported on a metal oxide carrier, such as Au/TiO_2–SiO_2, used in the presence of oxygen and the conditions indicated in the above reaction scheme. Methyl glycolate can be used as a solvent for semiconductor processes, as a building block for cosmetics and as a cleaner for boilers and metals. The Nippon Shokubai announcement also indicates that the catalyst technology will be used for other syntheses involving one-step esterification of carboxylic acids and lactones. One of their patents claims its use for the synthesis of methyl methacrylate.

6.5.3 Vinyl Chloride Synthesis

The manufacture of PVC is very important commercially [2, 18, 213, 248, 249] and the synthesis of the monomer, vinyl chloride is therefore an important step in this synthesis (see Sect. 6.4.3):

$$HC \equiv CH + HCl \rightarrow H_2C = CH.Cl \rightarrow PVC \tag{6.17}$$

Gold catalysts supported on activated carbon were found to be about three times more active than commercial mercuric chloride catalysts for vinyl chloride production and to deactivate much less rapidly than other supported metal catalysts. Deactivation can be minimised if high loadings of gold are used [18, 248, 249]. Also, Au catalysts can be reactivated by treatment offline with HCl or Cl_2, and by co-feeding NO with the reactants from the start of the reaction, deactivation could be virtually eliminated [259]. Gold is thus the catalyst of choice for this reaction [6, 7, 18, 213, 248, 249, 258, 259].

6.5.4 Gluconic Acid

The catalytic oxidation of D-glucose to D-gluconic acid is indeed an attractive target in industrial chemistry.

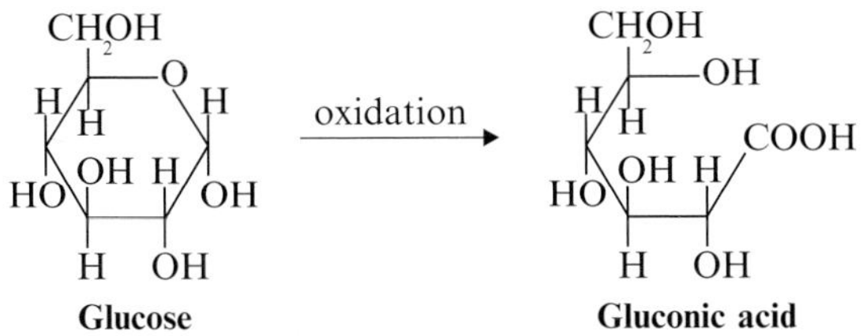

As discussed in Sect. 6.4.5, the oxidation of *glucose* or glucose-containing raw materials is of commercial importance since gluconic acid and its salts are important industrial products, used as additives in food and beverages and water-soluble cleansing agents.

Rossi and co-workers studies [360,361] showed that gold on carbon catalyst is a valid alternative to most of the investigated multi-metallic catalysts based on Pd and/or Pt metals. Moreover, gold has a unique property, i.e., it operates without the external control of pH, thus ensuring total conversion at all pH values, and total selectivity to gluconic acid.

6.5.5 Hydrogen Peroxide Production

The market for hydrogen peroxide is very large (ca. 1.9×10^6 tonnes per annum) and there is a big incentive to enable hydrogen peroxide to be synthesised, where it is to be used, and thus avoid the heavy transport costs for this hazardous material. Also, it is only economic to produce it on a large scale, using the sequential hydrogenation and oxidation of alkyl anthraquinone, whereas it is often required on a much smaller scale [103, 112]. As discussed in Sect. 6.4.7, theoretical calculations [470] and experimental results [112] have both shown that formation of H_2O_2 from hydrogen and oxygen is favoured over gold surfaces. Al_2O_3-supported Au catalysts are particularly effective for this reaction and Al_2O_3-supported Pd/Au alloys have now been shown to provide a significant improvement over the Pd catalysts used in industry to date [112].

6.5.6 Air Cleaning

Air purification is a very important issue. Air cleaning devices are needed for respiratory protection (gas masks) and for removing CO, CO_2, water vapour and trace amounts of VOCs and ozone from ambient air indoor office space (due to smoking, etc.) and in submarines or space crafts on long missions [2, 6, 7, 18, 31, 64, 71, 110, 125, 126, 132, 251, 260–264, 598].

Supported nanogold catalysts are promising materials to use in CO cleaning devices. And Haruta's studies on CO oxidation over Au catalysts (supported in Fe_2O_3 and TiO_2) in real air, confirms that they are useful for the removal of CO from both low concentrations (10–100 ppm) and high concentrations (10,000 ppm) [262].

Ozone decomposition can be a source of active oxygen for CO oxidation, and a system for simultaneous elimination of ozone and carbon monoxide has been devised by Engelhard Co. (USA). Their Prem AirTM [599] can be used to remove ozone and carbon monoxide directly from the ambient air. Installing such catalyst systems on all of the vehicles in Los Angeles (which experiences the highest ozone level in USA) would reduce ozone and carbon monoxide more effectively than other new control strategies such as introducing electric vehicles. Also, it was found that the ozone (0.25 ppm) and CO (25.0 ppm) conversion efficiencies decreased after prolonged use, the CO removal (being dependent on catalyst formulation) was a strong function of catalyst temperature. Virtually no CO conversion was observed at 303 K for all catalysts used, which included Pt/Al_2O_3, Pd/Al_2O_3, Ni/Al_2O_3 and $Pd/Ni/Al_2O_3$, although they were more effective at 373 K [600].

However, investigations of catalytic ozone decomposition and simultaneous CO oxidation [543] showed that Au/Fe_2O_3 was a good catalytic material for ozone decomposition and simultaneous elimination of ozone and carbon monoxide, at any O_3/CO concentration ratio, in the presence of rich oxygen, at ambient temperature. This new catalyst is also suitable for use in severe conditions, such as relatively high ozone concentration and large gas space velocity, without the stoichiometric limitations. It also shows a high room temperature activity and good resistance to moisture. This new type of Au catalytic material shows great potential as an environmental catalyst, particularly for indoor environmental pollution control.

Incinerator exhaust gases can contain a huge variety of pollutants such as dioxins, VOCs, odours, hydrocarbons, NO_x, CO_x and amine derivatives. Due to the variety, it would be unreasonable to expect a single-component catalyst to achieve high catalytic performance for all of the exhaust gas pollutants. However, Haruta and co-workers [516] have shown that if several supported single noble metal catalysts were encouraged to work in synergy, the overall catalytic activity would be greatly improved. Consequently, a ternary component noble metal catalyst (consisting of Au supported on Fe_2O_3–Pt/SnO_2–Ir/La_2O_3) achieving good results for purifying typical exhaust gases emanating from incinerators even at 423 K [28, 516]. This multicomponent gold-based catalyst is a promising material for cleaning of exhaust gases.

It has been shown also that other odour producing compounds, such as trimethylamine, can be oxidatively decomposed over $Au/NiFe_2O_4$ at temperatures below 373 K [7, 31] (see Fig. 6.26). Gold on Fe_2O_3, supported on a zeolite wash-coated honeycomb, has been used commercially as a deodoriser in modern Japanese toilets since 1992 [6, 7, 28].

6.5.7 Autocatalysts

Air pollution generated from mobile sources is the major problem in this field. In the last 60 years, the world vehicle fleet has increased from about 40 million vehicles to over 700 million, and this figure is projected to increase to 920 million by the year 2010 [490]. The three major pollutants emitted by internal combustion engines are carbon monoxide, non-methane hydrocarbons and nitrogen oxides (see Sects. 6.4.4, 6.4.9 and 6.4.11) [145]. Environmental legislation governing the emission of these three types of gases is becoming increasingly stringent. Catalysts that are capable of removing these pollutants simultaneously are referred to as TWCs and the design of these catalyst systems is continually evolving to meet lower emission requirements. Typically, these catalysts must operate in the presence of 10% H_2O and 10–60 ppm SO_2 at temperatures ranging from 623 to 1,273 K and gas hourly space velocities ranging from 10,000 to over 100,000 h^{-1} for the duration of 100,000 miles of operation [2, 145, 601]. All commercial TWCs in use at present are based on platinum, palladium and rhodium on a support comprised of zirconia-stabilised ceria, zirconia and α-alumina. Additives include barium oxide and zinc oxide.

Catalyst efficiency is usually evaluated under simulated driving conditions using the standardised federal test procedure (FTP). A key problem identified in FTP tests is the liberation of unburned non-methane hydrocarbons during the cold start mode of the test, when the catalyst monolith is at ambient temperature. As a consequence, the catalyst does not reach the hydrocarbon light-off temperature of about 573 K until approximately 2 min after the start of the test. During this delay, up to 50% of the total unburned hydrocarbons are emitted. Additionally, when the engine operates under prolonged idling conditions, the temperature at the inlet to the converter is typically 553 K, which is an accepted monolith temperature for a catalyst converter mounted approximately 80 cm from the exhaust manifold of a spark ignition engine of 1.8 l displacement [2, 145, 601]. This temperature is again below the light-off for hydrocarbons.

A gold catalyst with low-temperature activity towards CO and hydrocarbon oxidation could be suitable in such an application, especially during the low-temperature start-up and idling phases. Marsh and co-workers [145] formulated a gold-based catalyst for use as a TWC in gasoline and diesel applications. This catalyst, developed at AARL, consisted of 1-wt% Au supported on zirconia-stabilised–CeO_2, ZrO_2 and TiO_2, and contained 1 wt% CoO_x, 0.1 wt% Rh, 2 wt% ZnO and 2 wt% BaO as promoters. The catalytically active gold–cobalt oxide clusters were 40–140 nm in size. This catalyst was tested under conditions that simulated the exhaust gases of gasoline and diesel automobiles and survived 773 K for 157 h, with some deactivation [145].

A significant hurdle for the gold-based TWC is the high-operating temperature requirements imposed by gasoline engines. Typically, a catalyst must be able to withstand a temperature of 1,373 K for at least 12 h. The gold based TWC cannot survive under such conditions and it is accepted that gold will not be able to match the high-temperature performance of the PGM-based TWCs. However, a relatively simple system in which PGM- and gold-based catalysts operate in parallel or sequentially can be envisaged, where the gold catalyst is in use at low temperatures but is by-passed in favour of the PGM catalyst at higher temperatures. In this way, maximum conversion activity can be maintained both at low temperatures using the gold catalyst and at high temperatures using the PGM catalyst [145].

Under diesel conditions, CO and HC oxidation is favoured. Under the highly oxidising conditions encountered in the diesel gas stream, reduction of NO is not expected. A NO conversion window is observed at temperatures between 493 and 623 K, with a T_{50} value of 523 K. However, large NO absorption bands are observed at temperatures above and below the conversion window [145].

Other advances have been made by Pitchon and co-workers [45], who prepared Au catalysts by DAE on an alumina support. The catalyst durability was strongly improved by the complete removal of chloride using an ammonia washing procedure. This catalyst tested in various reactions of saturated and unsaturated hydrocarbons from C_1 to C_3 and oxidation of CO, revealed a good activity, which is in an appropriate range of temperature for automotive exhaust cleaning applications, and longer durability tests may demonstrate further promise. Datye and co-workers [65] have used a new impregnation method for supporting gold on alumina, with chloride removal, as described in Sect. 6.4. The reactivity for CO oxidation at room temperature was comparable with catalysts prepared by deposition precipitation. These 1% Au/Al_2O_3 2 nm nanoparticle catalysts are stable to hydrothermal sintering, in 10 mol% H_2O at 873 K for 100 h [65]. This could have important implications for their future use in autocatalyst and other pollution control applications.

Kung and co-workers [602] have studied the synergism between Pt/Al_2O_3 and Au/TiO_2 catalysts in the low-temperature oxidation of propene. It was shown that the light-off temperature for C_3H_6 over Pt/Al_2O_3 is significantly lower than for CO. However, in a C_3H_6 and CO mixture, the presence of CO greatly suppresses the oxidation of C_3H_6. The extent of suppression depends on the concentration of CO in the gas feed, since CO adsorbs strongly on Pt and inhibits its oxidative activity. Au/TiO_2 is significantly less active than Pt/Al_2O_3 for C_3H_6 oxidation, but has a high CO oxidation activity: Over a mixture of these two catalysts, the partial pressure of CO is lowered by oxidation over Au/TiO_2, reducing its suppression of C_3H_6 oxidation over Pt/Al_2O_3. Consequently, the C_3H_6 oxidation activity can be enhanced substantially. A similar enhancement can be observed when the two catalysts are placed in series in a two reactor-bed configuration, with the Au/TiO_2 up-stream of Pt/Al_2O_3. Thus, a good strategy for improving the light-off characteristics for automotive catalysts might be to use catalysts that contain both Au and Pt active sites. The challenge will be to make a catalyst containing Au that is durable in automotive exhaust conditions [602].

6.5.8 Fuel Cell Technology

Fuel cells are energy generators, which provide a combination of high efficiency combined with low pollutant emissions, based on the direct electrochemical oxidation of hydrogen or hydrocarbon fuels to provide an electric current, and at the same time forming water [101,188,189,191,202,213,233,334,603] which can be a useful by-product in space travel and submarines. Globally, fuel cells have been developed for a wide range of applications including stationary generators, cars and buses, industrial vehicles and small portable power supplies for mobile telephones and laptop computers, each of which represents a substantial market. After decades of development and large-scale trials, fuel cell generators are now entering commercial service to provide electricity and heat for individual homes, or district schemes. In addition, virtually all the world's major vehicle manufacturers are demonstrating electrically propelled passenger vehicles or municipal buses [603, 604].

Recent developments in supported gold catalysts make them suitable for a number of potential applications required for the application of fuel cell technology [2, 7, 104, 223, 334, 598, 603, 605–607]. These include generation of hydrogen by reforming hydrocarbons and use in the water-gas shift. A promising initial use for supported gold catalysts is in catalysing this WGS reaction, since they can be very active for it, as discussed already in Sect. 6.4.1. Another use for gold catalysts is the purification of the resulting gas mixture by selective oxidation of carbon monoxide (PROX). Encouraging results have been obtained with gold-based catalysts for this reaction also, as discussed in Sect. 6.4.4.

There could be opportunities to exploit the properties of gold and gold/PGM alloys in the form of catalysts, particularly for low-temperature PEM fuel cells, which are one of the most widely developed forms of generator and also the most susceptible to poisoning by impurities in hydrogen fuel [103, 334, 603]. PGMs are currently used for both anode and cathode catalysts in most commercial PEM fuel cells, but the tendency for poisoning of platinum by CO contaminant in the hydrogen feedstock gas remains a major technical issue [202]. Work is now emerging [608–610] investigating the potential benefits of alternative PGM/Au bifunctional catalysts. This is based on the premise that the alloying of gold with Pt in the nanosize range could generate unique bifunctional catalytic properties with Pt acting as the catalyst for the main fuel cell reaction and Au assisting via the removal of CO under the low operating temperatures of the fuel cell.

This approach would also potentially have two other advantages over current technology. Firstly, if Pt loadings could be reduced, it could produce a welcome reduction in the capital cost of fuel cell catalysts. Over the last 5 years, prices for Pt have significantly exceeded those for Au [233]. Cost remains a significant barrier to more widespread commercialisation of fuel cell technology and a reduction in catalyst cost is identified as a key objective throughout the industry. The second advantage from incorporation of gold into fuel cells would be the useful enhancement in electrical conductivity that could be derived, due to the lower electrical resistivity of gold compared to platinum [202, 233]. Industrial interest in the use of Au/Pt alloy nanoparticles as fuel cell catalysts is now emerging [611, 612].

A new improvement was made by Dumesic and co-workers [607]. They demonstrated the feasibility of making new generations of inexpensive fuel cells that operate with solutions of reduced polyoxometalate compounds. They produced electricity by catalytic oxidation of CO using gold nanoparticles on various supports such as carbon and titania at room temperature. The observed rates were faster than conventional processes operating at 500 K or higher for the conversion of CO with water to produce hydrogen and carbon dioxide through the WGS reaction. By eliminating this reaction, the need to transport and vaporise liquid water in the production of energy for portable applications is removed. This process can use CO-containing gas streams from the catalytic reforming of hydrocarbons to produce an aqueous solution of reduced polyoxometalate compounds that can be used to generate power. The reduced polyoxometalate can be re-oxidised in fuel cells that contain simple carbon anodes.

6.5.9 Sensors

The need for air-quality monitoring demands development of sensors which are selective for detection of individual pollutant gases. Gas sensors based on gold have therefore been developed for detecting a number of gases, including CO and NO_x. The use of gold is also particularly promising for colour-change sensors used for monitoring components of body liquids.

Sensors for *CO detection* are well established, using the use of Au nanoparticles doped on α–Fe_2O_3 [613–616] with particle sizes between 3.2 and 8.8 nm [615]. Also, nanogold particles, with diameters smaller than 5 nm, have been reported on ZnO [617] and on powdered cobalt oxide [618]. CuO/Au [619, 620], NiO [620] and Co_3O_4 [620] composite films (for optical CO sensors) are also reported, as well as nano-Au colloid modified on Pt disk microelectrode [621] and an Au–La_2O_3 loaded SnO_2 ceramic [622, 623]. The target of the last example was to develop an alcohol sensitivity depressed, steadily workable and widely available CO gas sensor, since conventional CO gas sensors had a poor selectivity to ethanol vapour, which co-exists very often in kitchens, causing false alarms; so ethanol absorbents such as activated carbon had to be used. Results showed that the sensitivity to CO was above 10 times as high as that to H_2, CH_4, i-C_4H_{10} and C_2H_4; while the sensitivity to C_2H_5OH alone remained high [623].

Sensors for *NO_x detection* have been reported. Porous silicon (PS) was catalysed by sputtering gold onto the surface and used as an NO_x sensor [624]. It was found that, as a result of Au catalysation, PS is suitable for sensing nitrogen oxides with negligible influence by interfering gases, such as CO, CH_4 or methanol; however, humidity appreciably affected the response [624].

WO_3 thin films activated by Au layers have been used [625]. This type of material possesses excellent sensitivity towards NO and NO_2 gases. An exhaust gas NO_x sensor that uses a platinum–gold alloy electrode to selectively remove oxygen but not NO was also reported [626].

Particular sensors for NO_2 detection using an Au/PVC composite prepared by dispersion of a fine gold powder in a highly plasticised PVC matrix containing a hydrophobic electrolyte (Au/PVC electrode) were studied [627]. It was found that the Au/PVC electrode is applicable to continuously operating sensors monitoring the gaseous environment under conditions of slow variation in NO_2 content. Also, metal oxide chemical sensors have been reported [628, 629]. This type of sensors is selective to NO_2 with NO, H_2 and CO as interfering gases.

Gold thin-film gas sensors can be used to detect parts per million of NO in air without the combined use of any semiconductor material [630]. Also, amperometric NO gas sensors prepared with Au/solid polymeric electrolytes can be used [631].

The detection of nitric oxide in biology and medicine is also important as it is in air, particularly to distinguish $NO^{\bullet}$ from its oxidation products, allowing in vivo measurements; assessment of chemical activity, making data available rapidly, ideally in real time. The techniques of electrochemistry are promising for use in sensors for biomedical applications. Nitric oxide oxidation signals were tested for gold and a nickel–porphyrin on graphite in de-aerated phosphate-buffered saline (pH 7.0) at 308 K and $NO^{\bullet}$ was detected amperometrically [632].

Also, third-generation biosensors for *superoxide anion* ($O_2^{\bullet -}$) have been developed based on superoxide dismutase (SOD) immobilised by thin silica–PVA sol–gel film on a gold electrode surface [633]. The preparation of SOD electrode is easy and simple. The uniform porous structure of the silica–PVA sol–gel matrix results in a fast response rate of immobilised SOD and is very efficient for stabilising the enzyme activity.

Hydrogen peroxide biosensors have been made by immobilising horseradish peroxidase (HRP) enzymes on colloidal Au modified ITO conductive glass supports [634], or on colloidal gold-modified carbon paste electrodes (Au–CPE) [635], nano-scaled particulate gold (nano-Au) monolayer modified chitosan-entrapped carbon paste electrodes (CCPE) [636], or by self-assembling gold nanoparticles on thiol-functionalised poly(styrene-*co*-acrylic acid) (St-*co*-AA) nanospheres [637]. In all cases, the immobilised HRP displayed excellent electrocatalytical response to the reduction of hydrogen peroxide. The resulting biosensors exhibited high sensitivity, good reproducibility, and long-term stability.

Glucose sensors have been developed using a bimetallic electrode composed of copper microparticles dispersed unto a gold surface (Au/Cu) to give an amperometric sensor for glucose [638]. Good mechanical stability was found, making this amperometric sensor very attractive for analytical applications in flowing streams.

Also, a glucose biosensor based on direct electron transfer of glucose oxidase immobilised on colloidal gold modified carbon paste electrode was reported [639], as well as a gold electrode modified with ferrocenylundecanethiol monolayer for the simultaneous detection of glucose and ascorbic acid [640]. A novel system with Au nanoparticles based on microdialysis and co-enzymes for continuous glucose monitoring has also been described [641]. Results of in vitro experiments show that this glucose sensor has a short response time, high sensitivity and good linearity. This is a demonstration that aqueous colloidal gold nanoparticles can enhance the activity of aqueous enzymes.

Gold-based *biosensors* for near-patient assessment were also described [642]. The new technology mimics sensing mechanisms found in biology, but is adapted to be more stable and directly compatible with electronic devices for reading and storing data from clinical measurements. The Ambri system works by measuring the changes in the ionic currents induced across a specially designed lipid membrane chemically tethered to a piece of gold-coated plastic. Being chemically tethered to the gold surface results in a very stable membrane whose conduction may easily be measured by simple electronics.

A nanocluster consisting of 11 gold atoms joined together, undecagold, is being manufactured by Nanoprobes Inc. in USA for use in some very *selective biomedical diagnoses* [643, 644]. The Ambri biosensor is being developed for both medical applications, such as the critical care pathology market, where valuable rapid results are produced, and non-medical applications, such as rapid analysis for food contamination and for individual components of wines and other drinks [643, 644].

Mesoporous gold has the unique property of being both metallic and highly resistant to oxidation. This material is relevant to potential applications in *biosensors* [645]. Also, the use of gold CDs as *biochemical sensors* has been described [646]. After removing the protective polymer film with nitric acid, species in body fluids containing –SH groups would react with the gold to form SAMs on the surface. It was also confirmed that these substrates are suitable not only for surface wetting measurements but also surface electrochemical studies.

Coordination polymers based on cyanoaurate derivatives ($M[Au(CN)_2]_2(Ligand)_x$) show large colour changes upon changing the ligand, thus illustrating a sensor-type application as *solvent and gas sensors* [647, 648]. Strongly luminescent heteropolynuclear complexes obtained by reaction between basic gold precursors and acid silver or thallium derivatives which are very good *sensors for VOCs*: In these cases, the change of colour and luminescent properties are due to interactions of the ligands with the silver or thallium centres [649]. An optical fibre sensor based on vapochromic gold complexes has been developed and used for the detection of some VOCs [650]. The design of luminescent polynuclear Au(I) complexes as *chemosensor towards metal ions* has also been referred [651]. Two-coordinate Au(I) complexes which display strong photoluminescence and participate in a number of unusual luminescence phenomena can be useful as *chemosensors* and others have potential applications as *sensors for gaseous organic compounds* [652].

Guth and co-workers [653, 654] developed Au-oxide composites as sensors for *hydrocarbon monitoring* in exhaust gases. Also, Au-loaded In_2O_3 ceramics were used at detection of NH_3 and other reducing gases [45] and Au promoted Li–Fe_2O_3 thin films for *humidity sensors* with high sensitivity, linearity and good stability have been developed [655].

A gold–solid polymer electrolyte sensor was developed for detecting *dissolved oxygen in water*, showing high selectivity [656]. The detection of *propene* under oxidising conditions using zirconia-based potentiometric sensor was also achieved [657]. Also, a nano-Au-assembled amperometric *SO_2 gas sensor* was developed, in which other co-present gases, such as CO, NO, NH_3 and CO_2, did not cause interference [658].

Many other examples include the work of Jian and co-workers [659–661], who reported colloidal Au for enhancement of the immobilisation capacity and ultimately detection limit of *DNA* using quartz crystal microbalance (QCM). These authors showed that Au nanoparticle films on the Au plate provide a novel means for the fabrication of DNA sensor [659–661].

6.6 Future Prospects

As we have seen in Sects. 6.4 and 6.5, research on gold nanocatalysts has indicated the potential of these materials for catalysing commercially viable chemical processes, as illustrated by the new pilot plant for the production of methyl glycolate, the vinyl acetate synthesis plants and the routes for the production of vinyl chloride and the oxidation of glucose to gluconic acid. Gold-containing fuel cell catalysts are being developed both for use in the cells themselves and for hydrogen generation via the WGS and purification via the PROX reaction. There are a variety of opportunities in pollution control applications, including cleaning living air space and vehicle exhausts.

Whilst gold nanocatalysis has clearly indicated this spectrum of new opportunities for the application of gold nanocatalysts, parallel developments in homogeneous catalysis by gold have clearly indicated new opportunities for unique organic syntheses and processes with high turnover numbers and high selectivities and TOF: Interested readers are referred to publications on these aspects [213, 662–680]. The recent developments in catalysis by gold, consequently, cover nanocatalysis by supported catalysts in both the gas and liquid phases and homogeneous catalysis in solution.

In the keynote lecture he delivered to the international conference on "New Industrial Applications for Gold", Dr. Haruta (Tokyo, Japan) [7] indicated that R&D efforts on nanogold catalysts are expected to expand in a number of directions. The first involves the discovery of new capabilities for Au particles larger than 2 nm that are stable up to 673 K. Combinations with other metals are already contributing to the further developments in the properties of Au nano-catalysts. There are already indications that gold catalysts are stable even at 873 K, so this may be an underestimate of the potential of catalysis by gold! In another keynote lecture, Professor Hutchings (Cardiff, UK) emphasised the importance of focussing research effort on the potential of gold nanocatalysis for a wider range of potential uses than has been studied to date [681].

Another direction for research could involve the development of cluster science for Au particles smaller than 2 nm. Fundamental work has led to an understanding of the unique catalysis of gold as well as its size and structure dependency. Gold itself has comparatively poor activity but this markedly increases when the Au particles have a specific structure and size. The advantage of studying Au model catalysts is that the activity can often be observed at lower temperatures, where designed structures can be maintained.

To increase the effectiveness of recent attempts to combine experimental work using real and model catalysts with theoretical calculations, the effect of moisture should also be considered. Most work in surface science occurs in UHV conditions, while measurements of the activity of real catalysts are conducted in fixed bed flow reactors containing between 1 and 10 ppm moisture. The tolerance of gold catalysis to water is an unexpected advantage, which is particularly valuable in developing new catalyst systems for pollution control and other applications.

Our overall conclusion is that the potential of gold catalysts for stimulating research is considerable and the results will lead to more practical, commercial applications, the full extent of which has still to be envisaged.

References

1. Thompson DT (2001) Chem. Br. 37:43
2. Corti CW, Holliday RJ, Thompson DT (2005) Appl. Catal. A 291:253
3. Dekkers MAP, Lippits MJ, Niewenhuys BE (1998) Catal. Lett. 56:195
4. Dekkers MAP, Lippits MJ, Niewenhuys BE (1999) Catal. Today 54:381
5. Dekkers MAP (2000) Ph.D. Thesis, Leiden University
6. Haruta M (2002) CATTECH 6:102
7. Haruta M (2003) Chem. Record 3:75
8. Wang D, Hao Z, Cheng D, Shi X, Hu C (2003) J. Mol. Catal. A 200:229
9. Kung HH, Kung MC, Costello CK, J. Catal. 216 (2003):425
10. Vinod CP, Niemantsverdriet JW, Nieuwenhuys BE (2005) Appl. Catal. A 291:93
11. Parker SC, Grant AW, Bondzie VA, Campbell CT (1999) Surf. Sci. 441:10
12. Syomin D, Koel BE (2001) Surf. Sci. 492:L693
13. Gottfried JM, Elghobashi N, Schroeder SLM, Christmann K (2003) Surf. Sci. 523:89
14. Kitchin J, Barteau MA, Chen JG (2003) Surf. Sci. 526:323
15. Rost MJ, van Albada SB, Frenken JWM (2002) Surf. Sci. 515:344
16. Mellor JW (1923) In: A Comprehensive Treatise on Inorganic and Theoretical Chemistry. Longmans, Green and Co., London

17. Faraday M (1857) Philos. Trans. R. Soc. London 147:145
18. Bond GC, Thompson DT (1999) Catal. Rev. Sci. Eng. 41:319
19. Mayya KS, Patil V, Sastry M (1997) Langmuir 13:2575
20. Patil V, Malvankar RB, Sastry M (1999) Langmuir 15:8197
21. Hwang Y-N, Jeong DH, Shin HJ, Kim D, Jeoung SC, Han SH, Lee J-S, Cho G (2002) J. Phys. Chem. B 106:7581
22. Saraiva M, de Oliveira JF (2002) J. Dispers. Sci. Technol. 23:837
23. Currao A, Reddy VR, Calzaferri G (2004) Chem. Phys. Chem. 5:720
24. Duff DG, Baiker A, Edwards PP (1993) Langmuir 9:2301
25. Tsunoyama H, Sakurai H, Ichikuni N, Negishi Y, Tsukuda T (2004) Langmuir 20:11293
26. Leff DV, Ohara PC, Heath JR, Gelbart WM (1995) J. Phys. Chem. 99:7036
27. Coluccia S, Martra G, Porta F, Prati L, Rossi M (2000) Catal. Today 61:165
28. Haruta M (2003) Proc. International Conference on 'New Industrial Applications for Gold' _www.GOLD2003.org_ (http://www.GOLD2003.org), Vancouver, Canada, 2003
29. Lee S-J, Gavriilidis A, Pankhurst QA, Kyek A, Wagner FE, Wong PCL, Yeung KL (2001) J. Catal. 200:298
30. Seker E, Gulari E (2002) Appl. Catal. A 232:203
31. Haruta M (1997) Catal. Today 36:153
32. Wolf A, Schüth F (2002) Appl. Catal. A 226:1
33. Zanella R, Giorgio S, Henry CR, Louis C (2002) J. Phys. Chem. B 106:7634
34. Centeno MÁ, Carrizosa I, Odriozola JA (2003) Appl. Catal. A 246:365
35. Haruta M, Ueda A, Tsubota S, Torres Sanchez RM (1996) Catal. Today 29:443
36. Chang CK, Chen YJ, Yeh C (1998) Appl. Catal. A 174:13
37. Grisel RJH, Kooyman PJ, Niewenhuys BE (2000) J. Catal. 191:430
38. Grisel RJH (2001) Ph.D. Thesis, Leiden University
39. Lee SJ, Gavriilidis A (2002) J. Catal. 206:305
40. Upshade BS, Akita T, Nakamura T, Haruta M (2002) J. Catal. 209:331
41. Meyer R, Lemire C, Shaikhutdinov ShK, Freund H-J (2004) _www.goldbulletin.org_ (http://www.goldbulletin.org) 37:72
42. Oh HS, Yang JH, Costello CK, Wang Y, Bare SR, Kung HH, Kung MC (2002) J. Catal. 210:375
43. Moreau F, Bond GC, Taylor AO (2005) J. Catal. 231:105
44. Kung HH, Kung MC, Costello CK, Yang JH (2003) Proc. International Conference on 'New Industrial Applications for Gold' _www.GOLD2003.org_ (http://www.GOLD2003.org), Vancouver, Canada, 2003
45. Cousin R, Ivanova S, Ammari F, Petit C, Pitchon V (2003) Proc. International Conference on 'New Industrial Applications for Gold' _www.GOLD2003.org_ (http://www.GOLD2003.org), Vancouver, Canada, 2003
46. Bethke GK, Kung HH (2000) Appl. Catal. A 194/195:43
47. Chen YJ, Yeh CT (2001) J. Catal. 200:59
48. Zanella R, Louis C, Giorgio S, Touroude R (2003) Proc. International Conference on 'New Industrial Applications for Gold' _www.GOLD2003.org_ (http://www.GOLD2003.org), Vancouver, Canada, 2003
49. Zanella R, Giorgio S, Shin C-H, Henry CR, Louis C (2004) J. Catal. 222:357
50. Zanella R, Delannoy L, Louis C (2005) Appl. Catal. A 291:62
51. Venezia AM, La Parola V, Nicolì V, Deganello G (2002) J. Catal. 212:56
52. Grisel RJH, Nieuwenhuys BE (2001) J. Catal. 199:48

53. Grisel R, Weststrate K-J, Gluhoi A, Nieuwenhuys BE (2002) _www.goldbulletin.org_ (http://www.goldbulletin.org) 35:39
54. Gluhoi AC, Dekkers MAP, Nieuwenhuys BE (2003) J. Catal. 219:197
55. Khoudiakov M, Gupta MC, Deevi S (2004) Nanotechnology 15:987
56. Khoudiakov M, Gupta MC, Deevi S (2005) Appl. Catal. A 291:151
57. Dobrosz I, Jiratova K, Pitchon V, Rynkowski JM (2005) J. Mol. Catal. A 234:187
58. Jia J, Haraki K, Kondo JN, Domen K, Tamaru K (2000) J. Phys. Chem. B 104:11153
59. Prati L, Martra G (1999) _www.goldbulletin.org_ (http://www.goldbulletin.org) 32:96
60. Lin JN, Wan BZ (2003) Appl. Catal. B 41:83
61. Mihut C, Descorme C, Duprez D, Amiridis MD (2002) J. Catal. 212:125
62. Soares JMC, Bowker M (2003) Proc. International Conference on 'New Industrial Applications for Gold' _www.GOLD2003.org_ (http://www.GOLD2003.org), Vancouver, Canada, 2003
63. Bowker M, Millard L, Greaves J, James D, Soares J (2003) Proc. International Conference on 'New Industrial Applications for Gold' _www.GOLD2003.org_ (http://www.GOLD2003.org), Vancouver, Canada, 2003
64. Bulushev DA, Kiwi-Minsker L, Yuranov I, Suvorova EI, Buffat PA, Renken A (2002) J. Catal. 210:149
65. Xu Q, Kharas KCC, Datye A (2003) Catal. Lett. 85:229
66. Parish RV (1998) _www.goldbulletin.org_ (http://www.goldbulletin.org) 31:14
67. Araki H, Fukuoka A, Sakamoto Y, Inagaki S, Sugimoto N, Fukushima Y, Ichikawa M (2003) J. Mol. Catal. A 199:95
68. Yuan Y, Kozlova AP, Asakura K, Wan H, Tsai K, Iwasawa Y (1997) J. Catal. 170:191
69. Koslova AP, Sugiyama S, Koslov AI, Akasura K, Iwasawa Y (1998) J. Catal. 176:426
70. Liu HC, Kozlov AI, Kozlova AP, Shido T, Iwasawa Y (1999) Phys. Chem. Chem. Phys. 1:2851
71. Liu H, Kozlov AI, Kozlova AP, Shido T, Asakura K, Iwasawa Y (1999) J. Catal. 185:252
72. Kozlova AP, Kozlov AI, Sugiyama S, Matsui Y, Asakura K, Iwasawa Y (1999) J. Catal. 181:37
73. Kozlov AI, Kozlova AP, Asakura K, Matsui Y, Kogure T, Shido T, Iwasawa Y (2000) J. Catal. 196:56
74. Koslov AI, Kozlova AP, Liu H, Iwasawa Y (1999) Appl. Cat. A 182:9
75. Okumura M, Tsubota S, Haruta M (2003) J. Mol. Catal. A 199:73
76. Brey LA, Wood TE, Buccellato GM, Jones ME, Chamberlain CS, Siedle AR (2005) 3M Innovative Properties Co. WO Patent 030382 A2
77. Veith GM, Lupini AR, Pennycook SJ, Ownby GW, Dudney NJ (2005) J. Catal. 231:151
78. Qiu S, Ohnishi R, Ichikawa M (1994) J. Phys. Chem. 98:2719
79. Kang Y-M, Wan B-Z (1995) Appl. Catal. A 128:53
80. Kang Y-M, Wan B-Z (1995) Catal. Today 26:59
81. Salama TM, Shido T, Minagawa H, Ichikawa M (1995) J. Catal. 152:322
82. Guillemot D, Polisset-Thfoin M, Fraissard J (1996) Catal. Lett. 41:143
83. Salama TM, Ohnishi R, Shido T, Ichikawa M (1996) J. Catal. 162:169

84. Kang Y-M, Wan B-Z (1997) Catal. Today 35:379
85. Guillemot D, Borovkov VYu, Kazansky VB, Polisset-Thfoin M, Fraissard J (1997) J. Chem. Soc. Faraday Trans. 93:3587
86. Mohamed MM, Salama TM, Ichikawa M (2000) J. Colloid Interf. Sci. 224:366
87. Horváth D, Polisset-Thfoin M, Fraissard J, Guczi L (2001) Solid State Ion. 153:141
88. Lin J-N, Chen J-H, Hsiao C-Y, Kang Y-M, Wan B-Z (2002) Appl. Catal. B 36:19
89. Riahi G, Guillemot D, Polisset-Thfoin M, Khodadadi AA, Fraissard J (2002) Catal. Today 72:115
90. Yang C-M, Kalwei M, Schüth F, Chao K-J (2003) Appl. Catal. A 254:289
91. Yang C-M, Liu P-H, Ho Y-F, Chiu C-Y, Chao K-J (2003) Chem. Mater. 15:275
92. Chao K-J, Cheng M-H, Ho Y-F, Liu P-H (2004) Catal. Today 97:49
93. Fisher J (2003) _www.goldbulletin.org_ (http://www.goldbulletin.org) 34:155
94. Kutsch B, Lyon O, Schmitt M, Mennig M, Schmidt H (1997) J. Non-Cryst. Solids 217:143
95. Sei T, Takatou N, Kineri T, Tsuchiya T (1997) Mater. Sci. Eng. B 49:61
96. Seker E, Cavataio J, Gulari E, Lorpongpaiboon P, Osuwan S (1999) Appl. Catal. A 183:121
97. Seker E, Gulari E, Hammerle RH, Lambert C, Leerat J, Osuwan S (2002) Appl. Catal. A 226:183
98. Claus P, Brueckner A, Mohr C, Hofmeister H (2000) J. Am. Chem. Soc. 22:11430
99. Schimpf S, Lucas M, Mohr C, Rodemerck U, Brückner A, Radnik J, Hofmeister H, Claus P (2002) Catal. Today 72:63
100. Mine E, Yamada A, Kobayashi Y, Konno M, Liz-Marzán LM (2003) J. Colloid Interf. Sci. 264:385
101. Luengnaruemitchai A, Osuwan S, Gulari E (2004) Int. J. Hydrogen Energy 29:429
102. Zhu H, Pan Z, Chen B, Lee B, Mahurin SM, Overbury SH, Dai S (2004) J. Phys. Chem. B 108:20038
103. Thompson DT (2004) Platinum Metals Rev. 48:169
104. Thompson DT (2005) Top. Catal. 38:231
105. Ponec V (2001) Appl. Catal. A 222:31
106. Ponec V, Bond GC (1996) In: Catalysis by Metals and Alloys. Elsevier, Amsterdam
107. Shibata M, Kawata N, Masumoto T, Kimura H (1985) Chem. Lett. 1605
108. Baiker A, Maciejewski M, Tagliaferri S, Hug P (1995) J. Catal. 151:407
109. Sanz N, Lomello-Tafin M, Valmalette J-C, Isa M, Galez Ph (2002) Mater. Sci. Eng. C 19:79
110. Haruta M, Daté M (2001) Appl. Catal. A 222:427
111. Boccuzzi F, Chiorino A, Manzoli M, Lu P, Akita T, Ichikawa S, Haruta M (2001) J. Catal. 202:256
112. Landon P, Collier PJ, Papworth AJ, Kiely CJ, Hutchings GJ (2002) Chem. Commun.:2058
113. Venezia AM, Liotta LF, Pantaleo G, La Parola V, Deganello G, Beck A, Koppány Zs, Frey K, Horváth D, Guczi L (2003) Appl. Catal. A 251:359
114. Venezia AM, La Parola V, Deganello G, Pawelec B, Fierro JLG (2003) J. Catal. 215:317

115. Kapoor MP, Ichihashi Y, Nakamori T, Matsumura Y (2004) J. Mol. Catal. A 213:251
116. Bowker M, Millard L, Greaves J, James D, Soares J (2004) _www.goldbulletin.org_ (http://www.goldbulletin.org) 37:170
117. Pawelec B, Cano-Serrano E, Campos-Martin JM, Navarro RM, Thomas S, Fierro JLG (2004) Appl. Catal. A 275:127
118. Scott RWJ, Wilson OM, Oh S-K, Kenik EA, Crooks RM (2004) J. Am. Chem. Soc. 126:15583
119. Pawelec B, Venezia AM, La Parola V, Thomas S, Fierro JLG (2005) Appl. Catal. A 283:165
120. Nieuwenhuys BE, Gluhoi AC, Rieks EDL, Westrate CJ, Vinod CP (2005) Catal. Today 100:49
121. Njoki PN, Luo J, Wang L, Maye MM, Quaizar H, Zhong C-J (2005) Langmuir 21:1623
122. Wang J-H, Kumar D, Yan Z, Goodman DW (2005) J. Catal. 232:467
123. Wang A-Q, Liu J-H, Lin SD, Linand T-S, Mou C-Y (2005) J. Catal. 233:186
124. van der Lingen E, Cortie MB, Schwarzer H, Roberts SJ, Pattrick G, Compton D (2003) Proc. International Conference on 'New Industrial Applications for Gold' _www.GOLD2003.org_ (http://www.GOLD2003.org), Vancouver, Canada, 2003
125. Bond GC, Thompson DT (2000) _www.goldbulletin.org_ (http://www.goldbulletin.org) 33:41
126. Hutchings GJ, Siddiqui MRH, Burrows A, Kiely CJ, Whyman R (1997) J. Chem. Soc. Faraday Trans. 93:187
127. Grisel RJH, Nieuwenhuys BE (2001) Catal. Today 64:69
128. Bond GC (2001) _www.goldbulletin.org_ (http://www.goldbulletin.org) 34:117
129. Bond GC (2002) Catal. Today 72:5
130. Okumura M, Akita T, Haruta M (2002) Catal. Today 74:265
131. Centeno MA, Paulis M, Montes M, Odriozola JA (2002) Appl. Catal. A 234:65
132. Cortie MB, van der Lingen E (2002) Mater. Forum 26:1
133. Costello CK, Kung MC, Oh H-S, Wang Y, Kung HH (2002) Appl. Catal. A 232:159
134. Carabineiro SAC (2002) Chemisch Weekblad 98:12
135. Uematsu T, Fan L, Maruyama T, Ichikuni N, Shimazu S (2002) J. Mol. Catal. A 182–183:209
136. Shaikhutdinov S, Meyer R, Lemire C, Freund H-J (2003) Proc. International Conference on 'New Industrial Applications for Gold' _www.GOLD2003.org_ (http://www.GOLD2003.org), Vancouver, Canada, 2003
137. Fu L, Wu NQ, Yang JH, Qu F, Johnson DL, Kung MC, Kung HH, Dravid VP (2005) J. Phys. Chem. B 109:3704
138. Iizuka Y, Fujiki H, Yamauchi N, Chijiiwa T, Arai S, Tosubota S, Haruta M (1997) Catal. Today 36:115
139. Iizuka Y, Tode T, Takao T, Yatsu K, Takeuchi T, Tsubota S, Haruta M (1999) J. Catal. 187:50
140. Biella S, Rossi M (2003) Proc. International Conference on 'New Industrial Applications for Gold' _www.GOLD2003.org_ (http://www.GOLD2003.org), Vancouver, Canada, 2003
141. Sanchez-Castillo MA, Cauto C, Kim WB, Dumesic JA (2004) Angew. Chem. Int. Ed. 43:1140

142. Fu Q, Saltsburg H, Flytzani-Stephanopoulos M (2003) Science 301:935
143. Fu Q, Deng W, Saltsburg H, Flytzani-Stephanopoulos M (2005) Appl. Catal. B 56:57
144. Deng W, de Jesus J, Saltsburg H, Flytzani-Stephanopoulos M (2005) Appl. Catal. A (in press)
145. Mellor JR, Palazov AN, Grigorova BS, Greyling JF, Reddy K, Letsoalo MP, Marsh JH (2002) Catal. Today 72:145
146. Kung HH, Kung MC, Costello CK, Yang JH (2003) Proc. International Conference on 'New Industrial Applications for Gold' _www.GOLD2003.org_ (http://www.GOLD2003.org), Vancouver, Canada, 2003
147. Cunningham DAH, Vogel W, Kageyama H, Tsubota S, Haruta M (1998) J. Catal. 177:1
148. Bore MT, Pham HN, Switzer RE, Ward TL, Fukuoka A, Datye AK (2005) J. Phys. Chem. B 109:2873
149. Gosh A, Patra CR, Mukherjee P, Sastry M, Kumar R (2003) Microporous Mesoporous Mater. 58:201
150. Zhu H, Lee B, Dai S, Overbury SH (2003) Langmuir 19:3974
151. Yan W, Petkov V, Mahurin SM, Overbury SH, Dai S (2005) Catal. Commun. 6:404
152. Schubert MM, Hackenberg S, van Veen AC, Muhler M, Plzak V, Behm RJ (2001) J. Catal. 197:113
153. Ilieva L, Sobczak JW, Manzoli M, Su BL, Andreeva D (2005) Appl. Catal. A 291:85
154. Nieuwenhuys BE (1999) Adv. Catal. 44:259, and references therein
155. Okumura M, Coronado JM, Soria J, Haruta M, Conesa JC (2001) J. Catal. 203:168
156. Feron J, Bakker JW, Nieuwenhuys BE (unpublished results)
157. Grisel RJH, Weststrate CJ, Goossens A, Crajé MWJ, van der Kraan AM, Nieuwenhuys BE (2002) Catal. Today 72:123
158. Lin SD, Gluhoi AC, Nieuwenhuys BE (2004) Catal. Today 90:3
159. Gluhoi AC, Lin SD, Nieuwenhuys BE (2004) Catal. Today 90:175
160. Gluhoi AC, Bogdanchikova N, Nieuwenhuys BE (2005) J. Catal. 229:154
161. Grunwaldt JD, Baiker A (1999) J. Phys. Chem. B 103:1002
162. Minicò S, Scirè S, Crisafulli C, Gavagno S (2001) Appl. Catal. B 34:277
163. Hodge NA, Kiely CJ, Whyman R, Siddiqui MRH, Hutchings GJ, Pankhurst QA, Wagner FE, Rajaram RR, Golunski SE (2002) Catal. Today 72:133
164. Hayashi T, Han LB, Tsubota S, Haruta M (1995) Ind. Eng. Chem. Res. 34:2298
165. Mul G, Zwijnenburg A, van der Linden B, Makkee M, Moulijn JA (2001) J. Catal. 201:128
166. Zwijnenburg A, Saleh M, Makkee M, Moulijn JA (2002) Catal. Today 72:59
167. Haruta M, Uphade BS, Tsubota S, Miyamoto A (1998) Res. Chem. Intermed. 24.329
168. Stangland EE, Stavens KB, Andres RP, Delgass WN (2000) J. Catal. 191:332
169. Hayashi T, Tanaka K, Haruta M (1998) J. Catal. 178:566
170. Qi C, Akita T, Okumura M, Kuraoka K, Haruta M (2003) Appl. Catal. A 253:75
171. Qi C, Akita T, Okumura M, Akita T, Haruta M (2004) Appl. Catal. A 263:19
172. van de Sandt EJAX, Wiersma A, Makkee M, van Bekkum H, Moulijn JA (1997) Catal. Today 35:163

173. Makkee M, van de Sandt EJAX, Wiersma A, Moulijn JA (1998) J. Mol. Catal. A 134:191
174. van de Sandt EJAX, Wiersma A, Makkee M, van Bekkum H, Moulijn JA (1998) Appl. Catal. A 173:161
175. Wiersma A, van de Sandt EJAX, den Hollander MA, van Bekkum H, Makkee M, Moulijn JA (1998) J. Catal. 177:29
176. Moulijn JA, Makkee M, Wiersma A, van de Sandt EJAX (2000) Catal. Today 59:221
177. Makkee M, Wiersma A, van de Sandt EJAX, van Bekkum H, Moulijn JA (2000) Catal. Today 55:125
178. Karpinski Z, Early K, d'Itri JL (1996) J. Catal. 164:378
179. Bonarowska M, Pielaszek J, Juszczyk W, Karpinski Z (2000) J. Catal. 195:304
180. Bonarowska M, Malinowski A, Juszczyk W, Karpinski Z (2001) Appl. Catal. B 30:187
181. Bonarowska M, Burda B, Juszczyk W, Pielaszek J, Kowalczyk Z, Karpinski Z (2001) Appl. Catal. B 35:13
182. Bonarowska M, Pielaszek J, Semikolenov VA, Karpinski Z (2002) J. Catal. 209:528
183. Haruta M, Yamada N, Kobayashi T, Iijima S (1989) J. Catal. 115:301
184. Rosd MTS, Atsushi U, Koji T, Haruta M (1997) J. Catal. 168 125
185. Okumura M, Nakamura S, Tsubota S, Nakamura T, Azuma M, Haruta M (1998) Catal. Lett. 51:53
186. Kahlich MJ, Gasteiger HA, Behm RJ (1999) J. Catal. 182:430
187. Schubert MM, Kahlich MJ, Gasteiger HA, Behm RJ (1999) J. Power Sources 84:175
188. Avgouropoulos G, Ioannides T, Papadopoulou Ch, Batista J, Hocevar S, Matralis HK (2002) Catal. Today 75:157
189. Choudhary TV, Goodman DW (2002) Catal. Today 77:65
190. Qiao B, Deng Y (2003) Chem. Commun.:2192
191. Zhang J, Wang Y, Chen B, Li C, Wu D, Wang X (2003) Energy Conv. Management 44:1805
192. Schubert MM, Venugopal A, Kahlich MJ, Plzak V, Behm RJ (2004) J. Catal. 222:107
193. Shou M, Tanaka K-I, Yoshioka K, Moro-oka Y, Nagano S (2004) Catal. Today 90:255–261
194. Liu Y, Fu Q, Stephanopoulos MF (2004) Catal. Today 93–95:241
195. Kandoi S, Gokhale AA, Grabow LC, Dumesic JA, Mavrikakis M (2004) Catal. Lett. 93:93
196. Margitfalvi JL, Hegedús M, Szegedi Á, Sajó I (2004) Appl. Catal. A 272:87
197. Manzoli M, Chiorino A, Boccuzzi F (2004) Appl. Catal. B 52:259
198. Schumacher B, Denkwitz Y, Plzak V, Kinne M, Behm RJ (2004) J. Catal. 224:449
199. Lomello-Tafin M, Chaou AA, Morfin F, Caps V, Rousset J-L (2005) Chem. Commun.:388
200. Rossignol C, Arrii S, Morfin F, Piccolo L, Caps V, Rousset J-L (2005) J. Catal. 230:476
201. Buffat P, Borel J-P (1976) Phys. Rev. A 13:2287
202. Pattrick G, van der Lingen E, Corti CW, Holliday RJ, Thompson DT (2003) Proc. 6th International Congress on Catalysis and Automotive Pollution Control (CAPOC 6), Brussels, Belgium

203. European Patent Application, EP1043059 A1
204. Adams KM, Cavataio JV, Hammerle RH (1996) Appl. Catal. B 10:157
205. Uphade BS, Okumura M, Tsubota S, Haruta M (2000) Appl. Catal. A 190:43
206. Liu G, Rodriguez JA, Jirsak T, Hrbek J, Chang Z, Dvorak J, Maiti A (2003) Proc. 18th Meeting of the North American Catalysis Society, Cancun, Mexico
207. Ruth K, Hayes M, Burch R, Tsubota S, Haruta M (2000) Appl. Catal. B 24:L133
208. Venezia AM, La Parola V, Pawelec B, Fierro JLG (2003) Proc. International Conference on 'New Industrial Applications for Gold' _www.GOLD2003.org_ (http://www.GOLD2003.org), Vancouver, Canada, 2003
209. Tada H, Soejima T, Ito S, Kobayashi H (2004) J. Am. Chem. Soc. 126(49):15952–15953
210. Kuperman A, Moir ME (2005) WO Patent 2005/005032
211. Andreeva D, Idakiev V, Tabakova T, Ilieva LI, Falaras P, Bourlinos A, Travlos A (2002) Cat. Today 72:51
212. Tabakova T, Boccuzzi F, Manzoli M, Andreeva D (2003) Appl. Catal. A 252:385
213. Thompson DT (2003) Appl. Catal. A 243:201
214. Kinne M, Leppelt R, Plzak V, Behm RJ (2003) Proc. International Conference on 'New Industrial Applications for Gold' _www.GOLD2003.org_ (http://www.GOLD2003.org), Vancouver, Canada, 2003
215. Tabakova T, Boccuzzi F, Manzoli M, Sobczak JW, Idakiev V, Andreeva D (2004) Appl. Catal. B 49:73
216. Schumacher N, Boisen A, Dahl S, Gokhale AA, Kandoi S, Grabow LC, Dumesic JA, Mavrikakis M, Chorkendorff I (2005) J. Catal. 229:265
217. Kim CH, Thompson LT (2005) J. Catal. 230:66
218. Sakurai H, Akita T, Tsubota S, Kiuchi M, Haruta M (2005) Appl. Catal. A (in press)
219. Andreeva D, Idakiev V, Tabakova T, Andreev A (1996) J. Catal. 158:354
220. Andreeva D, Idakiev V, Tabakova T, Andreev A, Giovanoli R (1996) Appl. Catal. A 134:275
221. Ilieva LI, Andreeva DH, Andreev AA (1997) Thermochim. Acta 292:169
222. Andreeva D, Tabakova T, Idakiev V, Christov P, Giovanoli R (1998) Appl. Catal. A 169:9
223. Andreeva D (2002) _www.goldbulletin.org_ (http://www.goldbulletin.org) 35:82
224. Sakurai H, Ueda A, Kobayashi T, Haruta M (1997) Chem. Commun. :271
225. Venugopal A, Aluha J, Mogano D, Scurrell MS (2003) Appl. Catal. A 245:149
226. Boccuzzi F, Chiorino A, Manzoli M, Andreeva D, Tabakova T (1999) J. Catal. 188:176
227. Boccuzzi F, Chiorino A, Manzoli M, Andreeva D, Tabakova T, Ilieva L, Iadakiev V (2002) Catal. Today 75:169
228. Tabakova T, Idakiev V, Andreeva D, Mitov I (2000) Appl. Catal. A202:91
229. Fu Q, Weber A, Flytzani-Stephanopoulos M (2001) Catal. Lett. 77:87
230. Fu Q, Kudriavtseva S, Saltsburg M, Flytzani-Stephanopoulos M (2003) Chem. Eng. J. 93:41
231. Jacobs G, Patterson PM, Williams L, Chenu E, Sparks D, Thomas G, Davis BH (2004) Appl. Catal. A 262:177
232. Li Y, Fu Q, Flytzani-Stephanopoulos M (2000) Appl. Catal. B 27:17

233. Cameron D, Holliday R, Thompson D (2003) J. Power Sources 118:298
234. Kim CH, Thompson LT (2003) Proc. International Conference on 'New Industrial Applications for Gold' _www.GOLD2003.org_ (http://www.GOLD2003.org), Vancouver, Can 2003
235. Kim CH, Thompson L (2003) ada, Preprints of Symposia—American Chemical Society, Division of Fuel Chemistry, vol 48. p.233
236. Idakiev V, Yuan Z-Y, Tabakova T, Su B-L (2005) Appl. Catal. A 281:149
237. Provine WD, Mills PL, Lerou JJ (1996) Stud. Surf. Sci. Catal. 101:191
238. Macleod N, Keel JM, Lambert RM (2004) Appl. Catal. A 261:37
239. Neurock M, Provine WD, Dixon DA, Couiston GW, Lerou JJ, van Santen RA (1996) Chem. Eng. Sci. 51:1691
240. (1967) US Patent No. 3658888
241. Bissot TC (1977) US Patent No. 4048096
242. Han Y-F, Wang J-H, Kumar D, Yan Z, Goodman DW (2005) J. Catal. 232:467
243. Sennewald (1971) US Patent 3631079
244. Barteley WJ, Jobson S, Harkreader GG, Kitson M, Lemanski M (1993) US Patent 5274181
245. Haley RD, Tikhov MS, Lambert RM (2001) Catal. Lett. 76:125
246. Abel R, Prauser G, Tiltscher H (1994) Chem. Eng. Technol. 17:112
247. Samanos B, Boutry P, Montarnal R (1971) J. Catal. 23:19
248. Thompson DT (1998) _www.goldbulletin.org_ (http://www.goldbulletin.org) 31:111
249. Hutchings GJ (2002) Catal. Today 72:11
250. Hutchings GJ, Grady D (1985) Appl. Catal. 16:441
251. Kulkarni GU, Vinod CP, Rao CNR (2002) Nanoscale catalysis by gold. In: Carley et al. (eds.) Surface Chemistry and Catalysis. Kluwer Academic/Plenum Publisher, New York
252. Hutchings GJ (2005) Catal. Today 100:55
253. Hutchings GJ (1996) _www.goldbulletin.org_ (http://www.goldbulletin.org) 29:123
254. Shinoda K (1975) Chem. Lett. :219
255. Hutchings GJ (1985) J. Catal. 96:292
256. Nkosi B, Coville NJ, Hutchings GJ (1988) Appl. Catal. 43:33
257. Nkosi B, Coville NJ, Hutchings GJ (1988) J. Chem. Soc. Chem. Commun.:71
258. Nkosi B, Coville NJ, Hutchings GJ, Adams MD, Friedl J, Wagner F (1991) J. Catal. 128:366
259. Nkosi B, Adams MD, Corville NJ, Hutchings GJ (1991) J. Catal. 128:378
260. Hoflund GB, Gardner SD, Schryer DR, Upchurch BT, Kielin EJ (1995) Appl. Catal. B 6:117
261. Epling WE, Hoflund GB, Weaver J, Tsubota S, Haruta M (1996) J. Phys. Chem. 100:9929
262. Sakurai H, Haruta M, Tsubota S (2003) Proc. International Conference on 'New Industrial Applications for Gold' _www.GOLD2003.org_ (http://www.GOLD2003.org), Vancouver, Canada, 2003
263. Wang GY, Zhang WX, Lian HL, Jiang DZ, Wu TH (2003) Appl. Catal. A 239:1
264. Mallick K, Scurrell MS (2003) Appl. Catal. A 253:527
265. Haruta M, Kobayashi T, Sano H, Yamada N (1987) Chem. Lett. :405

266. Haruta M, Tsubota S, Kobayashi T, Kageyama H, Genet MJ, Delmon B (1993) J. Catal. 144:175
267. Cunningham DAH, Kobayashi T, Kamijo N, Haruta M (1994) Catal. Lett. 25:257
268. Vogel W, Cunningham DAH, Tanaka K, Haruta M (1996) Catal. Lett. 40:175
269. Bamwenda GR, Tusobota S, Nakamura T, Haruta M (1997) Catal. Lett. 44:83
270. Tsubota S, Nakamura T, Tanaka K, Haruta M (1998) Catal. Lett. 56:131
271. Cunningham DAH, Vogel W, Haruta M (1999) Catal. Lett. 63:43
272. Cunningham DAH, Vogel W, Sanchez RMT, Tanaka K, Haruta M (1999) J. Catal. 183:183
273. Daté M, Haruta M (2001) J. Catal. 201:221
274. Daté M, Ichihashi Y, Yamashita T, Choirino A, Boccuzzi F, Haruta M (2002) Catal. Today 72:89
275. Daté M, Okumura M, Tsubota S, Haruta M (2003) Proc. International Conference on 'New Industrial Applications for Gold' _www.GOLD2003.org_ (http://www.GOLD2003.org), Vancouver, Canada, 2003
276. Choudhary TV, Goodman DW (2002) Topic. Catal. 21:25
277. Aderson JA (1992) J. Chem. Soc. Faraday Trans. 88:1201
278. Gardner SD, Hoflund SB, Schryer DR, Schryer J, Upchurch BT, Kielin EJ (1991) Langmuir 7:2135
279. Lin S, Vannice MA (1991) Catal. Lett. 10:47
280. Lin S, Bollinger M, Vannice MA (1993) Catal. Lett. 17:245
281. Bollinger MA, Vannice MA (1996) Appl. Catal. B 8:417
282. Yuan Y, Asakura K, Kozlova AP, Wan H, Tsai K, Iwasawa Y (1998) Catal. Today 44:333
283. Valden M, Pak S, Lai X, Goodman DW (1998) Catal. Lett. 56:7
284. Valden M, Lai X, Goodman DW (1998) Science 281:1647
285. Park ED, Lee JS (1999) J. Catal. 186:1
286. Grunwaldt J-D, Maciejewski M, Becker OS, Fabrizioli P, Baiker A (1999) J. Catal. 186:458
287. Grunwaldt J-D, Kiener C, Wögerbauer C, Baiker A (1999) J. Catal. 181:223
288. Fan L, Ichikuni N, Shimazu S, Uematsu T (2003) Appl. Catal. A 246:87
289. Soares JMC, Morrall P, Crossley A, Harris P, Bowker M (2003) J. Catal. 219:17
290. Arrii S, Morfin F, Renouprez AJ, Rousset JL (2004) J. Am. Chem. Soc. 216:1199
291. Yan W, Chen B, Mahurin SM, Dai S, Overbury SH (2004) Chem. Commun.:1918
292. Chou J, Franklin NR, Baeck S-H, Jaramillo TF, McFarland EW (2004) Catal. Lett. 95:107
293. Wu S-H, Zhong X-C, Wang S-R, Han D Z, Huang W-P, Zhang S-M (2004) Catal. Lett. 97:17
294. Moreau F, Bond GC, Taylor AO (2004) Chem. Commun.:1642
295. Stiehl JD, Kim TS, McClure SM, Mullins CB (2004) J. Am. Chem. Soc. 126:13574
296. Daté M, Okumura M, Tsubota S, Haruta M (2004) Angew. Chem. Int. Ed. 43:2129
297. Yan W, Chen B, Mahurin SM, Schwartz V, Mullins DR, Lupini AR, Pennycook SJ, Dai S, Overbury SH (2005) J. Phys. Chem. B 109:10676
298. Wagner FE, Galvagno S, Milone C, Visco AM, Stievano L, Calogero S (1997) J. Chem. Soc. Faraday Trans. 93:3403

299. Tripathi AK, Kamble VS, Gupta NM (1999) J. Catal. 187:332
300. Gupta NM, Tripathi AK (1999) J. Catal. 187:343
301. Hall MS, Landon P, Carley AF, Hutchings GJ, Herzing A, Kiely CJ (2003) Proc. International Conference on 'New Industrial Applications for Gold' _www.GOLD2003.org_ (http://www.GOLD2003.org), Vancouver, Canada, 2003
302. Waters RD, Weimer JJ, Smith JE (1995) Catal. Lett. 30:181
303. Guzman J, Gates BC (2003) J. Phys. Chem. B 107:2242
304. Guzman J, Carrettin S, Corma A (2005) J. Am. Chem. Soc. 127:3286
305. Venezia AM, Pantaleo G, Longo A, Di Carlo G, Casaletto M, Liotta FL, Deganello G (2005) J. Phys. Chem. B 109:2821
306. Carrettin S, Corma A, Iglesias M, Sánchez F (2005) Appl. Catal. A 291:247
307. Guzman J, Corma A (2005) Chem. Commun. 6:743
308. Chen J-H, Lin J-N, Kang Y-M, Yu W-Y, Kuo C-N, Wan B-Z (2005) Appl. Catal. A 291:162
309. Wu K-C, Tung Y-L, Chen Y-L, Chen Y-W (2004) Appl. Catal. B 53:111
310. Cheng W-H, Wu K-C, Lo M-Y, Lee C-H (2004) Catal. Today 97:145
311. Guczi L, Peto G, Beck A, Frey K, Geszti O, Molnár G, Daróczi C (2003) J. Am. Chem. Soc. 125:4332
312. Martra G, Prati L, Manfredotti C, Biella S, Rossi M, Coluccia S (2003) J. Phys. Chem. B 107:5453
313. Overbury SH, Ortiz-Soto L, Zhu H, Lee B, Amiridis MD, Dai S (2004) Catal. Lett. 95:99
314. Costello CK, Yang JH, Law HY, Wang Y, Lin J-N, Marks LD, Kung MC, Kung HH (2003) Appl. Catal. A 243:15
315. Bulushev DA, Yuranov I, Suvorova EI, Buffat PA, Kiwi-Minsker L (2004) Catal. 224:8
316. Gossner K, Mizera E (1979) J. Electroanal. Chem. 98:37
317. Lopez N, Janssens TVW, Clausen BS, Xu Y, Mavrikakis M, Bligaard T, Nørskov JK (2004) J. Catal. 223:232
318. Valden M, Lai X, Goodman DW (1998) Science 281:1647
319. Mavrikakis M, Stoltze P, Nørskov JK (2000) Catal. Lett. 64:101
320. Campbell CT (2004) Science 306:234
321. Bond GC (2004) CatGoldNews 6
322. Olea M, Iwasawa Y (2004) Appl. Catal. A 275:35
323. Gluhoi AC, Vreeburg HS, Bakker JW, Nieuwenhuys BE (2005) Appl. Catal. A 291:145
324. Lian H, Jia M, Pan W, Li Y, Zhang W, Jiang D (2005) Catal. Commun.6:47
325. Moreau F, Bond GC, Taylor AO (2005) J. Catal. 231:105
326. Milone C, Ingoglia R, Pistone A, Neri G, Frusteri F, Galvagno S (2004) J. Catal. 222:348
327. Gardner SD, Hoflund GB, Davidson MR, Laitinen HA, Schryer DR, Upchurch BT (1991) Langmuir 7:2140
328. Visco AM, Neri F, Neri G, Milone C, Galvagno S (1999) Phys. Chem. Chem. Phys. 1:2869
329. Chen MS, Goodman DW (2004) Science 306:252
330. Campbell CT (2004) Science 306:234
331. Boyen H-G, Kästle G, Weigl F, Koslowski B, Dietrich C, Ziemann P, Spatz JP, Riethmüller S, Hartmann C, Möller M, Schmid G, Garnier MG, Oelhafen P (2002) Science 297:1533

332. Kalvachev YA, Hayashi T, Tsubota S, Haruta M (1999) J. Catal. 186:228
333. Margitfalvi JL, Fasi A, Hegedus M, Lonyi F, Gobolos S, Bogdanchikova N (2002) Catal. Today 72:157
334. Cameron DS (2003) Proc. International Conference on 'New Industrial Applications for Gold' _www.GOLD2003.org_ (http://www.GOLD2003.org), Vancouver, Canada, 2003
335. Farrauto RJ (2005) Appl. Catal. B 56:3
336. Luengnaruemitchai A, Thoa DTK, Osuwan S, Gulari E (2005) Int. J. Hydrogen Energy 30:981
337. Rostrup-Nielson JR (1993) Catal. Today 18:305
338. Armor JN (1999) Appl. Catal. A 176:159
339. Hohlein B, von Andrian S, Grube T, Menzer R (2000) J. Power Sources 86:243
340. Freni S, Calogero G, Cavallaro S (2000) J. Power Sources 87:28
341. Oh SH, Sinkevitch RM (1993) J. Catal. 142:254
342. Kahlich MJ, Gasteiger HA, Behm RJ (1997) J. Catal. 171:93
343. Schubert MM, Gasteiger HA, Behm RJ (1997) J. Catal. 172:256
344. Igarashi H, Uchida H, Suzuki M, Sasaki Y, Watanabe M (1997) Appl. Catal. A 159:159
345. Teng Y, Sakurai H, Ueda A, Kobayashi T (1999) Int. J. Hydrogen Energy 24:355
346. Mariño F, Descorme C, Duprez D (2004) Appl. Catal. B 54:59
347. Schubert MM, Kahlich MJ, Feldmeyer G, Huttner M, Hackenberg S, Gasteiger HA, Behm RJ (2001) Phys. Chem. Chem. Phys. 3:1123
348. Zheng X, Wang S, Wang S, Zhang S, Huang W, Wu S (2004) Catal. Commun. 5:729
349. Choudhary TV, Goodman DW (2002) Top. Catal. 21:25
350. Lai XF, Goodman DW (2000) J. Mol. Catal. A 162:33
351. Kolmakov A, Goodman DW (2000) Catal. Lett. 70:93
352. Kolmakov A, Goodman DW (2001) Surf. Sci. Lett. 490:L597
353. Konova P, Naydenov A, Tabakova T, Mehandjiev D (2004) Catal. Commun. 5:537
354. Choudhary TV, Sivadinarayana C, Chusuei C, Datye AK, Fackler JP Jr, Goodman DW (2002) J. Catal. 207:247
355. Landon P, Ferguson J, Solsona BE, Garcia T, Carley AF, Herzing AA, Kiely CJ, Golunski SE, Hutchings GJ (2005) Chem. Commun. 27:3385
356. Biella S, Prati L, Rossi M (2003) Inorg. Chim. Acta 349:253
357. Prati L, Rossi M (1998) J. Catal. 176:552
358. Bianchi C, Porta F, Prati L, Rossi M (2000) Topic. Catal. 13:231
359. Porta F, Prati L, Rossi M, Scari G (2002) J. Catal. 211:464
360. Biella S, Prati L, Rossi M (2002) J. Catal. 206:242
361. Biella S, Castiglioni GL, Fumagalli C, Prati L, Rossi M (2002) Catal. Today 72:43
362. Biella S, Prati L, Rossi M (2003) J. Mol. Catal. A 197:207
363. Prati L, Porta F (2005) Appl. Catal. A 291:199
364. Comotti M, Della Pina C, Matarrese R, Rossi M, Siani A (2005) Appl. Catal. A 291:204
365. Mallat T, Baiker A (1994) Catal. Today 19:247
366. Porta F, Rossi M (2003) J. Mol. Catal. A 204–205:553
367. Biella S, Porta F, Prati L, Rossi M (2003) Catal. Lett. 90:23

368. Berndt H, Pitsch I, Evert S, Struve K, Pohl M-M, Radnik J, Martin A (2003) Appl. Catal. A 244:169
369. Berndt H, Martin A, Pitsch I, Prüsse U, Vorlop K-D (2004) Catal. Today 91–92:191
370. Carrettin S, McMorn P, Johnston P, Griffin K, Hutchings GJ (2002) Chem. Commun.:696
371. Carrettin S, McMorn P, Johnston P, Griffin K, Kiely CJ, Attard GA, Hutchings GJ (2004) Top. Catal. 27:131
372. Demirel-Gülen S, Lucas M, Claus P (2005) Catal. Today 102–103:166
373. Dimitratos N, Porta F, Prati L (2005) Appl. Catal. A 291:210
374. Bianchi CL, Canton P, Dimitratos N, Porta F, Prati L (2005) Catal. Today 102–103:203
375. Porta F, Prati L (2004) J. Catal. 224:397
376. Garcia R, Besson M, Gallezot P (1995) Appl. Catal. A 127:165
377. Dimitratos N, Porta F, Prati L, Villa A (2005) Catal. Lett. 99:181
378. Dimitratos N, Prati L (2005) _www.goldbulletin.org_ (http://www.goldbulletin.org) 38:73
379. Hermans S, Devillers M (2005) Catal. Lett. 99:55
380. Alardin F, Delmon B, Ruiz P, Devillers M (2000) Catal. Today 61:255
381. Alardin F, Ruiz P, Delmon B, Devillers M (2001) Appl. Catal. A 215:125
382. Milone C, Ingoglia R, Neri G, Pistone A, Galvagno S (2001) Appl. Catal. A 211:251
383. Milone C, Ingoglia R, Pistone A, Neri G, Galvagno S (2003) Catal. Lett. 87:201
384. Mirescu A, Preusse U, Vorlop K-D (2004) Proc. 13th International Congress on Catalysis, Paris, France
385. Prati L, Rossi M, Fumagalli C, Castiglioni G, Pirola C (1999) Italian Patent 99A 002611 for Lonza S.p.A.
386. Asahi Chemical Industry Co. Ltd. (1980) DE–OS2936652
387. Saito H, Ohnaka S, Fukuda S (1989) US Patent 4843173 for Kawaken Fine Chemical Corp. Ltd. and Kao Corp.
388. Deller K, Krause H, Peldszus E, Despeyroux B (1992) US Patent 5132452 for Degussa
389. Bronnimann C, Bodnar Z, Hug P, Mallat T, Baiker A (1994) J. Catal. 150:199
390. Besson M, Lahmer F, Gallezot P, Fuertes P, Fleche G (1995) J. Catal. 152:116
391. Abbadi A, van Bekkum H (1995) Appl. Catal. A 124:409
392. Besson M, Gallezot P (2000) Catal. Today 57:127
393. Claus P, Schimpf S, Onai Y (2003) Proc. 18th Meeting of the North American Catalysis Society, Cancun, Mexico
394. Comotti M, Della Pina C, Matarrese R, Rossi M (2004) Angew. Chem. Int. Ed. 43:5812
395. Lu G, Zhao R, Qian G, Qi Y, Wang X, Suo J (2004) Catal. Lett. 97:115
396. Zhao R, Ji D, Lv G, Qian G, Yan L, Wang X, Suo J (2004) Chem. Commun.:904
397. Pirutko LV, Khatitonov AS, Khramov MI, Uriate AK, US Patent 2004158103 A1
398. Corma A (private communication)
399. Hutchings GJ, Carrettin S, Landon P, Edwards JK, Enache D, Knight DW, Xu Y-J, Carley AF (2005) Top. Catal. 38:223
400. Zhu K, Hu J, Richards R (2005) Catal. Lett. 100:195
401. Uphade BS, Yamada Y, Akita T, Nakamura T, Haruta M (2001) Appl. Catal. A 215:137

402. Sinha AK, Seelan S, Akita T, Tsubota S, Haruta M (2003) Appl. Catal. A 240:243
403. Sinha AK, Seelan S, Tsubota S, Haruta M (2004) Top. Catal. 29:95
404. Yap N, Andres RP, Delgass WN (2004) J. Catal. 226:156
405. Stangland EE, Taylor B, Andres RP, Delgass WN (2005) J. Phys. Chem. B 109:2321
406. Merier R, Holderich U, Dingerdissen WF (1998) J. Catal. 176:376
407. Lu J, Luo M, Lei H, Li C (2002) Appl. Catal. A 237:11
408. Jenzer G, Mallat T, Maciejewski M, Eigemann E, Baiker A (2001) Appl. Catal. A 208:125
409. Laufer W, Hoelderich WF (2001) Appl. Catal. A 213:163
410. Zwijnenburg A, Makkee M, Moulijn JA (2004) Appl. Catal. A 270:49
411. Qi C, Akita T, Okumura M, Haruta M (2001) Appl. Catal. A 218:81
412. Kalvachev YA, Hayashi T, Tsubota S, Haruta M (1997) Stud. Surf. Sci. Catal. 110:965
413. Uphade BS, Tsubota S, Hayashi T, Haruta M (1998) Chem. Lett.:1273
414. Cumaranantunge L, Delgass WN (2005) J. Catal. 232:38
415. Taylor B, Lauterbach J, Delgass WN (2005) Appl. Catal. A 291:188
416. Bayer AG (2001) Patent Application WO200158887
417. Nippon Shokubai (2001) Patent Application US2001020105
418. Dow Patent, US 2004/0176629
419. Nijhuis TA, Huizinga BJ, Makkee M, Moulijn JA (1999) Ind. Eng. Chem. Res. 38:884
420. Swern D (1971) Organic Peroxide, vol. 2. Wiley-Interscience, New York
421. Patil NS, Jha R, Uphade BS, Bhargava SK, Choudhary VR (2004) Appl. Catal. A 275:87
422. Patil NS, Uphade BS, Jana P, Bharagava SK, Choudhary VR (2004) J. Catal. 223:236
423. Patil NS, Uphade BS, Jana P, Sonawane RS, Bhargava SK, Choudhary VR (2004) Catal. Lett. 94:89
424. Patil NS, Uphade BS, McCulloh DG, Bhargava SK, Choudhary VR (2004) Catal. Commun. 5:681
425. Coperet C, Adolfson H, Sharpless KB (1997) Chem. Commun.:1565
426. Kumar SB, Mirajkar SP, Pais GCG, Kumar P, Kumar R (1995) J. Catal. 156:163
427. Yang Q, Wang S, Lu J, Xiong G, Feng Z, Xin X, Li C (2000) Appl. Catal. A 194–195:507
428. Fu J, Yin D, Li Q, Zhang L, Zhang Y (1999) Microporous Mesoporous Mater. 29:351
429. Grieken VR, Sotelo JL, Martos C, Fierro JLG, Lopez Granados M, Mariscal R (2000) Catal. Today 61:49
430. Laha SC, Kumar R (2001) J. Catal. 204:64
431. Rudolph J, Reddy KL, Chiang JP, Sharpless KB (1997) J. Am. Chem. Soc. 119:6189
432. Sun J, Fujita S, Zhao F, Hasegawa M, Arai M (2005) J. Catal. 230:398
433. Augustine RL (1997) Catal. Today 37:419
434. Bailie JE, Hutchings GJ (1999) Chem. Commun.:2151
435. Bailie JE, Abdullah HA, Anderson JA, Rochester CH, Richardson NV, Hodge N, Zhang Z-G, Burrows A, Kiely CJ, Hutchings GJ (2001) Phys. Chem. Chem. Phys. 3:4113

436. Milone C, Tropeano ML, Gulino G, Neri G, Ingoglia R, Galvagno S (2002) Chem. Commun.:868
437. Okumura M, Akita T, Haruta M (2002) Catal. Today 74:265
438. Mohr C, Hofmeister H, Lucas M, Claus P (2000) Chem. Eng. Technol. 23:324
439. Mohr C, Hofmeister H, Claus P (2003) J. Catal. 213:86
440. Claus P (2005) Appl. Catal. A 291:222
441. Milone C, Ingoglia R, Tropeano ML, Neri G, Galvagno S (2003) Chem. Commun.:868
442. Bond GC, Sermon PA (1973) _www.goldbulletin.org_ (http://www.goldbulletin.org) 6:102
443. Bond GC, Sermon PA, Buchanan DA, Webb G, Wells PB (1973) J. Chem. Soc. Chem. Commun.:444
444. Sermon PA, Bond GC, Wells PB (1979) J. Chem. Soc. Faraday Trans. I 75:385
445. Sárkány A, Horváth A, Beck A (2002) Appl. Catal. A 229:117
446. Claus P, Hofmeister H, Mohr C (2003) Proc. International Conference on 'New Industrial Applications for Gold' _www.GOLD2003.org_ (http://www.GOLD2003.org), Vancouver, Canada, 2003
447. Claus P (1998) Topic. Catal. 5:51
448. Gallezot P, Richard D (1998) Catal. Rev.-Sci. Eng. 40:8
449. Gallezot P, Giroir-Fendler A, Richard D (1990) Catal. Lett. 5:169
450. Didillon B, El Mansour A, Candy JP, Bournonville JP, Basset JM (1991) Stud. Surf. Sci. Catal. 59:137
451. Claus P, Hofmeister H (1999) J. Phys. Chem. B 103:2766
452. Hutchings GJ, King F, Okoye IP, Padley MB, Rochester CH (1994) J. Catal. 148:453
453. Hutchings GJ, King F, Okoye IP, Padley MB, Rochester CH (1994) J. Catal. 148:464
454. Marinelli TBLW, Naarburs S, Ponec V (1995) J. Catal. 151:431
455. Marinelli TBLW, Ponec V (1995) J. Catal. 156:51
456. Ponec V (1997) Appl. Catal. A 149:27
457. Claus P, Hofmeister H, Mohr C (2003) J. Am. Chem. Soc. 125:1905
458. Ponec V, Bond GC (1995) Stud. Surf. Sci. Catal. 95:491
459. Duca D, Frusteri F, Parmaliana A, Deganello G (1996) Appl. Catal. A 146:269
460. Molero H, Bartlett BF, Tysoe WT (1999) J. Catal. 181:49
461. Zhang Q, Li J, Liu X, Zhu Q (2000) Appl. Catal. A 197:221
462. Molnár A, Sárkány A, Varga M (2001) J. Mol. Catal. 173:185
463. Guczi L, Schay Z, Stefler G, Liotta LF, Deganello G (1999) J. Catal. 182:456
464. Guzman J, Gates BC (2004) J. Catal. 226:111
465. Lopez-Sanchez JA, Lennon D (2005) Appl. Catal. A 291:230
466. Freemantle M (1987) Chemistry in Action. MacMillan Education
467. Słoczynski J, Grabowski R, Kozłowska A, Olszewski P, Stoch J, Skrzypek J, Lachowska M (2004) Appl. Catal. A 278:11
468. Sakuri H, Tsubota S, Haruta M (1993) Appl. Catal. 102:125
469. Sakurai H, Haruta M (1995) Appl. Catal. A 127:93
470. Paredes Olivera P, Patrito EM, Sellers H (1994) Surf. Sci. 313:25
471. Izumi Y, Miyazaki H, Kawahara S (1981) US Patent 4279883
472. Dalton AI Jr, Skinner RW (1982) US Patent 4335092
473. Chuang KI, Zhou B (1984) US Patent 5338531
474. Brill W (1987) US Patent 4661337

475. Gosser LW (1987) US Patent 4681751
476. Gosser LW, Schwartz JAT (1988) US Patent 4772458
477. Gosser LW, Schwartz JAT (1989) US Patent 4832938
478. Gosser LW (1989) US Patent 4889705
479. Chuang K (1992) US Patent 5082647
480. Hiramatsu Y, Ishiuchi Y (1992) US Patent 5132099
481. Hiramatsu Y, Ishiuchi Y, Nagashima H (1993) US Patent 5180573
482. van Waynbergh J, Schoebrichts J-P, Colery J-C (1995) US Patent 5447706
483. Zhou B, Lee L-K (2001) US Patent 6168775
484. Wanngard J (1998) European Patent 0816286 AI
485. Ishihara T, Ohura Y, Yoshida S, Hata Y, Nishiguchi H, Takita Y (2005) Appl. Catal. A 291:215
486. Haruta M (2002) Stud. Surf. Sci. Catal. 145:31
487. Carabineiro SAC (2000) Ph.D. Thesis. New University of Lisbon
488. Forzatti P (2001) Appl. Catal. A 222:221
489. Ueda A, Haruta M (1999) _www.goldbulletin.org_ (http://www.goldbulletin.org) 32:3
490. Kaspar J, Fornasiero P, Hickey N (2003) Catal. Today 77:419
491. Fritz A, Pitchon V (1997) Appl. Catal. B 13:1
492. Ueda A, Oshima T, Haruta M (1997) Appl. Catal. B 12:81
493. Ueda A, Haruta M (1998) Appl. Catal. B 18:115
494. Cant NW, Ossipoff NJ (1997) Catal. Today 36:125
495. Ohtsuka H (2001) Appl. Catal. B 33:325
496. Kung MC, Bethke KA, Yan J, Lee J-H, Kung HH (1997) Appl. Surf. Sci. 121–122:261
497. Jang BW-L, Spivey JJ, Kung MC, Kung HH (1997) Energy Fuels 11:299
498. Burch R, Watling TC (1997) Appl. Catal. B 11:207
499. Kung MC, Lee J-H, Chu-Kung A, Kung HH (1996) Stud. Surf. Sci. Catal. 101:701
500. Miyadera T (1993) Appl. Catal. B 2:199
501. Li Y, Armor JN (1994) J. Catal. 145:1
502. Tabata T, Kokitsu M, Okada O, Nakayama T, Yasumatsu T, Sakane H (1994) Stud. Surf. Sci. Catal. 88:409
503. Hirabayashi H, Yahiro H, Mizuno N, Iwamoto M (1992) Chem. Lett. :2235
504. Obuchi A, Ohi A, Nakamura M, Ogata A, Mizuno K, Ohuchi H (1993) Appl. Catal. B 2:71
505. Armor JN (1995) Catal. Today 26:147
506. Bamwenda GR, Obuchi A, Ogata A, Oi J, Kushiyama S, Yagita H, Mizuno K (1999) Stud. Surf. Sci. Catal. 121:263
507. Pattrick G, van der Lingen E, Schwarzer H, Roberts SJ (2003) Proc. International Conference on 'New Industrial Applications for Gold' _www.GOLD2003.org_ (http://www.GOLD2003.org), Vancouver, Canada, 2003
508. Kobylinsky TP, Taylor BW (1974) J. Catal. 33:376
509. Murakami H, Fujitani Y (1986) Ind. Eng. Chem. Prod. Res. Dev. 25:414
510. Schwartz JM, Schmidt LD (1994) J. Catal. 148:22
511. Skoglundh M, Torncrona A, Thormahlen P, Fridell E, Drewsen A, Jobson E (1998) Stud. Surf. Sci. Catal. 116:116
512. El Hamdaoui A, Bergeret G, Massardier J, Primet M, Renouprez A (1994) J. Catal. 148:47

513. Trillat JF, Massuadier J, Moraweck B, Praliaud H, Renouprez AJ (1998) Stud. Surf. Sci. Catal. 116:103
514. United States Environmental Protections Agency—Office of Research and Development. http://cfpub.epa.gov/ncea/cfm/dioxin.cfm
515. Hart JR (2004) Chemosphere 54:1539
516. Okumura M, Akita T, Haruta M, Wang X, Kajikawa O, Okada O (2003) Appl. Catal. B 41:43
517. Kajikawa O, Wang X-S, Tabata T, Okada O (1999) Organohalogen Compounds 40:581
518. Sakurai T, Iwasaki T, Shibuya E (1996) Zeolites 17:321; Sakurai T, Iwasaki T, Shibuya E (1995) Eur. Pat. Appl. 645172; Sakurai T, Iwasaki T, Shibuya E (1993) Jp. Appl.: for NE Chemcat Corp., NKK Corp
519. Minicò S, Scirè S, Crisafulli C, Maggiore R, Galvagno S (2000) Appl. Catal. B 28:245
520. Scirè S, Minicò S, Crisafulli C, Galvagno S (2001) Catal. Comm. 2:229
521. Scirè S, Minicò S, Crisafulli C, Satriano C, Pistone A (2003) Appl. Catal. B 40:43
522. McGraph M (1995) Appl. Catal. B 5:N25
523. Cordi EM, Falconer JL (1996) J. Catal. 162:104
524. Cordi EM, O'Neill PJ, Falconer JL (1997) Appl. Catal. B 14:23
525. Cordi EM, Falconer JL (1997) Appl. Catal. A 151:179
526. Quaranta NE, Soria J, Cortes Corberan V, Fierro JLG (1997) J. Catal. 171:1
527. Papaefthimiou P, Ioannides T, Verykios XE (1997) Appl. Catal. B 13:175
528. Gallardo-Amores JM, Armaroli T, Ramis G, Finocchio E, Busca G (1998) Appl. Catal. B 22:249
529. Baldi M, Finocchio E, Milella F, Busca G (1998) Appl. Catal. B 16:43
530. Luo J, Maye MM, Kariuki NN, Wang L, Njoki P, Lin Y, Schadt M, Naslund HR, Zhong C-J (2005) Catal. Today 99:291
531. Liu W, Flytzani-Stephanopoulos M (1995) J. Catal. 153:304
532. Andreeva D, Tabakova T, Ilieva L, Naydenov A, Mehanjiev D, Abrashev MV (2001) Appl. Catal. A: Gen. 209:291
533. Idakiev V, Ilieva L, Andreeva D, Blin JL, Gigot L, Su BL (2003) Appl. Catal. A 243:25
534. Andreeva D, Tabakova T, Idakiev V, Naydenov A (1998) _www.goldbulletin.org_ (http://www.goldbulletin.org) 31:105
535. Haruta M (1992) Now and Future 7:13
536. Miao S, Deng Y (2001) Appl. Catal. B 31:L1
537. Liu W, Flytzani-Stephanopoulos M (1995) J. Catal. 153:317
538. Grisel RJH, Slyconish JJ, Nieuwenhuys BE (2001) Topic. Catal. 16–17:425
539. Andreeva D (2003) Proc. International Conference on 'New Industrial Applications for Gold' _www.GOLD2003.org_ (http://www.GOLD2003.org), Vancouver, Canada, 2003
540. Ilieva LI, Nedyalkova RN, Andreeva DH (2002) Bulgarian Chem. Commun. 34:289
541. Andreeva D, Nedyalkova R, Ilieva L, Abrashev MV (2003) Appl. Catal. A 246:29
542. Andreeva D, Nedyalkova R, Ilieva L, Abrashev MV (2004) Appl. Catal. B 52:157
543. Hao Z, Cheng D, Guo Y, Liang Y (2001) Appl. Catal. B 33:217

544. Dhandapani B, Oyama ST (1997) Appl. Catal. B 11:129
545. Perry RH, Green D (1989) In: Perry's Chemical Engineer's Handbook. McGraw-Hill, New York, p. 147
546. Konova PM, Naydenov AI, Mehandjiev DR (2002) Bulgarian Chem. Commun. 34:437
547. Masafumi Y, Tadao N, Kazuhiko N, Tetsuo T (1991) Jpn Kokai Tokyo Koho 5
548. Katsuhiko M, Akira H, Tetsuo I (1992) Jpn Kokai Tokyo Koho 3
549. Horst M, Ruediger L, Manfred G, Matthias S, Friedrich F (1993) Ger. Offen. 7
550. Hao Z, An L, Wang H (2000) React. Kinet. Catal. Lett. 70:153
551. Petrov LA (2000) Stud. Surf. Sci. Catal. 130C:2345
552. Naydenov A, Mehanjiev D, Andreeva D, Tabakova T (2003) Oxid. Comm. 26:492
553. (August, 2002) Chem. Eng. News :19
554. Anderson DC, Anderson P, Galwey AK (1995) Fuel 74:1018
555. Haase JJ (1997) Phys. Condens. Matter 9:3647
556. DeBarr J, Lizzio AA (1997) Energy Fuels 11:267
557. Mochida I, Kuroda K, Kawano S, Matsumura Y, Yoshikawa M (1997) Fuel 76:533
558. Zhu T, Kundakovic L, Dreher A, Flytzani-Stephanopoulos M (1999) Catal. Today 50:381
559. Rodriguez JA, Jirsak T, Chaturvedi S, Hrbek JA (1998) J. Am. Chem. Soc. 120:11149
560. Jirsak T, Rodriguez JA, Chaturvedi S, Hrbek J (1998) Surf. Sci. 418:8
561. Rodriguez JA, Hrbek J (1999) Acc. Chem. Res. 32:719
562. Rodriguez JA, Jirsak T, Chaturvedi S (1999) J. Chem. Phys. 110:3138
563. Rodriguez JA, Ricart JM, Clotet A, Illas F (2001) J. Chem. Phys. 115:454
564. Rodriguez JA, Jirsak T, Gonzalez L, Evans J, Perez M, Maiti A (2001) J. Chem. Phys. 115:10914
565. Rodriguez JA, Liu G, Jirsak T, Hrbek J, Chang Z, Dvorak J, Maiti A (2002) J. Am. Chem. Soc. 124:5242
566. Rodriguez JA, Perez M, Jirsak T, Evans J, Hrbek J, Gonzalez L (2003) Chem. Phys. Lett. 378:526
567. Carabineiro SAC, Ramos AM, Vital J, Loureiro JM, Órfão JJM, Silva IF (2003) Catal. Today 78:203
568. Paik SC, Chung JS (1995) Appl. Catal. B 5:233
569. Jirsak T, Rodriguez JA, Hrbek J (1999) Surf. Sci. 426:319
570. Shelef M, McCabe RW (2000) Catal. Today 62:35
571. Kauffmann TG, Kaldor A, Stuntz GF, Kerby MC, Ansell LL (2000) Catal. Today 62:77
572. Solomon JL, Madix RJ, Wurth W, Stöhr J (1991) J. Phys. Chem. 95:3687
573. Tsai S-H, Liu Y-H, Wu P-L, Yeh C-S (2003) J. Mater. Chem. 13:978
574. Yam CM, Cho J, Cai C (2003) Langmuir 19:6862
575. Shi F, Zhang Q, Ma Y, He Y, Deng Y (2005) J. Am. Chem. Soc. 127:4182–4183
576. Naka K, Itoh H, Tampo Y, Chujo Y (2003) Langmuir 19:5546
577. Tai Y-W, Chen J-S, Yang C-C, Wan B-Z (2004) Catal. Today 97:95
578. Takata T, Takata A, Hara M, Kondo JN, Domen K (1998) Catal. Today 44:17
579. Lee JK, Lee K-B, Kim DJ, Choi IS (2003) Langmuir 19:8141
580. Ojima I, Li Z, Zhu J (1998) In: Z Rappoport, Y Apeilog (eds.) The Chemistry of Organic Silicon Compounds. Wiley, New York

581. Caporusso AM, Aronica LA, Schiavi E, Martra G, Vitulli G, Salvadori P (2005) J. Organomet. Chem. 690:1063
582. Manzer LE (1992) Catal. Today 13:13
583. Manzer LE, Rao VNM (1993) Adv. Catal. 39:329
584. Ocal M, Maciejewski M, Baiker A (1999) Appl. Catal. B 21:279
585. Ramos AL, Schmal M, Aranda DAG, Somorjai GA (2000) J. Catal. 192:423
586. Chen B, Bai C, Cook R, Wright J, Wang C (1996) Catal. Today 30:15
587. Farrauto RJ, Heck RM, Speronello BK (1992) Chem. Eng. News 7:34
588. Yuan G, Lopez JL, Louis C, Delannoy L, Keane MA (2005) Catal. Commun. 6:555
589. Potapenko DV, Horn JM, Beuhler RJ, Song Z, White MG (2005) Surf. Sci. 574:244
590. Furimsky E, Massoth FE (1999) Catal. Today 52:381
591. Qian W, Yoda Y, Hirai Y, Ishihara A, Kabe T (1999) Appl. Catal. A 184:81
592. Caldwell TE, Abdelrehim IM, Land DP (1996) Surf. Sci. 367:L26
593. Johansson Å, Stafström S (2000) Chem. Phys. Lett. 322:301
594. Bond GC, Rawle AF (1996) J. Mol. Catal. A 109:261
595. Venezia AM, La Parola V, Deganello G, Pawelec B, Fierro JLG (2003) Proc. International Conference on 'New Industrial Applications for Gold' _www.GOLD2003.org_ (http://www.GOLD2003.org), Vancouver, Canada, 2003
596. http://www.bp.com/company_overview/technology/frontiers/fr04aug02/fr04leapavada.asp
597. Chemical Engineering (New York) (September, 2004) 111:20
598. Corti CW, Holliday RJ (2004) _www.goldbulletin.org_ (http://www.goldbulletin.org) 37:20
599. Hoke J, Anderson D, Heck R, Poles T, Steger J (1996) SAE Paper 960800
600. Wu MC, Kelly NA (1998) Appl. Catal. B 18:79
601. Farrauto RJ, Bartholomew CH (1997) Fundamentals of Industrial Catalytic Processes. Chapman & Hall, London
602. Kim DH, Kung MC, Kozlova A, Yuan SD, Kung HH (2004) Catal. Lett. 98:11
603. Cameron DS (2003) _www.goldbulletin.org_ (http://www.goldbulletin.org) 36:17
604. Katikaneni SP, Gaffney AM, Ahmed S, Song C (2005) Catal. Today 99:255
605. Maye MM, Luo J, Han L, Nancy, Kariuki N, Zhong C-J (2003) _www.goldbulletin.org_ (http://www.goldbulletin.org) 36:75
606. Boccuzzi F, Chiorino A, Manzoli M (2003) J. Power Sources 118:304
607. Kim WB, Voitl T, Rodriguez-Rivera GJ, Dumesic JA (2004) Science 305:1280
608. Maye MM, Lou Y, Zhong C-J (2000) Langmuir 16:7520
609. Zhong CJ, Maye M (2001) Adv. Mater. 13:1507
610. Luo J, Maye MM, Lou Y, Han L, Hepel M, Zhong CJ (2002) Catal. Today 77:127
611. Electrode Catalyst for Fuel Cells, Jap. Patent Appl. JP2002305001
612. Worldwide Patent 9424710 for Johnson Matthey Plc (1994)
613. Funazaki N, Hemmi A, Ito S, Asano Y, Yamashita S, Kobayashi T, Haruta M (1993) Sens. Actuators B 14:536
614. Neri G, Bonavita A, Galvagno S, Caputi L, Pacilè D, Marsico R, Papagno L (2001) Sens. Actuators B 80:222
615. Neri G, Bonavita A, Milone C, Galvagno S (2003) Sens. Actuators B 93:402

616. Neri G, Bonavita A, Rizzo G, Galvagno S, Donato N, Caputi LS (2004) Sens. Actuators B 101:90
617. Boccuzzi F, Chiorino A, Tsubota S, Haruta M (1995) Sens. Actuators B 24–25:540
618. Wu R-J, Hu C-H, Yeh C-T, Su P-G (2003) Sens. Actuators B 96:596
619. Ando M, Kobayashi T, Haruta M (1995) Sens. Actuators B 2425:851
620. Ando M, Kobayashi T, Haruta M (1997) Catal. Today 36:135
621. Chen C, He J, Xu D, Tan X, Zhou X, Wang X (2005) Sens. Actuators B 107:866
622. Fukui K, Nakane M, Sens. Actuators B 24–25 (1995):486
623. Fukui K, Nishida S (1997) Sens. Actuators B 45:101
624. Baratto C, Sberveglieri G, Comini E, Faglia G, Benussi G, La Ferrara V, Quercia L, Di Francia G, Guidi V, Vincenzi D, Boscarino D, Rigato V (2000) Sens. Actuators B 68:74
625. Penza M, Martucci C, Cassano G (1998) Sens. Actuators B 50:52
626. Skelton DC, Tobin RG, Lambert DK, DiMaggio CL, Fisher GB (2003) Sens. Actuators B 96:46
627. Hoheréáková Z, Opekar F (2004) Sens. Actuators B 97:379
628. Filippini D, Weiss T, Aragón R, Weimar U (2001) Sens. Actuators B 78:195
629. Filippini D, Fraigi L, Aragón R, Weimar U (2002) Sens. Actuators B 81:296
630. Toda K, Ochi K, Sanemasa I (1996) Sens. Actuators B 32:15
631. Do J-S, Wu K-J (2000) Sens. Actuators B 67:209
632. Allen BW, Piantadosi CA, Coury LA Jr (2000) Nitric Oxide: Biol. Chem. 4:75
633. Di J, Bi S, Zhang M (2004) Biosens. Bioelectron. 19:1479
634. Wang L, Wang E (2004) Electrochem. Comm. 6:225
635. Liu S-Q, Ju H-X (2002) Anal. Biochem. 307:110
636. Lei C-X, Hu S-Q, Shen G-L, Yu R-Q (2003) Talanta 59:981
637. Xu S, Han X (2004) Biosens. Bioelectron. 19:1117
638. Casella IG, Gatta M, Guascito MR, Cataldi TRI (1997) Anal. Chim. Acta 357:63
639. Liu S, Ju H (2003) Biosens. Bioelectron. 19:177
640. Sato Y, Kato D, Niwa O, Mizutani F (2005) Sens. Actuators B 108:617
641. Pan M, Guo X, Cai Q, Li G, Chen Y (2003) Sens. Actuators A 108:258
642. Cornell BAA (2003) Proc. International Conference on 'New Industrial Applications for Gold' _www.GOLD2003.org_ (http://www.GOLD2003.org), Vancouver, Canada, 2003
643. http://www.nanoprobes.com; http://www.ambri.com
644. Corti CW, Holliday RJ, Thompson DT (2002) _www.goldbulletin.org_ (http://www.goldbulletin.org) 35:111
645. Cortie MB, van der Lingen E (2003) Proc. International Conference on 'New Industrial Applications for Gold' _www.GOLD2003.org_ (http://www.GOLD2003.org), Vancouver, Canada, 2003
646. Yu H-Z (2003) Proc. International Conference on 'New Industrial Applications for Gold' _www.GOLD2003.org_ (http://www.GOLD2003.org), Vancouver, Canada, 2003
647. Lefebvre J, Batchelor RJ, Leznoff DB (2003) Proc. International Conference on 'New Industrial Applications for Gold' _www.GOLD2003.org_ (http://www.GOLD2003.org), Vancouver, Canada, 2003
648. Leznoff DB, Shorrock CJ, LeFebvre J, Xue B-Y, Batchelor RJ (2003) Proc. International Conference on 'New Industrial Applications for Gold' _www.GOLD2003.org_ (http://www.GOLD2003.org), Vancouver, Canada, 2003

649. Laguna A, Fernández EJ, López-de-Luzuriaga JM, Monge M, Montiel M, Olmos ME, Pérez J (2003) Proc. International Conference on 'New Industrial Applications for Gold' _www.GOLD2003.org_ (http://www.GOLD2003.org), Vancouver, Canada, 2003
650. Bariain C, Matías IR, Fdez-Valdivielso C, Elosúa C, Luquin A, Garrido J, Laguna M (2005) Sens. Actuators B 108:535
651. Yam VW-W (2003) Proc. International Conference on 'New Industrial Applications for Gold' _www.GOLD2003.org_ (http://www.GOLD2003.org), Vancouver, Canada, 2003
652. Balch AL (2003) Proc. International Conference on 'New Industrial Applications for Gold' _www.GOLD2003.org_ (http://www.GOLD2003.org), Vancouver, Canada, 2003
653. Zosel J, Westphal D, Jakobs S, Müller R, Guth U (2002) Solid State Ion. 152–153:525
654. Schmidt-Zhang P, Guth U (2004) Sens. Actuators B 99:258
655. Neri G, Bonavita A, Milone C, Pistone A, Galvagno S (2003) Sens. Actuators B 92:326
656. Chou TC, Ng KM, Wang SH (2000) Sens. Actuators B 66:184
657. Hibino T, Wang S, Kakimoto S, Sano M (1998) Sens. Actuators B 50:149
658. Li H, Wang Q, Xu J, Zhang W, Jin L (2002) Sens. Actuators B 87:18
659. Lin L, Zhao H-q, Li J-r, Tang J-a, Duan M-x, Jiang L (2000) Biochem. Biophys. Res. Commun.274:817
660. Liu T, Tang J-a, Jiang L (2002) Biochem. Biophys. Res. Commun.295:14
661. Liu T, Tang J, Jiang L (2004) Biochem. Biophys. Res. Commun.313:3
662. Teles JH, Brode S, Chabanas M (1998) Angew. Chem. Int. Ed. 37:1415
663. Hashmi ASK, Frost TM, Bats JW (2000) J. Am. Chem. Soc. 122:11553
664. Hashmi ASK, Frost TM, Bats JW (2002) Catal. Today 72:19
665. Mizushima E, Hayashi T, Tanaka M (2002) Angew. Chem. Int. Ed. 41:4563
666. Hashmi ASK (2003) _www.goldbulletin.org_ (http://www.goldbulletin.org) 36:3
667. Weyrauch JP, Hashmi ASK (2003) Proc. International Conference on 'New Industrial Applications for Gold' _www.GOLD2003.org_ (http://www.GOLD2003.org), Vancouver, Canada, 2003
668. Cinellu MA, Minghetti G, Bacciu D, Stoccoro S, Zucca A (2003) Proc. International Conference on 'New Industrial Applications for Gold' _www.GOLD2003.org_ (http://www.GOLD2003.org), Vancouver, Canada, 2003
669. Mizushima E, Hayashi T, Tanaka M (2003) Proc. International Conference on 'New Industrial Applications for Gold' _www.GOLD2003.org_ (http://www.GOLD2003.org), Vancouver, Canada, 2003
670. Hashmi ASK (2004) _www.goldbulletin.org_ (http://www.goldbulletin.org) 37:51
671. Yao X, Li C-J (2004) J. Am. Chem. Soc. 126:6884
672. Stephen A, Hashmi K, Weyrauch JP, Rudolph M, Kurpejovi E (2004) Angew. Chem. Int. Ed. 43:6545
673. Cinellu MA, Minghetti G, Stoccoro S, Zucca A, Manassero M (2004) Chem. Commun. :1618
674. Roembke P, Schmidbaur H, Cronje S, Raubenheimer H (2004) J. Mol. Catal. A 212:35

675. González-Arellano C, Corma A, Iglesias M, Sánchez F (2005) Chem. Commun. 15:1990; 27:3451
676. Carrettin S, Guzman J, Corma A (2005) Angew. Chem. Int. Ed. 44:2
677. Hashmi ASK, Salathé R, Frost TM, Schwarz Lr, Choi J-H (2005) Appl. Catal. A 291:238
678. Zhang L, Kozmin SA (2005) J. Am. Chem. Soc. 127:6962
679. Yang C-G, He C (2005) J. Am. Chem. Soc. 127:6966
680. Ferguson G, Gallaghor JF, Kelleher A-M, Spalding TR, Deeney FT (2005) J. Organomet. Chem. 690:2888
681. Hutchings GJ (2004) _www.goldbulletin.org_ (http://www.goldbulletin.org) 37:3

Index

Printing: Krips bv, Meppel
Binding: Stürtz, Würzburg